T0202542

Lecture Notes in Computer Science   11596

Commenced Publication in 1973
Founding and Former Series Editors:
Gerhard Goos, Juris Hartmanis, and Jan van Leeuwen

Editorial Board Members

David Hutchison
  Lancaster University, Lancaster, UK
Takeo Kanade
  Carnegie Mellon University, Pittsburgh, PA, USA
Josef Kittler
  University of Surrey, Guildford, UK
Jon M. Kleinberg
  Cornell University, Ithaca, NY, USA
Friedemann Mattern
  ETH Zurich, Zurich, Switzerland
John C. Mitchell
  Stanford University, Stanford, CA, USA
Moni Naor
  Weizmann Institute of Science, Rehovot, Israel
C. Pandu Rangan
  Indian Institute of Technology Madras, Chennai, India
Bernhard Steffen
  TU Dortmund University, Dortmund, Germany
Demetri Terzopoulos
  University of California, Los Angeles, CA, USA
Doug Tygar
  University of California, Berkeley, CA, USA

More information about this series at http://www.springer.com/series/7409

Heidi Krömker (Ed.)

HCI in Mobility, Transport, and Automotive Systems

First International Conference, MobiTAS 2019
Held as Part of the 21st HCI International Conference, HCII 2019
Orlando, FL, USA, July 26–31, 2019
Proceedings

 Springer

Editor
Heidi Krömker
Technische Universität Ilmenau
Ilmenau, Germany

ISSN 0302-9743   ISSN 1611-3349   (electronic)
Lecture Notes in Computer Science
ISBN 978-3-030-22665-7   ISBN 978-3-030-22666-4   (eBook)
https://doi.org/10.1007/978-3-030-22666-4

LNCS Sublibrary: SL3 – Information Systems and Applications, incl. Internet/Web, and HCI

© Springer Nature Switzerland AG 2019
This work is subject to copyright. All rights are reserved by the Publisher, whether the whole or part of the material is concerned, specifically the rights of translation, reprinting, reuse of illustrations, recitation, broadcasting, reproduction on microfilms or in any other physical way, and transmission or information storage and retrieval, electronic adaptation, computer software, or by similar or dissimilar methodology now known or hereafter developed.
The use of general descriptive names, registered names, trademarks, service marks, etc. in this publication does not imply, even in the absence of a specific statement, that such names are exempt from the relevant protective laws and regulations and therefore free for general use.
The publisher, the authors and the editors are safe to assume that the advice and information in this book are believed to be true and accurate at the date of publication. Neither the publisher nor the authors or the editors give a warranty, expressed or implied, with respect to the material contained herein or for any errors or omissions that may have been made. The publisher remains neutral with regard to jurisdictional claims in published maps and institutional affiliations.

This Springer imprint is published by the registered company Springer Nature Switzerland AG
The registered company address is: Gewerbestrasse 11, 6330 Cham, Switzerland

Foreword

The 21st International Conference on Human-Computer Interaction, HCI International 2019, was held in Orlando, FL, USA, during July 26–31, 2019. The event incorporated the 18 thematic areas and affiliated conferences listed on the following page.

A total of 5,029 individuals from academia, research institutes, industry, and governmental agencies from 73 countries submitted contributions, and 1,274 papers and 209 posters were included in the pre-conference proceedings. These contributions address the latest research and development efforts and highlight the human aspects of design and use of computing systems. The contributions thoroughly cover the entire field of human-computer interaction, addressing major advances in knowledge and effective use of computers in a variety of application areas. The volumes constituting the full set of the pre-conference proceedings are listed in the following pages.

This year the HCI International (HCII) conference introduced the new option of "late-breaking work." This applies both for papers and posters and the corresponding volume(s) of the proceedings will be published just after the conference. Full papers will be included in the *HCII 2019 Late-Breaking Work Papers Proceedings* volume of the proceedings to be published in the Springer LNCS series, while poster extended abstracts will be included as short papers in the HCII 2019 *Late-Breaking Work Poster Extended Abstracts* volume to be published in the Springer CCIS series.

I would like to thank the program board chairs and the members of the program boards of all thematic areas and affiliated conferences for their contribution to the highest scientific quality and the overall success of the HCI International 2019 conference.

This conference would not have been possible without the continuous and unwavering support and advice of the founder, Conference General Chair Emeritus and Conference Scientific Advisor Prof. Gavriel Salvendy. For his outstanding efforts, I would like to express my appreciation to the communications chair and editor of *HCI International News,* Dr. Abbas Moallem.

July 2019   Constantine Stephanidis

HCI International 2019 Thematic Areas
and Affiliated Conferences

Thematic areas:

- HCI 2019: Human-Computer Interaction
- HIMI 2019: Human Interface and the Management of Information

Affiliated conferences:

- EPCE 2019: 16th International Conference on Engineering Psychology and Cognitive Ergonomics
- UAHCI 2019: 13th International Conference on Universal Access in Human-Computer Interaction
- VAMR 2019: 11th International Conference on Virtual, Augmented and Mixed Reality
- CCD 2019: 11th International Conference on Cross-Cultural Design
- SCSM 2019: 11th International Conference on Social Computing and Social Media
- AC 2019: 13th International Conference on Augmented Cognition
- DHM 2019: 10th International Conference on Digital Human Modeling and Applications in Health, Safety, Ergonomics and Risk Management
- DUXU 2019: 8th International Conference on Design, User Experience, and Usability
- DAPI 2019: 7th International Conference on Distributed, Ambient and Pervasive Interactions
- HCIBGO 2019: 6th International Conference on HCI in Business, Government and Organizations
- LCT 2019: 6th International Conference on Learning and Collaboration Technologies
- ITAP 2019: 5th International Conference on Human Aspects of IT for the Aged Population
- HCI-CPT 2019: First International Conference on HCI for Cybersecurity, Privacy and Trust
- HCI-Games 2019: First International Conference on HCI in Games
- MobiTAS 2019: First International Conference on HCI in Mobility, Transport, and Automotive Systems
- AIS 2019: First International Conference on Adaptive Instructional Systems

Pre-conference Proceedings Volumes Full List

1. LNCS 11566, Human-Computer Interaction: Perspectives on Design (Part I), edited by Masaaki Kurosu
2. LNCS 11567, Human-Computer Interaction: Recognition and Interaction Technologies (Part II), edited by Masaaki Kurosu
3. LNCS 11568, Human-Computer Interaction: Design Practice in Contemporary Societies (Part III), edited by Masaaki Kurosu
4. LNCS 11569, Human Interface and the Management of Information: Visual Information and Knowledge Management (Part I), edited by Sakae Yamamoto and Hirohiko Mori
5. LNCS 11570, Human Interface and the Management of Information: Information in Intelligent Systems (Part II), edited by Sakae Yamamoto and Hirohiko Mori
6. LNAI 11571, Engineering Psychology and Cognitive Ergonomics, edited by Don Harris
7. LNCS 11572, Universal Access in Human-Computer Interaction: Theory, Methods and Tools (Part I), edited by Margherita Antona and Constantine Stephanidis
8. LNCS 11573, Universal Access in Human-Computer Interaction: Multimodality and Assistive Environments (Part II), edited by Margherita Antona and Constantine Stephanidis
9. LNCS 11574, Virtual, Augmented and Mixed Reality: Multimodal Interaction (Part I), edited by Jessie Y. C. Chen and Gino Fragomeni
10. LNCS 11575, Virtual, Augmented and Mixed Reality: Applications and Case Studies (Part II), edited by Jessie Y. C. Chen and Gino Fragomeni
11. LNCS 11576, Cross-Cultural Design: Methods, Tools and User Experience (Part I), edited by P. L. Patrick Rau
12. LNCS 11577, Cross-Cultural Design: Culture and Society (Part II), edited by P. L. Patrick Rau
13. LNCS 11578, Social Computing and Social Media: Design, Human Behavior and Analytics (Part I), edited by Gabriele Meiselwitz
14. LNCS 11579, Social Computing and Social Media: Communication and Social Communities (Part II), edited by Gabriele Meiselwitz
15. LNAI 11580, Augmented Cognition, edited by Dylan D. Schmorrow and Cali M. Fidopiastis
16. LNCS 11581, Digital Human Modeling and Applications in Health, Safety, Ergonomics and Risk Management: Human Body and Motion (Part I), edited by Vincent G. Duffy

17. LNCS 11582, Digital Human Modeling and Applications in Health, Safety, Ergonomics and Risk Management: Healthcare Applications (Part II), edited by Vincent G. Duffy
18. LNCS 11583, Design, User Experience, and Usability: Design Philosophy and Theory (Part I), edited by Aaron Marcus and Wentao Wang
19. LNCS 11584, Design, User Experience, and Usability: User Experience in Advanced Technological Environments (Part II), edited by Aaron Marcus and Wentao Wang
20. LNCS 11585, Design, User Experience, and Usability: Application Domains (Part III), edited by Aaron Marcus and Wentao Wang
21. LNCS 11586, Design, User Experience, and Usability: Practice and Case Studies (Part IV), edited by Aaron Marcus and Wentao Wang
22. LNCS 11587, Distributed, Ambient and Pervasive Interactions, edited by Norbert Streitz and Shin'ichi Konomi
23. LNCS 11588, HCI in Business, Government and Organizations: eCommerce and Consumer Behavior (Part I), edited by Fiona Fui-Hoon Nah and Keng Siau
24. LNCS 11589, HCI in Business, Government and Organizations: Information Systems and Analytics (Part II), edited by Fiona Fui-Hoon Nah and Keng Siau
25. LNCS 11590, Learning and Collaboration Technologies: Designing Learning Experiences (Part I), edited by Panayiotis Zaphiris and Andri Ioannou
26. LNCS 11591, Learning and Collaboration Technologies: Ubiquitous and Virtual Environments for Learning and Collaboration (Part II), edited by Panayiotis Zaphiris and Andri Ioannou
27. LNCS 11592, Human Aspects of IT for the Aged Population: Design for the Elderly and Technology Acceptance (Part I), edited by Jia Zhou and Gavriel Salvendy
28. LNCS 11593, Human Aspects of IT for the Aged Population: Social Media, Games and Assistive Environments (Part II), edited by Jia Zhou and Gavriel Salvendy
29. LNCS 11594, HCI for Cybersecurity, Privacy and Trust, edited by Abbas Moallem
30. LNCS 11595, HCI in Games, edited by Xiaowen Fang
31. LNCS 11596, HCI in Mobility, Transport, and Automotive Systems, edited by Heidi Krömker
32. LNCS 11597, Adaptive Instructional Systems, edited by Robert Sottilare and Jessica Schwarz
33. CCIS 1032, HCI International 2019 - Posters (Part I), edited by Constantine Stephanidis

http://2019.hci.international/proceedings

First International Conference on HCI in Mobility, Transport, and Automotive Systems (MobiTAS 2019)

Program Board Chair(s): **Heidi Krömker,** *Germany*

- Angelika C. Bullinger-Hoffmann, Germany
- Bertrand David, France
- Marco Diana, Italy
- Cyriel Diels, UK
- Riender Happee, The Netherlands
- Christophe Kolski, France
- Lutz Krauss, Germany
- Josef Krems, Germany
- Lena Levin, Sweden
- Peter Mörtl, Austria

- Gerrit Meixner, Germany
- Philipp Rode, Germany
- Matthias Roetting, Germany
- Thomas Schlegel, Germany
- Ulrike Stopka, Germany
- Jacques Terken, The Netherlands
- Alejandro Tirachini, Chile
- Andree Woodcock, UK
- Fang You, P.R. China
- Xiaowei Yuan, P.R. China

The full list with the Program Board Chairs and the members of the Program Boards of all thematic areas and affiliated conferences is available online at:

http://www.hci.international/board-members-2019.php

HCI International 2020

The 22nd International Conference on Human-Computer Interaction, HCI International 2020, will be held jointly with the affiliated conferences in Copenhagen, Denmark, at the Bella Center Copenhagen, July 19–24, 2020. It will cover a broad spectrum of themes related to HCI, including theoretical issues, methods, tools, processes, and case studies in HCI design, as well as novel interaction techniques, interfaces, and applications. The proceedings will be published by Springer. More information will be available on the conference website: http://2020.hci.international/.

General Chair
Prof. Constantine Stephanidis
University of Crete and ICS-FORTH
Heraklion, Crete, Greece
E-mail: general_chair@hcii2020.org

http://2020.hci.international/

Contents

Driving Experience

Mobility and Transport

Interaction in Autonomous and Semiautonomous Vehicles

Turmoil Behind the Automated Wheel

An Embodied Perspective on Current HMI Developments in Partially Automated Vehicles

Anika Boelhouwer[1]([⊠]), Jelle van Dijk[2], and Marieke H. Martens[1,3]

[1] Centre for Transport Studies, University of Twente,
Drienerlolaan 5, 7522 Enschede, NB, The Netherlands
a.boelhouwer@utwente.nl
[2] Department of Design, Production and Management, University of Twente,
Drienerlolaan 5, 7522 Enschede, NB, The Netherlands
[3] TNO Traffic and Transport,
Anna van Buerenplein 1, 2595 The Hague, DA, The Netherlands

Abstract. Cars that include combinations of automated functions, such as Adaptive Cruise Control (ACC) and Lane Keeping (LK), are becoming more and more available to consumers, and higher levels of automation are under development. In the use of these systems, the role of the driver is changing. This new interaction between the driver and the vehicle may result in several human factors problems if not sufficiently supported. These issues include driver distraction, loss of situational awareness and high workload during mode transitions. A large conceptual gap exists on how we can create safe, efficient and fluent interactions between the car and driver both during automation and mode transitions. This study looks at different HMIs from a new perspective: Embodied Interaction. The results of this study identify design spaces that are currently underutilized and may contribute to safe and fluent driver support systems in partially automated cars.

Keywords: Embodied interaction · Human computer interaction ·
Human Machine Interface · Review · Self-driving vehicles ·
Automated vehicles

1 Introduction

While cars with basic automated functions, such as Adaptive Cruise Control (ACC) and Lane Keeping (LK), are becoming more widely available to consumers, higher levels of automation such as level 3 and 4 [1] are under development. These highly automated systems take over the longitudinal and lateral control of the car. In case of a level 2 system, drivers still need to monitor the driving situation continuously. With a level 3 system, drivers no longer need to continuously monitor the driving situation, but still have to be able to take back control when requested within a given time frame. A level 4 system includes a minimal risk maneuver in case the driver does not take back control after a request. As the systems have an Operational Design Domain (ODD) and do not function in all possible situations, drivers still need to take

© Springer Nature Switzerland AG 2019
H. Krömker (Ed.): HCII 2019, LNCS 11596, pp. 3–25, 2019.
https://doi.org/10.1007/978-3-030-22666-4_1

back control occasionally. In this interaction, the HMI plays a crucial role to help drivers understand their automated vehicle (Carsten and Martens 2018).

Automated cars can provide multiple benefits for both the driver and society as a whole. These include improved traffic safety, potentially reduced fuel consumption and accompanying costs reductions, co2 emission reductions, and improved driver comfort [2–4]. In case drivers still have to monitor the situation continuously, they are at least relieved from some of the physical efforts in driving. When they do not need to monitor the situation, they can engage in non-driving tasks while traveling. Studies like those of [5] have already shown that drivers engage in tasks ranging from reading to playing games on a tablet.

However, recent studies show that besides the potential benefits, automated cars may create safety issues in the driver-car interaction [6–8]. Expected issues are driver distraction, automation surprise, loss of situational awareness and high workload [6, 9–11] when the driver needs to take over. The role of drivers shifts from operator to supervisor. This new role of supervisor, that is required with level 2 systems, is shown to be difficult for humans [12, 13]. Distraction towards non-driving activities with loss of situational awareness is in this case expected. Even more so, shifting from the distraction back to the driving task can be challenging. Especially in level 3 or 4 vehicles when the driver is temporarily not required to monitor the driving situation and is immersed in a non-driving task. Drivers have to disengage both physically and mentally from the non-driving task before resuming manual control.

Studies have shown lowered situational awareness in drivers that were engaged in non-driving tasks for long periods of time [14]. Using the commonly used definition by [15], the situational awareness of drivers can be described as: perceiving the driving situation, understanding this situation, and projecting the status of this situation in the future. When drivers are requested to take-back control, they first need to be able to regain their situational awareness to a level on which they are capable of safely resuming control. To avoid negative effects on safety, acceptance and driver comfort, the car Human Machine Interface (HMI) should be taking these human factors into consideration. In case of distraction or emersion in non-driving activities, the HMI should be able to support the driver in smoothly returning to the driving task and regaining situational awareness efficiently. It can also support during the automated phase to, for example, retain a certain level of situational awareness in the driver. The interesting thing here is that very often, solutions are found in improving system reliability. The more reliable the system will be, the less human factors issues will arise. However, as Carsten and Martens (2018) already indicated, this is not correct. With improving system reliability, comfort and trust will increase, but automation surprise and response times will also increase, and situational awareness, attention and trust calibration will decrease. Therefore, instead of focusing on improving system reliability, we believe that the primary focus should be on a proper interaction between the vehicle and the user, irrespective of the ODD or the system level.

Until now, the development and research on driver support through in-car HMIs has been mainly addressed from a traditional cognitive psychology perspective and human centered design. In this traditional perspective, cognition is considered to be "the mental action or process of acquiring knowledge and understanding through thought, experience, and the senses" [16]. Although specific perspectives of course differ from each other, the mode transition from automation to manual control is commonly described as a sequence of consecutive mental processes and physical actions.

This study investigates driver support in partially automated vehicles from a different perspective: embodied interaction [17–19]. This may allow us to identify unused design spaces. Embodied interaction proposes that all knowledge and sense-making of the world emerges from a continuous and simultaneous interaction with the world [20]. In this perspective, cognition is not strictly designated as sequential processes of the brain. Rather, cognition is the entire system of interaction between mind, body and world. As [19] stated "cognition is a highly embodied or situated activity […] thinking beings ought be considered first and foremost as acting beings". Furthermore, embodied cognition states that all abstract symbols (including words) only gain meaning through embodied experiences and physical aspects. In this embodied perspective, the emphasis of gaining knowledge is thus more focused on the physical acting of a person in a specific situation. By discarding the idea that cognition only occurs in the mind, new design spaces may be discovered. More emphasis can be put on the combination of mind and body making sense of situations in ongoing interactions with the environment.

This study reviews current HMI feedback systems of partially automated cars during two phases. One phase consists of the Take-Over Request (TOR) by the car. This includes messages from the car that the driver needs to take back control from the automation. The second phase that is reviewed is the general HMIs during automation. This phase also includes any Hands on Wheel Warnings (HOWW). These warnings indicate that the drivers have to put their hands on the wheel (or ideally eyes on the road). In most systems, the automation disengages if the driver does not comply with the HOWW. Since this is not a formal request to take over and is not linked to system limitations, this will be described as feedback during automation.

The HMI systems are reviewed in the light of three important characteristics of embodied interaction: suppleness [21], bodily experience [18] and situatedness [19]. These characteristics include for example: the fluency with which TORs are introduced, in- and output modalities, and whether the feedback systems are adaptive to the situation. Further details of the review protocol are discussed in the methodology section. While most review papers only discuss academic papers and patents, this review includes currently commercially available systems and systems that are being studied in experiments but are not yet on the market.

The goal of this review is to identify the current state of HMI support during the TOR phase and the automation phase in both literature and commercially available systems. We want to examine how they consider the main characteristics of embodied interaction in their design. This will allow us to identify unexplored design spaces, and new opportunities for the design of HMI systems of partially automated cars. Concluding, this study investigates two main research questions: (1) What embodied design elements are currently used in driver support during TORs and automated driving? (2) What are the unused embodied design spaces for designing HMI support for TORs and automated driving in partially automated cars?

2 Methods

2.1 Data Collection

The materials gathered for this literature review consisted of the following types: journal papers, conference papers, work-in-progress papers, technical reports and product documentation of commercial cars. It was decided to include technical reports and product documentation of commercial cars as the current development of HMI in automated car systems is proceeding fast. Including these material types allowed the study to review the latest developments in both industry and academia. Both the commercial car systems as the concepts in literature were reviewed on (1) the HMI during TORs, and (2) the HMI during automated mode. As the majority of the gathered materials does not specify the level of automation nor the exact Operational Design Domain (ODD), it was decided that the requirement for inclusion was that the system automated both lateral and longitudinal control simultaneously.

The literature papers and reports had to be written in English, and published after 2008. Although other studies have conducted reviews on shorter periods, we believe that it is necessary to include sources of a 10 year period. Condensing this work into a short snapshot would undermine the continuous progress within the field. Literature reviews and meta-analysis studies were excluded. The following leading research databases were used to collect the journal- and conference papers: Web of Science, IEEE, Scopus, Google Scholar.

For the TOR reviews, we solely considered systems that indicate a take-over request due to system limitations. Therefore Hands on Wheel Warnings (HOWWs), which prompt the driver to keep their hands on the wheel without the need to disengage automation, were *not* reviewed among the TORs. However, the HOWWs *were* included in the review of the general HMI during automation. These warnings are often included in car systems both for legal reasons and with the intention to keep the driver ready to take back control instantly.

Literature Concepts – TOR. For the TOR review on literature concepts, the literature papers had to be specifically focusing on the design or testing of HMI support during TOR. Studies that only used HMI as a means to perform their experiment on a different topic were excluded. The following keywords were used in the research databases: (("Autonomous" OR "Self-driving" OR "Automated") AND ("HMI" OR "Human machine interaction") AND ("Design" OR "Feedback") AND ("Take-over" OR "Take over" OR "Transition" OR "Warning" OR "Request") AND ("Car")).

Literature Concepts – HMI During Automation. For the review on general HMI feedback during automated mode, only studies that specifically address the development and testing of an HMI design were included. Studies that use an HMI purely as a means to perform their experiment on a different topic were excluded. The search entry for materials on HMI systems during automated mode contained the following keywords. (("Autonomous" OR "Self-driving" OR "Automated") AND ("HMI" OR "Human machine interaction") AND ("Interface" OR "Feedback") AND ("Car")).

Commercial Cars – TOR and HMI During Automation. The selection of commercial car brands was done based on their official user manuals and websites. The car system had to be available for purchase at the time of this review. To avoid an incomplete review, only systems that included all necessary information for the categorization were considered. Of the current available systems, only two formally include a TORs [22, 23]. Therefore the TOR review included just these two commercial car systems. (As mentioned before, the systems do include HOWWs. These are reviewed in the 'general HMI during automation' section.)

2.2   Data Coding

The materials gathered were labelled on three main use qualities of embodied interaction: suppleness [21, 24, 25], bodily experience [18, 20] and situatedness [19]. Although not exhaustive, these are discussed frequently within the embodied interaction domain and are generally excepted to portrait (some of) the core elements. Each quality will be discussed briefly with their respective measures. Some of the specific variables were used from the study by [26] who created a categorization framework for control transition interfaces. Tables 1 and 2 show all variables that were examined, respectively for the TORs and general HMI during automation.

Suppleness. [24, 25] introduced the use quality of suppleness. They stressed designing for supple back and forth interaction between a user and system, which can be seen as a fluent 'dance' [25]. The Webster dictionary definition of supple is considered the base for this use quality: "easy and fluent without stiffness or awkwardness". In this study, we categorized the TORs on three supple qualities. The first was whether the transfer is introduced abrupt or gradually: **Temporal Output Mode** [26]. The TORs could be categorized as being shown once, several times, or incremental. It was specified whether the support was given: before/during deactivation of the system, or before a hazard. It was important to take this into account as the time to take-over would be either the time before a collision or deactivation of the system. The second variable was the amount of time the driver has to take back control: **Time to take over**. More specifically, how much time does the driver have after the TOR until the system disengages or the car crashes? The third item entailed the use of **Social cues**. The research and design area of embodied interaction is increasingly focusing on incorporation of natural social interactions in artificial intelligent systems [27]. As we are social beings, we engage in continuous social interactions to understand and act on the world [28]. Therefore we investigated whether there is any use of social cues that we use daily in human to human communication in the HMI systems. These could for example facial expressions and gestures.

Bodily experience. Inclusion of the body in making sense of the situation is at the core of embodied interaction [18, 20]. Our entire body and all our senses are included in learning, and creating an understanding of the world. By including multiple senses in a feedback system, overload may be reduced or prevented. Therefore, the way the driver has to disengage automation (**Input**) was included in this review as well as the modality of the TOR itself (**Output**). For the input, we used a similar classification as [26] which included physical, touchscreen, gesture and speech. However, touchscreen was made into a sub classification of physical and we additionally included the options for activity recognition and 'other' input. Activity recognition includes all forms of system initiated recognition such as eye movement recognition or posture recognition. The physical class contains input through buttons, the steering wheel, the pedals and touchscreen. For the output modalities, we included all five basic modalities: visual, auditory, haptic, smell, taste. As directional forces such as acceleration and deceleration are a large part of the driving experience, the vestibular sense is also included.

Situatedness. As the name would make one suspect, the situatedness [19] describes how the meaning of interactions with technology cannot be seen in isolation from the context in which it occurs: interaction is always situated. Cognition relies on embodied interactions that take place within a specific situation. For example, a symbol or gesture can have a very different meaning in different contexts and for different people. In this study, TORs were investigated on whether or not they are **Adaptive to the driver and driving situation**. Is the feedback the same for all drivers and all their driver states? Also, is the feedback the same in all driving situations?

Table 1. Data coding scheme for TOR feedback.

TOR						
Suppleness			Bodily experience		Situatedness	
Temporal output mode	Time to take over	Social cues	Output of TOR	Input to disengage automation	Adaptive to driver	Adaptive to driving situation
(O)nce, before deactivation (O)nce, during deactivation (O)nce, before hazard (S)everal (I)ncremental	Seconds before deactivation	(Y)es (N)o	(V) Visual (A) Auditory (C) taCtile (S) Smell (T) Taste (E) vEstibular	(P) Physical (S) Speech (G) Gesture (A) Activity recognition (O) Other	(Y)es (N)o	(Y)es (N)o

Table 2. Data coding scheme for HMI during automation.

HMI during automation			
Suppleness	Bodily experience	Situatedness	
Social cues	Output of HMI system	Adaptive to driver	Adaptive to driving situation
(Y)es, .. (N)o	(V) Visual (A) Auditory (H) Haptic (S) Smell (T) Taste (E) vEstibular	(Y)es, .. (N)o	(Y)es, .. (N)o

3 Results

3.1 Reviewed Materials

An overview of the results can be found in Table 3 until 6 in Appendix A. Seven different commercial car brands were selected for this review. All systems have the option to simultaneously activate the automation of lateral and longitudinal control. As the systems have different names across brands (and sometimes even within the brand) they will be addressed by their company assigned name on their official websites and or user manuals. The included brands and systems are: (1) Audi – AI Traffic Jam Pilot [22, 29], (2) Tesla – Autopilot [30], (3) Cadillac - Super Cruise [23], (4) BMW – Steering and Lane Guidance Assistant [31], (5) Volvo – Pilot Assist, (6) Mercedes – Drive Pilot [32, 33]. All commercial systems will be reviewed on their general HMI during automation (including HOWW). However only two of these systems included a formal TOR since they allow the driver to be temporarily out of the loop due to a traffic jam assist. Therefore only these two system could be reviewed on its TOR [22, 23].

A total of 20 literature papers were reviewed on their TOR concepts in this analysis. Some papers discussed multiple concepts within the same paper. These were considered as individual concepts, resulting in a total of 31 concepts that were reviewed. 15 papers were selected for the general HMIs during automation. Again, as some papers presented multiple concepts, a final total of 17 concepts were reviewed. None of the literature concepts contained HOWWs, therefore these could not be included in the general HMI review.

3.2 Take-Over Request (TOR)

The result tables are situated in appendix A. Table 3 shows the results for TORs in commercial cars. Table 4 shows the results for TORs in literature concepts.

Commercial Cars. Formally, only two of the assessed systems issue a TOR [22, 23]. Therefore only these two commercial systems will be reviewed here. The remaining systems all require the driver to continuously keep their hands or eyes on the road.

Suppleness. Audi AI traffic jam pilot provides multiple TORs before the system disengages due to system limitations. Cadillac Super Cruise provides one TOR before deactivation. During the second warning, at the end of the take-over period, the system already deactivates. Both systems provide a social cue in the form of a symbol in which hands are holding a steering wheel (or an animation that the hand grab the steering wheel).

Bodily Experience. The TORs are in both cases visually displayed on a screen. Cadillac Super Cruise uses additional use of color and illumination in the steering wheel. The visual cues are complemented in both systems with auditory beep(s) and/or a spoken take-over message. Cadillac Super Cruise includes vibrations in the seat as a TOR. Audi includes a short brake jerk and tightens the safety belt three times during the second warning. Drivers can disengage the automation in both systems by turning the steering wheel, pressing one of the pedals or pushing a button.

Situatedness. The reviewed systems are not adapted to the driver. This means that the same message is given regardless of the current driver (state) or activity he is currently performing. None are adapted to the driving situation. The feedback does not change according to, for example, the reason that the car needs to transfers back control.

In conclusion, we found that the two reviewed commercial car TORs are very similar on the reviewed embodiment aspects. The Audi system is slightly more supple as it provides multiple TORs before the system disengages. Both TOR systems provide visual and auditory cues. These are complemented with vibrations (Cadillac), seatbelt tightening or vestibular feedback through braking (Audi). The situatedness of the TOR feedback is lacking as they did not change their form to the driver, nor to the specific driving situation.

Literature Concepts. Suppleness. The majority of the concepts (N = 20) consist of a single TOR before a detected hazard (without the automation deactivating). Two concepts are similar to the commercial car systems as they provide one take-over request during which the system is immediately deactivated [34, 35]. Eight of the reviewed concepts give several warnings before deactivation. Five of these warnings increase in intensity and cue modalities over time. The time that drivers have to take back control before deactivation or impact ranges widely from 10 s to 'a few minutes'. Two studies only report that the drivers had 'sufficient' time to take back control [36, 37]. It is not stated how much this specifically is. As a social cue, a few of the concepts (N = 5) include a symbol with hands on the steering as is also seen in the commercial car systems (Fig. 1). One concept uses a distressed voice in a verbal message in order to portrait urgency [38].

Bodily Experience. 23 of the TOR concepts give auditory feedback. This feedback is divided into abstract beeps (N = 16), and verbal messages (N = 2), while the rest of the concepts combines both (N = 5). The majority of the concepts (N = 17) use a display. These include standardized symbols, text and use of color or flashers. The color red is used in all cases to indicate an immediate required take-over. Of the display messages, thirteen are complimented with auditory or haptic feedback. Four of the concepts include lighting. While two concepts have a simple LED on the dashboard, the concept by [39] has a LED strip on the steering wheel that can light in directional patterns. This way, it hints towards the required steering direction after take-over. Two studies included mechanical transformations in their concepts. In the concept by [40], part of the steering wheel was replaced with grips that change direction during the TOR depending on the required steering direction. In the concept by [41], the upper part of the steering wheel moves backwards during automation and is shifted back during the TOR. This is mainly done to emphasize the need to take back control. Eight of the concepts include vibration feedback. This is mainly applied in either the driver seat or steering wheel. However, the concept in [38] gives vibration feedback in a wristband. The vibration feedback in the driver seat is either static or dynamic. In case of dynamic feedback, the vibration shifts along rows, creating the 'illusion' of motion or direction. Besides three papers, drivers can take back control in all concepts by engaging with the steering wheel or pedals.

Fig. 1. Examples of TORs that include a 'hands on wheel' symbol in literature. Left by [36] and right by [42].

Situatedness. One of the literature TORs is adaptive to the driver. The concept by [35] shows the TOR on the driver's mobile device if he is using this. More than half of the concepts (N = 16) adapt to the driving situation. Most of these concepts contain a suggested (steering) action based on the situation. The way in which this is done ranges widely. Some provide a suggested steering direction through vibration or lighting direction while others adapt the color or symbol accordingly. [40] even adapts the shape of the steering wheel according to the suggested steering direction. Some concepts do suggest a direct action but rather provide boundaries in which the driver can operate. For example, the concept by [37] shows the intent and expected actions of other road users, while the concept by [43] shows an overlay on the driving lane whether it is safe to continue driving there. Two concepts show the upcoming situation visually and why the driver needs to take-over, for example dense fog or roadworks.

Concluding, the majority of the literature concepts present one or multiple TORs before deactivation of the automation. This is expected as it is easier to implement warnings before deactivation of automation as a pre-set in an experimental setting, compared to in a car driving on the road. The variety of social cues is scarce. More variety is found in the bodily experience but only in the output. The variety consists of physical shape changes, verbal messages, dynamic vibrations and lighting. However, the main outputs are still displays and auditory beeps. Only one of the concepts adapts to the driver. However, more concepts adapt to the specific driving situation. In these cases they mainly provide a suggested action, boundaries after the transfer of control, or reasons for the TOR.

3.3   HMI During Automation

The result tables are situated in Appendix A. Table 5 shows the results for general HMI during automation in commercial cars. Table 6 shows the results for general HMI during automation in literature concepts.

Commercial Cars. Suppleness. Most of the commercial systems (N = 5) include a Hands on Wheel Warning. While it is not indicated exactly how long these warnings continue before the system disengages or stops the car, all systems provide these warnings several time while increasing the intensity (in any form). All systems use a 'hands on wheel' symbol as a social cue to indicate that the driver needs to keep their hands on the wheel.

Bodily experience. All systems use a visual display on a screen with illustrations, symbols, text and changing colors to provide feedback. If drivers keep their hands or eyes to long of the road they will receive auditory beeps as a warning and vibrations in the steering wheel or seat. Cadillac Super Cruise includes illumination and changing colors in the steering wheel as additional feedback on the automation state. Drivers get visual feedback of the current car actions as they see the turning of the steering wheel. Besides the visual feedback, drivers can feel the car's actions through the turning of the wheel.

Situatedness. All HMI systems during automation are partially adapted to the driver as they sense whether the driver has their hands on the wheel, or their eyes on the road, and prompts a HOWW accordingly. There is some variation to the extend in which the systems are adapted to the driving situation. However, all of them show a combination of automation mode, detected vehicles, lane markings and speed limit.

Conclusion. The general HMI during automation of commercial cars is very much the same across the systems on the investigated aspects. The suppleness with regard to social cues is limited to 'Hands on the wheel' symbols. The output is mainly given through displays, auditory beeps and vibrations in the steering wheel. The feedback is partially adaptive to the driver as it issues a 'hands on wheel' (or eyes on road) message in case the car detects that the hands are not on the wheel (or the eyes are not on the road). The feedback is adaptive to the driving scenario as all systems present the detected vehicles, obstacles, speed limit and/or lane markings. Figure 2 shows examples of HMI during automation in the Audi (A8) and Tesla systems.

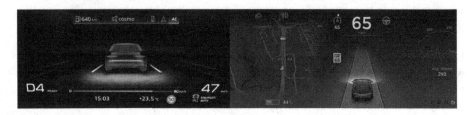

Fig. 2. Examples of HMI during automation. The left dashboard is by Audi for their A8 [44], the right dashboard is by Tesla [45] (copyright Tesla.com).

Literature Concepts. Suppleness. Two concepts [36, 46] use the social cue of showing 'hands on a steering wheel'. While the concept by [36] uses this to indicate manual driving mode, [46] uses it as a soft warning in case of potential hazards. Two concepts [47, 48] use facial expressions in emoticons as social cues to indicate the confidence of the automated system. [49] uses the tendency to engage in joint attention/gaze to redirect the driver's attention. Their concept contains three physical mini robots on the dashboard that turn their head from and towards the road ahead. The concept by [50] uses small talk to engage with the driver, which consisted of sentences that were either driving related or not.

Bodily experience. Six of the concepts provide multimodal feedback. These are combinations of auditory, visual and/or haptic stimuli. Eleven of the concepts include visual feedback, most of which are on displays (N = 9). Two concepts use lighting in their feedback [46, 49]. In [49] this is used to intensify the movement of the physical dashboard robots (as described above). [46] uses light in the windscreen as a soft warning to direct the driver's attention towards potential hazards in the driving environment. [49] is the only concept to use movement of mechanical objects in their HMI. Two concepts use tactile stimuli. [51] uses vibrations in different parts of the driver seat to indicate approaching vehicles. The concept by [52] consists of a high resolution haptic surface the driver can touch with his fingers. The authors report that the concept may be used for visually impaired passengers of automated cars, but an exact function of the device is not specified. The use of auditory feedback is split evenly between beeps and verbal statements. The study by [53] uses auditory icon sounds. They describe these sounds as "non-speech sounds that bear some ecological relationship to their referent processes". An example is a water gurgle sound to represent the message that fuel is running low.

Situatedness. Four of the reviewed concepts are adaptive to the driver in some form. The concept by [49] tries to engage the driver in looking at the road by personification (small robot looking at the driver and then looking at the road) when the driver is inattentive. Similarly, the concept by [51] only starts the vibration feedback if the driver is not looking at the road, to provide information about the surrounding traffic. The concept by [54] shows adaptive information on the driver's condition during automation. What this information exactly entails is not specified. While the study by [52] (15) is directed specifically at visually impaired drivers, the feedback is not dynamically adapted to the driver during automation. Almost all concepts are adapted to some degree to the driving situation (N = 13). They use a variety of combined methods to show adaptive feedback about the driving situation. Five of the concepts show the currently detected elements of the driving situation through a display, such as road users, lane markings and traffic signs. All of these five concepts also include the planned next action of the car, such as an upcoming turn or brake. [22] and [32] change the location of their feedback, which are respectively vibration and illumination, according to the detected hazards. While [50] uses casual remarks and questions about the driving situation to engage the driver, [55] adapts the verbal level of information according to the situation. For example, in some situations the system only mentions the current action "the car is braking", while in other situations it gives the reason why it is performing this action "the car is braking because a traffic jam is coming up ahead". [43] uses a direct overlay on the windscreen to show whether it is safe to continue in that lane after deactivation of the automation. Thee of the concepts display the confidence of the system to continue in automated mode [47, 48, 56].

Conclusion. The general HMI during automation in literature concepts shows a variety of supple social cues. These cues mainly include facial expressions, shared gaze, a 'hands on wheel' symbol and small talk. The bodily experience of literature HMI concepts is shows some variation. Only a small part of the literature concepts is adaptive to the driver. The ones that are, mainly show the driver condition or provide feedback if the driver is not paying attention. The feedback is adaptive to the driving situation in most concepts. In these concepts the feedback show the confidence of the automated system, the detected environment and detected hazards. Some concepts change the location of their feedback according to the environment and next actions of the car.

4 Discussion

The goal of this study was firstly to identify the current state of embodied design elements in driver support in partially automated cars. This way, new design spaces may be discovered to guide the design of innovative driver support in automated vehicles. To achieve this, partially automated car systems from literature and industry were reviewed from an embodied perspective. More specifically, we reviewed TOR feedback and the general HMI during automation on suppleness, bodily experience and situatedness.

Several opportunities for new designs were found in the current TOR feedback systems. Firstly, most commercial car systems do not provide a formal TOR since the driver is considered to continuously monitor the road. Rather the system disengages when it can no longer function with only a simple visual or auditory cue. While we recognize that this is most likely a technical limitation, implementing multiple incremental TORs before system disengagement may greatly improve the suppleness [57]. Especially as it can be very difficult for drivers to recognize themselves when the system reaches its limits, the system should indicate its limits as clearly as possible [11, 58]. Second, the use and variety of social cues was very limited in both commercial cars and literature concepts. Social cues may create more easily understood, fluent and accepted car-driver interactions. These may for example include social behaviour such as facial expressions, or gestures such as pointing or turning towards a joint interest [59]. Third, while literature showed an increasing variety of TOR output methods, TORs of commercial cars mainly kept to displays, beeps and steering wheel vibrations. It is important to transfer this development into commercial cars as dividing feedback to different senses may prevent overloading of the driver during take-over. Alternative output modes may be useful as drivers are engaged in non-driving tasks and not holding the steering wheel or looking at the dashboard. Lastly, both literature and commercial cars lacked situated feedback to the driver. This leaves a large opportunity to design driver adaptive feedback systems. The request may for example take the current activity of the driver into consideration. This is especially relevant in higher level automated cars where the driver may be immersed in different activities such as work. In order to create a safe mode transition the system should take the driver into consideration and adapt the feedback accordingly. This can be done not only by timing, but also by changing the location, intensity or modality of the information according to the driver's activity.

Design opportunities for general HMI during automation were also identified. First, although a few concepts with social cues were presented in literature, only 'hands on wheel' symbols were present in the commercial cars. Again, there is an opportunity to transfer more variety of social cues to commercial cars. The literature concepts already included facial expressions, small talk and mutual gaze. It is encouraged to expand the development of these and new cues to aid the driver in understanding the car through continuous fluent interactions [60, 61]. Second, the bodily experience in general HMI during automation mainly consists of visuals, audio and vibrations. This holds for both the commercial cars and literature concepts. An opportunity is found to include other senses that may be less obvious at a first sight such as smell [62, 63], taste and the vestibular sense [64, 65]. Although a few concepts use braking as vestibular feedback, it can be explored further as the lateral and longitudinal forces make up such a large part of the driving experience, and seems to be a natural cue for passengers of vehicles to respond so. Developing other forms of vestibular feedback may improve the situational awareness of drivers in automated mode while they perform non-driving activities [64]. Third, with regard to the situatedness, the HMI in commercial cars and literature are mainly adapting the timing of their message to the driver state. The form or message content however does not change. As previously stated, it may be necessary to design driver dependent feedback due to the different activities the driver may be engaged in during automation.

Some limitations of this review have to be taken into account. We recognize that we may have missed papers or car systems that would have been relevant to this review. The search terms described in the method section were carefully chosen, however they may still not include all relevant papers. New commercial or industrial concepts may have been missed in particular as the development of automated cars is currently proceeding so fast. Another limitation is that, as mentioned before, the three reviewed embodied characteristics (suppleness, bodily experience, situatedness) do not represent every aspect of embodied interaction. No method to review interactional systems on their embodiment exists. However, we chose to take these key elements of embodied interaction as a guideline to explore HMI in partially automated car systems, as they represent the main concepts.

In conclusion, we firmly believe that embodied interaction holds a great promise for all next generation automated vehicles. While often the industry aims to fight human factors issues by improving vehicle technology, we believe that this may even enlarge some classic human factors issues. Therefore, the role of self-explaining and supportive feedback will even become more important as technology improves. Embodied interaction holds a great promise for both the TOR feedback and general HMI during automated. For TOR feedback, new embodied designs are encouraged to focus espe- cially on the development of social cues, in- and output methods and adaptivity to the driver. For the general HMIs during automation, new embodied design opportunities are in the output methods and adaptivity to the driver. By including these embodied elements, we can create HMI designs that foster a more fluent and natural interaction movement between automation and manual driving, reducing the need to invest in extensive training. This entails keeping drivers in the loop during automation so they are not overwhelmed at transfer of control, and support fluent transfer back to manual driving. Including the key characteristics of embodied interaction in future HMI may create safer, more efficient and effective car-driver interactions in automated cars.

Acknowledgements. This research is supported by the Dutch Domain Applied and Engineering Sciences, which is part of the Netherlands Organization for Scientific Research (NWO), and which is partly funded by the Ministry of Economic Affairs (project number 14896).

Declaration of Conflicting Interests. The authors declared no potential conflicts of interest with respect to the research, authorship, and/or publication of this article.

Appendix A

Table 3. TOR feedback - commercial cars.

TOR – commercial cars								
		Suppleness			Bodily experience		Situatedness	
Car brand	System	Temporal output mode	Time to take over	Social cues	Output of TOR	Input to disengage automation	Adaptive to driver	Adaptive to driving situation
Audi [22, 44]	Audi AI traffic jam pilot	S	Up to 10 s	Y (Hands on wheel icon)	V (Display), A (Beep), C/E (Brake, Belt tightening)	P (Wheel, Pedals, Button)	N	N
BMW [31, 66, 67]	Steering & Lane Control Assist including Traffic Jam Assist	–	–	–	–	–	–	–
Cadillac [23, 68]	Super Cruise	Once, before deactivation	N/A	Y (Hands on wheel icon)	V (Display, Lighting), A (Beep), H (Vibration in seat)	P(Wheel, Pedals, Button)	N	N
Mercedes [32, 33, 69]	Drive Pilot	–	–	–	–	–	–	–
Tesla [30, 45, 70, 71]	Traffic aware cruise control and Autosteer	–	–	–	–	–	–	–
Volvo [72, 73]	Pilot Assist	–	–	–	–	–	–	–

Table 4. TOR feedback - literature concepts.

TOR – literature concepts							
	Suppleness			Bodily experience		Situatedness	
Source	Temporal output mode	Time to take over	Social cues	Output of TOR	Input to disengage automation	Adaptive to driver	Adaptive to driving situation
[41]	O, before hazard	D	N	V (Display), A (Beep), V/H (Shape change)	P (Wheel, Pedals, Button)	N	N

(*continued*)

Table 4. (*continued*)

TOR – literature concepts							
	Suppleness			Bodily experience		Situatedness	
[40]	O, before hazard	D	N	A (Beep), C (Vibration)	P (Wheel, Pedals, Button)	N	Y (Vibration location changes according to the required turn direction)
	O, before hazard	D	N	A (Beep), V/C (Shape change)	P (Wheel, Pedals, Button)	N	Y (Wheel shape changes according to the required turn direction)
[36]	S,I	"Sufficient time to take over"	Y(Hands on wheel)	V (Display)	P (Wheel, Pedals)	N	Y (Visual and text shows upcoming situation)
[38]	O, before hazard	D	Y (Use of female distressed voice)	V (Display), A (Verbal, Beep), C (Vibration)	P (Wheel, Pedals)	N	N
[37]	S	"Sufficient time to take over"	N	V (Display)	P (Button, Pedals)	N	Y (Intent and action of other road users, suggested action in this situation)
[50]	O, before hazard	D	N	A (Beep)	P (Wheel, Pedals)	N	N
	O, before hazard	D	N	A (Verbal)	P (Wheel, Pedals)	N	N
[74]	S,I	50 s	N	V (Display, Lighting)	P (Wheel, Pedals)	N	N
	S,I	50 s	N	V (Display, Lighting), C (Vibrations)	P (Wheel, Pedals)	N	N
[34]	S,I	20 s	N	A (Beeps)	P (Wheel, Pedals)	N	N
	O, during deactivation	D	N	A (Beeps, verbal)	P (Wheel, Pedals)	N	N
[75]	S	"A few minutes"	N	V (Display), A (Beeps)	P (Wheel, Pedals)	N	N
	S	"A few minutes"	N	V (Display), A (Verbal)	P (Wheel, Pedals)	N	N
[76]	O, before hazard	D	N	A (Beeps)	P (Wheel, Pedals, Button)	N	Y (Feedback is directional)
	O, before hazard	D	N	H (Vibration)	P (Wheel, Pedals, Button)	N	Y (Feedback is directional)
	O, before hazard	D	N	A (Beeps), C (Vibration)	P (Wheel, Pedals, Button)	N	Y (Feedback is directional)

(*continued*)

Table 4. (*continued*)

TOR – literature concepts

	Suppleness			Bodily experience		Situatedness	
[77]	O, before hazard	D	N	V (Display), A (Beeps, verbal)	P (Wheel, Pedals)	N	Y (TOR changes to the weather condition)
	O, before hazard	D	N	V (Display), A (Beeps, verbal)	P (Wheel, Pedals)	N	N
	O, before hazard	D	N	V (Display), A (Beeps, verbal)	P (Wheel, Pedals)	N	N
[78]	O, before hazard/deactivation	60 s	N	C (Vibration)	N/A	N	N
[79]	O, before hazard	D	N	C (Vibration)	P (Lever)	N	Y (Direction pattern corresponds to the suggested/required action of the driver)
[43]	O, before hazard	D	N	V (Display), A (Beeps)	P (Wheel, Pedals)	N	Y (Color of lane overlay AR adapts to the safety to continue on that lane)
[80]	S,I	10 s	Y (Hands on wheel)	V (Display),	P (Button)	N	Y (Lines show restrictions of trajectory after to)
[81]	O, before hazard	D	N	V (Display), A (Beeps)	P (Wheel, Pedals)	N	N
[39]	O, before hazard	D	N	V (Lighting), A (Beeps)	P (Wheel, Pedals)	N	Y (Location/movement direction of the lighting indicated suggested steering direction)
[46]	O, before hazard	D	N	A (Beeps)	P (Wheel, Pedals)	N	Y (Sound/type of TOR depends on the required actions)
	O, before hazard	D	Y (Hands on wheel)	V (Display), A (Beeps)	P (Wheel, Pedals)	N	Y (Color/type of TOR depends on the required actions)
	O, before hazard	D	N	V (Lighting), A (Beeps), C (Vibration)	P (Wheel, Pedals)	N	Y (Color/type of TOR depends on the required actions)
[42]	O, before deactivation	10 s	Y (Hands on wheel)	V (Display), A (Beeps), E (Brake Jerk)	P (Wheel, Pedals)	Y (Request also appears on phone when the driver is using it)	N
[35]	O, during deactivation	D	Y (Hands on wheel)	V (Display), A (Beeps)	P (Wheel, Pedals, Button)	N	N

Table 5. HMI during automation - commercial cars.

Feedback HMI during automation – commercial cars

Car brand	System	Suppleness			Bodily experience	Situatedness	
		Temporal output of HOWW	Time to place hands on wheel after HOWW before system disengages	Social cues	Output of HMI system	Form is adaptive to driver	Form is adaptive to driving situation
Audi [22, 44]	Audi AI traffic jam pilot	-	-	Y (Use of 'hands on wheel' symbol)	V (Display), C (Vibration), A (Beeps)	Y (Full TOR is started if driver is performing unallowed activities or keeps their eyes closed for long periods)	Y (Detected lane markings, detected vehicles, detected speed limit)
BMW [31, 66, 67]	Steering & Lane Control Assist including Traffic Jam Assist	S,I	N/A	Y (Use of 'hands on wheel' symbol)	V (Display), C (Vibration), A (Beeps)	Y (HOWW, if hands are off)	Y (Detected lane markings, detected vehicles, detected speed limit)
Cadillac [23, 68]	Super Cruise	S,I	N/A	Y (Use of 'hands on wheel' symbol)	V (Display, Lighting), C (Vibration), A (Beep, Verbal)	Y (Request to look back at the road, if eyes are off road)	Y (Detected lane markings, detected vehicles)
Mercedes [32, 33, 69]	Drive Pilot	S,I	N/A	Y (Use of 'hands on wheel' symbol)	V (Display), C (Vibration), A (Beeps)	Y (HOWW, if hands are off)	Y (Detected lane markings, detected vehicles, detected speed limit)
Tesla [30, 45, 70, 71]	Traffic aware cruise control and Autosteer	S,I	N/A	Y (Use of 'hands on wheel' symbol)	V (Display), C (Vibration), A (Beeps)	Y (HOWW, if hands are off)	Y (Detected lane markings, detected vehicles, detected speed limit)
Volvo [72, 73]	Pilot Assist	S,I	N/A	Y (Use of 'hands on wheel' symbol)	V (Display), C (Vibration), A (Beeps)	Y (HOWW, if hands are off)	Y (Detected vehicles, detected speed limit)

Table 6. HMI during automation - literature concepts.

	Suppleness	Bodily experience	Situatedness	
Source	Social cues	Output of HMI system	Form is adaptive to driver	Form is adaptive to driving situation
[47]	Y (Use of social gesture "uncertainty gesture", face and hands)	V (Display)	N	Y (An uncertainty symbol -face and hands- appears according to the situation)
[54]	N	V (Display)	Y (Shows detected driver condition)	Y (Detected environment. Next action of the car within the situation)
[50]	Y (Asks non-task related questions, small talk)	A (Verbal)	N	Y (Questions and remarks on the current driving situation)
[56]	N	V (Display)	N	Y (Level of certainty adapts to the situation and car capabilities)
[55]	N	A (Verbal), V (Display, VR)	N	Y (Level of information is adapted to the situation)
[36]	N	A (Verbal), V (Display)	N	Y (Detected environment. Next action of the car within the situation)
[49]	Y (Use head movement to redirect gaze, shared gaze)	C (Movement), V(Lighting)	Y (Trying to engage in mutual gaze is started when the driver is not paying attention enough to the road)	N
[82]	N	A (Verbal)	N	Y (Next action of the car within the situation)
[80]	N	V(Display)	N	Y (Detected environment. Next action of the car within the situation)
[43]	N	V(Display)	N	Y (Color of lane overlay AR adapts to the safety to continue on that lane)
[53]	N	A (Verbal, Auditory icon sound)	N	Y (Car state. Detected hazard)
[48]	Y (Frown or smile to suggest system confidence)	V (Display)	N	Y (Detected environment. Level of certainty adapts to the situation and car capabilities)
[51]	N	C (Vibrations)	Y (Message is only provided if driver is looking away from task)	Y (Location of vibration in seat changes according to detected vehicles)
[52]	N	C (Vibrations)	Y (Designed for visually impaired drivers)	NA
[46]	N	A (Beeps)	N	N
	Y (Use of 'hands on wheel' symbol)	V (Display), A (Beeps)	N	N
	N	V (Lighting), A (Beeps)	N	Y (Location and color of illumination changes to place and type/degree of hazard)

References

1. SAE International: Taxonomy and Definitions for Terms Related to Driving Automation Systems for On-Road Motor Vehicles (2016)
2. Fagnant, D.J., Kockelman, K.: Preparing a nation for autonomous vehicles: opportunities, barriers and policy recommendations. Transp. Res. Part A Policy Pract. **77**, 167–181 (2015). https://doi.org/10.1016/j.tra.2015.04.003

3. Tientrakool, P., Ho, Y.C., Maxemchuk, N.F.: Highway capacity benefits from using vehicle-to-vehicle communication and sensors for collision avoidance. In: IEEE Vehicular Technology Conference, pp. 0–4 (2011). https://doi.org/10.1109/vetecf.2011.6093130
4. Van Wee, B., Annema, J.A., Banister, D.: The Transport System and Transport Policy, an Introduction. Edward Elgar Publishing Limited, Cheltenham (2013)
5. Merat, N., Jamson, A.H., Lai, F.C.H., Carsten, O.: Highly automated driving, secondary task performance, and driver state. Hum. Factors J. Hum. Factors Ergon. Soc. **54**, 762–771 (2012). https://doi.org/10.1177/0018720812442087
6. Martens, M.H., Van Den Beukel, A.P.: The road to automated driving: dual mode and human factors considerations. In: IEEE Conference on Intelligent Transportation Systems, pp 2262–2267 (2013)
7. Endsley, M.R., Kaber, D.B.: Level of automation effects on performance, situation awareness and workload in a dynamic control task. Ergonomics **42**, 462–492 (1999). https://doi.org/10.1080/001401399185595
8. Saffarian, M., de Winter, J.C.F., Happee, R.: Automated driving: human-factors issues and design solutions. Proc. Hum. Factors Ergon. Soc. Annu. Meet. **56**, 2296–2300 (2012). https://doi.org/10.1177/1071181312561483
9. Banks, V.A., Stanton, N.A., Harvey, C.: Sub-systems on the road to vehicle automation: Hands and feet free but not "mind" free driving. Saf. Sci. **62**, 505–514 (2014). https://doi.org/10.1016/j.ssci.2013.10.014
10. De Winter, J.C.F., Happee, R., Martens, M.H., Stanton, N.A.: Effects of adaptive cruise control and highly automated driving on workload and situation awareness: a review of the empirical evidence. Transp. Res. Part F Traffic Psychol. Behav. **27**, 196–217 (2014). https://doi.org/10.1016/j.trf.2014.06.016
11. Carsten, O., Martens, M.H.: How can humans understand their automated cars ? HMI principles, problems and solutions. Cogn. Technol. Work, 1–18 (2018). https://doi.org/10.1007/s10111-018-0484-0
12. Brookhuis, K.A., de Waard, D., Janssen, W.H.: Behavioural impacts of advanced driver assistance systems–an overview. Eur. J. Transp. Infrastruct. Res. **1**, 245–253 (2001)
13. Farrell, S., Lewandowsky, S.: A connectionist model of complacency and adaptive recovery under automation. J. Exp. Psychol. Learn. Mem. Cogn. **26**, 395–410 (2000)
14. Stanton, N.A., Young, M.S.: Driver behaviour with adaptive cruise control. Ergonomics **48**, 1294–1313 (2005). https://doi.org/10.1080/00140130500252990
15. Endsley, M.: Situation awareness. In: Salvendy, G. (ed.) Handbook of Human Factors and Ergonomics, pp. 553–568. Wiley (2012). Chapter 19
16. Cognition (n.d.). In: Oxford Living Dictionaries. https://en.oxforddictionaries.com/definition/cognition. Accessed 21 Sept 2018
17. Wachsmuth, I., Lenzen, M., Knoblich, G.: Introduction to embodied communication: why communication needs the body. In: Embodied Communication in Humans and Machines, pp. 1–34 (2012). https://doi.org/10.1093/acprof:oso/9780199231751.003.0001
18. Clark, A.: Embodiment and explanation. In: Calvo, P., Gomila, T. (eds.) Handbook of Cognitive Science: An Embodied Approach, pp. 41–56. Elsevier (2008)
19. Anderson, M.L.: Embodied cognition: a field guide. Artif. Intell. **149**, 91–130 (2003). https://doi.org/10.1016/S0004-3702(03)00054-7
20. van Dijk, J., van der Lugt, R., Hummels, C.: Beyond distributed representation: embodied cognition design supporting socio - sensorimotor couplings. In: 8th International Conference on Tangible, Embedded and Embodied Interaction, pp. 181–188 (2014). https://doi.org/10.1145/2540930.2540934

21. Isbister, K., Höök, K.: Supple interfaces: designing and evaluating for richer human connections and experiences. In: Extended Abstracts on Human Factors in Computing Systems, CHI 2007, pp. 2853–2856 (2007). https://doi.org/10.1145/1240866.1241094
22. Audi MediaCenter: TechDay piloted driving. The traffic jam pilot in the new Audi A8, pp. 1–14 (2017)
23. Cadillac: CT6 SUPER CRUISE ™ Convenience & Personalization Guide (2018)
24. Isbister, K., Höök, K.: On being supple: in search of rigor without rigidity in meeting new design and evaluation challenges for HCI practitioners. In: Proceedings 27th International Conference on Human Factors in Computing Systems, pp. 2233–2242 (2009). https://doi.org/10.1145/1518701.1519042
25. Sundström, P., Höök, K.: Hand in hand with the material: designing for suppleness. In: Proceedings 28th International Conference on Human Factors in Computing Systems, CHI 2010, pp. 463–472 (2010). https://doi.org/10.1145/1753326.1753396
26. Mirnig, A.G., et al.: Control transition interfaces in semiautonomous vehicles: a categorization framework and literature analysis. In: Proceedings of the 9th International Conference on Automotive User Interfaces and Interactive Vehicular Applications (AutomotiveUI 2017), pp. 209–220 (2017)
27. Dourish, P.: Where The Action Is: The Foundations of Embodied Interaction. The MIT Press, Cambridge (2004)
28. De Jaegher, H., Di Paolo, E.: Participatory sense-making: an enactive approach to social cognition. Phenomenol. Cogn. Sci. 6, 485–507 (2007). https://doi.org/10.1007/s11097-007-9076-9
29. Audi: Owner's Manual (2018)
30. Tesla: Model S Owner's Manual (2018)
31. BMW (Bayerische Motoren Werke): THE BMW 7 SERIES (2015)
32. Mercedes-Benz: E-Class Sedan and Wagon Operator's Manual (2016)
33. Mercedes-Benz: Mercedes-Benz Intelligent Drive (2017). https://www.mercedes-benz.com/en/mercedes-benz/innovation/mercedes-benz-intelligent-drive/. Accessed 7 Nov 2017
34. van der Heiden, R.M.A., Iqbal, S.T., Janssen, C.P.: Priming drivers before handover in semi-autonomous cars. In: Proceedings 2017 CHI Conference on Human Factors in Computing Systems - CHI 2017, pp. 392–404 (2017). https://doi.org/10.1145/3025453.3025507
35. Naujoks, F., Mai, C., Neukum, A.: The effect of urgency of take-over requests during highly automated driving under distraction conditions. In: Proceedings 5th International Conference on Applied Human Factors and Ergonomics, AHFE, pp. 2099–2106 (2014)
36. Naujoks, F., Forster, Y., Wiedemann, K., Neukum, A.: A human-machine interface for co-operative highly automated driving. Adv Hum. Asp. Transp., 585–595 (2017). https://doi.org/10.1007/978-3-319-41682-3_49
37. Zimmermann, M., Bengler, K.: A multimodal interaction concept for cooperative driving. In: IEEE Intelligent Vehicles Symposium Proceedings, pp. 1285–1290 (2013). https://doi.org/10.1109/ivs.2013.6629643
38. Politis, I., Brewster, S., Pollick, F.: Using multimodal displays to signify critical handovers of control to distracted autonomous car drivers. Int. J. Mob. Hum. Comput. Interact 9, 1–16 (2017). https://doi.org/10.4018/ijmhci.2017070101
39. Borojeni, S.S., Chuang, L., Heuten, W., Boll, S.: Assisting drivers with ambient take-over requests in highly automated driving. In: AutomotiveUI 2016, pp. 237–244 (2016). https://doi.org/10.1145/3003715.3005409
40. Borojeni, S.S., Wallbaum, T., Heuten, W., Boll, S.: Comparing shape-changing and vibro-tactile steering wheels for take-over requests in highly automated driving. In: Proceedings 9th International Conference Automotive User Interfaces and Interactive Vehicular Applications, AutomotiveUI 2017, pp. 221–225 (2017). https://doi.org/10.1145/3122986.3123003

41. Kerschbaum, P., Lorenz, L., Bengler, K.: A transforming steering wheel for highly automated cars. In: IEEE Intelligent Vehicles Symposium, pp. 1287–1292, August 2015. https://doi.org/10.1109/ivs.2015.7225893

42. Melcher, V., Rauh, S., Diederichs, F., Widlroither, H., Bauer, W.: Take-over requests for automated driving. Procedia Manuf. **3**, 2867–2873 (2015). https://doi.org/10.1016/j.promfg. 2015.07.788

43. Lorenz, L., Kerschbaum, P., Schumann, J.: Designing take over scenarios for automated driving: how does augmented reality support the driver to get back into the loop? In: Proceedings of the Human Factors and Ergonomics Society Annual Meeting, pp. 1681–1685, January 2014. https://doi.org/10.1177/1541931214581351

44. Chnl, C.: 2018 Audi A8 Traffic Jam Pilot Test Drive (2017). https://www.youtube.com/ watch?v=BkcZ2OPmIq0. Accessed 21 Sept 2018

45. Tesla: Uw Autopilot is er (2015). https://www.tesla.com/nl_NL/blog/your-autopilot-has-arrived?redirect=no. Accessed 12 Dec 2018

46. van den Beukel, A.P., van der Voort, M.C., Eger, A.O.: Supporting the changing driver's task: exploration of interface designs for supervision and intervention in automated driving. Transp. Res. Part F Traffic Psychol. Behav. **43**, 279–301 (2016). https://doi.org/10.1016/j.trf. 2016.09.009

47. Beller, J., Heesen, M., Vollrath, M.: Improving the driver-automation interaction: an approach using automation uncertainty. Hum. Factors J. Hum. Factors Ergon. Soc. **55**, 1130–1141 (2013). https://doi.org/10.1177/0018720813482327

48. Rezvani, T., Driggs-campbell, K., Sadigh, D., Sastry, S.S., Seshia, S.A., Bajcsy, R.: Towards trustworthy automation: user interfaces that convey internal and external awareness. In: IEEE 19th International Conference on Intelligent Transportation Systems (2016)

49. Karatas, N., Yoshikawa, S., Tamura, S., Otaki, S., Funayama, R., Okada, M.: Sociable driving agents to maintain driver's attention in autonomous driving. In: 26th IEEE International Symposium on Robot and Human Interactive Communication, pp. 143–149 (2017)

50. Hester, M., Lee, K., Dyre, B.P.: "Driver take over": a preliminary exploration of driver trust and performance in autonomous vehicles. In: Proceedings of the Human Factors and Ergonomics Society Annual Meeting, pp. 1969–1973, October 2017. https://doi.org/10. 1177/1541931213601971

51. Telpaz, A., Rhindress, B., Zelman, I., Tsimhoni, O.: Haptic Seat for Automated Driving: Preparing the Driver to Take Control Effectively (2015). https://doi.org/10.1145/2799250. 2799267

52. Wi, D., Sodemann, A., Chicci, R.: Vibratory haptic feedback assistive device for visually-impaired drivers. In: 2017 IEEE SmartWorld, Ubiquitous Intelligence & Computing, Advanced & Trusted Computed, Scalable Computing & Communications, Cloud & Big Data Computing, Internet of People and Smart City Innovation, pp. 1–5 (2017)

53. Nees, M.A., Helbein, B., Porter, A., College, L.: Speech auditory alerts promote memory for alerted events in a video-simulated self-driving car ride. Hum. Factors (2014). https://doi. org/10.1177/0018720816629279

54. Gowda, N., Kohler, K., Ju, W.: Dashboard design for an autonomous car. In: Proceedings 6th International Conference on Automotive User Interfaces and Interactive Vehicular Applications, AutomotiveUI 2014, pp. 1–4 (2014). https://doi.org/10.1145/2667239. 2667313

55. Hock, P., Kraus, J., Walch, M., Lang, N., Baumann, M.: Elaborating feedback strategies for maintaining automation in highly automated driving. In: Proceedings of the 8th International Conference on Automotive User Interfaces and Interactive Vehicular Applications, pp. 105–112 (2016)

56. Helldin, T., Falkman, G., Riveiro, M., Davidsson, S.: Presenting system uncertainty in automotive UIs for supporting trust calibration in autonomous driving. In: Proceedings of the 5th International Conference on Automotive User Interfaces and Interactive Vehicular Applications, AutomotiveUI 2013, pp. 5:210–5:217 (2013). https://doi.org/10.1145/2516540.2516554

57. Gold, C., Damböck, D., Lorenz, L., Bengler, K.: "Take over!" how long does it take to get the driver back into the loop? Proc. Hum. Factors Ergon. Soc. Annu. Meet. **57**, 1938–1942 (2013). https://doi.org/10.1177/1541931213571433

58. Boelhouwer, A., van den Beukel, A.P., van der Voort, M.C., Martens, M.H.: Should I take over? Does system knowledge help drivers in making take-over decisions while driving a partially automated car? Transp. Res. Part F Traffic. Psychol. Behav. **60**, 669–684 (2019). https://doi.org/10.1016/j.trf.2018.11.016

59. Cassell, J., Bickmore, T., Vilhjálmsson, H., Yan, H.: More than just a pretty face. In: Proceedings 5th International Conference Intelligent User Interfaces, IUI 2000, pp. 52–59 (2000). https://doi.org/10.1145/325737.325781

60. Flemisch, F.O., Adams, C.A., Conway, S.R., Goodrich, K.H., Palmer, M.T., Schutte, P.C.: The H-metaphor as a guideline for vehicle automation and interaction (2003)

61. Damböck, D., Kienle, M., Bengler, K., Bubb, H.: The H-metaphor as an example for cooperative vehicle driving. In: Jacko, J.A. (ed.) HCI 2011. LNCS, vol. 6763, pp. 376–385. Springer, Heidelberg (2011). https://doi.org/10.1007/978-3-642-21616-9_42

62. Dmitrenko, D., Maggioni, E., Thanh Vi, C., Obrist, M.: What did i sniff? Mapping scents onto driving-related messages. In: Proceedings 9th International Conference on Automotive User Interfaces and Interactive Vehicular Applications, pp. 154–163 (2017). https://doi.org/10.1145/3122986.3122998

63. Kaye, J.J.: Making scents: aromatic output for HCI. Interactions **11**, 48–61 (2004). https://doi.org/10.1145/962342.964333

64. Peterson, B., Wells, M., Furness, T.A., Hunt, E.: The effects of the interface on navigation in virtual environments. In: Proceedings of the Human Factors and Ergonomics Society Annual Meeting, vol. 42, no. 21, pp. 1496–1500, 1696 (1998). https://doi.org/10.1177/154193129804202107

65. Chance, S.S., Gaunet, F., Beall, A.C., Loomis, J.M.: Locomotion mode affects the updating of objects encountered during travel: the contribution of vestibular and proprioceptive inputs to path integration. Presence Teleoperators Virtual Environ. **7**, 168–178 (1998). https://doi.org/10.1162/105474698565659

66. BMW (Bayerische Motoren Werke): DE BMW 7 SERIE. ASSISTENTIESYSTEMEN (2018). https://www.bmw.nl/nl/modellen/7-serie/sedan/ontdek/assistentiesystemen.html. Accessed 21 Sept 2018

67. eGearTv: BMW Driving Assistant Plus (7-Series) - POV Test Drive (2017). https://www.youtube.com/watch?v=JAfr8NXpJuI&t=417s. Accessed 21 Sept 2018

68. SlashGear: Cadillac Super Cruise First Drive on the 2018 Cadillac CT6 (2017). https://www.youtube.com/watch?v=T90LPU_JT7Q. Accessed 21 Sept 2018

69. MercBenzKing: 2018 Mercedes S Class Long - NEW Full Review Drive Pilot Assist Lights Distronic Plus Lane Keeping (2018). https://www.youtube.com/watch?v=SW3OfMlGrwA. Accessed 21 Sept 2018

70. Nick's Tesla Life: Tesla 8.0 New Dashboard View (2016). https://www.youtube.com/watch?v=GzwpCVzYwg0. Accessed 21 Sept 2018

71. Tesla Family: "Take Over Immediately" Tesla Autopilot Warning Test Model X (2017). https://www.youtube.com/watch?v=daHgi5qNUHA. Accessed 21 Sept 2018

72. Volvo: S90 Owner's Manual (2018)

73. Volvo Cars: Volvo Cars How-To: Pilot Assist (2017). https://www.youtube.com/watch?v=N5gmgqXY5FI. Accessed 21 Sept 2018
74. Blanco, M., et al.: Automated vehicles: take-over request and system prompt evaluation. In: Meyer, G., Beiker, S. (eds.) Road Vehicle Automation 3. LNM, pp. 111–120. Springer, Cham (2016). https://doi.org/10.1007/978-3-319-40503-2_9
75. Toffetti A, et al.: CityMobil: human factor issues regarding highly automated vehicles on eLane. Transp. Res. Rec. J. Transp. Res. Board. 1–8 (2009). https://doi.org/10.1258/itt.2010.100803
76. Petermeijer, S., Bazilinskyy, P., Bengler, K., de Winter, J.: Take-over again: investigating multimodal and directional TORs to get the driver back into the loop. Appl. Ergon. **62**, 204–215 (2017). https://doi.org/10.1016/j.apergo.2017.02.023
77. Walch, M., Lange, K., Baumann, M., Weber, M.: Autonomous driving: investigating the feasibility of car-driver handover assistance. In: Proceedings of the 7th International Conference on Automotive User Interfaces and Interactive Vehicular Applications, AutomotiveUI 2015, pp. 11–18 (2015). https://doi.org/10.1145/2799250.2799268
78. Duthoit, V., Sieffermann, J.M., Enrègle, É., Michon, C., Blumenthal, D.: Evaluation and optimization of a vibrotactile signal in an autonomous driving context. J. Sens. Stud. **33**, 1–10 (2018). https://doi.org/10.1111/joss.12308
79. Petermeijer, S.M., Cieler, S., de Winter, J.C.F.: Comparing spatially static and dynamic vibrotactile take-over requests in the driver seat. Accid. Anal. Prev. **99**, 218–227 (2017). https://doi.org/10.1016/j.aap.2016.12.001
80. Langlois, S., Soualmi, B.: Augmented reality versus classical HUD to take over from automated driving: an aid to smooth reactions and to anticipate maneuvers1. In: IEEE Conference on Intelligent Transportation Systems, Proceedings, ITSC, pp. 1571–1578 (2016). https://doi.org/10.1109/itsc.2016.7795767
81. Kim, N., Jeong, K., Yang, M., Oh, Y., Kim, J.: "Are You Ready to Take-over?" An exploratory study on visual assistance to enhance driver vigilance figure. In: Proceedings of the 2017 CHI Conference Extended Abstracts on Human Factors in Computing Systems, CHI EA 2017, pp. 1771–1778 (2017). https://doi.org/10.1145/3027063.3053155
82. Koo, J., Kwac, J., Ju, W., Steinert, M., Leifer, L., Nass, C.: Why did my car just do that? Explaining semi-autonomous driving actions to improve driver understanding, trust, and performance. Int. J. Interact. Des. Manuf. **9**, 269–275 (2014). https://doi.org/10.1007/s12008-014-0227-2

Self-driving Vehicles: Do Their Risks Outweigh Their Benefits?

Peng Liu[1(✉)] and Zhigang Xu[2]

[1] Tianjin University, Tianjin 300072, People's Republic of China
pengliu@tju.edu.cn
[2] Chang'an University, Xi'an 710064, Shaanxi, People's Republic of China

Abstract. For self-driving vehicles (SDVs), do their benefits to society outweigh their risks? Or their risks outweigh their benefits? Public responses to these questions were not yet surveyed previously. A total of 1032 participants in China were asked this question. Their answers showed that 42.4% thought that the benefits of SDVs are higher than their risks. However, more than 50% participants held other opinions: 20.3% participants believed that the risks of SDVs are higher than their benefits, and 37.3% participants thought that the benefits of SDVs are equal to their risks. Four demographic characteristics were found to affect participants' the benefit-risk perception of SDVs. Those who have heard about SDVs before the survey, male, older (≥ 40), and highly educated tended to perceive higher benefits versus risks. Our findings imply that the public do not show optimism toward SDVs. Effective risk communication is necessary to prevent SDVs from becoming another controversial technology in society.

Keywords: Self-driving vehicle · Benefit-risk perception ·
Demographic factors

1 Introduction

Mass adoption of autonomous vehicles (AVs) has the great potential to improve traffic and mobility. More than 70% of traffic crashes were thought to be directly or indirectly related to human error [1], which can be largely reduced through widely adopting AVs [2]. It also can increase the mobility for those currently unable to drive. It, with electrification, shared mobility, and connectivity innovations, may have profound environmental benefits in terms of reducing the greenhouse gas emissions and energy use [3–5]. On the other side, it also poses a new set of challenges about safety, security, legal liability, and regulation issues [6, 7]. For example, current road tests [8, 9] did not confirm the safety of AVs over conventional vehicles. AVs usually refer to those vehicles that can run in an automated-driving mode. Self-driving vehicles (SDVs) refer to the AVs with the full automation (the highest automation) and without any human intervention, which are also called fully AVs (FAVs).

A common question for any emerging technology is, do its benefits to society outweigh risks? It would be not a real question to the scientists and engineers working on SDVs in the vehicle industry. If the benefits of SDVs cannot greatly outweigh their risks, there is no reason to design, develop, and deploy them. From the viewpoints of

© Springer Nature Switzerland AG 2019
H. Krömker (Ed.): HCII 2019, LNCS 11596, pp. 26–34, 2019.
https://doi.org/10.1007/978-3-030-22666-4_2

the lay public, however, they may have different responses to this question. The history of certain controversial technologies such as nuclear power, genetically modified organisms, and nanotechnologies shows that technical experts/scientists and the lay people have conflicting responses to this raised question on the risks versus benefits of these technologies. Understanding public sentiment towards any emerging technology is pivotal because, historically, public perceptions and attitudes have shaped the direction and pace of technology development [10].

This raised question related to the relative risks versus benefits of SDVs is not yet discussed by previous studies, which is the focus in the current study. The risk perception literature on other technologies [11] tells us that public benefit and risk perceptions are driven by many internal and external factors, including cognitive and affective factors, media coverage and information exposure, risk communication and management-related factors. This paper investigates the associations between certain demographic factors (i.e., whether participants have heard about SDVs, gender, age, and education) and the benefit-risk perception.

1.1 Demographic Factors and the Benefit-Risk Perception

Here we simply operationalize familiarity as whether participants have heard about SDVs before the survey or not. The "familiarity hypothesis" in psychology argues that support for an emerging technology will likely grow as awareness of it expands [11–13]. For example, higher level of knowledge was found to positively correlate with more beliefs that benefits of nanotechnologies exceed their risks [11, 13] and with higher benefit perception and lower risk perception [12]. The AV literature does not have any studies on the relationship between familiarity and the benefit-risk perception. But, our previous work [6] observed that those participants who have heard of SDVs before the survey perceived higher benefits and lower perceived risk than those who have not heard of SDVs. In this study, we will examine the relationship between familiarity and the benefit-risk perception.

Gender difference in beliefs and attitudes of risks has been noted in the risk perception literature [14, 15]. Usually, females hold less positive attitudes toward emerging technologies and perceive higher risks from these technologies than males. This difference is weak but systematic. The current literature notices that female participants were less willing to use the AV technology [16] and less willing to pay extra money for adding partial and full automation on their next vehicle [17], but does not touch the gender difference in the benefit-risk perception. In this study, we assume that females perceive lower benefits versus risks than males.

Older adults expect to benefit the most from self-driving in mobility. However, due to their limitations in physical and cognitive functions, using the new technology would be a challenge for them. Younger people are more likely to desire the in-vehicle technologies and options [18], and express more interested in full automation and more willingness to pay for full automation [17, 19]. Another survey [6], however, showed that older participants perceived higher benefits of SDVs than younger participants. In this study, we suspect that age influences people's benefit-risk perception.

Education may affect people's perception and preference related to the SDV technology. An international survey [20] showed that higher educated respondents expressed more concerns about their vehicle transmitting data; however, another survey [6] did not find education affects the surveyed participants' perceived benefit and perceived risk of SDVs.

2 Methodology

The methodology can be found in our previous work [6]. The data used in this study was extracted from our current large-scale surveys on the general Chinese public's responses to SDVs in the future. Next we brief the methodological issues related to the current study.

2.1 Participants

An offline survey was conducted in Tianjin (a tier-2 city in China). Participants were approached through direct intercept by trained interviewers while at recreational areas. A total of 1032 participants submitted valid data for further analysis. Among them, 79.4% ($n = 819$) have heard about SDVs before the survey, 47.2% ($n = 487$) were female. Other demographic information is shown in Fig. 1. Our participants were skewed toward younger and highly educated compared to the general Chinese population.

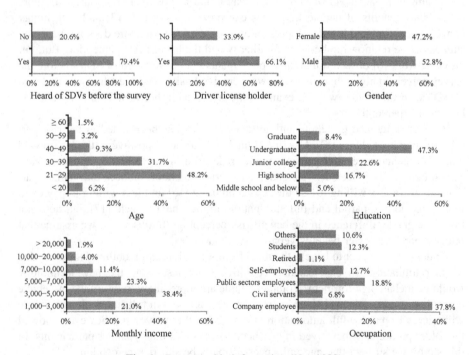

Fig. 1. Demographic information ($N = 1032$).

2.2   Procedure and Measure Design

After participants agreed to participate in the survey, they were instructed to read a short description and graphic scenario of SDVs on the cover page (to reduce response biases when participants were unfamiliar with this technology), respond to a series of items about SDVs, and give their demographic information [6]. The verbal description was as below [20]: "The automated driving system takes over speed and steering control completely and permanently, on all roads and in all situations. The driver or passenger sets a destination via a touchscreen. The driver or passenger cannot drive manually and perform interventions, because the vehicle does not have a steering wheel" (p. 131). A graphic scenario [21], for illustrating the application and utility of SDVs, showed the driver and passengers in rearward-facing seating arrangements having face-to-face interactions and read as follows: "The picture shows one possible application scenario of fully automated driving. Fully automated driving enables the driver (i.e., passengers) to perform more non-driving activities, such as reading a book, watching a film, surfing the Internet, playing their phones, dealing with their working affairs, sleeping, and so on and so forth. The driver and front seat passenger are able to swivel their seats and have a face-to-face communication and conversation."

In this study, we only focused on public responses to the benefit-risk perception. Participants were required to weigh the risks and benefits of SDVs to society on a five-point scale with the following five options: "risks far outweigh benefits", "risks outweigh benefits", "risks and benefits are the same", "benefits outweigh risks" and "benefits far outweigh risks" [22].

3   Results

Those who chose "risks far outweigh benefits" (2.4%) or "risks outweigh benefits" (17.8%) held the position of "risks > benefits" (20.3%; $n = 209$) (see Fig. 2); those who chose "benefits outweigh risks" (36.1%) or "benefits far outweigh risk" (6.3%) held the position of "risks < benefits" (42.4%; $n = 438$). Thus, those who perceived greater benefits outnumbered those who perceived greater risks by 2 to 1. The left 37.3% participants ($n = 385$) thought "risks = benefits".

We coded the five responses from "risks far outweigh benefits" to "benefits far outweigh risks" as 1–5 and then conducted a regression analysis to examine the relationships between the benefit-risk perception (as a dependent variable) and demographic factors (as predictors). Four demographic factors affected the benefit-risk perception (see Table 1). Those who have heard about SDVs before the survey, male, older (≥ 40), and highly educated (with a college degree or higher) tended to believe that the benefits of SDVs outweigh their risks.

We summarized the percentage of the three opinion positions—risks > benefits, risks = benefits, and risks < benefits—among different segments of participants on the basis of the four demographic factors that associated with the benefit-risk perception (see Table 2). The relative percentage of participants holding the two polarized positions was usually reversed between groups in terms of each demographic factor. For instance, in the "risks > benefits" position, the percentage of participants holding this

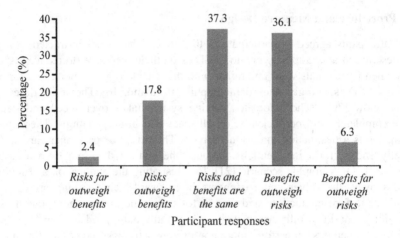

Fig. 2. Distribution of participant responses.

Table 1. Results of regression on benefit-risk perception.

Predictors	B	SE
Heard about SDVs (no = 0, yes = 1)	0.34***	0.07
Gender (male = 0, female = 1)	−0.24***	0.06
Age (<40 = 0, ≥40 = 1)	0.24**	0.08
Education (college and above = 1, others = 0)	0.21**	0.07
Occupation (civil servants and public sectors = 1, others = 0)	0.06	0.06
Income (>CNY5,000 = 1, others = 0)	0.03	0.06
Driver (driver license holder = 1, others = 0)	−0.10	0.06
$F = 9.17, p < 0.001, R^2 = 0.059$		

Note: B, unstandardized coefficients; SE, standard error. **$p < 0.01$; ***$p < 0.001$. 1 CNY ≈ 0.145 USD.

position in the male group (17.1%) was lower than that in the female group (23.8%). It was reversed in the "risks < benefits" position that the percentage of participants holding this position in the male group (49.0%) was greater than that in the female group (35.1%). This qualitative finding was also true for other three demographic factors.

Table 2. Benefit-risk perception by demographic factors

Perception	Heard about SDVs		Gender		Age		Education	
	No	Yes	Male	Female	<40	≥40	< college	≥ college
Risks > Benefits	33.3%	16.8%	17.1%	23.8%	20.8%	16.7%	26.0%	18.3%
Risks = Benefits	37.1%	37.4%	33.9%	41.1%	38.2%	31.9%	35.7%	37.9%
Risks < Benefits	29.6%	45.8%	49.0%	35.1%	41.0%	51.4%	38.4%	43.8%

4 Discussion

This analysis is one part of a multifaceted research that aims to understand public responses to SDVs. More than 50% participants (57.6%) did not believe that the benefits of SDVs outweigh their risks to society, and the left 42.4% thought benefits higher than risks. It might be an unwanted finding for promoters of SDVs.

SDVs might be a controversial technology in the future. We can see similar public responses to existing controversial technologies. For instance, Cobb and Macoubrie [13] surveyed the general American public's benefit-risk perception and found that sizeable percentage (38%) thought risks and benefits would be about equal, and slightly more (40%) thought that nanotechnology would produce more benefits than risks, while half that many (22%) said risks would outweigh the benefits.

Our finding would invoke findings on other kinds of public responses to SDVs. For example, a survey by Pew Research Center [23] showed that more Americans express worry than enthusiasm about the development of driverless vehicles: 40% were at least somewhat enthusiastic about this development, while 54% expressed some level of worry; 44% Americans say they would want to ride in a driverless vehicle if given the opportunity, whereas 56% Americans say they would not. We think, currently, the public do not show optimism toward SDVs.

Understanding the relationship between individual characteristics and the benefit-risk perception may provide practical insights for segmenting the future SDV market. Four demographic factors might influence the public's benefit-risk judgment. Participants who have heard about SDVs tended to perceive greater benefits versus risks. This finding might invoke the "familiarity hypothesis" in the risk-perception literature, which argues that higher level of familiarity forms more positive attitudes. Our previous study [29] also reported that those who have heard about SDVs were willing to pay more extra money for adding the SDV technology. Regarding the benefit-risk judgment, similar findings can be found in studies on nanotechnologies [11, 13]. The positive influence of familiarity is usually seen in the early period of a technology when media coverage may be more likely to communicate the benefits of this technology. When the public perceive more risks from the deployment of this technology, familiarity could induce a negative influence on the benefit-risk judgment.

Female participants tended to perceive greater risks versus benefits. This finding might invoke the gender difference identified in the risk-perception literature. Facing with emerging technologies, females usually express more concerns and worry for the same risks, perceive risks as more serious, take less risks, than males [14, 15, 24]. Different affective responses to technologies between females and males [12] might account for the gender difference in the benefit-risk perception. Hohenberger et al. [16] found that female participants expressed higher anxiety when they imaged to ride in an AV than male participants, which led them to have lower willingness to use AVs. Note that this gender difference might not be seen in other measures [25, 26, 29]. For example, we did not find the gender difference in the acceptable risk of SDVs [25] and the willingness to pay for the SDV technology [29].

Findings on the negative relationship between age and public acceptance of SDVs [17, 19, 29] may suggest a negative relationship between age and the benefit-risk perception. Unexpectedly, a positive relationship was observed: older participants (age $\geq$ 40] tended to believe higher benefits versus risks than younger participants. Similarly, our previous work [6] found that older participants perceived higher benefits of SDVs than younger participants and these two groups did not report different risk concerns. The relationship between age and the benefit-risk perception deserves more explorations. SDVs promise to largely increase the mobility of senior populations who cannot drive themselves. Their benefit-risk perceptions will largely determine whether the society gains this mobility benefit.

Highly educated participants tended to perceive greater benefits versus risks. We do not have clear and specific accounts for this finding. We guess that these participants have a higher expectation that this technology can assure the purported benefits or they have more resources to seek out and to credit benefit information of the SDV technology. We should note certain studies did not find the significant influence of education on people's benefit perception and risk perception [6] and on intention to use [27] related to SDVs. However, we noted that education was a positive predictor to people's willingness to pay for the SDV technology [29].

5 Conclusions

Our survey focused on the public's benefit-risk perception of SDVs that was not addressed in past studies. Our participants were asked whether they think the risks of SDVs to society far outweigh their benefits to society, their risks outweigh their benefits, their risks are equal to their benefits, their benefits outweigh their risks, or their benefits far outweigh their risks. In sum, we found that 37.3% Chinese participants perceived SDVs' benefits to society are equal to their risks to society, 20.3% thoughts SDVs' risks higher than their benefits, and the left 42.4% thought SDVs' benefits higher than their risks. Thus, more than 50% participants did not have optimistic perceptions about SDVs. Four individual characteristics significantly influenced the benefit-risk perception. Those who have heard about SDVs before the survey, male, older ($\geq$ 40), and highly educated (with a college degree or higher) tended to believe the benefits of SDVs higher than their risks. Reasoned public opinion does not spontaneously emerge from accumulating scientific information about the risks and benefits of a technology [28]. We suggest active, effective risk communication to prevent the SDV technology from becoming another controversial technology in society.

Acknowledgements. This study was supported by the Seed Foundation of Tianjin University (No. 2018XRG-0026).

References

1. Dhillon, B.S.: Human Reliability and Error in Transportation Systems. Springer, London (2007). https://doi.org/10.1007/978-1-84628-812-8
2. NHTSA: Federal automated vehicles policy: accelerating the next revolution in roadway safety. National Highway Traffic Safety Administration (NHTSA), U.S. Department of Transportation, Washington, D.C. (2016)

3. Anderson, J.M., Kalra, N., Stanley, K.D., Sorensen, P., Samaras, C., Oluwatola, O.A.: Autonomous Vehicle Technology: A Guide for Policymakers. RAND Corporation, Santa Monica (2016)
4. Arbib, J., Seba, T.: Rethinking transportation 2020–2030: the disruption of transportation and the collapse of the internal-combustion vehicle and oil industries. RethinkX (2017)
5. Greenblatt, J.B., Saxena, S.: Autonomous taxis could greatly reduce greenhouse-gas emissions of us light-duty vehicles. Nat. Clim. Change 5, 860–864 (2015)
6. Liu, P., Yang, R., Xu, Z.: Public acceptance of fully automated driving: effects of social trust and risk/benefit perceptions. Risk Anal. (in press). https://doi.org/10.1111/risa.13143
7. Marchau, V., Zmud, J., Kalra, N.: Editorial for the special issue–autonomous vehicle policy. Transp. Res. Part A Policy Pract. (in press). https://doi.org/10.1016/j.tra.2018.04.017
8. Favarò, F.M., Nader, N., Eurich, S.O., Tripp, M., Varadaraju, N.: Examining accident reports involving autonomous vehicles in California. PLoS ONE 12(9), e0184952 (2017)
9. Banerjee, S.S., Jha, S., Cyriac, J., Kalbarczyk, Z.T., Iyer, R.K.: Hands off the wheel in autonomous vehicles? A systems perspective on over a million miles of field data. In: Proceedings of the 48th Annual IEEE/IFIP International Conference on Dependable Systems and Networks (DSN), Luxembourg City, Luxembourg, pp. 586–597 (2018)
10. Currall, S.C., King, E.B., Lane, N., Madera, J., Turner, S.: What drives public acceptance of nanotechnology? Nat. Nanotechnol. 1, 153–155 (2016)
11. Satterfield, T., Kandlikar, M., Beaudrie, C.E.H., Conti, J., Harthorn, B.H.: Anticipating the perceived risk of nanotechnologies. Nat. Nanotechnol. 4, 752–758 (2009)
12. Kahan, D.M.: The evolution of risk perceptions. Nat. Nanotechnol. 4, 705–706 (2009)
13. Cobb, M.D., Macoubrie, J.: Public perceptions about nanotechnology: risks, benefits and trust. J. Nanopart. Res. 6, 395–405 (2004)
14. Gustafson, P.E.: Gender differences in risk perception: theoretical and methodological perspectives. Risk Anal. 18, 805–811 (1998)
15. Slovic, P.: Trust, emotion, sex, politics, and science: surveying the risk-assessment battlefield. Risk Anal. 19, 689–701 (1999)
16. Hohenberger, C., Spörrle, M., Welpe, I.M.: How and why do men and women differ in their willingness to use automated cars? The influence of emotions across different age groups. Transp. Res. Part A Policy Pract. 94, 374–385 (2016)
17. Bansal, P., Kockelman, K.M., Singh, A.: Assessing public opinions of and interest in new vehicle technologies: an Austin perspective. Transp. Res. Part C Emerg. Technol. 67, 1–14 (2016)
18. Owens, J.M., Antin, J.F., Doerzaph, Z., Willis, S.: Cross-generational acceptance of and interest in advanced vehicle technologies: a nationwide survey. Transp. Res. Part F Traffic Psychol. Behav. 35, 139–151 (2015)
19. Abraham, H., et al.: Autonomous vehicles and alternatives to driving: trust, preferences, and effects of age. In: 96th Annual Meeting of the Transportation Research Board. Washington, D.C. (2017)
20. Kyriakidis, M., Happee, R., de Winter, J.C.F.: Public opinion on automated driving: results of an international questionnaire among 5000 respondents. Transp. Res. Part F Traffic Psychol. Behav. 32, 127–140 (2015)
21. Diels, C., Bos, J.E.: Self-driving carsickness. Appl. Ergon. 53, 374–382 (2016)
22. Poortinga, W., Pidgeon, N.F.: Trust in risk regulation: cause or consequence of the acceptability of GM food? Risk Anal. 25, 199–209 (2005)
23. Smith, A, M. Anderson, M.: Automation in everyday life. Pew Research Center, Washington, D.C. (2017)
24. Harris, C.R., Jenkins, M., Glaser, D.: Gender differences in risk assessment: why do women take fewer risks than men? Judgment Decis. Making 1, 48–63 (2006)

25. Liu, P., Yang, R., Xu, Z.: How safe is safe enough for self-driving vehicles? Risk Anal. (in press). https://doi.org/10.1111/risa.13116
26. Xu, Z., Zhang, K., Min, H., Wang, Z., Zhao, X., Liu, P.: What drives people to accept automated vehicles? Findings from a field experiment. Transp. Res. Part C Emerg. Technol. **95**, 320–334 (2018)
27. Zmud, J., Sener, I.N., Wagner, J.: Self-driving vehicles: determinants of adoption and conditions of usage. In: 95th Annual Meeting of the Transportation Research Board, Washington, D.C. (2016)
28. Kahan, D.M., Braman, D., Slovic, P., Gastil, J., Cohen, G.: Cultural cognition of the risks and benefits of nanotechnology. Nat. Nanotechnol. **4**, 87–90 (2008)
29. Liu, P., Guo, Q., Ren, F., Wang, L., Xu, Z.: Willingness to pay for self-driving vehicles: influences of demographic and psychological factors. Transp. Res. Part C Emerg. Technol. (accepted)

Methodologies to Understand the Road User Needs When Interacting with Automated Vehicles

Evangelia Portouli[1(✉)], Dimitris Nathanael[1], Angelos Amditis[1],
Yee Mun Lee[2], Natasha Merat[2], Jim Uttley[2], Oscar Giles[2],
Gustav Markkula[2], André Dietrich[3], Anna Schieben[4],
and James Jenness[5]

[1] Institute of Communication and Computer Systems, 15773 Zografou, Greece
v.portouli@iccs.gr
[2] Institute for Transport Studies, University of Leeds, Leeds, UK
[3] Chair of Ergonomics, Technical University of Munich,
85748 Garching, Germany
[4] German Aerospace Center, Brunswick, Germany
[5] Westat, 1600 Research Blvd., Rockville, MD 20850, USA

Abstract. Interactions among road users play an important role for road safety and fluent traffic. In order to design appropriate interaction strategies for automated vehicles, observational studies were conducted in Athens (Greece), Munich (Germany), Leeds (UK) and in Rockville, MD (USA). Naturalistic behaviour was studied, as it may expose interesting scenarios not encountered in controlled conditions. Video and LiDAR recordings were used to extract kinematic information of all road users involved in an interaction and to develop appropriate kinematic models that can be used to predict other's behaviour or plan the behaviour of an automated vehicle. Manual on-site observations of interactions provided additional behavioural information that may not have been visible via the overhead camera or LiDAR recordings. Verbal protocols were also applied to get a more direct recording of the human thought process. Real-time verbal reports deliver a richness of information that is inaccessible by purely quantitative data but they may pose excessive cognitive workload and remain incomplete. A retrospective commentary was applied in complex traffic environment, which however carries an increased risk of omission, rationalization and reconstruction. This is why it was applied while the participants were watching videos from their eye gaze recording. The commentaries revealed signals and cues used in interactions and in drivers' decision-making, that cannot be captured by objective methods. Multiple methods need to be combined, objective and qualitative ones, depending on the specific objectives of each future study.

Keywords: Automated vehicles · Interactions among road users · Methodologies

© Springer Nature Switzerland AG 2019
H. Krömker (Ed.): HCII 2019, LNCS 11596, pp. 35–45, 2019.
https://doi.org/10.1007/978-3-030-22666-4_3

1 Introduction

Interactions among road users play an important role for road safety and fluent traffic [1]. A typical case is when Driver A wishes to turn left at a junction with oncoming traffic. The traffic in the oncoming lane may be so dense that Driver A is uncertain when it is safe to turn left. Driver A turns on the left indicator and waits. One of the oncoming drivers, Driver B, notices the left indicator and slows down while flashing the vehicle headlights. Driver A perceives this and starts turning left, since they anticipate that this will now be safe. Through similar communicative interactions, drivers in a way purposefully agree or settle on a common future motion plan, each one adapting their own planned future trajectory, so as to enable the safe execution of a manoeuvre.

The above example is a typical case of how humans use multiple means of implicit cues, such as approach speed, and explicit communication, such as eye contact and gestures, as well as vehicle signals, to anticipate the intention of the other road users. Previous research has identified a number of factors influencing both pedestrian-vehicle interactions and vehicle-vehicle interactions in different settings. Drivers can engage in explicit communication with other road users through the use of eye contact, hand gestures, flashing lights and indicator signals, or implicit communication strategies such as speed reduction [2]. Mutual eye-contact has been identified as a factor in facilitating safe interactions between vehicles and Vulnerable Road Users (VRUs) [3], with some research suggesting that establishing eye contact with a driver increases the likelihood that the driver will yield to a pedestrian [4]. Interview data [5] showed that drivers make use of a variety of techniques to force pedestrians to yield, including refusing to decelerate, speeding up, and driving more in the centre of the road to avoid hitting a pedestrian while not stopping for them. Finally, environmental factors such as traffic volume [6], darkness and weather conditions [7, 8], are also likely to affect crossing behaviour. Although the exact means of such interactions may vary across different regions and cultures, it is through such means that effective coordination of future motion plans between different road users is achieved. The phenomenon has not been studied in detail yet, especially as regards interactions among drivers.

Automated vehicles currently lack such interaction capabilities and their behaviour is mostly dominated by the rational principle of collision avoidance. This results in non-human-like, (robotised) behaviour of the automated vehicles, whose actions are not predictable by other road users, and can actually be quite frustrating. Therefore, to safely integrate automated vehicles in complex, mixed, traffic environments, in the future, one must ensure that the automated vehicles can interact with other road users in an intuitive, expectation-conforming manner. This will allow the surrounding road users to correctly interpret the intentions of the automated vehicles, and coordinate their planned actions accordingly.

In order to design appropriate interaction strategies for AVs, observational studies were conducted in Athens (Greece), Munich (Germany) and Leeds (UK), as part of the interACT project "Designing cooperative interaction of automated vehicles with other road users in mixed traffic environments", funded from the European Union's Horizon 2020 research and innovation programme, and in Rockville, MD (USA), as part of the

NHTSA-sponsored project, "Automated Vehicle Communication and intent with Shared Road Users." Both projects are connected by a twinning partnership organised by the EU and the US funding organisations. The aim of the studies was to identify interaction-demanding situations, and understand how road users resolve these in current traffic, focussing in particular on the explicit and implicit forms of communication. This paper presents the research objectives of each study, outlines the data collection methods used and provides an overview of the advantages and disadvantages of each method and of the main research purpose served by each method.

2 Observing Vehicles and Pedestrians' Interaction: Cameras and LiDAR Observations at European Test Sites

Cameras were placed at elevated locations (e.g. upper floor or roof of multi-storey building) in Athens, Munich and Leeds to record interactions at predefined use cases. pedestrian-vehicle and vehicle-vehicle interaction at unregulated intersections and shared space parking areas were chosen to identify how road users interact in these differently regulated areas. All videos were recorded in accordance to the data privacy policies of the individual countries. Overall around 600 h of video data was recorded across all locations using GoPro cameras in Athens and Munich and an HD wireless IP camera in Leeds.

The recorded videos served two purposes: (a) to review interactions in traffic that were manually observed and (b) to extract the positions of observed road users in each frame. Computer Vision algorithms can be used to extract kinematic information from videos. As no plug and play open source solution was found that was able to detect, track and classify road users in the recorded videos, existing algorithms were adapted and evaluated. These tracking algorithms ranged from simple blob tracking with background reduction to Histogram of Oriented Gradients [9] to convolutional neural networks trained on open source datasets (Fig. 1).

Fig. 1. Example images of blob tracking

The kinematic data extracted from the videos will help to understand which situations actually require explicit interactions between road users and which situations can be resolved by adjusting the approach velocity.

To have a more accurate account of traffic participant position and velocity, a ground-based LiDAR was utilized to receive synchronized quantitative measurements [10]. The LUX LiDAR sensor provides an object tracking with object properties position, size and velocity of traffic objects. The sensor was integrated in a housing with power supply, a hard disk storage and a GNSS receiver, to synchronize the LiDAR recordings with the video observations. Overall about 20 h of LiDAR data was recorded across all locations.

The point clouds generated by the LiDAR are merged to objects, classified and tracked using Python scripts. Polygons are manually generated used to recreate the road geometry and allow the re-identification of objects that were lost due to short time occlusions. The generated data will be used to understand how the kinematic behaviour from yielding vehicles differentiates from not yielding ones and the condition for cooperative traffic encounters.

3   Observing Pedestrians and Driver Behaviour: Manual Observations at European Test Sites

In addition to the video recordings, three researchers were positioned at each location in Athens, Munich and Leeds, to manually observe the vehicle-vehicle and pedestrian-vehicle interactions. The main purpose of the manual observations was to capture the presence and sequence of any explicit (e.g. hand gestures, signals, honking) and implicit event types (e.g. decelerated for pedestrian, stopped for traffic, accelerated) that was used between these observed road users while interacting with each other at the junctions.

During the data collection for pedestrian-vehicle interactions, three researchers positioned themselves at the designated location, where they were close enough to observe the interaction without interfering in the process. One of the researchers observed the behaviour of the pedestrian and one observed the behaviour of the driver/vehicle. The researchers also spoke out aloud about any event types that was being observed, and this material was recorded. After the end of the interaction (i.e. after a pedestrian had crossed the road), the two researchers then completed an HTML application that was specifically created to record any of these observed behaviours, demographic data of pedestrians observed, as well as the weather and infrastructure details of the observation site. The app also allowed an illustration of the trajectories of the observed road users, if required. The same procedure was conducted for vehicle-vehicle interactions, where one researcher observed the behaviour of one vehicle and one researcher observed the behaviour of the other vehicle.

For the pedestrian-vehicle interactions, a third researcher approached the pedestrian after they had completed their crossing, and asked if they wished to complete a short questionnaire, to provide a subjective measurement of their decision making while crossing the road. This questionnaire included questions about the types of information portrayed by the vehicle and driver that assisted in the crossing decision; how pedestrians themselves indicated their crossing intention; whether the presence of other pedestrian affected their crossing decision, and their familiarity of that particular crossing. These individuals were also asked to complete the Adolescent Road User Behaviour Questionnaire [11]. The data collected from the observation protocol was used to investigate which of these factors predict whether vehicle drove passed the pedestrian or whether pedestrian managed to cross in front of the vehicle, as well as the sequence of behaviours which led to a crossing.

4 Driving with an Eyeglass Mounted Gaze Sensor and Retrospective Commentary: An On-Road Study in Athens

An on-road, video-assisted observational study with retrospective commentary by drivers was designed and conducted so as to collect empirical evidence relevant to drivers' interactions with other drivers and pedestrians.

Twenty-one experienced drivers were asked to drive their own passenger car in a predefined urban course, while wearing an eye glass mounted gaze sensor. This system records the traffic scene from the driver's point of view and identifies the driver's eye-fixations points with a 50 Hz sampling frequency and gaze position accuracy of 0.5°. The course consisted of a circular route of 0.75 km which was driven 5 times by each driver. The total course length was 3.75 km and the mean driving duration was 18 min. The course included left turning from a two-way street and right turning from a smaller to a two-way street. Turns were not regulated by a traffic light and given the traffic density it was expected that there would be a lot of interactions between drivers relevant to the left and right turns. Example traffic scenes are shown in Fig. 2.

Fig. 2. Examples of eye gaze video recording relevant to left turn from two-way street with oncoming traffic (left) and right turn to two-way street (right)

After arriving at the lab, participants were introduced to the general setup and were calibrated on the eye-tracker, while seated on driver's seat their own passenger car, with a five-point procedure. Then they were instructed to drive at the selected site in their normal style and to repeat the selected course five times in a row. The driving duration was estimated to approximately 15 min.

Immediately following the driving session, participants returned to the lab and were asked to watch their eye-gaze video recording while commenting aloud on their behaviour and decision making for each case of interaction with another driver or pedestrian. The commentary was recorded trough video and voice capture software. Verbal protocols offer a way to record the human thought process [12] and have been used in driving studies [1].

Afterwards, an analyst watched the participant's eye gaze and scene video as well as his/her retrospective commentary, and labelled the interactions between the participant and another driver. An interaction start with another driver was defined as the time point when (i) the participant had to wait for a gap in the oncoming traffic before turning or (ii) the participant started turning knowing that the oncoming driver would have to modify his/her vehicle motion. For each interaction, the analyst labelled the type of the interacting vehicle and whether the other driver reacted. The signals or cues by the participant and his/her vehicle and by the other driver and his/her vehicle and their sequence were labelled for each interaction.

An interaction case with another pedestrian was defined when a pedestrian in the vicinity of the participant driver (i) affected the car movement and/or the driver's behaviour in an observable manner and (ii) received at least one eye-fixation from the driver. The starting point for each interaction case was defined by the observers according to the following criteria: either (i) the drivers' first fixation towards to the pedestrian or (ii) the first cue from the pedestrian interpreted as intention to cross. For each interaction case with a pedestrian, the video data were analysed by labelling the following indices: (i) participant-drivers' eye-fixations on the pedestrians, (ii) eye-contacts between pedestrian and participant-driver, (iii) cues denoting a pedestrian's projected direction (i.e. pedestrian's head orientation, body movement/orientation), (iv) cues denoting pedestrians awareness of the participant's vehicle (i.e. pedestrian's eye-gazes towards to the participant's vehicle). In addition, based on the video-assisted retrospective commentary (v) participants' expressed confidence about the future intended action of a pedestrian was noted when mentioned.

5 Real-Time Commentary Used to Study Shared Road Users' Interactions in Rockville, Maryland

The objective of the Rockville study was to determine the cues that drivers, pedestrians, and bicyclists frequently use when interacting with traffic to perceive drivers' intent and to predict vehicle manoeuvres.

Forty study participants (automobile drivers, pedestrians, and bicyclists) were recruited and trained to perform verbal commentary procedures while engaged in travel through intersections, merge lanes, parking lots, and other situations where interactions between road users occur. For each participant, data collection took place over two sessions including a supervised session where a researcher communicated with the participant, and a naturalistic, unsupervised session where the participant travelled independently without any communication with researchers.

Participating drivers wore a head-mounted GoPro 6 video camera that captured their approximate field of view and recorded audio of both their comments and the accompanying researcher's follow-up questions. Participants drove their own vehicles in this study.

Pedestrians in the study wore a head-mounted GoPro 6 video camera that captured their approximate field of view and recorded audio of their comments. During the supervised data collection session, the accompanying researcher wore a chest-mounted GoPro video camera to capture a view of the participant within the traffic environment and to record audio of follow-up questions. During the supervised data collection session, both the researcher and participant also used cell phones with wireless earpieces to maintain communications when they were not immediately next to each other.

For the bicyclist participants, two GoPro 6 video cameras were mounted on the participant's bicycle, one was pointed in the forward direction and one was pointed in the reverse direction. During the supervised data collection session, a smartphone was also mounted to the participant's bicycle handlebars with its face camera pointing toward the forward roadway. During the supervised data collection session, a researcher remotely viewed the forward scene and communicated with the participant through a cell phone application that provided a live video phone call. The participant wore a Bluetooth earpiece and microphone to hear and speak to the researcher. The cell phone was not used during the unsupervised data collection session.

For all participants data collection included a supervised session and a naturalistic, unsupervised session. In the first session (supervised) the participant travelled for approximately one hour along a predefined route that was chosen to include traffic situations where road user to road user communication may be necessary to avoid or resolve conflicts. Supervised sessions were scheduled on weekdays during time periods with greater vehicular traffic volumes including morning and afternoon commute hours and midday lunch hours.

At the start of the first data collection session, each participant was trained to use the video cameras and to perform the verbal commentary procedure. The researcher showed a video example of the verbal commentary procedure and then the participant engaged in approximately 10 min of practice traveling and commenting prior to starting data collection. The researcher provided navigation instructions. As needed, the researcher also prompted the participant to do more talking aloud, and reminded them to focus their comments on the cues that they were using to determine the actions of nearby vehicles and the intent of nearby drivers. Following traffic interactions where the participant commented, the researcher sometimes asked open-ended follow-up questions to elicit more information such as, "How did you know it was safe for you to cross the street?" or "You mentioned that the driver was going to stop for you, how did you know that?" In all interactions with participants, the researcher was careful to avoid biasing the participant toward reporting any particular vehicle-based or driver-based cues.

In the second data collection session (naturalistic, unsupervised), participants video recorded at least one hour of additional verbal commentary data as they travelled anywhere that they choose to go on public roads. Participants were instructed to restrict their travel to daylight hours and to travel during times of the day with moderate to heavy traffic volume. Pedestrians were urged to find routes that included many street crossings. Prior to conducting unsupervised session, the researcher reviewed instructions for conducting the verbal commentary procedure.

The predefined routes used in this study were located in and around Rockville, Maryland; an urban/suburban city with approximately 65,000 residents that is close to Washington, DC. Different routes were defined for drivers, pedestrians, and bicyclists.

Drivers. The 15-mile route involved driving on both local roads and highways. It took approximately one hour to complete. The route included merges, lane changes, navigating roundabouts, stop signs, right/left turns, U-turns, navigating parking lots, and periods of driving straight. Along the route there were sixteen scenario locations where participants were prompted to engage in verbal commentary. The researcher rode in the front seat of the participant's vehicle throughout the entire drive.

Pedestrians. The 1.3-mile walking route designed for the supervised data collection session involved signalized intersections, controlled intersections with stop signs, mid-block crossings, crossings with pedestrian signals, driveways, entrances and exits to retail establishments, and parking lots. In total there were eighteen planned scenario locations on the route where participants were prompted to engage in verbal commentary. The researcher walked near the participant throughout the session. However, at certain times, the participant was separated from the researcher by a short distance, for example, standing on opposite sides of a street. During these times communication was maintained using hands-free cell phones.

Bicyclists. The riding route designed for the supervised data collection session involved signalized intersections, controlled intersections with stop signs, mid-block crossings, right/left turns, driveways, entrances and exits to retail establishments, and bike paths. The route was a loop, approximately 3.5 miles long and participants were instructed to complete the route twice, once in each direction. There were fifteen planned scenario locations along the route where participants were prompted to engage in verbal commentary. Although the researcher did not physically accompany the participant during the ride, communication was maintained using hands-free cell phones running a video calling application.

Video and audio data were downloaded from the GoPro cameras onto a computer where the files were edited using Adobe Premiere software. For pedestrians and bicyclists, when two cameras were used for data collection, the two videos were synchronized and composited into a single split screen view.

Data were analysed separately for drivers, pedestrians, and bicyclists, and for supervised and unsupervised sessions. Researchers reviewed and manually coded the video data from both supervised and unsupervised sessions using Morae Manager software. Participants' comments concerning the cues that they used to assess vehicle manoeuvres and drivers' intentions were categorized and marked with their time-referenced position in the video. For supervised sessions, the comment markers were also referenced to the appropriate set of predefined scenario locations, such as "Roundabout 1" or "Midblock crossing 1." Similarly, cues identified in the data from unsupervised sessions were assigned to generic scenario locations, such as "midblock crossing," or "driving straight."

For the supervised sessions, data collected at each planned scenario location were combined across participants. For the unsupervised sessions, data were combined within the generic scenario locations. Subsequent analyses focused on determining the frequencies for use of different types of cues and how these frequencies differed by types of traffic scenarios. Implicit cues that signal intent such as vehicle movements were compared to explicit cues, such as use of a turn signal. Of special interest was determining how frequently cues coming directly from the driver, such as eye contact, or gestures were used because these cues will not be available or reliable from occupants of highly automated vehicles.

6 Discussion

The objective of the presented studies was (i) to support the development of safe AVs by developing kinematic models of behaviour during interactions and (ii) to inform system designers about important existing cues available from human-driven vehicles that may need to be replicated, or replaced by cues from the AV's operational behaviours and/or explicit external signals from an external human machine interface designed to communicate with nearby humans. It is assumed that to ensure safe and efficient interactions between AVs and shared road users, the AV's intent, for example, must be clearly legible, and compatible with shared road users' expectations. Communication through such cues is important for safe and efficient mixed traffic. There is little evidence, however, concerning the signals and cues used by the drivers to infer the future intention of other road users.

Naturalistic behaviour was studied, as it may expose interesting scenarios not encountered in controlled conditions.

A first type of methods used in the studies was video and LiDAR recordings. The aim of these recordings was to extract kinematic information of all road users involved in an interaction and to develop appropriate kinematic models that can be used to predict other's behaviour or plan the behaviour of an AV. While a stationary ground-based LiDAR suffers heavily from occlusion, it directly records points in space with relative positions matching the real world. This is very beneficial compared to the video recordings, as the video data loses accuracy due to distortion and homography. Furthermore, the LiDAR does not generate any personal data and can be used in situations, where installing an elevated camera is unfeasible. On the other hand, video data is simpler to understand and offers more information, such as head rotation or posture of pedestrians. For stationary high angle videos, blob tracking with background reduction works well in different lighting conditions but suffers from ID loss when a traffic participant stands still or gets too close to another road user. Furthermore, classifying tracked objects using the blob size is inaccurate. Detectors using deep learning methods will enhance the tracking results but require large training datasets.

Manual on-site observations of interactions provided additional behavioural information that may not have been visible via the overhead camera or LiDAR recordings. For instance, the camera would not be able to detect any hand movements from pedestrians who were facing away from the camera. On the other hand, these observations could be used to confirm the presence of certain behaviours, recorded by the videos. Using two different methodologies to collect the same data could therefore provide a redundancy gain. The manual on-site observations posed the difficulty that the researchers needed to complete a protocol very quickly, as the interaction frequently evolved very quickly. To assist this process, extensive effort was invested in creating a standardised observation protocol that could be used in different countries, with repeated piloting by members of the team. Following the pilot studies, a list of 98 observable event types was drawn up. Examples of these event types are: drivers' and pedestrians' looking behaviour, observed hand gestures, as well as signals used and movements observed, during an "approaching phase" (approaching the junction) and "crossing phase" (at the junction). At first, the protocols were developed in Microsoft (MS) Excel and tested by using a pen on printouts. To simplify the data extraction from the observation protocols, enable measurements synchronized in time and reduce the amount of paper used within the observation, the protocols were transferred into an HTML app that was programmed and usable on a variety on smartphones and tablets. The app enabled the researchers to quickly record any of these observed behaviours, as well as demographic data of pedestrians observed, as well as the weather and infras-tructure details of the observation site. The app also supported a sketching of the trajectories of the observed road users, if needed. Furthermore, the app enabled the synchronization of the different observation methods, by displaying the device's UNIX time to the observing cameras and logging every input by the ground observers with a timestamp and a sequence number. To be usable in areas with low mobile reception, the HTML app was specifically designed to work offline, saving each observed traffic interaction in a .csv file.

The data from observation protocol and questionnaires also provided an overview of the most common behaviours observed by pedestrians and car drivers during crossings at un-signalised junctions, across the three European cities, allowed the exploration on the frequency and to what extent that a particular event type occurred when the pedestrian and vehicle are at the junction and parking space.

To get a more direct recording of the human thought process, verbal protocols were also applied. Real-time verbal reports seem to provide a more complete and richer representation of pre-reflective cognition and deliver a richness of information that is inaccessible by purely quantitative data [13]. Still, they may pose excessive cognitive workload and remain incomplete. Participants in Maryland were trained to perform the verbal commentary in real time and were encouraged as much as possible to report what they were attending to in real time. In some of the data collection sessions, a researcher also asked participants follow-up questions to clarify and expand upon what had just been said. Such real-time commentaries may remain incomplete in complex traffic environments. This was the case in the Athens study, where due to the density of traffic, participants in the trial runs very frequently did not perform well in the real-time commentary. For this reason, a retrospective commentary was applied, which however carries an increased risk of omission, rationalization and reconstruction. This is why it was applied while the participants were watching videos from their eye gaze recording. The commentaries revealed signals and cues used in interactions and in drivers' decision-making, that cannot be captured by objective methods.

Understanding interactions and behaviour is a complex process and multiple methods need to be combined, objective and qualitative ones, depending on the specific objectives of each study.

Acknowledgement. This article is developed from the twinning partnership between the EU project interACT and the US project AVintent. Some work is a part of the interACT project. interACT has received funding from the European Union's Horizon 2020 research & innovation programme under grant agreement no 723395. Content reflects only the authors' view and European Commission is not responsible for any use that may be made of the information it contains. Funding for the study conducted in Rockville, MD was provided by the National Highway Traffic Safety Administration under contract DTNH2214D00329L, Task Order 0007.

References

1. Portouli, E., Nathanael, D., Marmaras, N.: Drivers' communicative interactions: on-road observations and modelling for integration in future automation systems. Ergonomics (2014). https://doi.org/10.1080/00140139.2014.952349

2. Fuest, T., Sorokin, L., Bellem, H., Bengler, K.: Taxonomy of traffic situations for the interaction between automated vehicles and human road users. In: Stanton, N.A. (ed.) AHFE 2017. AISC, vol. 597, pp. 708–719. Springer, Cham (2018). https://doi.org/10.1007/978-3-319-60441-1_68

3. Schneemann, F., Gohl, I.: Analyzing driver-pedestrian interaction at crosswalks: a contribution to autonomous driving in urban environments. In: 2016 IEEE Intelligent Vehicles Symposium (IV), pp. 38–43 (2016)

4. Guéguen, N., Meineri, S., Eyssartier, C.: A pedestrian's stare and drivers' stopping behavior: a field experiment at the pedestrian crossing. Saf. Sci. **75**, 87–89 (2015)

5. Šucha, M.: Road users' strategies and communication: driver pedestrian interaction. In: Proceedings 5th Conference Transport Solutions from Research to Deployment, Transport Research Arena, Paris (2014)

6. Harrell, W.A.: Factors influencing pedestrian cautiousness in crossing streets. J. Soc. Psychol. **131**, 367–372 (1991)

7. Klop, J.R., Khattak, A.J.: Factors influencing bicycle crash severity on two-lane, undivided roadways in North Carolina. Transp. Res. Rev. **1674**, 78–85 (1999)

8. Sayed, T., Zaki, M.H., Autey, J.: Automated safety diagnosis of vehicle–bicycle interactions using computer vision analysis. Saf. Sci. **59**, 163–172 (2013)

9. Dalal, N., Triggs, B.: Histograms of oriented gradients for human detection. In: 2005 IEEE Computer Society Conference on Computer Vision and Pattern Recognition (CVPR 2005), San Diego, CA, USA, vol. 1, pp. 886–893 (2005)

10. Dietrich, A., Ruenz, J.: Observing traffic – utilizing a ground based LiDAR and observation protocols at a T-junction in Germany. In: Bagnara, S., Tartaglia, R., Albolino, S., Alexander, T., Fujita, Y. (eds.) IEA 2018. AISC, vol. 823, pp. 537–542. Springer, Cham (2019). https://doi.org/10.1007/978-3-319-96074-6_56

11. Elliott, M.A., Baughan, C.J.: Developing a self-report method for investigating adolescent road user behaviour. Transp. Res. Part F **7**, 373–393 (2004)

12. Ericsson, K.A., Simon, H.A.: Protocol Analysis: Verbal Reports as Data. MIT Press, Cambridge (1993)

13. Banks, V.A., Stanton, N.A., Harvey, C.: What the drivers do and do not tell you: using verbal protocol analysis to investigate driver behaviour in emergency situations. Ergonomics (2014). https://doi.org/10.1080/00140139.2014.884245

Multi-methods Research to Examine External HMI for Highly Automated Vehicles

Melissa Cefkin[✉], Jingyi Zhang, Erik Stayton, and Erik Vinkhuyzen

Renault-Nissan-Mitsubishi Alliance Innovation Lab,
Sunnyvale, CA 94089, USA
`Melissa.Cefkin@nissan-usa.com`

Abstract. W*ill other road users know how to interact safely and effectively with highly automated and driverless vehicles (HAV's), especially in situations where there may be uncertainty, such as turn-taking and stop intersections?* Focusing on interaction-rich, urban contexts, we studied roadway practices among todays' drivers, pedestrians, bicyclists, and motorcyclists. Based on results from these field studies, we have been developing and testing concepts for *using external Human-Machine Interfaces* to signal intent. In this paper we provide a summary of this work based on a multi-methods program of research.

Keywords: Highly automated vehicle · Roadway interactions · External HMI

1 Introduction

In this paper, we report on a research program designed to explore the following questions: will other road users know how to interact safely and effectively with *highly automated vehicles (HAVs) when drivers are not in control*, especially in situations where there may be uncertainty, such as when two vehicles arrive at nearly the same time at a stop intersection? What will be the cumulative impact of HAVs on traffic flow? How will people's feelings of safety, comfort and trust evolve in traffic that includes HAVs?

Specifically, we explore the potential of additional signaling on the exterior of the vehicle as a means to communicate information about the HAVs to other road users. The aim of this paper is to recount the approach we have taken that has led to our current perspective on external signaling, which is that signals to indicate HAV vehicle state and intent hold value and should be implemented. We argue that a range of research methods and approaches are necessary to assess the problem and potential solutions for future roadway interactions with HAV solutions, and moreover, that the problem must be approached as both a scientific and design problem. This is due to the indeterminate and speculative state of knowledge from today's world.

The problem of roadway interaction is fundamentally a problem of inter-subjective interactions. Therefore, to understand and interpret how these interactions happen and how varying technological changes will affect interactions on the road requires social analysis. In this paper we describe the program of empirical examination and testing we have performed to examine these questions. Our backgrounds in the theories and methods of anthropology and ethnomethodology have significantly informed this program of research.

© Springer Nature Switzerland AG 2019
H. Krömker (Ed.): HCII 2019, LNCS 11596, pp. 46–64, 2019.
https://doi.org/10.1007/978-3-030-22666-4_4

Below we briefly describe a prototype we have developed for communicating intent and operational state by an AV. We then elaborate the investigations we have performed that have helped us evolve our prototype solutions and have led us to conclude that AV state and intent markers hold promise for traffic interactions in a future with highly automated vehicles. Nonetheless, we cannot expect to derive definitive answers from today's world to the questions of whether road users will know how to interact effectively with autonomous vehicles in urban settings, trust having them in their presence, or benefit from having additional signaling. The conditions for testing such eventualities, which rely on broad-scale enduring and collective practices, do not exist. We conclude with a discussion of the limitations of this research in light of that reality.

1.1   The Problem of Communicating Intent

In order to navigate through traffic, particularly in urban environments, road users participate in a steady stream of fleeting interactions with others on the road. Road users must anticipate and judge such things as: whether others are going before or after them; whether another road user is maintaining or changing speeds; whether a road user is going right away or pausing first; or whether a stopped body/vehicle will stay stopped or is about to go. Experienced drivers make these judgments constantly and typically without great difficulty.

Road users also express their own actions and intentions through both deliberate and inadvertent signaling. For instance, a driver may adjust his or her speed upon approach to a mid-block crosswalk where a pedestrian is waiting to cross. The pedestrian would likely interpret that movement as a signal that the driver was yielding. Someone driving beside a driver who turns his or her head to look over their shoulder, would likely assume that the driver is looking for a gap to perform a lane change.

These actions become *inter*-actional when two or more participants coordinate to establish who is doing what. Establishing this agreement may be almost instantaneous and imperceptible to the naked eye. Further, people may bring assumptions about a situation into their actions, such as varying behaviors in a small, rural town versus a large city. A crucial problem for HAVs is to be able to participate effectively in such *interactional* moments.

HAVs present two broad challenges to this set of practices. First, perception systems and reasoning systems are far from sensitive enough to be able to observe and interpret the movement and actions of others in communicative terms. Second, the kinematic cues that the HAV gives off will be different, though perhaps only very subtly. These differences will potentially increase the predictability of HAVs over time, as people learn what to expect about how they do and do not move, and thus have positive potential. Such differences, nonetheless, will limit their ability to engage actively in micro-negotiations as people do. Without the control of a human driver to manage those critical and fairly common situations in which actual interaction is required to negotiate with other road users, participants may be confounded.

1.2   The Intention Indicator

Building the systems to support HAVs in taking socially appropriate actions on the road is one of the key challenges for autonomous vehicle development. Signaling has the potential to assist in meeting this challenge by providing information that is difficult or impossible to observe otherwise, and by reinforcing other vehicle signals. We have devised a communication signal, which we call an "Intention Indicator", with the aim to provide other road users with information about the HAV so as to help mitigate and manage difficult roadway situations.

Based on our research, we are proposing a simple light signal viewable from 360° that expresses the state of the HAV in three conditions: (1) when it is stopping or stopped, (2) when it is going, and (3) when it is about to go. Importantly, the Intention Indicator is designed to communicate only about its own state, and not to instruct others directly. And it is designed to offer a distinct signal that is viewable at a glance/out the corner of the eye. The goal of new signaling would be to affect people's confidence in interacting with an autonomous vehicle. It is important to focus on generating the *appropriate* level of confidence. The goal of intention indication would be to provide sufficient information about the HAV's actions to help people establish foreknowledge that they can apply to their own decision-making and actions. It is important to avoid communicating information that is not valid (misinformation), or may not be accurate (leading to over-trust).

1.3   Background Research

There is a growing body of literature on studies relevant to the problem of interaction with HAV. These studies range from foundational studies on roadway interactions to studies of drivers' interactions with autonomous systems to reports on testing of potential signaling solutions. Vanderbilt (2009) cites a large number of relevant studies, including studies on eye contact, traffic flow, etiquette, and trust. The survey studies of Parkin et al. (2016), Merat and Madigan (2017), and (Schieben et al. 2018) more specifically focus on the question of road user to HAV interaction and the potential of markings and signals to impact those interactions. A few highlights from this literature follow.

Risto et al. (2017) identify "movement gestures" as the primary means by which vehicle intent is expressed. Clamann et al. (2017) tested prototype signals for communicating information about vehicle state and intent, and concluded that pedestrians relied primarily on "legacy motions" of vehicles (e.g., accelerating, decelerating) for their decision making when interacting with a single Wizard-of-Oz automated vehicle. Respondents nonetheless expressed a preference for additional signals from the vehicle.

However, research has shown that movement is not a sufficient guide to action in all circumstances. In instances of uncertainty and where priority is unclear, body motion, such as advancing and retreating, waves, and eye gaze may also play a role. Kitazaki and Myhre's (2017) examination of unmarked intersections found that in certain cases, such as cross-traffic turns, the use of hand gestures in addition to slowing motion has a considerable impact on decision-making effectiveness. At the same time, Vanderbilt (2009) reminds us that there are many occasions in which eye context is not used:

traffic may be moving too fast, there may be an obstruction, or it may simply be unnecessary or inappropriate. Indeed, Dey and Terken (2017) have concluded that eye contact is rare. Vanderbilt argues, nonetheless, that it is in part *because* of this rarity that the significance of eye contact, when it does occur, is heightened. When eye contact is present, agents are more likely to engage in cooperative behavior, a reason, he conjectures, that people avoid offering eye contact when they do not want to give way in their actions. Eye contact, then, could be seen as an escalation of a situation in which the other is ignoring a (perhaps more subtle) communicative act, suggesting, for instance, 'you act like you haven't seen me.'

"Ghost driver" testing, where the driver is disguised or pretends to be otherwise occupied (e.g., reading a newspaper) has been used to try to directly examine what will happen in roadway interactions with driverless vehicles. (Ford 2017; Rothenbucher et al. 2016). Ghost driver tests have shown that in a majority of cases people pay little meaningful regard to the lack of a driver. Rothenbucher et al. (2016) found that in only two of 67 instances did pedestrians seemed to hesitate and walked around the back of a ghost driven car as it exited a parking lot.

Other researchers have tested concepts for communicating vehicle state, intent and mode of operation externally, in preparation of interacting with autonomous vehicles. Madigan (2016) found that the primary signal desired by other road users was confirmation that they had been seen by the autonomous vehicle. Langstrom and Lundgren (2015) found that signals for communicating intent could be understood. Studies by Bockle et al. (2017); Habibovic et al. (2019) and Clamann (2017) found that knowing operational mode and intent improved pedestrian comfort.

In the remainder of this paper we summarize the investigations we performed that led to this concept.

2 Research Questions and Approach

Based on an assumption that HAVs will need to move through mixed-use environments in which they will intermingle with road users, including manual drivers, bicyclists, pedestrians and motorcyclists, we have been engaged in a robust program of research aimed at understanding roadway interactions and testing concepts for communicating intent. Since our aim is for HAVs to engage effectively with other road users we not only considered new signaling but also the development of vehicle 'behavior'.

2.1 Research Questions

Three overarching questions have framed our research program.

Q1: How can road users interact and communicate with HAVs safely and effectively?

- How do people communicate intent in traffic currently? How are interactions on the road achieved as a social practice?
- Are there local and regional variances to driving and communication practices?
- What happens in particular roadway configurations such as 4-way stops?

- When and how do interactions on the road affect people's sense of comfort or unease on the road?

Q2: Could additional external communication signals support road user-to-autonomous vehicle interactions?

- When, if ever, is movement insufficient for communicating intent?
- How are ambiguous situations in roadway interactions resolved?
- What factors generate complexity, and how does the complexity of the situation affect what transpires?
- Would additional signaling add distraction or confuse other road users?

Q3: What broader effects could a signal to communicate intent have?

- Would new visual signals help people resolve ambiguous traffic situations more quickly or slow them down? Would it improve or hinder the quality of the decisions made?
- Would indicating intent make road users feel more comfortable?
- Would the effectiveness of signaling intent increase or decrease in more complicated traffic situations? (e.g., many road users, many HAVs)
- Would intent signaling vary in its effectiveness by settings (e.g., regions, nations) or situations (e.g., stop intersections versus signaled crossing), and if so which settings or situations are relevant?
- Would intent indication signals create added distraction or confusion?

3   Research Activities and Results

Roadway interactions are more than just the sum total of cognitive and motor-body action of individual actors. They occur as a part of interactional dynamics shaped by psychological, social and cultural expectations. We must not only enable HAVs to proceed down the road, but also ensure comfortable, fluid, and socially acceptable road use for all.

It is essential that we consider the everyday understandings and expectations that people themselves bring to varied situations. We therefore strongly favor field study with observational and participatory methods, as well as iterative prototyping and regular participant feedback[1]. Many kinds of people other than passengers and drivers are affected by vehicle actions, not just pedestrians, bicyclists but also workers like school crossing guards and garbage collectors. We have aimed to remain sensitized to the voices of these different stakeholders during our research.[2]

---

[1] This work has been supported by collaborations with researchers at the DesignLab at the University of California, San Diego (Don Norman, Jim Hollan, Colleen Emmenegger and others), the University of Edinburgh (Eric Laurier and students) and Stockholm University (Barry Brown and students).

[2] Student groups at both San Jose State University (in courses taught by Dr. Jan English-Lueck) and the University of North Texas (in courses taught by Dr. Christina Wasson) have conducted class projects that consider these, and other topics, and have been welcome thought partners in this process.

We advanced our research from several angles, engaging in a range of activities as a part of a holistic approach to the topic. These have consisted of (1) exploratory research, (2) ideation and concept development, and (3) testing.

3.1 Empirical Field Studies

Our research began with a series of field observations of road use in the Bay Area of California.

Stop Intersection Observations. We started by examining traversals in un-signalized intersections such as 4-way stops. We identified two locations for comparative observations on 4-way stops. In each location, two researchers gathered data by conducting two to three ~45 min observations at both peak and non-peak hours. We gathered video recordings from both fixed-point cameras (3rd person perspective) and from cameras mounted on researchers as they crossed the intersection as pedestrians. We also conducted "intercept" interviews, stopping pedestrians who traversed the intersection for brief interviews.

One location was a 4-way stop in the commercial district of a small town in the Silicon Valley area. We portray this as a "destination" intersection, with the following characteristics:

- The intersection of a 2-lane road
- Marked crosswalks
- On-road parking near-by
- Wide sidewalks and bulb-out corners (designed for pedestrian comfort)
- Cafes and stores immediately at the corners and lining the streets
- Stop-intersections at each (short) block in each direction

The second location was a 4-way stop on a college campus in the Silicon Valley area. In contrast to above, this setting appears to have been designed to support the flow of a higher density road. We portrayed this as a "pass-through" intersection with the following characteristics:

- A 4-lane road with additional dedicated left and right-turn lanes in one direction.
- An intersecting 4-lane road with a dedicated right-turn lane, and merging into a single lane for continued traffic following the intersection.
- Considerable distance between stops.

Travel Alongs. A second round of study focused on first-person examinations using "travel-along" methods. We recruited six participants, three to observe as drivers and three to observe as pedestrians. With each participant we:

- Conducted an interview about their driving/walking history and experience with transportation;
- Performed a drive/walk of about 45 min in an area familiar to the subject;
- Video recorded the journeys from at least two perspectives, (1) the participant's view of the journey and (2) from a secondary observation perspective following the participant;

- Reviewed the video recordings with the participant in the days just following the drive/walk with the participant.

Comparative and International Studies. We have gathered further data in a range of locations. These include observations in multiple locations in San Francisco; observation in Beijing, Paris, Sao Paulo, and Yokohama.

Findings. Our primary analytical activity has been in-depth interaction analysis of the video-recorded material. Interaction analysis is an empirical method for investigating the interaction of human beings with each other and with their environments and is useful in "identifying routine practices and problems and the resources for their solutions" (Jordan and Henderson 1995). We have reviewed our interviews to identify themes and perceptions that arise from people's experiences on the road.

Our research has reconfirmed that movement is the primary way in which action and intent are communicated on the road today. Understandings of movement are socially and culturally embedded in people's expectations of behavior. This means that, for example, a pedestrian crossing at a marked crosswalk but where the traffic must yield is likely to interpret a car's slowing as a meaningful action: "that car is slowing down *for me*".

HAVs, as we asserted at the start, may disrupt those expectations and be unable to participate with similar degrees of social awareness in the kinds of micro-negotiations that go on today in performances of roadway coordination. In cases of uncertainty, people *inter*-act in order to resolve troubles and clarify actions. Inadvertent communications sent by body motion, acceleration or deceleration, and start/stop actions are a key part of this process. In addition, people sometimes communicate directly using eye contact, gaze, speech and hand gestures to clarify order and actions.

3.2 Testing

Working in an iterative manner, we created initial concepts after our first round of field study. The concept we developed was to assist in this negotiation process. Early on we identified lighting configurations which would vary in terms of representation (icon, colors, words, movement patterns), and in terms of timing and position (forward or side facing, front grill, rooftop) as promising directions. We have continued to evolve these concepts through many iterations into our current 3-state Intention Indicator concept.

Our first goal in testing has been to determine the value of the *concept* of intent indication, and only after to determine the design of the concept. Nonetheless, in the case of communicative technologies, form and function are tightly coupled, and therefore it has been necessary to elaborate somewhat the specific form and function of the Intention Indicator.

A further step was to consider in more detail how an intention indicator would function. Should it function like a turn signal which can emit highly contextual information – for instance, waiting until after passing through an intersection to turn on a right-turn signal meant to signal an impending right turn into a driveway – or like a brake-light, activated mechanically whenever the brake pedal is pressed? This question has remained as an on-going focus of our research.

Lo-fi Prototyping. For early lo-fi prototyping, we outfitted a small remote-control car with strips of lights on the front. Our primary goal in this early round of testing was to evaluate the promise of a lighting concept for expressing vehicle motion and intent to other road users. We were interested to gather feedback on the impact of the signal on both decision-making and perception. Secondarily we aimed to gather early feedback on the strengths and weaknesses of different lighting configurations. We tested this concept by having people interact with the vehicle in both controlled and uncontrolled encounters. Our aims in this test were:

- To see if the concept is understandable
- To examine participants' reactions to the prototype in various interaction contexts
- To explore participants' reactions to various concept implementations of signals on an LED light strip

We recruited participants in our work facility who had only a passing under-standing of the concept and told them to imagine the remote-controlled car as a full-sized, autonomous vehicle. Researchers controlled the remote-control car from a distance while the participants interacted with it in both controlled settings (simulated street configurations) and random settings (simulating an uncontrolled encounter, such as a sudden jaywalker, on the road) (Fig. 1).

Fig. 1. Remote-control car and 'random' experiment setting for lo-fi prototyping

Findings. The first result of the lo-fi prototype testing was to reconfirm that vehicle motion provides context for interpretability. When the car was moving and would slow and stop, participants could understand the meaning of the light signal. When the car was stopped and the lights on the LED strip changed, the meaning was less clear to participants on their first interaction, additional interactions were required for the meaning of the signal to become clearer.

Second, the concept was understandable and had implications on the actions that participants took. Participants interpreted the signal as communicating their safety to cross, the next action of the car, or its detection of a pedestrian. Some participants used the lights to predict the next action of the car before the car motion changed.

Third, we tested both signal color lightings that changed in hue or intensity, as well as a variety of colors and/or movement patterns (blinking and light-strip motions forward/back). Participants preferred the Red-Yellow-Green lighting and commented that red, yellow, and green signals were already familiar on the road and had meaning in roadway contexts. (This configuration, however, is disallowed by motor vehicle regulations.)

Finally, we tested both relative signals (e.g. color spectrums or frequency variations) and discrete state signals (e.g. color changes or solid vs. blinking). Discrete states were easier to understand than spectrum states. Discrete states were easier to see and "felt more decisive" to participants. Frequency and brightness spectrums were unfavorable as they needed a comparison point for reference, and small changes were difficult to detect. This has been further validated in real world testing where glancing behavior makes relative states difficult to assess. What the discrete color changes leave out, however, is any indication of the direction of an impending change: that is, if the light on a moving vehicle is yellow–meant to convey "slowing", – is it slowing to stop or about to accelerate again, or in other words, is the vehicle about to become more aggressive or more yielding?

Survey. We used a web-based survey to collect public feedback on the Intention Indicator concept. We tested for overall comprehension of the signals' purpose—were they understood to be communicating about the vehicle's action or state—as well as of their specific meanings and configurations. The survey included video clips of a Nissan Leaf outfitted with a strip of LED lights operating to communicate five vehicles states: Slowing Down, Waiting, Planning to Go, Starting to Go and Going. Various light configurations (light color and motion) of each vehicle state were tested (Fig. 2).

The survey consisted of single-choice questions on general demographics, concept identification, concept configuration comparison, as well as open-ended questions to collect additional feedback. A total of 145 participants accessed the survey, among which 92 complete responses were gathered.

Findings. Results (Zhang et al. 2018) confirmed that the Intention Indictor should be self-referential, that is, it should signal its own state/intent and not instruct others. Despite strong associations with the red-yellow-green, the precise meaning of the colors was more difficult to disambiguate than movement. Green was slightly more frequently associated with a moving/accelerating vehicle, but the question remained whether the color expressed the vehicle's movement or was intended as a signal to other road users about their movement. Finally, the five-mode configuration was

Fig. 2. Nissan Leaf equipped with Intention Indicator for survey study

unclear. Participants were able to recognize Slowing Down, Waiting, and Starting to Go, but were confused by Planning to Go (confused with Waiting) and Going (confused with Starting to Go)[3].

On-Road in Test Environment Using Wizard-of-Oz Test Method. Neither the remote-controlled car experiment nor the survey provided realistic interactions with the Intention Indicator concept. We thus developed a manually operable Wizard-of-Oz test car (a human-driven vehicle made to look like an autonomous vehicle). The vehicle was outfitted with an LED strip across the front and sides and operated in different modes and colors. We marked the car with decals suggesting it was an autonomous vehicle, and drove the vehicle around the NASA Ames campus where Nissan performs some AV testing and where people are used to seeing the test vehicles. The roads on campus are two-way with speed limits of either 15 or 25 miles per hour. There are marked and unmarked crosswalks and stop sign regulated intersections.

The human operator wore a visor to reduce the likelihood of eye contact and drove with his hands positioned low on the steering wheel so as to minimize visible evidence of steering tasks. The LED strip was controlled by an operator in the backseat of the car. We mounted cameras on the vehicle facing forward and to the side, and inside the cabin towards the driver and passenger. The whole test was approved by both the NASA Safety and Human Subjects Institutional Review committees, and both signage and a broadcast announcement were used to announce the testing to participants on campus.

---

[3] This confusion may have been compounded by testing biases. The participants were only presented with videos of the signals on a stopped vehicle, and there is nothing inherent in the meaning of the signals. Rather, as observed in the lo-fi prototype testing, the signals only had meaning in context of motion on the road.

There were three parts to the test:

1. Driving the vehicle marked as an HAV, but without turning the Intention Indicator on to establish a baseline
2. Driving the vehicle marked as an HAV while running the Intention Indicator manually
3. Intercepting pedestrians who interacted with the vehicle for brief interviews

Findings. Results from these initial road tests were limited. Traffic was minimal and we completed but few interviews. We determined that the lights were difficult to see and there was no clear evidence that they were noticed or understood to be signals about the vehicle actions. With only one-time viewings and without introduction to the concept, we did not have any validation that people understood the meaning of the signals. Rather, vehicle motion was the primary basis of communicating vehicle actions and intent. At the conclusion of the NASA testing, we were left with the question of whether more complex settings might result in different outcomes.

Road Test in Public Environment Using Wizard-of-Oz Test Method. Next we sought to test in real-world conditions that were more densely populated and where we would be likely to encounter more frequent interactions. We also aimed to involve knowing participants so that we could gather feedback from people who had actually noticed encountered the vehicle with the mock-Intention Indicator.

For the second road test we selected a busy commercial area with short blocks, wide sidewalks and stop signs on every block of a 10 square block area. We conducted the experiment in the summer when the area is active with people strolling, gathering for drinks or dinner, sitting in cafes, and shopping. The vehicle was outfitted with an enhanced prototype for improved visibility and aesthetic appeal. Although the vehicle was manually controlled, it was marked as an autonomous vehicle with a technician operating the Intention Indicator signals from the back seat. (This arrangement of safety driver plus technician mimics current testing practice of the actual HAV.)

The Intention Indicator was positioned over the A-pillar of the vehicle (Fig. 3). The design we tested used green lights with the following configuration:

- Solid: "go" mode
- Pulse (slow fade in and out): "stopping" or "stopped" mode
- Flashing: "about to go" mode, a few moments prior to the visible initiation of movement

We recruited 21 participants, including seven drivers, five bicyclists, nine walkers. Each participant was given a route to walk, drive or bike. The aim was to try to multiply their encounters with the test vehicle. Two walkers and one bicyclist recorded video as they followed their route (mounted via a hat, chest harness, or handlebar mount) and three drivers had front-facing cameras mounted inside their windshields. The research vehicle was equipped with three front-facing cameras mounted on the roof (and simulating HAV vehicle sensors) and a mounted inside the vehicle pointed at the driver. Four static cameras on tripods were positioned at intersections for recording data from third-person perspectives.

Fig. 3. Intention Indicator prototype used in public road test

We provided an introduction to the test, in which we introduced the goal of the experiment, described the plan for the testing, and introduced the Intention Indicator.[4] We emphasized that our core interest was in how people perceived and would interact with HAV, and whether additional signaling would assist. With the WoZ car parked nearby, we showed the three states of Intention Indicator and defined what they were designed to mean. This brief introduction (total of about 10 min) was conducted in an open-air setting in a near-by park. All participants were instructed to put their own comfort and safety first, and could stop participation at any time if they needed to.

The test ran for about 50 min on a Thursday evening following work hours. During the 50 min the research vehicle passed through the two selected intersections 38 times while the participants continued on their route.

At the conclusion of the test, participants gathered in a café for focus groups discussion, divided into groups of three. The focus groups were relatively informal, designed to hear from the subjects about their experiences. Audio records were made. A survey was also administered.

Following the experiment, the video data was processed and reviewed in an effort to identify whether any discernible differences could be seen in the way road users interacted with the research vehicle. We looked for evidence of hesitation, glance behavior, and deference or assertiveness towards the research vehicle.

In a final step, we conducted post-interviews with eight participants: one driver, one bicyclist, and six pedestrians. These participants returned to our lab several days later to

---

[4] We referred to the vehicle as "the research vehicle". Subjects took this to mean that the car was autonomous, though it was in fact manually controlled. We encouraged them to act as they normally would on the road, while recognizing that participants had had no prior "normal" interactions with HAV.

review video data of their interactions with the research vehicle together with a researcher, and share further reflections. These post-interviews were video recorded.

Findings. During the 50-min test, the following actions were observed:

- 38 crossings of the recorded intersections by the research vehicle
- 334 'interactions' (148 with cars, 163 with pedestrians, 23 with bicycles)
- 237 instances when another road user went ahead of the research vehicle (58% of cars, 85% of pedestrians, 52% of bicyclists) (counted from the moment when the research vehicle was the first vehicle approaching the intersection)
- 97 instances when another road user yielded to the research vehicle (42% of the cars, 15% of pedestrians, 48% of bicyclists)
- 19 instances when either another road user or the research vehicle went out of turn (5.6%)

Our analysis of the video-recorded data did not show that people behaved differently in interactions with the research vehicle as compared to similar analysis of road users interacting with non-HAVs.[5] During the experiment there was a lot of traffic and the pedestrian activity quite high, which, combined with this being a downtown and destination, contributed to the sense that this was a pedestrian area, with most cars yielding to pedestrians patiently.

From our recruited participants we gathered the following insights:

- Most subjects reported that they treated the research vehicle as a regular car.
- One participant reported that her observation of the Intention Indicator did influence her decision-making in one instance. However, we could not corroborate her story from the video data[6].
- A number of participants reported paying little attention to the Intention Indicator. The vehicle's movement and location in time and space with respect to the other road users already provided salient and familiar resources for interpreting vehicle intent[7].
- When asked, a majority of participants indicated they liked the idea of the Intention Indicator.
- A number of participants commented that they were unaccustomed to looking for the Intention Indicator and that in case they had been, they might have paid closer attention to it. This is in contrast to having been trained to look for brake lights and turn signals—observing these signals had become second nature.

---

[5] At one point during the experiment, a man purposely tried to stop the vehicle in the middle of the intersection by walking out and stopping in the crosswalk in front of the vehicle. He repeated this gaming behavior for several more crossings, until one of the experimenters spoke with him and took him aside. This incident shows that HAVs may encounter some harassment by members of the public, as has been reported on occasion in the press.

[6] It is unclear if one out of 163 interactions per 50 min of road use represents a low or high impact rate. A similar question is raised by the results of Rothenbucher et al's (2016) Ghost Driver study. Baseline measures do not yet exist to assess the relative significance of the small number of participants observed to take a different action vis-à-vis HAV.

[7] The fact that the research vehicle was driven manually may have further contributed to this feeling.

- Participants commented that a Red-Yellow-Green signal system would be more familiar and easier to understand than our design.
- The most noticeable state of the Intention Indicator was when it was flashing "about to go".
- The pulsing "not my turn" state was not reported to have helped participants decide whether to could cross in front of the stopped research vehicle.

In response to the survey question *"Did you feel the light strip provided any value to your encounter?"* responses were distributed with a strong majority responding that it would be of some or great value. In the focus groups many people expressed that an autonomous vehicle *should* communicate. On the survey question *"Should Autonomous Vehicles communicate what they are doing?"* a majority of respondents replied "yes". It is interesting that even though video analysis did not yield a discernible direct effect of the Intention Indicator and many respondents reported not paying attention to the signals, the sentiment after-the-fact remained strong that automated vehicle communications are a good idea.

In conclusion, the WoZ-car road test resulted in a strong positive reaction to the *idea* of the Intention Indicator concept, while results on its effectiveness remain inconclusive. The following considerations should be noted:

- The Intention Indicator lights were not that noticeable. The lights were not that bright and the "not my turn" state was especially difficult to notice.
- Subjects had only just been introduced to the Intention Indicator concept, and their encounter was brief. There was thus minimal time to learn the meaning of the signal or to adapt their behavior accordingly.
- Vehicle motion provided a more familiar means of communication than a new signal.
- The research vehicle was driven in a careful and polite manner, and in a way that was consistent with other vehicles in the setting. The HAV's behavior may have generated confidence in the subjects and other road users that it was a normal car. While an actual HAV could also be expected to drive carefully as well, differences in movement patters (such as slow accelerations through an intersection traversal, as have been commonly reported in today's HAV) could yield different results.
- Though the streets were busy, all traffic had to stop every block and speeds remained low. Given the compact space and 4-way stop configuration, forfeiting turns had little cost to participants as there would be minimal delay before they could take another spot in the flow.

What this setting did not allow us to observe, therefore, is what would happen in the occasions where such practices are challenged or break-down, and where vehicle movement gestures, which an HAV would not be guaranteed to exhibit, would be needed to effectively resolve the conflict.

Multi-HAV Simulation Testing. A further question we have examined is how people might respond if most vehicles they encounter simultaneously were equipped with Intention Indicator. (A limit of road tests to date is that participants often encounter just one car at a time.) We created a semi-immersive simulation to examine manual drivers' responses to interactions with multiple HAVs at once. Results of this study are being prepared for publication. Here we provide a brief introduction to the study.

The multi-AV study involved using a semi-immersive testing set up consisting of monitors at 180° and a steering plus pedal set up to simulate a driving context. Videos of actual driving scenes were edited in post-production to add different Intention Indicator markings. We tested three configurations, as well as control group with no signals. The street scenes involved intersections where the driver would encounter up to five simulated HAV at a time. Participants wore eye-tracking glasses, and were tested for their ability to understand the signals, for any variance in their actions, and for their opinions.

The multi-HAV simulation study provided insight on (1) the ability of participants to recognize and make sense of the Intention Indicator, (2) participants' preferences regarding Intention Indicator configurations, and (3) and the potential for the Intention Indicator either to create distraction, and hence disruption in traffic flow, or to enhance traffic flow. A majority of the participants reported that the Intention Indicator would be useful or very useful for future interactions with autonomous vehicles. Crucially, the results did not indicate that the Intention Indicator would introduce additional distractions in any of the conditions tested. In fact, results indicated that the Intention Indicator with a correct configuration could potentially improve traffic flow at 4-way stop sign intersections with multiple AVs present.

4 Discussion and Conclusions

Each of our research activities has deepened our understanding of how people interact on the road, and how they perceive their roadways experiences. Based on our research to date, we have concluded that signaling state and intent has the potential to provide a valuable contribution to future roadway interactions with highly automated vehicles when driver is not present or not in control. While behavioral effects have yet to be observed in real life testing, our results reveal that:

1. Signaling intent will have the potential to improve flow when multiple HAVs are encountered at once;
2. Users have expressed strong expectations for signaling from HAVs;
3. The value of signaling is likely to increase with familiarity and training.

Moreover, vehicle manufacturers are responsible for communicating vehicle operational state. Today this is achieved internally through standardized displays of machine operational states directed at the driver. In the future, we must increasingly consider how other road users will understand HAVs' operational states.

Intention indication will not be useful in all occasions, just as today the value of turn signals and brake lights varies by context. For instance, if a vehicle is in a turn lane, the turn signal is at that moment redundant, as are brake lights when a vehicle is approaching a red light. Rather, intention indication would make generally available to other road users information about a vehicle's state and intentions that could be useful in those occasions in which uncertainty requires micro-negotiations with others.

The primary goal of this research to date has been to assess the value of the *idea* of using a signal to communicate vehicle state and intent, rather than to test executions of the concept, which should follow from the results of the concept testing. Nonetheless, as with many other HMI, form and function are highly coupled and the idea cannot be assessed separate from its execution. Thus we have also surmised some important design considerations for how the Intention Indicator should be executed. The Intention Indicator should:

- Communicate about the HAV's own state rather than explicitly instructing others
- Be visible at eye level of pedestrians, bicyclists, and drivers
- Consist of discrete, rather than continuous, signal-states
- Avoid textual or iconic messages that require directed gaze to be seen or understood
- Be designed in such a way that the messages are discernible at a glance without sustained or focused attention; it must function and be discernible in highly dynamic, 'noisy' environments
- Color use will need to coordinate with regulatory standards globally.

Finally, a large area of consideration is how the Intention Indicator pairs with vehicle motion. Without careful pairing of signaling with actual HAV actions, there is a high risk of creating over-trust, confusion or other effects equally detrimental to future road use. Signaling performance should not be designed separate from the vehicle behavior programming.

4.1   Beyond Limits to This Study: Anticipating Future Effects

In the program of research described herein we have attempted to explore possible effects of having HAVs on the road, and the potential need for and impact of new kinds of signaling. We explored the issues from a range of perspectives, using a range of research methods. Each research and testing method is limited.

Simulation studies are useful for controlling variables, but are limited in at least two key ways: they lack real consequences for the actions taken by participants, and they tend to project unavoidably artificial scenes. *Real-world testing* has the benefit of realistic settings and consequences, but it too has at least two important restrictions: it is difficult to perform across sufficient cases and settings, and what can be ascertained about people's experiences is partial—both externally observable and reported data are only a subset of what people actually experience. From *video-based interaction analysis* we can learn a tremendous amount about the actions people take and make strong assertions about why. For instance, if in a video we see a pedestrian start to walk out in front of a vehicle but then retreat back to the curb while turning her head, we can fairly safely assert that upon seeing the approaching traffic, she determined that that moment was not the best time to cross. But we may not be able to know for certain whether the pedestrian felt unsafe (e.g., "it is risky for me to go now"), or whether she thought it would be more polite to wait (e.g., "there is a gap behind that car so why should I make it slow down and wait for me when I can go soon anyway"?). We may not know if they felt they had 'miscalculated' to begin with, or whether they felt a change in the environment precipitated a change in their action. From *eye-tracking analysis* we can get a robust and accurate understanding of what is in road users' visual field, but we may not be able to capture enough contextual elements needed to study road user interactions.

In other words, there is no perfect testing method. This is why we have used a host of different approaches in the effort to build our understanding. Further, and perhaps more consequentially, the practices and conventions that do emerge will take shape iteratively, over time, in real-world encounters between people and vehicles. This is why research to date, including the research program recounted here, cannot provide conclusive answers to what the future of roadway interactions with HAV holds, nor the significance of intention indication. We conclude this paper with an extended treatment of these challenges because we hope to encourage HMI researchers, designers and developers to recognize that they must inhabit multiple roles, one part scientist, one part designer.

Results from testing in today's world remain approximations and inevitably involve a degree of speculation. The reasons for this are numerous. Firstly, and most profoundly, people do not yet have a meaningful conceptual appreciation of *autonomous* vehicles. In form and in behavior, they will be as any other car on the road. The behavioral differences between HAVs and regular cars will likely be subtle and perhaps infrequent, and people will therefore likely discover such differences over time. Consider, for comparison, the early days of mobile phones. It took some time for conventions to develop around the use of phones in public, and new technologies emerged to address changing expectations (e.g., ear buds and directional sound). Further, over time the mobile phone has become less a "phone" and more a miniature-computer. This has giving rise, for instance, to the action of someone standing still at the curb while interacting with their device (punching in mapping information? searching a restaurant recommender app?). This has altered not only what we consider acceptable etiquette in face-to-face interactions—we now tolerate the many buzzes and rings that may emit from our interlocutor's device—but further how we "read" people's actions on the road. In short, practices evolve as technologies change and familiarity grows. While the actions of the phone user in public may be mostly a matter of etiquette, actions of people vis-à-vis HAV in the future could have implications to safety and trust in the highly consequential condition of traffic.

Relatedly, other aspects of HAV exterior form and visual state continue to evolve. Will HAV be marked with painting or signage, or will sensors remain visible in ways that make it evident that the vehicle is an HAV? The sensor suites on today's test vehicles tend to make them highly visible, while production vehicles with sensors capable of lower-level automated driving are not (advanced ADAS). In other in-progress research we are performing, we are finding that people tend to be aware of the HAV test vehicles as "different" vehicles.

People may expect to interact with HAVs the same as they do with any other car on the road today. Or they may note that it is a different type of vehicle and therefore expect to interact differently. Indeed, testing in public environments such as Mountain View, CA, Pittsburg, PA and Phoenix, AZ has in fact revealed that both of the above relations coexist: there are occasions in which people act as they would around regular cars, and occasions where they react to HAVs with extreme caution or curiosity. A consequence of this dynamic situation is that it will be some time before the metrics and measures for assessing signaling effectiveness can be determined. Attention to the histories of vehicle signaling reminds us that conventions around signal use and effectiveness remain dynamic and evolving.

Further, only a small portion of the population has had regular and repeated interactions with autonomous vehicles on the road, and those vehicles are all test vehicles. Few companies are testing large numbers of HAVs in real world settings. We should not yet expect to see 'typical' behavior and instead need to account for the novelty effect of having HAVs on the road. Behaviors are likely to shift as people encounter HAVs under more circumstances. People may well become both more trusting and less trusting over time when dealing with HAVs, depending on the contexts of those interactions.

Finally, effects may depend greatly on the concentration of HAVs in a particular area. As an unavoidably dynamic, multi-agent context, and given that what one road user does influences another, which can impact a third and so on, the impacts of HAVs at scale could vary. When there is uncertainty in the interactions that contribute to unnecessary hesitation, braking or positioning changes by individual road users, there is a risk of introducing minor disruptions to the overall traffic flow[8]. Today, the best we can do is simulate the anticipated chain reaction effect of interactions (Millard-Ball 2018).

The introduction of HAV to the roadways has the potential to impact society in a variety of ways, not the least of which is from the standpoint of what it feels like and what behaviors people will encounter in the everyday process of going about their lives. Given the wide range of factors that need to be considered, it is important to recognize there is no single test method that can answer all the questions from today's standpoint. Our goal has been to show the value of working iteratively from a variety of standpoints to examine the possibilities of both what 'will' be and what 'should' be, using both science and design.

References

Bockle, M.P., Brenden, A.P., Klingegard, M., Habibovic, A., Bout, M.: SAV2P: Exploring the Impact on an Interface for Shared Automated Vehicles on Pedestrians' Experience. In: Proceedings of AutomotiveUI (2017)

Clamann, M., Aubert, M., Cummings, M.L.: Evaluation of vehicle-to-pedestrian communication displays for autonomous vehicles. In: Proceedings of 96th Annual Transportation Research Board Meeting (2017)

Dey, D., Terken, J.: Pedestrian Interaction with Vehicles: Roles of Explicit and Implicit Communication. In: Proceedings of AutomotiveUI (2017)

Ford: Ford, Virginia Tech go undercover to develop signals that enable autonomous vehicles to communicate with people (2017). https://media.ford.com/content/fordmedia/fna/us/en/news/2017/09/13/ford-virginia-tech-autonomous-vehicle-human-testing.html. Accessed 24 Feb 2019

---

[8] Chapter 4 of Vanderbilt's "Traffic: Why We Drive the Way We Do" (2009) summarizes a large number of studies looking at the effect braking and uneven movement and acceleration causing others to brake on overall traffic flow, showing that even slight disturbances can magnify into traffic problems.

Habibovic, A., Andersson, J., Malmsten Lundgren, V., Klingegård, M., Englund, C., Larsson, S.: External vehicle interfaces for communication with other road users? In: Meyer, G., Beiker, S. (eds.) AUVSI 2018. LNM, pp. 91–102. Springer, Cham (2019). https://doi.org/10.1007/978-3-319-94896-6_9

Jordan, B., Henderson, A.: Interaction analysis: foundations and practice. J. Learn. Sci. 4(1), 39–103 (1995)

Kitazaki, S., Myhre, N.: Effects of non-verbal communication cues on decisions and confidence of drivers at an uncontrolled intersection. In: Proceedings of the Ninth International Driving Symposium on Human Factors in Driving Assessment Conference (2017)

Langstrom, T., Lundgren, V.: AVIP-autonomous vehicles interaction with pedestrians: an investigation of pedestrian-driver communication and development of a vehicle external interface. Master thesis, Chalmers University of Technology, Gothenborg, Sweden (2015)

Madigan, R., Louw, T., Wilbrink, M., Schieben, A., Merat, N.: What influences the decision to use automated public transport? Using UTAUT to understand public acceptance of automated road transport systems. Transp. Res. Part F Traffic Psychol. Behav. 50, 55–64 (2017)

Madigan, R.: CityMobile2 project. In: Proceedings of the First External Human Machine Interface (e-HMI) Harmonization Workshop (2016)

Merat, N., Madigan, R.: Human Factors, user requirements, and user acceptance of ridesharing in automated vehicles. International Transport Forum (2017)

Millard-Ball, A.: Pedestrians, autonomous vehicles, and cities. J. Plann. Educ. Res. 38, 6–12 (2018)

Parkin, J., Clark, B., Clayton, W., Ricci, M., Partkhurst, G.: Understanding interactions between autonomous vehicles and other road users: a literature review. Project Report. University of the West of England, Bristol (2016)

Risto, M., Emmenegger, C., Vinkhuyzen, E., Cefkin, M., Hollan, J.: Human-vehicle interfaces: the power of vehicle movement gestures in human road user coordination. In: Proceedings of the Ninth International Driving Symposium on Human Factors in Driving Assessment Conference 2017

Rothenbucher, D., Li, J., Sirkin, D., Mok, B., Ju, W.: Ghost driver: a field study investigating the interaction between pedestrians and driverless vehicles. In: 25th IEEE International Symposium on Robot and Human Interactive Communication (RO-MAN) (2016)

Schieben, A., Wilbrink, M., Kettwich, C., Madigan, R., Louw, T., Merat, N.: Designing the interaction of automated vehicles with other traffic participants: design considerations based on human needs and expectations. Cogn. Technol. Work 21, 69–85 (2018)

Vanderbilt, T.: Traffic: Why We Drive the Way We Do. Vintage Press, New York (2009)

Zhang, J., Vinkhuyzen, E., Cefkin, M.: Evaluation of an autonomous vehicle external communication system concept: a survey study. In: Stanton, N.A. (ed.) AHFE 2017. AISC, vol. 597, pp. 650–661. Springer, Cham (2018). https://doi.org/10.1007/978-3-319-60441-1_63

The Hexagonal Spindle Model for Human Situation Awareness While Autonomous Driving

Daehee Park[1] and Wanchul Yoon[2(✉)]

[1] Samsung Electronics, 56, Seongchon-gil, Seocho-gu, Seoul, Republic of Korea
daehee0.park@samsung.com
[2] KAIST, 291, Daehak-ro, Yuseong-gu, Daejeon, Republic of Korea
wcyoon@kaist.ac.kr

Abstract. Over the years, many automobiles have been installed with auto-mated functions (Debernard et al. [5]). Traditional automotive manufacturers in addition to newly emerging IT companies have been recently developing autonomous driving functions. Common sensor technologies have recently been discussed, although human situation awareness regarding autonomous driving systems has not. According to Endsley, the development of autonomous vehicles reveals three major problems [8] that may affect drivers' overall safety, as the autonomous driving system is not perfect; thus, if any problems occur that disable the driver and autonomous system from maintaining safe driving, a serious car crash becomes a possibility. Therefore, this paper suggests that the hexagonal spindle model be employed for human situation awareness during autonomous driving. Through the model, Human Machine Interface (HMI) designers may be expected to consider various aspects of human situation awareness during autonomous driving to help drivers implement driving strategies when facing unanticipated events on the road.

Keywords: Situation awareness · Cognitive engineering model

1 Introduction

1.1 Development of Autonomous Driving

In recent years, autonomous driving has been discussed as a significant topic and has obtained broad public attention (Brenner and Herrmann [1]). Over the years, many automobiles have been installed with automated functions (Debernard et al. [5]). Many traditional automotive manufacturers such as Ford, Toyota, and Hyundai as well as newly emerging IT companies such as Google, Apple, and Samsung have recently been developing autonomous driving functions. Common sensor technologies have recently been discussed, although, in regard to the cognitive engineering aspect, human situation awareness designed alongside autonomous driving systems has not yet been sufficiently discussed. This means most manufacturers are generally concentrating on improving the functions or quality of their automated systems but consequently neglecting manual functions of cars equipped with autonomous driving systems.

© Springer Nature Switzerland AG 2019
H. Krömker (Ed.): HCII 2019, LNCS 11596, pp. 65–75, 2019.
https://doi.org/10.1007/978-3-030-22666-4_5

1.2 Definition of Autonomous Driving

National Highway Traffic Administration (NHTSA) categorised automated driving into six levels [11].

- No-Automation (Level 0): There is no automation. The driver performs all driving tasks; brake, steering, throttle, and motive power.
- Driver Assistance (Level 1): Basically, the vehicle is controlled by the driver. However, some specific functions involved in the vehicle, that helps the control of the driver.
- Partial Automation (Level 2): The vehicle consists of some automated functions such as acceleration and steering wheel control, but the driver should be engaged with the driving tasks and monitoring the driving environment.
- Conditional Automation (Level 3): The driver should present in the vehicle, but the driver does not need to monitor the driving environment. The driver should be ready to take control of the vehicle all times with notice.
- High Automation (Level 4): The vehicle can control all functions under certain conditions. The driver may have option to control the vehicle.
- Full Automation (Level 5): The vehicle can control all functions under all conditions. The driver may have option to control the vehicle.

In this paper, we believe Levels 3 and 4 possess powerful situation awareness systems. In addition, although full automation (Level 5) does not require a human driver, humans drive the vehicles that possess conditional automation (Level 3) and high automation (Level 4) due to several reasons that will be discussed in the following section. Therefore, we discuss the cognitive engineering model to consider how the powerful situation awareness systems inform drivers when situational events that may influence the driving strategy unexpectedly occur.

1.3 Human Driving with Autonomous Driving Vehicle

Although the autonomous driving system has been developing dramatically, many cases can be expected wherein drivers continue operating autonomous driving vehicles manually. There are several examples of humans manually driving autonomous driving vehicles; for example, the driver may prefer driving by him/herself for fun, as many drivers prefer manually driving for fun. Cai's and Lin's research suggests that drivers potentially express various emotions while driving, such as calmness, pleasure, happiness, and fear [3]; in addition, they may enjoy speedy driving, controlling the vehicle, and even improving their driving skills. On the other hand, many people do not trust the autonomous driving system. Various types of driver support systems have been recently developing in the automotive domain to improve autonomous driving systems that may be proactively aware of the road's situational environment. Although autonomous driving systems provide a variety of benefits to the human driver, they might introduce some human factor issues; at times, even small human factor issues might be associated with a serious car crash. It is assumed that human factor errors such as human trust, system acceptance, behaviour adaptation, situation awareness, and main agent to the control may be regarded as negative effects of automated driving systems [3]. Thus, a human driver can drive a highly automated system vehicle by oneself. At this time, it is important to consider how these systems effectively inform human drivers regarding situation awareness.

2 Situation Awareness

2.1 General Definition

Endsley stated that 'situation awareness is being aware of what is happening around you and understanding what that information means to you now and in the future' [7]. The formal definition of situation awareness can be divided into three separate levels:

Level 1: perception of the elements in the environment;
Level 2: comprehension of the current situation;
Level 3: projection of future status.

Level 1 indicates that the driver perceives the status, attributes, and dynamics of relevant elements in his/her environment (Endsley [7]). Information can be processed through human sensory perception by way of visual, auditory, and/or tactile functions. In Level 2, the driver understands what the perceived data and cues mean in relation to relevant goals. At this level, the driver should understand how the situational information influences safe driving to a specific destination. In Level 3, the driver predicts elements perceived during previous levels that may be relevant in the future. This projection leads the driver to proactively make decisions when events arise.

2.2 Situation Awareness in Autonomous Driving

Situation awareness is a critical factor of a driver's ability to make decisions to avoid hazards, plan routes, and maintain safe travel. With the advent of automation, many current techniques used to assess driver performance, workload, or behaviour become useful only after a transition from autonomous to manual driving has occurred [8, 9, 14, 16]. However, situation awareness is also important to assess prior to transition due to concerns about driver performance. Such awareness can also be an indicator of an individual's trust in automation and will hence be an increasingly important element in future automotive studies. In addition, according to Endsley, the development of autonomous vehicles can reveal three major problems [8]: firstly, poor vigilance when humans become monitors, often coupled with increased trust or over-reliance on the automation; secondly, limited information regarding the behaviour of the automation, relevant system, and environment due to either intentional or unintentional design decisions; thirdly, a reduced level of cognitive engagement that originates from one being a passive rather than an active processor of information. The autonomous driving system is not perfect; thus, if any problems occur that disable the driver and autonomous system from maintaining safe driving, a serious car crash becomes a possibility. Therefore, this paper suggests that the hexagonal spindle model be employed for human situation awareness during autonomous driving. The general version of situation awareness typically provides three levels of situations: perception, comprehension, and projection (Endsley [6]). Along the aspect of driving, human drivers should perceive any signs that induce irregular situations. Then, the human driver should understand the signs' meanings. Lastly, the human driver should expect which situations will soon occur. Those three levels are general descriptions of situation awareness as provided by Endsley [6]. On the other hand, it is necessary that drivers examine various factors occurring during autonomous driving in order to discuss situation awareness within the cognitive engineering aspect because autonomous driving should consider all safe driving aspects to get the driver to his/her destination effectively. Furthermore, each situation that occurs during autonomous driving should be connected and considered in regard to safe driving.

2.3   SPIDER Framework

In order to analyse situation awareness, the mental model of the driving environment should be considered (e.g., Durso et al. and Endsley [4, 6]). Many researchers suggest that working memory is regarded as a significant element of the driver's situation awareness (e.g., Jo-hannsdottir and Herdman [10]). According to Strayer et al., driving is related to several cognitive processes, as it requires that the driver visually scan the driving environment, predict where potential hazards might appear if they are not already visible, identify hazards and objects in the driving environment, decide which action is necessary and when it should be executed, and execute proper responses [15]. Strayer et al. named those cognitive processes SPIDER [15] and suggested that situation awareness is notified and updated through the SPIDER processes, which include scanning, predicting, and identifying [15]. Strayer et al.'s SPIDER framework involves Endsley's three levels of situation awareness [15]. Scanning is similar to Level 1 of situation awareness (perception of environmental elements), predicting is connected to Level 3 of situation awareness (projection of future status), and identification is connected to Level 2 of situation awareness (comprehension of current situation). In addition, the level of situation awareness influences the driver's responses. According to Strayer et al., there is even a small decrease in likelihood that a driver will successfully complete a SPIDER-related activity, which may influence the driver's situation awareness and driving performance [15]. Thus, even small errors in situation awareness can induce poor performance (Endsley [6]).

3   Previous Hexagonal Model

3.1   Benedyk's Model

The hexagonal spindle model was originally suggested by Benedyk et al. in order to clearly consider design activity based on personal, organisational, and contextual sectors within education areas [2]. Benedyk et al.'s paper suggested that, from an ergonomic perspective, learning—as the transformation and extension of the learner's knowledge and/or skills—can be regarded as work. Its 'workplace' is the educational environment in which the learning tasks are performed, and the 'learning work' is composed of a series of learning tasks [2]. The model of Benedyk et al. was designed based on a concentric ring model of ergonomics (informed by Kao's earlier model) to propose a new model for educational ergonomics: the hexagonal spindle model [2]. Different from Kao's model (1976), Benedyk et al.'s model involves the concept of time—as a spindle for serial and simultaneous tasks—and space shared by multiple learners that highlights areas wherein conflicts may occur [2]. The authors proposed a generic, high-level, holistic model of educational ergonomics that could influence the design of more effective learning environments. They also anticipated that its generic nature and transferability would encourage its use for the design and evaluation of different forms of learning materials, aids, devices, and environments with consideration of the requirements of different types of learners and learning tasks [2] (Fig. 1).

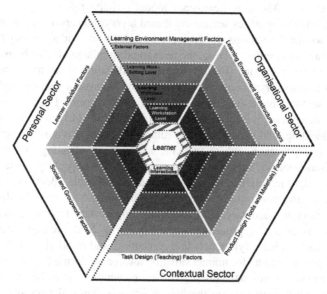

Fig. 1. A hexagonal spindle model for education from Benedyk et al. (2009)

4 New Hexagonal Model for Human Situation Awareness

4.1 Cognitive Engineering Approach

The cognitive engineering approach is considered in order for the new hexagonal model to describe human situation awareness. The general concept of cognitive engineering concerns the analysis, design, and evaluation of complex sociotechnical systems (Vincente and Wickens [16, 18]). Such sociotechnical systems are composed of several layers, and traditionally, many disciplines have viewed their technical cores as comprising the whole system [16]. The reason the cognitive engineering approach is applied to the new hexagonal model is that driving tasks require complex types of cognitive and environmental constraints. We additionally believe that, if the autonomous driving system is properly developed, complex situation information should be effectively conveyed to the driver. Such information processing involves many factors that influence the driver's driving strategy. Many different types of work analysis techniques that have been proposed should be categorised such that they may be used effectively. Rasmussen suggested three generic models that can be applied to group together work analysis techniques into *normative*, *descriptive*, and *formative* models [13]. Normative models describe how a system should behave, descriptive models depict how a system behaves in practice, and formative models clarify the requirements that must be satisfied. The new hexagonal model can be categorised as a normative model since the model describes how the situation awareness system conveys situation information to the driver in each layer.

Unlike the model introduced by Benedyk et al. (2009), this paper presents the hexagonal spindle model's different approaches toward applying situation awareness in autonomous driving. The new hexagonal model's shape refers to Benedyk et al.'s (2009) model, which comprises several combined layers and partitions that consider the environment within the various dimensions. In order to consider various factors and situations during autonomous driving, the hexagonal spindle model for human situation awareness has been suggested (Fig. 2). Similar to education areas, this paper suggests the following four sectors that may be applied to the hexagonal spindle model: vehicle sector, traffic sector, navigation sector, and other vehicle sector. Each sector possesses five situation awareness levels referred to in Strayer's study [15]: scanning, predicting, identifying, deciding, and executing. According to Strayer, driving is dependent upon several cognitive processes, which may include visually scanning the driving environment for indications of irregular events, predicting and expecting where potentially unsafe events might materialise if they are not already visible, identifying events and objects in the driving environment when they are in the field of view, deciding whether and which action is necessary, and executing appropriate responses [15]. Strayer defined those six cognitive processes while driving as SPIDER cognitive processes (scanning, predicting, identifying, deciding, executing, and responding) [15]. However, this paper proposes a model for human situation awareness during autonomous driving. Thus, if emergency situations suddenly occur, 'responding' may be the autonomous system's responsibility rather than the human driver's. Thus, the responding process was not included in the hexagonal spindle model for human situation awareness during autonomous driving.

4.2   Four Divisions of Sectors

In general, the hexagonal spindle model can be divided into the abovementioned four sectors to analyse human situation awareness while driving a vehicle with an autonomous system. In the vehicle sector, there are many types of vehicle controls, such as a steering wheel, throttle, brake, mirrors, and even the infotainment system. The human driver should control various devices and perform tasks according to each situation to change his/her driving strategy Controlling steering wheel, accelerator, and brake regarded as primary driving tasks that changed simultaneously as change of situation awareness. In the traffic sector, there exist various external factors, such as accidents, heavy traffic, road environments, and weather, among others, that influence human situation awareness. For example, if an accident occurs ahead, heavy traffic may follow behind, and the driver should revise his/her driving strategy to drive the vehicle

Fig. 2. A hexagonal spindle model for human situation awareness

effectively and safely. In the other vehicle sector, the movement of both nearby and distant cars can influence the driving strategy. Finally, navigation—considered part of the manual control—is an important factor of driving strategy, as route guidance to a destination can be changed according to changes in the traffic sector and the other vehicle sector.

4.3   Five Levels of Cognitive Task in Each Sector

Each sector comprises five layers that constitute a modified version of Strayer's cognitive processes [15]. Essentially, Strayer proposed six levels of cognitive process regarding situation awareness [15]. In this model, we regarded Strayer's cognitive process as tasks within layers [15]. Initially, a human drives a vehicle within his/her own driving scenario. During the first stage, the driver must scan the situation information regarding the driving scenario. If something happens, the autonomous driving system scans the situation information and informs the driver about what is happening. During the second stage, the autonomous driving system calculates any hazards or issues based on the previous data and scanned information. Thus, the driver can easily predict the volume of hazards and/or issues. During the third stage, the system informs the driver about related information to assist his/her decision making, and the driver can consider his/her driving strategy based on information provided by the system. During the fourth stage, the human driver decides whether to drive the vehicle effectively, safely, or economically (i.e., driving strategy). During the fifth stage, the driver executes several actions according to his/her driving strategy to avoid any events scanned during the first stage (Table 1).

Table 1. Five situation awareness levels

Segment	Objectivities
Scanning	Scan situation regarding driving scenario
Predicting	Predict hazards or issues related to driving scenario
Identifying	Identify road environment related to driving scenario
Deciding	Decide when the driver completes the driving strategy
Executing	Execute the driving strategy

4.4   Interaction with Situation

Figure 3 indicates several separate hexagons connected to one another. Each hexagon is regarded as a separate situation that can be connected according to the context of each situation. For example, if a car accident occurs ahead, cars behind the accident begin driving slowly, which influences traffic negatively. Therefore, each situation should be connected and considered together. The purpose of this paper is to provide a framework that considers and designs human situation awareness during autonomous driving. Thus, each situation can be analysed in detail, and a human decision model can be developed. Furthermore, each analysed situation can be linked in a specific sequence.

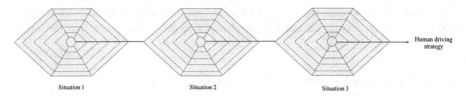

Fig. 3. Each spindle regarded as separate situation

5   Application of the Model

5.1   How the Model Can Be Applied

Table 2 indicates the detailed context for each model's sectors and levels that should be considered for efficient human situation awareness within driving environments. Each sector can be regarded as a driving environment, and the event occurring in each sector should be considered a different level of cognitive process. Each sector constitutes a driving environment involving several cognitive tasks that can be divided into various levels of situation awareness and considered when certain types of situations occur. According to Vincente and Rasmussen, events in complex human machine systems can be categorised according to their degree of novelty as familiar events, unfamiliar/ anticipated events, or unfamiliar/unanticipated events [17]. Firstly, a familiar event is regarded as a routine in that operators encounter it frequently. Secondly, an unfamiliar/anticipated event can be described as only occurring infrequently, although the operator can expect how it progresses and then easily determine a solution. Thirdly, an unfamiliar/unanticipated event is also unfamiliar to operators because it rarely occurs. Unlike the second event, the third is not anticipated by designers, and hence, it is not possible for operators to rely on their own solutions—they must improvise instead [17]. Therefore, if a situation occurs while driving, the intelligent system can provide situation information specifically for each type of event, meaning information within each sector's cognitive task should be provided to the driver at the proper time. Thus, the system designer can use the model to consider how information should be designed to handle the various situations that may occur while driving.

Table 2. Contents of each section of the model

Segment	Vehicle sector	Other vehicle sector	Traffic sector	Navigation sector
Scanning	Scan situation awareness around driving vehicle	Scan situation awareness around other vehicle (nearby own vehicle)	Scan situation awareness about traffic (sudden accident ahead, traffic jam, road status etc.)	Scan the current navigation route
Predicting	Predict hazards or issues which might come to the vehicle	Predict hazards or issues which might come from the other vehicle (nearby own vehicle)	Predict hazards or issues which might come from the traffic change	Predict hazards or issues which might come on the navigation route
Identifying	Identify road environment around vehicle what is happening	Identify road environment around the other vehicle what is happening (nearby own vehicle)	Identify hazards or issues which might come from the traffic change	Identify hazards or issues which might come on the navigation route
Deciding	Decide driver's action to avoid hazards or issues for safe driving	Decide driver's action to avoid hazards or issues for safe driving	Decide driver's action to avoid hazards or issues for safe driving	Decide driver's action to change route to avoid hazards or issues for safe driving
Executing	Execute driver's action to avoid hazards or issues for safe driving	Execute driver's action to avoid hazards or issues for safe driving	Execute driver's action to avoid hazards or issues for safe driving	Execute driver's action to change the route to avoid hazards or issues for safe driving

In addition, the designer can consider the abstraction level of information according to each situation. This model is expected to help apply the concept of the ecological interface design when the designer creates situation awareness support systems. According to Vincente and Rasmussen, the *ecological interface design* is a kind of framework that proposes a set of principles for designing interfaces that provide fundamental properties of human cognition using a specific method [17]. Thus, it can be expected that operators adapt to the complexity and events unanticipated during system design. Along the concept of ecological interface design, it is recommended that the environment be analysed with a holistic view before operators' work, tasks, and knowledge are analysed [17]. Thus, it is not easy to comprehend human behaviours without simultaneously understanding the environment in which operators work. Through this model, the designer considers the driving environment under the holistic view. Then, the designer deliberates what kind of information can be provided to the driver according to each situation in order to assist human situation awareness and avoid complexity of information.

Furthermore, information should be provided according to Rasmussen's SRK framework [12], which includes skills, rules, and knowledge taxonomy. The SRK framework depicts three qualitatively different ways in which operators interact with their environments [12]. Skill-based behaviours and direct behavioural interaction with the world occur without conscious control. Rule-based behaviours involve a sequence of subroutines and a familiar, worldly, perceptual cue with an action or intent without intervening cognitive processing. Knowledge-based behaviours include serial, analytical problem solving based on an analysis of the whole environment and a symbolic mental model [12]. If the SRK framework is applied during information provision, the designer can consider which information should be chosen and provided through the model.

6   Conclusion

In summary, although many automobile manufacturers concentrate on autonomous driving, some human situation awareness research has been recently conducted. The conveyance of information from the system to the human becomes an important issue in autonomous driving system. As autonomous driving systems develop, the situation that is collected from the system will be complex, and thus, conveying that information to the human driver is also difficult. Thus, situation information that can be conveyed from the system to the human driver should be considered under the holistic view in order to reduce the cognitive load and provide information at the proper time. This model was designed to consider the detailed cognitive process along the different perspectives when an unanticipated event occurs on the road. Through the model, the designer can consider four different sectors, five different levels of cognitive processes, and several interrelated situations whose relationships induce specific events. This model can be used to help analyse different situations of each event as well as how cognitive processes in each division are connected.

Therefore, this paper suggests that the hexagonal spindle model be employed for human situation awareness during autonomous driving. Through the model, it can be expected that HMI designers are capable of considering various aspects of human situation awareness in regard to autonomous driving to help the driver implement his/her own driving strategy when facing unanticipated events on the road.

References

1. Brenner, W., Herrmann, A.: An overview of technology, benefits and impact of automated and autonomous driving on the automotive industry. In: Linnhoff-Popien, C., Schneider, R., Zaddach, M. (eds.) Digital Marketplaces Unleashed, pp. 427–442. Springer, Heidelberg (2018). https://doi.org/10.1007/978-3-662-49275-8_39
2. Benedyk, R., Woodcock, A., Harder, A.: The hexagon-spindle model for educational ergonomics. Work **32**(3), 237–248 (2019)
3. Cai, H., Lin, Y.: Modeling of operators' emotion and task performance in a virtual driving environment. Int. J. Hum.-Comput. Stud. **69**(9), 571–586 (2011)
4. Durso, F., Rawson, K., Girotto, S.: Comprehension and situation awareness. In: Durso, F.T., Nickerson, R., Dumais, S.T., Lewandowsky, S., Perfect, T. (Eds.) The Handbook of Applied Cognition, 2nd ed., pp. 163–194). Wiley, Chichester (2007)
5. Debernard, S., Chauvin, C., Pokam, R., Langlois, S.: Designing human-machine interface for autonomous vehicles. IFAC-PapersOnLine **49**(19), 609–614 (2016)
6. Endsley, M.R.: Toward a theory of situation awareness in dynamic systems. Hum. Factors **37**(1), 32–64 (1995)
7. Endsley, M.R.: Designing for Situation Awareness: An Approach to User-Centered Design. CRC press, Boca Raton (2016)
8. Endsley, M.R.: Situation awareness in future autonomous vehicles: beware of the unexpected. In: Bagnara, S., Tartaglia, R., Albolino, S., Alexander, T., Fujita, Y. (eds.) IEA 2018. AISC, vol. 824, pp. 303–309. Springer, Cham (2019). https://doi.org/10.1007/978-3-319-96071-5_32
9. Hart, S.G., Staveland, L.E.: Development of NASA-TLX (Task Load Index): results of empirical and theoretical research. Adv. Psychol. **52**, 139–183 (1988)
10. Johannsdottir, K.R., Herdman, C.M.: The role of working memory in supporting drivers' situation awareness for surrounding traffic. Hum. Factors **52**, 663–673 (2010)
11. NHTSA: https://www.nhtsa.gov/technology-innovation/automated-vehicles-safety. Accessed 24 Dec 2018
12. Rasmussen, J.: Skills, rules, and knowledge; signals, signs, and symbols, and other distinctions in human performance models. IEEE Trans. Syst. Man Cybern. **3**, 257–266 (1983)
13. Rasmussen, J.: Merging paradigms: decision making, management, and cognitive control. In: Third International NDM Conference, pp. 67–85. Ashgate (1996)
14. Sirkin, D., Martelaro, N., Johns, M., Ju, W.: Toward measurement of situation awareness in autonomous vehicles. In: Proceedings of the 2017 CHI Conference on Human Factors in Computing Systems, pp. 405–415. ACM, May 2017
15. Strayer, D.L., Fisher, D.L.: SPIDER: a framework for understanding driver distraction. Hum. Factors **58**(1), 5–12 (2016)
16. Vicente, K. J.: Cognitive Work Analysis: Toward Safe, Productive, and Healthy Computer-Based Work. CRC Press, Boca Raton (1999)
17. Vicente, K.J., Rasmussen, J.: Ecological interface design: theoretical foundations. IEEE Trans. Syst. Man Cybern. **22**(4), 589–606 (1992)
18. Wickens, C.D., Hollands, J.G., Banbury, S., Parasuraman, R.: Engineering Psychology & Human Performance. Psychology Press, Routledge (2015)

Future Personalized Autonomous Shared Car Design Based on User Experience

Yufei Xie and Ting Han[(✉)]

School of Design, Shanghai Jiao Tong University, Shanghai, China
xyfdesign@163.com, hanting@sjtu.edu.cn

Abstract. In view of the social problems such as traffic congestion, energy shortage, parking difficulties caused by the high ownership and low utilization of urban private cars, this paper aims to explore sustainable future travel solutions with good user experience. Car-sharing emerges as a sustainable mode of transportation and become one of the trends of the future cars. Using research and interview, Persona model, Experience Map and Kano model, we study and analyze the user experience of car-sharing, excavate the pain points and user's needs, explore the effective user experience optimization scheme, and apply it to car conceptual design. The combination of autonomous driving and car-sharing will effectively improve road safety, realize the independent travel of special population and improve transportation efficiency. According to the research results of pain points, user's needs and opportunities, future autonomous shared car will help to provide users with a better personalized travel experience, and there will be more possibilities for the interaction between people and cars.

Keywords: Car-sharing · Autonomous driving · User experience ·
User's need analysis · Kano model

1 Introduction

1.1 Background

Private car ownership revealed a trend of dramatic increase. According to the official statistics of the Ministry of Public Security, China's car ownership had reached 217 million vehicles by the end of 2017, an increase of 11.85%, 23.44 million vehicles compared with 2016. The total number of passenger cars is 185 million, of which 170 million are small and mini passenger cars registered in the name of individuals, accounting for 91.89% of passenger cars. With the growth of private car ownership, urban traffic congestion, parking difficulties and other issues are becoming increasingly serious.

The major city traffic analysis report, published by Gaode Map in 2017, shows that during commuting peak 26% of the cities in China fell into traffic congestion, 55% of the cities' traffic are in slow-moving condition [1]. Statistics indicates that urban road congestion is serious and commuting time is wasted. Road congestion has spread from first-tier cities and some big cities to second-tier and third-tier small-mediumcities. The direct cause of traffic congestion is that the car ownership is far more than the capacity of roads.

© Springer Nature Switzerland AG 2019
H. Krömker (Ed.): HCII 2019, LNCS 11596, pp. 76–91, 2019.
https://doi.org/10.1007/978-3-030-22666-4_6

Parking difficulties and low utilization of parking spaces lead to urban traffic congestion and waste of resources to a large extent. According to the "Big Data Report on China's Intelligent Parking Industry 2017", 30% of the traffic congestion problems are caused by parking difficulties, 48% of the vehicles have to queue in the parking yard; in 2017, the total number of traditional parking spaces reached more than 80 million, while the demand for parking spaces exceeded 280 million, and the contradiction between supply and demand further increased. However, the utilization rate of parking space is less than 50% in more than 90% of the cities in China [2].

These social problems caused by the high ownership and low utilization of private cars have greatly affected the efficiency of road transportation, resulting in serious economic losses and social costs, and the energy shortage and environmental pollution problems are highlighted. Automobile exhaust has caused serious environmental pollution, become the main source of air pollution in many cities, and has become a "killer" threatening the lives and health of urban residents.

1.2 Car-Sharing and Autonomous Driving

Car-sharing emerges as a sustainable mode of transportation. Car-sharing, referring to multiple people sharing a car, is a traffic innovation mode between private cars and public transport. Compared with the traditional private car, the car-sharing not only provides users with free and flexible travel, but also meets users' needs of long-distance and fast travel and achieves resource sharing and full utilization. Car-sharing increases mobility for community members to reach destinations otherwise inaccessible by public transit, walking or biking, while increasing the citizens' awareness about the social and environmental impact of using private cars [3]. As an advanced mode of car service, car-sharing can effectively reduce the amount of private cars, improve the utilization rate of the car and the road, and to alleviate the traffic congestion and environment pollution.

Autonomous vehicle relies on artificial intelligence, visual computing, radar, monitoring devices and GPS to cooperate with each other, so that the intelligent driving system can operate smoothly and safely without the initiative of the human driver. Autonomous driving technology can reduce traffic accidents and alleviate traffic congestion; it can effectively use road space, increase the capacity of roads, improve the efficiency of transportation system. Auto-driving technology improves people's safety in the car, liberates drivers' hands, and makes full use of driving time for social or leisure activities. It can also facilitate the travel of special groups.

The existing papers on car-sharing mainly include the study of the business model, operation model, sharing mode and user research of shared cars. Guido Perboli, Francesco Ferrero, Stefano Musso and Andrea Vesco finished the first comprehensive analysis of business models in car-sharing services and made a comparison of 4 car-sharing companies [4]. There are three specific modes of car sharing. Including Two-way (station based) mode, One-way (station based) mode and Free-floating mode. Richard Mounce and John D. Nelson believe one-way car-sharing has the potential to become important components of future city transport systems [5]. The study of Henrik Becker, Francesco Ciari and Axhausen shows that free-floating car-sharing both complements and competes with station-based car-sharing [6]. The research of

Morency Catherine shows that users are segmented in two large groups: the regular and occasional ones, and most users belong to the occasional group [7].

The innovation of this study is to mine the pain points and user needs of shared cars at the present stage based on user experience research, and to propose new solutions for future travel.

2 Automobile Development Trend Analysis

With the application of artificial intelligence, big data, vehicle networking, new energy and other new generation technologies in the automotive field, tremendous changes will take place in the way people travel in the future. The future vehicles will develop rapidly in the direction of intellectualization, sharing, electrification and personalization.

- Intellectualization. Developments of various driver assistance technologies, such as cruise control, onboard Wifi and vehicle-to-vehicle communication, drive the changes in mobility [5]. The application of big data, artificial intelligence, cloud computing and vehicle networking technology promotes the development of advanced driving assistance system and Intelligent Connected Vehicle products are constantly upgraded and iterated. In the future, a new generation of driverless travel mode may be formed, which simplifies the complexity of people's operation and provides users with excellent comfort and safety.
- Electrification. With the concept of "light assets" becoming more and more popular, automobile users will pay more attention to the right of use than ownership in the future. Car-sharing is being globally implanted in over 1,100 cities and has become an important part of the sharing economy [8]. People have a low utilization rate of automobiles in daily life, and sharing can make efficient use of idle social resources.
- Electrification. Automobile electrification is one of the effective ways to cope with the depletion of petroleum resources. In the future, electric vehicles will become the main means of transportation. The rise of the all-electric vehicle (EVs) has a major influence on the future of the car, and the overall cost of EVs will be less [9]. Driven by policy incentives and industry competition, all major auto manufacturers are investing heavily in EVs and its charging systems [9]. The technology of new energy vehicles has been developed rapidly, and the construction of charging piles and other infrastructure supporting electric vehicles will also be greatly developed.
- Personalization. Once people's basic needs are met, higher-level needs will emerge. People begin to have the needs of respect and self-realization. Products become more and more information densely (i.e. smart) and personalized (i.e. low volume in high variety) to meet user's individual requirements [10]. People's high quality and personalized needs need to be met urgently. Therefore, personalization will become one of the development trends of automobiles in the future.

3   Aim

In view of the current unhealthy urban traffic situation, this study aims to explore a reasonable and sustainable way of urban residents' travel in the future by analyzing the way of urban residents' travel, the choice of means of transportation and the development of sharing economy. Taking the shared car as the carrier, we analyze the pain points and excitement points in the existing shared car use flow, excavates the travel needs and personalized expectations of the target population, seeks for the design points and innovation points of future shared car, and explores the feasible scheme of combining individual travel and sharing mode. Combining with the development trend of automobile industry, a future concept vehicle is designed, which not only improves the current situation of high private car ownership, but also provides excellent driving experience for target users.

4   Methodology

The study methods were (1) Survey and interview, (2) Persona model, (3) User Experience Map and (4) Kano model. Through surveys and interviews, we obtain the data such as the user preferences and user performances to explore users' needs. Persona models are created to segment user groups and analyze the usage scenarios. User Experience Map is used to sort out the use flow of shared car and key touch points, analyze user experience and emotions at each stage, discover user's pain points and satisfaction points, and look for design opportunities. The Kano model is used to classify the user's needs. The analysis results are used to guide and improve the products and services of car-sharing at the present stage, and to provide innovative ideas for future shared car design.

4.1   Survey and Interview

Jiguang Data published a research report on car-sharing industry. The data shows that as of November 2018, 95.24 million users installed car-sharing apps, an increase of 1.4 times over the same period last year, and the industry penetration rate has not yet reached 1%. The proportion of males and females is 76.2% and 23.8%, and users aged 26–35 accounted for 71.9% and 18-25 accounted for 19.9%. The majority of users are in the first-tier and second-tier cities, and the top three cities are Shanghai, Chongqing and Chengdu. Overall, there are tide phenomena in the use of shared cars, including morning peak and evening peak [11].

This indicates that there is still a huge space for the development of car-sharing industry. At present, the first and second-tier cities are in the lead, and the user groups are relatively young. The main travel scenario is commuting to work.

According to the survey results, we recruited four typical users of shared cars in Shanghai for interviews. The basic information of the respondents is shown in Table 1.

Table 1. Basic information of the respondents.

Names (Alias)	Gender	Age	Occupation	Usage frequency	Scenario
Chenyu Wang	Male	34	Photographer	2–5 times/week	Multiple trips
Chufan Zhang	Female	26	Sales manager	3–4 times/week	Commuting
Tianze Zhang	Male	21	College student	2–3 times/month	Travel/Gathering
Ziyu Li	Male	32	Programmer	1–3 times/month	Travel

Through interviews, we collected the specific experience information of the respondent in the process of using shared cars. Respondents' words included some of the advantages of shared cars and pain points in using them. Here are some examples of their original interview fragments.

- "… Driving 30 km, 50 min, the cost is 24-yuan, 20 yuan is deducted with coupons, and only 4 yuan is spent when paying…"
- "… The subway is too crowded, and there are two transfers. It's fast and don't need to take a detour to drive a shared car…"
- "… It feels like the air inside the car is not very good, the environment is also a little bad, and there are oil-rich plastic bags on the co-driver's seat…"
- "… The number of shared cars is relatively small. It's too difficult to find cars… The last user left too little vehicle location information…"

We use text analysis to collate and summarize the content of user interviews. The advantages of car-sharing include energy saving, environmental protection, comfortable driving, economical cost, free parking space, flexible and free travel, etc. The summary of the main pain points is shown in Table 2.

Table 2. Summary of main pain points.

Main pain points	Insight	Opportunities
Shared cars are idle in some areas, but it is difficult to find vehicles in some areas	Distribution of shared car resources is unreasonable	Adjust vehicle distribution and increase vehicle scheduling frequency
Parking spaces are often unavailable and designated parking spaces are often occupied by private cars	There are insufficient parking spaces and lack of parking space management	Increase flexible parking spaces and strengthen the management of designated parking spaces
Some vehicles have a bad smell and environment	Vehicle is difficult to be cleaned in time	Clean vehicle regularly and increase the frequency of cleaning
It's difficult to find charging stations; users are anxious due to fear of insufficient electricity	Adequate power and easy charging are key factors affecting user experience	Remind when power is insufficient and mark nearby charging stations in APP

(*continued*)

Table 2. (*continued*)

Main pain points	Insight	Opportunities
The description of vehicle location is inaccurate	Information of vehicle location needs to be clear	Define the floor of the vehicle in the stereo parking lot; increase the navigation function of car search
Most of the vehicles need to be locked when users leave temporarily. They fear that the vehicles will be borrowed by others	Users are anxious when they leave temporarily; complex operations affect user experience	Optimize temporary leaving scenarios, such as adding suspension function in APP

4.2 Persona Model

Previous research results show that the scale of shared car users is growing, and its unique convenience and accessibility make more urban residents willing to try car-sharing as a way of travel. Car-sharing users are mainly young and middle-aged. They can accept new things quickly and earn considerable income. With the improvement of consumption level, people also have higher requirements for the quality of life. They are keen to pursue personalization and fashion and get self-satisfaction from it. The users of shared cars are subdivided into the following groups according to the specific travel scenarios.

- **White-collar workers.** They have driver's license but no car. They have a certain economic base and have a long commuting journey.
- **Urban short-distance travel.** Use shared car to travel between business circles, go to the suburbs or nearby cities to play, achieve flexible self-driving travel.
- **College students.** They are willing to try new things. They drive shared cars with friends to downtown shopping or entertainment (from remote campus).
- **Multiple travel.** Travel/work trips require multiple turns.

User groupsare basically determined, so we create Persona models to elaborate in detail. Combined with specific scenarios, we analyze the reasons why different user groups choose car-sharing and the needs under different travel scenarios.

- **Persona A**

Zhang Feng, a 26-year-old software engineer from a company in downtown Shanghai, has no car and rents a house in the suburbs (Jiading District). He likes playing video games with friends or relaxing at home after work. He usually commutes to work by subway, after the popularity of car-sharing, he chooses to use shared cars for about 1–2 times a week.

Travel Scene: Feng Zhang woke up in the morning and found that he has overslept. He was late to take the subway and taxi was expensive. There was a shared car parking station downstairs of his apartment. Through simple operation, the car was rented immediately, He arrived at the company on time.

- **Persona B**

Zhang Xi, a 27-year-old female civil servant, is married and lives in Beijing. On weekends and holidays, she likes to have small gatherings with her friends, or staying with her children and husband.

Travel Scene 1: Zhang Xi made an appointment with his friend to go shopping at the weekend. They bought a lot of things and it is not convenient to take these things by bus or subway. There were many shared cars near the shopping mall, so Zhang Xi rented a shared car. They put the purchased goods in the trunk, Zhang Xi started the car, first sent her friend home, and then drove home by herself.

Travel Scene 2: Zhang Xi and her family went to Nanjing for a trip on National Day holiday. It was 8 p.m. after a day's trip, as a result of exhaustion, they planned to take a taxi back to the hotel, but it was still difficult to get a taxi after a long wait. Opening APP, they found that there was just an idle shared car nearby, they found and rented the shared car and quickly returned to the hotel.

- **Persona C**

Xu Ye, 22 years old, is a graduate student in a university (far from the city center) in Shanghai, who has no car but has obtained a driver's license. He is willing to try new things. He likes to socialize through Microblog and Instagram. On weekends, he and his friends occasionally go to design exhibitions, shop or eat in the city, or visit other interesting places in the city.

Travel Scene: Xu Ye plans to go to the Exhibition Center with his classmates at the weekend, and after the exhibition they can eat and shop in the mall nearby. It's more expensive to take a taxi or a DidiTaxi, the subway requires two transfers, and it's hard to get a taxi when they come back. So, Xu Ye and his classmates discussed renting a shared car, which is less expensive and convenient, and they can stay in the city for a little longer and return to school later.

- **Persona D**

Liu Yang, a 31-year-old photographer living in Shanghai, is fond of traveling. He often has multiple travels to collect local customs and finish photographic works at the same time.

Travel Scene: Liu Yang needed to go to Hangzhou to complete his photographic work. His one-day journey included four destinations, which is time-consuming and laborious to travel by bus. When Liu Yang arrived in Hangzhou, he rented a shared car and drove flexibly between the four locations to complete the task efficiently.

4.3   User Experience Map

User experience map is used to decompose the use flow of shared cars. And the user satisfaction points, pain points, mood and opportunities are analyzed in each stage. Through research and interviews, we collect enough user behavior, experience, feelings in the process of using shared cars. We refine the key task flow of using shared cars, find the goals of users to accomplish each key task, and sort out the specific behavior paths and touch points in each stage. The next steps are sorting out the user pain points and satisfaction points obtained in the research process, judging the emotional level of each stage and connecting them to form emotional curve, and searching for opportunities behind satisfaction points and pain points.

The key task flow of using shared cars can be concisely refined into five stages: renting, finding, driving, returning and paying.

- **Renting.** Set pick-up and return location in the APP and select the desired car.
- **Finding.** Go to the pick-up location, check the vehicle and unlock it.
- **Driving.** Drive the car to the destination from pick-up location to return location.
- **Returning.** Arrive at the designated parking lot and return the car.
- **Paying.** Finish payment after the vehicle is returned.

The "renting" and "finding" stages can be interchanged in different scenarios. Persona A and Persona B travel scenarios are to find the target shared car first, and then complete the rental; Persona C and Persona D travel scenarios are to reserve the vehicle in advance, and then go to the shared car pick-up station to rent the car.

The User Experience Map of car-sharing (see Fig. 1) reveals that the users are relatively satisfied in the renting and driving stages, while the user experience in the car-finding and car-returning stages is poor. User's needs are analyzed and sorted out based on User Experience Maps, and the results are shown in Table 3.

Stages	Renting	Finding	Driving	Returning	Paying
Goals	Renting is simpler, faster and easier to operate	Easier to find the target vehicle.	• Environment in car is more comfortable • Driving more smoothly	Finding parking lot faster	Payment is simpler and free from worry
Actions & Touchpoints	Setting pick-up p, return location point and vehicle type (APP)	Getting Vehicle Position Information (APP) - Finding Vehicles (shared car)	Driving the car (shared car)	Finding Parking space (APP & Parking lot) - Stopping Travel with the return button(APP & Automobile)	Paying on mobile phones (APP)
Happy Moment	• Recommend the nearest pick-up location • Display battery level and driving kilometers • Shared car models are optional • High credit score can be exempted from deposit	• Provide navigation	• Save commuting time • Easy to operate • More comfortable, flexible and well-equipped (compared with public transport) • Easy to carry luggage	• Free parking in designated parking spaces	• Save money with coupons • Giving coupons after payment
Pain Point	• Pick-up locations are far away • Small number of shared cars nearby • First registration steps are complex	• Fuzzy Location Information of Some Vehicles • Poor identification of reserved vehicles (license plate) • Surface of the car is dusty	• The interior environment is occasionally poor • inconvenient to leave the car temporarily during the journey • difficult to charge when the electricity is low.	• Parking spaces are often occupied by private cars • Designated parking spaces are hard to find • Parking space is far from the destination	• The process of dealing with traffic violations is complex • Additional deduction for unknown reasons
Emotion	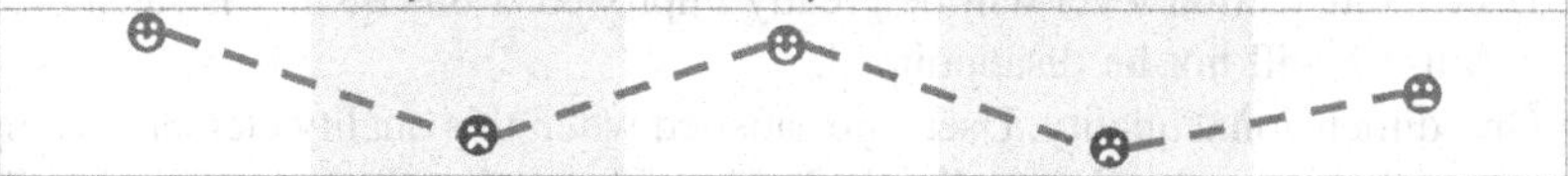				
Opportunity	• Arranging pick-up locations reasonably • Increasing vehicle scheduling frequency	• Providing vehicle specific location information • Improving vehicle identification by flashing lights or body marks • Increasing vehicle cleaning frequency	• Product optimization for user's "temporary leaving" behavior • Reminding users of power shortage and recommending the nearest charging station • Increasing vehicle cleaning frequency	• Rational Planning of Parking Space • Increasing vehicle scheduling frequency	• Attracting price-sensitive users with coupons • Optimizing the Procedure of Disposal of Violation and Vehicle Damage

Fig. 1. User Experience Map of car-sharing.

Table 3. Summary of user's needs in each stage.

Stages	User's needs
Renting	Increase the number of pick-up stations Increase the number of shared cars Increase more types of Vehicle Provide more personalized choices
Finding	Define the specific location of pick-up place Improve the identifiability of reserved vehicle Complete car exterior cleaning in time
Driving	Guarantee driving safety Enhance the comfort of environment and vehicle quality Improve the quality of automobiles Complete car interior cleaning in time Increase the number of charging facilities Define the location of charging pile
Returning	Increase the number of return stations Increase the number of free parking spaces
Paying	Reduce the price of car-sharing travel Guarantee payment security Optimize the solution of traffic violations and accidents

The opportunities we get from the user experience map are arranging pick-up stations and return stations reasonably, increasing vehicle scheduling frequency, reminding users of power shortage and recommending the nearest charging station, increasing vehicle cleaning frequency, providing vehicle specific location information, improving vehicle identifiability by flashing lights or body marks, designing around User's "temporary leaving" behavior, etc.

4.4 Kano Model

Kano model was proposed by Professor Noriaki Kano. The Kano model of customer satisfaction is a useful tool to classify and prioritize customer needs based on how they affect customer's satisfaction [12]. In his model, user's needs for something are sub-divided into five qualities (see Fig. 2): Must-be quality, One-dimensional quality, Attractive quality, Indifferent quality and Reverse quality.

- Attractive quality. Attractive Quality is the quality beyond user's expectation. Users' satisfaction level will be greatly improved if this quality is satisfied. Even if not, users will not be disappointed.
- One-dimensional quality. Users are satisfied when this quality element is ample, and their satisfaction level is in liner relation with quality element adequacy [13].
- Must-be quality. Optimizing this quality will not increase user satisfaction, but if it is not satisfied, user satisfaction will be greatly reduced.
- Indifferent quality. Whether this quality is provided or not, it has no effect on user experience and satisfaction.

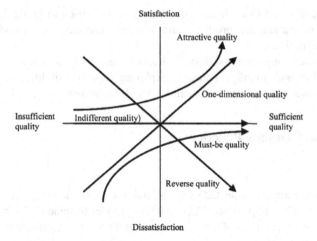

Fig. 2. Kano model of user satisfaction (Kano, 1984).

- Reverse quality. Many users do not have this need, the level of user satisfaction will decrease after providing this quality.

We use Kano model to categorize and classify the user's needs that were mined earlier. Both Indifferent quality and Reverse quality need to be avoided. Therefore, we mainly analyze Must-be quality, One-dimensional quality and Attractive quality, and the results are shown in Table 4.

Table 4. Summary of user's needs in each stage.

Quality	User's needs
Must-be quality	Safe driving of shared car Definite location of pick-up station and return station Sufficient oil and electricity supply Adequate charging facilities Complete vehicle infrastructure Clean exterior and interior environment
One-dimensional quality	Enough and spacious interior space Adequate pick-up and return stations Good comfort while driving Easy and fast operation of renting and returning cars Enhanced the Identifiability of Target Vehicles Simplified user driving operations
Attractive quality	Autonomous driving of shared cars Optimized interior layout of shared cars Personalized Services and Facilities

Must-be quality and One-dimensional quality will be used to guide the improvement of current car-sharing products and services, enhance user satisfaction and improve user experience.

We obtain new opportunities through Kano model analysis, such as simplifying driving operation with intelligent means, exploring more possibilities of car layout, providing personalized services and facilities for different users, etc.

5 Result and Design Practice

5.1 Result

In the previous section, we collected users' behavior and experience information of car-sharing and used User Experience Map and Kano model to mine the pain points and opportunities. The application of auto-driving technology will effectively solve some of the main pain points of sharing automobile at current stage and improve user experience and satisfaction.

- **Solving the difficulties of finding and parking shared cars.** The existing shared cars require users to pick up and return their cars by themselves. The location distribution of pick-up stations and return stations can not provide users with the greatest convenience, resulting in waste of time and vehicle resources. Auto-driving shared car can complete automatic cruise, arrive at the user's designated boarding position and head for destination automatically.
- **Reducing the demand for parking spaces.** After arriving at the destination, the autonomous shared car can directly navigate to the location of the next user, further reducing the idle time of the shared car.
- **Solving the difficulties of charging and relieving range anxiety.** With the application of autonomous driving, users no longer worry about the lack of electricity during the journey. When the power is insufficient, the shared car will automatically cruise to the nearest charging station to recharge. When the power is full, it will resume its working state.
- **Regular vehicle cleaning.** The auto-driving shared car can cruise to the maintenance center regularly and automatically for cleaning and maintenance by batch, which can effectively ensure the vehicle environment and improve the comfort of the car.
- **Completing vehicle dispatch automatically and scientifically.** Shared cars can be dispatched automatically according to the demand for vehicles. Vehicles are more distributed in areas where the need is concentrated, so as to improve the utilization rate of vehicles.

The application of autonomous driving technology can effectively improve vehicle safety, making users no longer confined to the driving seat, providing more possibilities for vehicle layout and human-computer interaction. It also facilitates the travel of special groups (such as the elderly, the disabled, etc.).

According to the analysis of users' needs, opportunities, and the future automobile development trend, the design concept of future automobile sharing is formed.

- Making full use of the advantages of autonomous driving technology to optimize the operation mode of shared vehicles and completing the task of picking up and carrying passengers, charging, cleaning and dispatching by automatic cruise.
- Expanding the interior space of the car and exploring new possibilities of in-car layout and human-computer interaction. The future shared cars will become mobile living space and interactive terminals.
- Providing personalized services and facilities to meet the user's personalized needs. Following the concept of environmental protection and improving the utilization of space in the car.
- Improving the identifiability of shared vehicles to make users easily identify their reserved vehicles as soon as possible, which also improves the efficiency of using vehicles.

5.2 Design Practice

This design aims at making use of autonomous driving technology and sharing mode to make future cars become mobile living space and interactive terminals. It provides reasonable solutions to negative problems caused by urban cars while optimizing users' travel experience. The design includes automobile styling, human-computer interaction and operation mode.

The styling of future autonomous sharing car is shown in Figs. 3, 4 and 5, simple and futuristic. According to Kano model analysis, One-dimensional quality includes adequate and spacious interior space. Therefore, the wheelbase of the car body is lengthened, the front and rear suspensions become shorter, which provides more space for passengers and can meet the travel needs of 5–6 people at the same time. Two sides of the body are embedded with parameterized LED display screen elements, which are used to display the personalized logo set by the user. It is convenient for car identification and solves the pain point of poor identification of the reserved vehicle. The layout and size of the shared car concept are shown in Fig. 6.

The future shared car will become an intelligent interactive terminal and mobile living space. Users can interact with automobiles through the three-dimensional interactive interface of holographic projection, which helps to improve the interaction efficiency between users and automobiles and bring users a pleasant immersive experience. Users can enjoy entertainment, online social networking, online shopping, work and other forms of activities during the ride.

The analysis results of car-sharing User Experience Map and Kano model show that user pain points such as unreasonable distribution of pick-up and return stations, difficulties in finding charging piles and poor vehicle environment seriously affect user experience. Shared car operation mode based on autonomous driving technology can help users to improve their driving experience, enhance safety, comfort and utilization of shared cars. The vehicles can cruise automatically, realizing the automatic operation such as picking up and delivering passengers, charging, regular cleaning, short-distance connection of "the last kilometer" and vehicle dispatching. It can solve the user's pain points while reducing human costs and improving operation efficiency.

Users reserve cars through APP on the mobile phone before travel. They are able to set the time, starting point, destination, number of peers and whether they are willing to

Fig. 3. The future autonomous sharing car (a).

Fig. 4. The future autonomous sharing car (b).

carpool in this APP. The vehicle cruises automatically to the starting point, picks up the passengers and drives automatically to the destination. After the passengers get off safely, the vehicle automatically goes to the starting point of the next passenger. Vehicles regularly go to the maintenance center for cleaning and maintenance. When the power is insufficient, they automatically drive to the charging station to replenish the power.

Based on user's needs for personalized services and facilities, this design focuses on personalization. Users set up their own personalized logos, and the car body shows the pre-designed logo when the vehicle arrives (see Fig. 7). To meet the special needs of users, modular seats and tools (such as children's seats, barrier-free facilities, etc.) in the car can be flexibly combined and moved. Based on the user's demand for the optimization of the interior layout, the car interior is wraparound layout. All passengers

Fig. 5. The future autonomous sharing car (c).

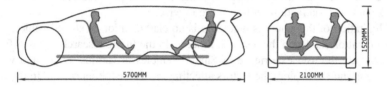

Fig. 6. The layout and size of the shared car concept.

can face to the center, which facilitates the communication between passengers and their companions, and provides the possibility for the communication between carpool passengers as well.

Fig. 7. Example of personalized logo.

The future autonomous shared car design based on user experience meets various needs of users and provides better personalized experience. It can also effectively alleviate urban problems and improve the utilization rate of shared vehicles and road transport efficiency.

6 Conclusion

In this paper, we study the user experience in the existing shared car use flow. We use survey and interview, Persona model, User Experience Map and Kano model to excavate user pain points, satisfaction points and user's needs, find design opportunities and use them to guide creative design. The application of autonomous driving technology can solve many pain points of existing shared cars. By combining the future development trend of automobiles, we can provide users with more high-quality personalized travel experience and effectively improve user satisfaction. Making full use of autonomous driving can effectively solve users' pain points such as range anxiety and difficulties in finding a car, parking and charging. The operating mode of shared cars based on autonomous driving will make it easier to realize automatic scheduling and regular cleaning, and improve the utilization rate of Shared vehicles and the comfort level of users in ride experience. In the future, cars will become mobile living space and interactive terminals. Shared vehicles will have personalized service and the interior layout and interaction will be more diverse.

The limitation of this paper is that there is no classification of shared vehicles with multiple operation modes in the research process. In the future research, we will further refine and study the user experience of Shared vehicles in various modes and vehicle scenarios, continue to explore the factors affecting the user experience, and improve the user satisfaction of Shared vehicles.

References

1. Traffic analysis report of major Chinese cities in 2017. https://report.amap.com/share.do?id=8a38bb8660f9109101610835e79701bf. Accessed 07 Feb 2018
2. Big Data Report on China's Intelligent Parking Industry 2017, https://www.cbndata.com/report/505/detail. Accessed 30 Jan 2018
3. Ferrero, F., Perboli, G., Rosano, M., Vesco, A.: Car-sharing services: an annotated review. Sustain. Cities Soc. **37**, 501–518 (2018)
4. Perboli, G., Ferrero, F., Musso, S., Vesco, A.: Business models and tariff simulation in car-sharing services. Transp. Res. Part A Policy Pract. **115**, 32–48 (2018)
5. Mounce, R., Nelson, J.D.: On the potential for one-way electric vehicle car-sharing in future mobility systems. Transp. Res. Part A Policy Pract. **120**, 17–30 (2019)
6. Becker, H., Ciari, F., Axhausen, K.W.: Measuring the car ownership impact of free-floating car-sharing – a case study in Basel, Switzerland. Transp. Res. Part D Transp. Environ. **65**, 51–62 (2018)
7. Morency C., Martin, T., Bruno, A., Basile, M., Joel, Q.: Car sharing system: What transaction datasets reveal on users' behaviors. In: The 10th International IEEE Conference on Intelligent Transportation Systems, pp. 284–289. IEEE, Bellevue (2007)
8. Hamari, J., Sjöklint, M., Ukkonen, A.: The sharing economy: Why people participate in collaborative consumption. J. Assoc. Inf. Sci. Technol. **67**, 2047–2059 (2016)
9. Webb, J.: The future of transport: literature review and overview. Econ. Anal. Policy **61**, 1–6 (2019)
10. Zheng, P., Shiqiang, Yu., Wang, Y., Zhong, R.Y., Xu, X.: User-experience based product development for mass personalization: a case study. Procedia CIRP **63**, 2–7 (2017)

11. Research Report on Shared Automobile Industry. https://www.jiguang.cn/reports/367. Accessed 30 Jan 2019

12. Kano, N., Seraku, N., Takahashi, F., Tsuji, S.: Attractive quality and must-be quality. J. Jpn. Soc. Qual. Control **14**(2), 39–48 (1984)

13. Yao, M.-L., Chuang, M.-C., Hsu, C.-C.: The Kano model analysis of features for mobile security applications. Comput. Secur. **78**, 336–346 (2018)

HMI Design for Autonomous Cars: Investigating on Driver's Attention Distribution

Weizhe Chen and Wei Liu[✉]

Tongji University, Shanghai 200092, China
liuweiinnovation@126.com

Abstract. This paper provides an overview of the predictable possibilities and opportunities of autonomous human-computer interaction design offered by studying the driver's attention distribution. We examined attention distribution from novice to advanced drivers and a visual experiment was conducted using stickers and camera to both quantify and qualify attention while participants were in a simulation auto-driving scenario. Thus, an consciousness hot-map could be determined by stickers' distribution and eye-tracking data, which could contribute to interface design and its structure in autonomous driving vehicles.

Keywords: Autonomous driving · UX · User interface

1 Introduction and Related Works

1.1 Autonomous Cars

Autonomous cars and its relevant technologies have been greatly developed in the past few years, and they are expected to occupy a large presence in the market, with a massive market demand and a promising future [1, 2]. Currently, automobile systems vary from those which assist with driving tasks (e.g. automatic transmission) to those which replace manual operations [3]. The National Highway Traffic Safety Administration (NHTSA) defines five degrees of car autonomy which vary in the integration of cars with Advanced Driver Assistance Systems (ADAS), and the extent to which a car is taken over by autonomous systems [4]. Autonomous and semiautonomous vehicles are currently being developed by over 14 companies [32], most of which stay in navigating in complex environment. Thus, this paper and its relevant experiment will focus on cars capable of driving automatically for a certain distance, which is L4, the current study has been conducted.

Different levels of driver engagement are involved in each degree of autonomous driving [5]. The term 'Attention Distribution' [6] is used to describe what drivers focus on and their engagement. In autonomous cars, drivers are relieved of the cognitive load from monitoring traffic and executing driving tasks [7] – which forms a clear contrast against traditional driving practices – and the driver's attention is free to stray towards secondary tasks instead of focusing on the primary task of driving. Thus, the higher the degree of automation, the more tendency for the driver's attention to shift from driving to secondary tasks.

© Springer Nature Switzerland AG 2019
H. Krömker (Ed.): HCII 2019, LNCS 11596, pp. 92–102, 2019.
https://doi.org/10.1007/978-3-030-22666-4_7

This change in attention distribution (AD) raises new requirements and challenges of Human Machine Interface (HMI) for autonomous driving in two perspectives:

1. How can the non-driving-related interface (for entertainment and recreation purposes) be designed to be capable of functioning without shifting attention from the driving interface? For example, as an expansion of the Internet of Things, autonomous cars should be capable of accessing the internet, communicating with smart devices as well as other cars and road infrastructures, and collecting and dealing with data for drivers [8]. The innovation of the mobile phone had changed the world, and the similar changes occurring in autonomous cars would do so again. This pivotal development would require these functions be well organized according to user experiences (UX) on varying levels of autonomy.

2. Related study had investigated automated steering systems and found that they improve driver satisfaction and performance but also increase the time to recover from a system shutdown, demonstrating an out-of-the-loop problem. [33] How the driving interface can better transmit information with minimal input attention, increasing the efficiency of information as to decrease the frequency of driver distractions. Autonomous cars relieve drivers from the most stressful operations needed for driving [9], however, the sense of control during driving should still exist despite the decrease in driving tasks. Therefore, how the driving task interface conveys its reliability and safety during driving is another important factor.

1.2 Attention and Cognition

Attention has been described as the allocation of limited cognitive processing resources. [10] Generally speaking, you can think of attention as a highlighter [11]. For an example when you read through a section of text, the highlighted section where you are aiming at stands out, causing you to focus your interest in that area. In other words, with attention, some sensory inputs will process faster and deeper [12, 13].

It was pointed out that the pattern of eye fixations (visual attention) that a given observer produces was influenced by properties of the scene as well as the goals and interests of the perceiver [14]. Present a major categorical boundary of visual attention is the distinction between bottom-up (stimulus-driven) and top-down (goal-directed) attention. The point is that the attention function processing may depend on the properties of an image (e.g. a striking colors or a sudden movement) as well as goal driven (e.g. hungry people looking for food).

1.3 Top-Down Attention Review

The ability that distribution of attention can be affected by intentions of the observer was first noted by Helmholtz [15]. And its relevant perceptual consequences were studied thoroughly by Mertens [16]. Eriksen and his colleagues started a seminal series of studies about the quantitative understanding of the top-down deployment of attention [17–22]. Further exams were conducted by Posner and his colleagues for top-down attention control processing [23–27].

1.4   Bottom-Up Attention Review

Meanwhile, evidence and researches concerning captured attention (Bottom-up attention) are more recent and could be differed to two major categories based on stimulus properties: feature singletons and abrupt visual onsets [12]. Feature singletons can be identified easily in visual research. (Neisser found that curved letters distinguish themselves among straight letters [28]). Same as this, dynamic, and colorful road vision are much more attractive than stationary S-IVIS (In-vehicle information system) in a simulation auto-driving experiment. So we decided participants to allocate their attention points after the driving video and tell them to focus on S-IVIS.

1.5   Driver's Attention

Psychologists have confirmed that top-down attention and bottom-up attention processes work together [12], as is under natural condition. In a typical experiment [29], participants were asked to search for and identify a red target in a display containing several white distractors, while on other trials a "cue" display (only containing red color singleton) precede the search display. The cue singleton involuntarily attracted top-down attention and affected participants' response to upcoming targets even though participants were told to ignore the cue display and draw down-up attention as much as possible.

In driving processing, regardless of autonomous levels, drivers attention are distributed in both Top-down way and Bottom-up way. Drivers are responsible for constantly changing views (bottom-up) and naturally allocating their attention to driving task interface when drivers should operate it, or just watch it on the highly autonomous vehicle (top-down). Meanwhile, S-IVIS tends to capture drivers' attention through attractive visual elements (shining screen, sudden movements and sounds). So driving process involves both two kinds of attention and its complex combination.

Despite this, we had refined opportunities and points associated with singleton attention from the two perspectives we mentioned before.

In perspective 1, when the drivers actively focus on non-driving-related tasks, the proper spatial location for secondary tasks interface and how to guide the attention shifts quickly and accurately could be the points for top-down attention research, out of drivers' initiative intention to take secondary tasks. On the other side, visual elements on the non-driving task interface should also be limited in case bottom-up attention plays a negative role. (e.g. unnecessary massage distract drivers in driving and weak reminder may cause drivers to miss important messages.)

In perspective 2, involved with the driving-task interface, how to express driving information (speed; cars conditions; etc.) efficiently since top-down attention deployment inevitably decline because of developed automation and how emergency alert capturing driver's bottom-up attention immediately could be meaning for the research on attention distribution on driving-task interface.

1.6   Research Approach

Conducting a user evaluation with a real self-driving car is difficult. Some studies have provided alternatives that a fake self-driving car [30, 34], a VR simulation [35, 36] and a video experiment [31, 37] can be appropriate for such research. This study therefore adopted a video experiment method to extract both quantitative and qualitative data on top-down and Bottom-up attention distribution on the driving process. Both driver's view and co-driver's view were taken into consideration because there will be no driver in a highly-autonomous car. And passengers' experiment could not only be a control experiment but a reference for fully-auto driving experiments as well. These data were used to frame an attention heat map for HMI (human machine interface) in autonomous cars. Thus, the research and conclusion of this hot map may act as a guideline for the functionality planning and interaction disciplines of interactive interface design for autonomous automobiles.

2   Experiment

In the video visual experiment. 12 adults were selected and invited into a mimicked typical driving environment in terms of the sample selection criteria as below:

(1) The age of them strictly ranged from 22 to 46, which contained 90% drivers in the range of age.
(2) Gender was logically taken into account, therefore, we introduced 6 male drivers and 6 female drivers.
(3) The experience of autopilot also affects the feedback from the participants. We have chosen drivers who have at least a low level of assisted driving experience.
(4) Both novice drivers and experienced drivers were introduced to this experiment. We also noticed that the participants' driving experience would definitely affect their feedback. The novices would not be expected to show signs of distraction in these studies, of course, because they had an observer with them at all times and so they might be expected to behave as they thought they should behave [31]. And expert, naturally, could be more proficient in driving behavior, and easier to immerse in driving experiments.

2.1   Driving Video

For our experiments, the video we showed to participants is transformed from a driving recording video. We edited the recording video to exclude the original driving interface and to include several common driving scenarios (reversing, turning, sharp turns, changing lanes and overtaking, accelerating or braking). We edited video out of two reasons: first, the video should be as inclusive as possible, to ensure that the experiment can completely replicate the real scene; second, different driving operations will lead to different attention distribution. The fluency after video editing was also considered, as we wanted to provide the complete driving experience for the subjects. Meanwhile, driving video was slightly blurred according to perspective.

2.2   Participants

During the test, while watching a piece of driving video provided by a projector on the window area of the driving interface paper, participants were asked to evaluate and rank their active attention (goal-directed) as playing a role of a real driver and then add attention stickers (10 in total) to the relevant area. In this way, we measured top-down attention distribution through the stickers added by participants. At the same time, a camera, above the participant's head, would monitor where the participant was looking at. Camera data would be evidence of framing bottom-up attention. Thus, both stimulus-driven and goal-directed attention on the driving interface could be framed out in this experiment.

In addition, we also introduced the passenger's perspective for the control experiment: Participants were asked to play a passengers' role in co-driver seat location, and allocate their attention stickers to the same interface. Data from these passenger-view experiments could be fundamental support for both passengers' attention and drivers' attention in L5 autonomous driving scenario.

2.3   Potential Error Source

As we mentioned before, conducting a user evaluation with a real self-driving car is difficult, therefore we applied video experiments as an alternative. The lack of an unrelated driving environment, caused by simulation, can make the participants in the experiment unable to immerse themselves in driving imagination. Apparently, the biggest source of error in this test is determined by the simulation of the simulated driving environment.

To ensure the effect of driving simulation during the experiment, we used a projector that could provide a real size driving interface instead of a limited size digital screen, which had led us to be unable to use the eye tracker to monitor changes in visual attention precisely. To address this problem, a camera monitored participants eye's moving was introduced. Therefore, inaccurate positioning caused by the manual approach leads to experimental error.

Although repeatedly told that the participants should not be attracted too much from the dynamic video, the objective existence of the dynamic image tends to affect the subjective evaluation and ranking after the video.

3   Data Collecting and Analysis

Unlike previous psychology visual experiments that gather eye-tracking data while participants are viewing static pictures, web pages, or documents, every operation in driving environment is rather complex, even though we have refined the visual elements related to driving attention in this experiments. Therefore, it is hard to say which attention process (top-down or bottom-up) dominates a certain attention shift during a single driving action. And we decided to focus on one of them, separately.

3.1   Top-Down Attention (Goal-Driven)

For driver's goal-driven attention, we rely on the points (stickers) to figure out how much attention is applied to and where it is applied to. Through studying the extracted data, we present the data as a heat map lying on the graphic driving interface (see Fig. 1).

Fig. 1. Drivers' top-down attention distribution

Also passenger's top-down attention is presented in a heat map too, as an addition to drivers' heat map (see Fig. 2).

As we can see from these two graphs, when participants assume driving tasks and do not undertake driving tasks, their top-down attention distribution is clearly biased: the driver's attention is relatively concentrated on several points which are almost completely cover the whole driving interface; meanwhile the passenger's attention tends to be evenly dispersed to the core of the driver's interface, and part of it is concentrated in the middle of the lower half of the driver's interface.

This result can be explained by the driver's driving habits: the driver controlled the car by focusing on several control areas, just like how they usually drove. Once liberated from the driving task, their attention will still be placed on the driving interface, but there will be no specific concentration points. So our point of the whole driving interface is: the driving interface should be as flat as possible and decentralized, since users without driving tasks no longer pay top-down attention to a few points but the whole interface. And the perception and cognition could be better developed on a flat interface without visual elements interference. ("Flat", is not to say be 2D, but to say nothing distinguish itself and be simple.)

Both driver's attention and passenger's attention show relativeness to their aerial location: Steering wheel The steering wheel and its vicinity occupy the most attention of driver's attention, which is exactly the center of driver's normal sight. And the most

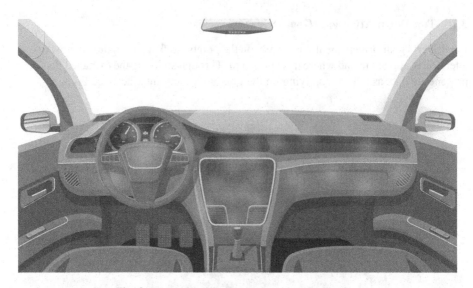

Fig. 2. Passengers' top-down attention distribution

attractive area for passengers is side control panel where passengers naturally turn their heads when communicating with drivers.

The results show that the aerial location has an influence on the attention distribution. However, we have proposed in the above that the driving interface should try to avoid protruding parts. Therefore, we believe that the driving interface should be changed to curved if conditions permit, reducing visual differences in different parts in the driving interface due to changes in spatial distance.

Also, we put their attention heat map together, to see how the attention of these two transforms and conflicts (see Fig. 3).

Obviously, we can see that the attention of passengers and drivers is barely coincident in the whole driving interface except side control panel in the middle of driving interface, which means drivers and passengers still tends to behave differently under auto-driving scenario, even though they were thought to be almost similar in high-level automatic driving.

At a autonomous cars, user's attention re-distribute when their identity is transformed from driver to passenger. Therefore, the area, to which both drivers and passengers pay mainly top-down attention, should be the places that information be expressed. And the control panel, the only conflict area, is being fully developed and used in Tesla, which is consistent with the phenomenon observed from our experiments.

3.2   Bottom-Up Attention (Stimulus-Driven)

We also counted the heat map of a distribution from the video recording the participants's (driver) eye-gazing area (see Fig. 4).

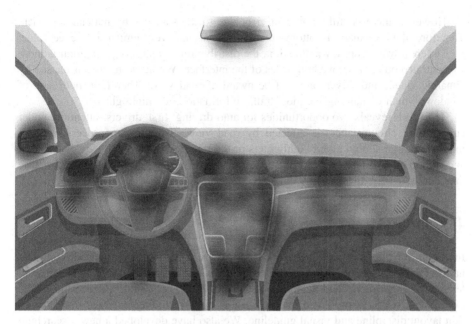

Fig. 3. Drivers and Passengers' top-down attention distribution

Fig. 4. Drivers' bottom-up attention distribution

However, there is still significant error due to the statistics by manual approach. Therefore, the heat map of bottom-up attention only has very limited reference value.

Different from Top-down attention, the distribution of bottom-up attention is more concentrated on areas above both sides of the interface. We insist that this is caused by dynamic video, and drivers are still be aware of road view. Therefore, both side of window, which containing the most traffic information, are highlighted.

The result reveals two opportunities for auto-driving: first, drivers' attention, under auto-driving circumstance, is affected by road view in bottom-up process; second, interface design should avoid highlighters or different texture on both side of interface and important messages should not been presented on the side of interface, because they can be easily distracted by dynamic road view.

4 Conclusion

In conclusion, we have conducted a visual experiment to simulate auto-driving scenario. Top-down attention has become our main research object, due to the characteristics of experiments and autonomous driving. Based on visualization results of the experiment, we have presented interface suggestions including interface frame, function layout discipline and visual guideline. We also have developed a new researching approach for analyzing attention distribution.

Although studies on spatial characteristics of attention distribution is available, there is still a need to conduct detailed functional division studies on the driving interface. This paper roughly considered the driving interface as several parts. The actual driving interface may need more specific distinction as different driving approaches involve different driving actions, and therefore the attention distribution will be slightly different. The future research may apply experiments in specific areas for certain driving actions.

References

1. McKerracher, C., et al.: An integrated perspective on the future of mobility. McKinsey & Company and Bloomberg New Energy Finance (2016)
2. Borojeni, S.S., Chuang, L., Heuten, W., Boll, S.: Assisting drivers with ambient take-over requests in highly automated driving. In: Proceedings of the 8th International Conference on Automotive User Interfaces and Interactive Vehicular Applications, pp. 237–244. ACM, October 2016
3. Rödel, C., Stadler, S., Meschtscherjakov, A., Tscheligi, M.: Towards autonomous cars: the effect of autonomy levels on acceptance and user experience. In: Proceedings of the 6th International Conference on Automotive User Interfaces and Interactive Vehicular Applications, pp. 1–8. ACM, September 2014
4. National Highway Traffic Safety Administration. Preliminary statement of policy concerning automated vehicles, Washington, DC, pp. 1–14 (2013)
5. Taxonomy SAE. Definitions for terms related to on-road motor vehicle automated driving systems. Technical report, SAE International (2014)

6. Hughes, P.K., Cole, B.L.: What attracts attention when driving? Ergonomics **29**(3), 377–391 (1986)

7. Eckoldt, K., Knobel, M., Hassenzahl, M., Schumann, J.: An experiential perspective on advanced driver assistance systems. IT-Information Technology Methoden und innovative Anwendungen der Informatik und Informationstechnik, **54**(4), 165–171 (2012)

8. Coppola, R., Morisio, M.: Connected car: technologies, issues, future trends. ACM Comput. Surv. (CSUR) **49**(3), 46 (2016)

9. Damböck, D., Weißgerber, T., Kienle, M., Bengler, K.: Evaluation of a contact analog head-up display for highly automated driving. In: 4th International Conference on Applied Human Factors and Ergonomics, San Francisco, USA (2012)

10. Anderson, J.R.: Cognitive Psychology and Its Implications, 6th edn, p. 519. Worth Publishers, New York (2004). ISBN 978-0-7167-0110-1

11. James, W.: The Principles of Psychology. Read Books Ltd., Redditch (2013)

12. Egeth, H.E., Yantis, S.: Visual attention: control, representation, and time course. Ann. Rev. Psychol. **48**(1), 269–297 (1997)

13. Posner, M.I.: Attention: the mechanisms of consciousness. Proc. Nat. Acad. Sci. **91**(16), 7398–7403 (1994)

14. Yarbus, A.L.: Eye movements during perception of complex objects. In: Eye Movements and Vision, pp. 171–211. Springer, Boston (1967). http://doi.org/10.1007/978-1-4899-5379-7_8

15. Helmholtz, H.: Treatise on physiological optics. III. In: Southall, J.P.C. (ed.) The Perceptions of Vision. Optical Society of America, New York (1925)

16. Mertens, J.J.: Influence of knowledge of target location upon the probability of observation of peripherally observable test flashes. JOSA **46**(12), 1069–1070 (1956)

17. Eriksen, B.A., Eriksen, C.W.: Effects of noise letters upon the identification of a tar- get letter in a nonsearch task. Percept. Psychophys. **16**, 143–149 (1974)

18. Eriksen, C.W., Collins, J.F.: Temporal course of selective attention. J. Exp. Psychol. **80**, 254–261 (1969)

19. Eriksen, C.W., Hoffman, J.E.: Temporal and spatial characteristics of selective encoding from visual displays. Percept. Psychophys. **12**, 201–204 (1972)

20. Eriksen, C.W., Hoffman, J.E.: The extent of processing noise elements during selective encoding from visual displays. Percept. Psychophys. **14**, 155–160 (1973)

21. Eriksen, C.W., Murphy, T.D.: Movement of attentional focus across the visual field: a critical look at the evidence. Percept. Psychophys. **42**, 299–305 (1987)

22. Eriksen, C.W., Rohrbaugh, J.W.: Some factors determining efficiency of selective at- tention. Am. J. Psychol. **83**, 330–342 (1970)

23. Posner, M.I.: Orienting of attention. Q. J. Exp. Psychol. **32**, 3–25 (1980)

24. Posner, M.I., Cohen, Y.: Components of visual orienting. In: Bouma, H., Bouwhuis, D.G. (eds.) Attention and Performance, 10th edn, pp. 531–555. Erlbaum, Hillsdale (1984)

25. Posner, M.I., Marin, O. (eds.): Attention and Performance, 11th edn. Erlbaum, Hillsdal (1985)

26. Posner, M.I., Rafal, R.D., Choate, L., Vaughan, J.: Inhibition of return: neural basis and function. Cognit. Neuropsychol. **2**, 211–228 (1985)

27. Posner, M.I., Snyder, C.R.R., Davidson, B.J.: Attention and the detection of signals. J. Exp. Psychol. Gen. **10**, 160–174 (1980)

28. Neisser, U.: Cognitive Psychology, p. 351. Appleton-Century-Crofts, New York (1967)

29. Folk, C.L., Remington, R., Johnston, J.C.: Involuntary covert orienting is contingent on attentional control settings. J. Exp. Psychol.: Hum. Percept. Perform. **18**, 1030–1044 (1992)

30. Rothenbücher, D., Li, J., Sirkin, D., Mok, B., Ju, W.: Ghost driver: a platform for investigating interactions between pedestrians and driverless vehicles. In: Adjunct Proceedings of the 7th International Conference on Automotive User Interfaces and Interactive Vehicular Applications, pp. 44–49 (2015)
31. Underwood, G.: Visual attention and the transition from novice to advanced driver. Ergonomics **50**(8), 1235–1249 (2007)
32. BMWNEWS. Autonomous cars to be in production by 2017, July 2016. https://news.bmw. co.uk/article/autonomous-cars-to-be-in-production-by-2021/. Accessed Apr 2017
33. Lagstrom, T., Lundgren, V. M.: AVIP-Autonomous vehicles interaction with pedestrians. Doctoral Dissertation, Chalmers University of Technology, Gothenborg (2015)
34. Debargha, D., Martens, M., Eggen, B., Terken, J.: The impact of vehicle appearance and vehicle behavior on pedestrian interaction with autonomous vehicles. In: Proceedings of the 9th International Conference on Automotive User Interfaces and Interactive Vehicular Applications, pp. 158–162, ACM (2017)
35. Marc-Philipp, B., Brenden, A.P., Klingegård, M., Habibovic, A., Bout, M.: SAV2P: exploring the impact of an interface for shared automated vehicles on pedestrians' experience. In: Proceedings of the 9th International Conference on Automotive User Interfaces and Interactive Vehicular Applications Adjunct, pp. 136–140. ACM (2017)
36. Chang, C.-M., Toda, K., Sakamoto, D., Igarashi, T.: Eyes on a car: an interface design for communication between an autonomous car and a pedestrian. In: Proceedings of the 9th International Conference on Automotive User Interfaces and Interactive Vehicular Applications, pp. 65–73. ACM, September 2017
37. Matthias, B., Witzlack, C., Krems, J.F.: Gap acceptance and time-to-arrival estimates as basis for informal communication between pedestrians and vehicles. In: Proceedings of the 9th International Conference on Automotive User Interfaces and Interactive Vehicular Applications, pp. 50–57. ACM (2017)

Investigating Temporal Changes of Behavioral Adaptation and User Experience During Highly Automated Driving

Dominique Stimm[✉], Arnd Engeln[✉], Julia Schäfer,
and Holger Schmidt

Stuttgart Media University, Stuttgart, Germany
{stimm, engeln, schmidtho}@hdm-stuttgart.de

Abstract. Sleepiness and micro-sleep as a consequence of the monotony of moving in queues as well as a very stressful daily routine of truck drivers put a serious risk on traffic safety (National Transportation Safety Board 1995). The automation of heavy traffic provides an opportunity to enhance traffic safety and drivers' convenience and allows the safe use of integrated infotainment and communication systems. The research project TANGO (German abbreviation for "Technologie für automatisiertes Fahren nutzergerecht optimiert", English equivalent "Technology for autonomous driving, optimized to user needs") is funded by the German Federal Ministry of Economic Affairs and Energy. It takes place in cooperation with Robert Bosch GmbH, Volkswagen Aktiengesellschaft, MAN Truck & Bus, University of Stuttgart and Stuttgart Media University. The project aims at improving user experience and acceptance of (highly) automated driving functions for trucks. The project focuses the user-centered development of an Attention and Activity Assistance system (AAA) which provides the truck driver with a variance of non-driving-related activities based on current traffic situation, automation level up to SAE level 3 (SAE international 2018), and the driver's current attentional state. While behavioral adaptation of drivers to the first use of highly automated systems has already been considered in a number of studies, little is known about the development of these behavioral changes over time, when familiarity with the system increases. In order to address these issues, a long term static driving simulator study will be conducted in spring 2019. The central research subject is the adaptation of drivers' behavior in take-over scenarios with low time budgets, which require an immediate reaction by the driver. The study will run from March to June, 2019. First research results will be presented at the HCI International Conference in July.

Keywords: Automated driving · Behavioral adaptation · User experience

1   Introduction

1.1   Autonomous Driving

While only a few decades ago, self-driving vehicles were more of science fiction than realistic prospect, the topic now has become a determinate and very near future. In fact,

© Springer Nature Switzerland AG 2019
H. Krömker (Ed.): HCII 2019, LNCS 11596, pp. 103–114, 2019.
https://doi.org/10.1007/978-3-030-22666-4_8

partly automated vehicles allowing the driver to temporarily pass the driving task to the automated system and only monitor the traffic and system state are commonplace on today's roads. Tesla firstly released the Model S equipped with an "Autopilot" in 2015 (Your Autopilot has arrived 2015), quickly followed by other car manufacturers (e.g., The Mercedes-Benz Intelligent Drive system offers compelling tech advances 2015). Automated driving systems have also been released for heavy traffic: In 2015, the partly automated Freightliner Inspiration was introduced in North America (Stromberg 2015).

Driver distraction and, especially in heavy traffic, sleepiness due to monotony of driving state serious risks for traffic safety (e.g., Young and Reagan 2007; Sagberg et al. 2004). The increasing automation of the driving task is supposed to reduce these risks by decreasing the driver's workload and by overcoming natural human limitations of attention, readiness to react and reaction quality. Simultaneously, the driver's convenience is addressed with the proceeding of technical development, new opportunities for entertainment, productivity, and recreation. For instance, the safe accomplishment of watching movies, playing games, writing e-mails and reading will be possible in the near future.

The SAE International taxonomy of autonomous driving functions (SAE 2018) describes six levels of motor vehicle automation and has reached mainstream and official recognition.

Level 0 (no driving automation) vehicles refer to 'conventional' cars and trucks which are fully operated by the driver.

In level 1 vehicles (driver assistance), the driver is supported by advanced driver assistance systems (ADAS) which temporarily either take over the lateral or the longitudinal subtask.

Level 2 refers to the already mentioned partial driving automation, which requires the driver to permanently monitor the traffic environment as well as the automated driving system which completes both, longitudinal and lateral vehicle control in definite situations.

Vehicles with level 3 automation systems (conditional driving automation) further release the driver from his monitoring task, allowing him to get involved in non-driving-related activities. However, the readiness to resume vehicle control after a sufficient time budget has to be ensured at any time. Note that the opportunity to attend to non-driving-related activities reflects only the system design but not necessarily the legal situation. The current research and developmental focus of automotive manufacturers lies on the imminent market launch of conditionally automated cars and trucks. Besides the technical realization of level 3 automation, the psychological aspects of driver behavior and experience during automation are of major research interest (see Sects. 1.2 and 1.3).

In SAE level 4 (high driving automation), the automated systems fully completes the driving task in a particular range of situations without any expectation that the driver will intervene. The driver is only obliged to intervene in foreseen tasks in specific driving situations which are not included into the particular range of automated situations.

The limitation of application to specific use cases is suspended in level 5. Here, the automated system is required to accomplish the driving task anytime and in any situation. No driver is needed any longer.

Contrary to the SAE taxonomy and nomenclature, many researchers use the term 'highly automated driving' when they refer to SAE level 3 (conditional automation). This stems from alternative taxonomies as by the American National Highway Traffic Safety Administration (NHTSA 2013) or the German Federal Highway Research Institute (BASt; Gasser and Westhoff 2012) which have been established earlier than the SAE standard. Also in the current study, 'highly automated driving' denotes SAE level 3.

1.2 Behavioral Adaptation

While a few decades ago cars equipped with assistance systems were rare occurrences, advanced driver assistance systems are commonly used these days and drivers rely on them even in situations where a system is at risk to fail (Itho 2012). Manual driving without any assistance systems involves the risk of driver distraction and sleepiness due to monotony of driving state, whereas assistance systems are supposed to reduce these risks by decreasing the driver's workload and by overcoming potential attentional deficits. They promise relief and relaxation to the driver, they inform them in case of critical traffic situations and release them from control tasks (Sullivan et al. 2016). But with regard to traffic safety, assistance systems can also bear a risk as driver's familiarity with the systems increases. According to Sullivan et al. (2016), behavioral changes in driver's behavior take place as they get more and more familiar with the systems. This might have a negative impact on safety issues because drivers sometimes rely on the systems too much. These behavioral changes based on individual experiences over time are defined as behavioral adaptation (Sullivan et al. 2016).

Behavioral adaptation can be triggered by different factors such as drivers' reduced risk perception, overtrust in the capability of automated systems, false estimation concerning the systems' competencies, and potential loss of engagement in the driving task (e.g., Itoh 2012; Rudin-Brown and Parker 2004; Sullivan et al. 2016). Reliance to automated systems even in critical situations where the system might become inactive increases the risk of not being prepared to take over vehicle control in the right moment and with the reaction quality required. For example later braking in case of ACC use compared to manual driving was shown by Nilsson (1995). Larsson et al. (2014) found similar effects in experienced as well as in inexperienced ACC drivers. These results show that slower reaction does not seem to be only a cause of unfamiliarity with the system but might be caused by behavioral adaptation. Given these facts, behavioral adaptation seems to play an important role for the evaluation of traffic safety in the context of automated systems (Mehler et al. 2014).

According to Manser et al. (2013), behavioral adaptation may be structured into three temporal stages: immediate (the first learnings when the driver initially interacts with the system), short term (over days or weeks) and long term (over weeks or months) (Sullivan et al. 2016). While immediate behavioral adaptation of drivers to the use of highly automated systems has already been considered in a number of studies, little is known about the development of these behavioral changes over time, when familiarity with the system increases. To address this issue, a long term static driving simulator study will be conducted in spring 2019, based on the research project TANGO.

1.3   Research Project TANGO and User-Centered Development

The current study is based on the research project TANGO (German abbreviation for Technologie für automatisiertes Fahren nutzergerecht optimiert, English equivalent Technology for autonomous driving, optimized to user needs), funded by the German Federal Ministry of Economic Affairs and Energy, in cooperation with Robert Bosch GmbH, Volkswagen Aktiengesellschaft, MAN Truck & Bus, and University of Stuttgart. The project aims at developing an Attention and Activity Assistance system (AAA; see Fig. 1) which shall enable truck drivers to engage in secondary tasks during automated driving phases. The AAA provides the drivers with diverse non-driving-related activities based on current traffic situation, automation level up to SAE level 3 (SAE international 2018), and the driver's current attentional state. It is designed by combining proven environment sensors with novel cabin-interior sensors and novel HMI-concepts (projekt-tango-trucks 2017).

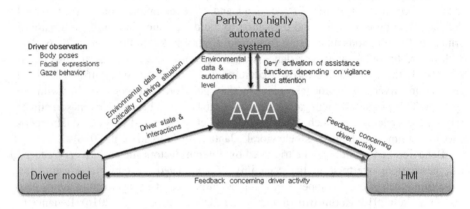

Fig. 1. The Attention and Activity Assistance System (AAA)

For market success, automated driving systems to enhance traffic safety have to be accepted by the user. This can be accomplished by a positive user experience which can be achieved by a user-centered development (DIN EN ISO 9241-210 2010). That for the AAA is developed in a user-centered way.

User-Centered Development of the AAA. The user-centered development of the AAA starts with various empirical studies for a deep understanding of the current working situation likes and dislikes of truck drivers, as well as their attitude towards automated driving. Based on the studies' results, functionalities, interactions and other prototypes of an AAA are realized in an iterative development process: In first iteration, a great number of very easy to realize "paper prototypes" are created and then evaluated in focus groups and online studies in form of user stories. In second iteration, the most promising ideas are provided in simulation and tested in laboratory environment. Finally, the optimized concept is implemented in a truck for evaluation in realistic environment. But how can the construct user experience be described?

Six Facets of User Experience. User experience is a complex phenomenon. In most situations, only some experiences are selected consciously for decision making, whereas the majority of experiences influences the user acceptance on a subconscious level. For the systematic design of a good user experience of products or services it is important to know as much as possible about the complex phenomenon. For a broad understanding of user experience (Engeln 2013; Engeln and Engeln 2015) published a six facets model (see Fig. 2).

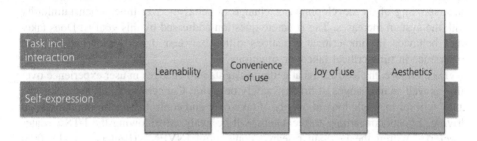

Fig. 2. Six facets of user experience

The facets are not selective but influence each other. In particular, the facets task (with interaction) and self-expression are influenced by comprehensibility, convenience, joy, and aesthetics of use.

Task (incl. interaction) addresses the experienced utility of an offer or product. Does it help to reach one's goals, reduce the effort or improve the quality of the result?

Self-expression means the (mostly implicit) thoughts about what others would think about the user when he owns or uses the offer or product. Does the user identify himself with the product, is he proud of it or rather ashamed?

Learnability runs with the investment of effort for competent use. Is the use intuitive, self-explaining or do I have to learn for it?

Convenience of use is relevant in case of dominant extrinsic motivation, when the user mainly wants to gain the results of an action. A good convenience of use leads to relaxation, a bad one to pressure.

Joy of use becomes important in case of mainly intrinsic motivation, when the action itself dominantly motivates the user. A good joy leads to happiness, a bad one to boredom.

Aesthetics focusses impressions of all senses. People adore good views, sounds, tastes, smells or touches. They avoid negative ones.

Asked for their user experience, people typically talk about the task but seldom about the self-expression. Very often, people stop reflecting when they have found "one good argument", although their experience in fact is affected by several factors. Because of that, the whole phenomenon needs to be taken into consideration in order to design a good user experience.

This report focuses on a specific long-term simulator study based on TANGO, which looks on changes over time in user experience and behavioral adaptation to automated driving and interaction with the automated system. As the TANGO project also deals with the similarities and differences between truck and car drivers and wants to give insights concerning the transfer between truck and car context, this study is focused on car drivers' behavioral adaptation.

1.4   Objectives

The central objective of this study is to learn more about the development and change of drivers' adaptation behaviors in time-critical take-over scenarios with low time budgets which require an immediate reaction by the driver. The main interest is a better understanding of the development or change in behavior over time, when familiarity with the system increases. The concrete question addressed by this goal is: Does take-over behavior in time-critical situations differ between drivers with and without experience in time-critical takeover situations?

Furthermore, the study aims at getting insights into changes in user experience over time as well as in changes of drivers' safety behavior. Concerning safety behavior, the study will for example look at quality of take-over and control gazes during automated driving. Effects concerning traffic safety will be analyzed following the I-TSA scales developed within the German research consortium INVENT (Böttcher et al. 2005; Glaser et al. 2005).

2   Method

In order to access the development of behavioral adaptation and user experience over time, a long-term static driving simulator experiment will be conducted in spring 2019.

2.1   Driving Simulator and Driving Scenario

For the purpose of the experiment, the driving simulator at Stuttgart Media University (see Fig. 3) will be used. It is based on a platform with multiple interacting high-end computers. The simulation is presented on three large displays in the participant's main field of view. In addition, we placed two smaller side-displays for showing virtual mirrors of the rear scenery. A realistic seating position is provided by the combination of actual automotive components, i.e., driver seat, steering wheel, and pedals. The equipment is mounted within a steel frame to ensure a consistent environment for all participants. The haptic feedback from the input gear resembles real automotive technology in case of the gear shifter, the pedal stiffness/response, and an actuator behind the steering wheel for active steering feedback controlled by the simulation. Surround audio gear also gives the participant a more holistic experience.

The simulator mockup is placed in a tiny laboratory room without window but ventilation. This ensures a calm and reproducible environment for the experimental situation. The simulation is controlled from a separate control computer outside the laboratory room. The participant is able to communicate with persons outside via speaker.

The SILAB simulation software (WIVW - Würzburg Institute of Traffic Science n.d.) is used. The modular software provides deep configuration capabilities for customizing the simulation process. The configuration of the simulation for this specific experiment uses a single road course with alternating events over the course of the full experiment. As the driving situation is intended to simulate a daily way to work the same route is used for all rides. Participants will start aside the road and drive manually towards a highway entrance. On the highway, participants will be asked to transfer vehicle control to the automated driving system. Within the autonomous driving situation, several take-over requests are placed. Some take-over requests are critical some are uncritical according to the time given to take over control. After the experiment-related events will be completed, participants will drive manually off the highway and park the car.

On runtime, the simulation measures several driving related data. The following list contains the most relevant of them:

- Ego-car speed and acceleration (longitudinal and lateral)
- Other vehicle speeds and accelerations (longitudinal and lateral)
- Distances between the ego car and other vehicles or objects
- Runtime of the simulation
- Position of ego-car and other cars (absolute and lane-dependent)
- Inputs of the participant, i.e. steering angle, pedal offset, gear shifter

Collecting these metrics enables to calculate further simulation related values like time to collision with other vehicles as part of the scientific questions behind the experiment and for analyzing effects on traffic safety.

2.2 Experimental Design

A 5 × 2 mixed design with factors driving session (within) and experience with timecritical take-over situations (between) will be used. The factor driving session has five levels (ride1/ride 2/.../ride 5). Rides should take place once a week for five weeks à 45 min each ride. The second factor (experience with time-critical takeover situations) has two levels (time-critical situation: yes/no). The experimental group will experience time-critical take-over situations in rides 1 to 4 whereas a control group will only experience non-time-critical take-over situations in rides 1 to 4. In ride 5, both groups will experience a time-critical take-over situation.

Fig. 3. The driving simulator at Stuttgart, Media University

2.3   Procedure

The study will be conducted in the static driving simulator (see Sect. 2.1). For first trial, participants will be welcomed in the laboratory and will get written as well as oral instructions concerning the experiment. Participants will be instructed that the experiment aims at evaluating the user experience of secondary tasks under realistic conditions. They will be asked to imagine being on their way to work while driving. Furthermore, participants will be asked to complete a demographic survey and a data protection declaration.

After filling in all documents, participants will start with a training session of about five to ten minutes to get familiar with the simulator and to exclude participants who suffer from motion sickness (e.g., Brooks et al. 2010). For this training session a drive through a small piece of landscape including curvy land roads and a part of a small town is chosen. During the training session, participants will learn how to activate the autonomous driving and how to take over vehicle control. Before starting the test ride they will be told that they can stop the experiment anytime they do not feel comfortable anymore.

After the training session, the eye-tracking system will be applied and calibrated. The ride will start with a manual driving session for about three minutes. Participants will be instructed not to overtake during this session. After three minutes, participants will be asked to start automation. This is done by pressing a button located to the right of the steering wheel. During the automated driving session, the ego car follows a lead vehicle without performing takeover maneuvers. This enables a better control of events within the simulation scenario for example surrounding traffic.

Participants in both groups will be asked to perform a secondary Sudoku task on a tablet during the automated driving session. After 20 min of autonomous driving, participants of both groups will be faced with a take-over situation. While control group participants will experience a non-time-critical take-over request because of road works, experimental group participants will be faced either with a time-critical or a non-time-critical request. The time budget in the time-critical take-over situation will be seven seconds (according to Gold et al. 2016) the budget for the time-uncritical take-over requests will be 60 s. Participants will be instructed to take-over vehicle control immediately. The following events are chosen to define the time-critical take-over situation:

- Deer crossing
- Vehicle going into lane from the right side
- Unsecured accident site
- Sudden emergency braking of lead vehicle because of traffic jam

Control group participants and those of the experimental group who will be faced with the non-time-critical request do not have to take-over immediately. They will be informed that they will have to resume vehicle guidance in one minute because of road works. There will be a reminder 30 s before they should take over. In case participants do not take over, there will be a safety stop.

After take-over, participants will drive manually for about two minutes. Then participants will be asked to give the vehicle control back to the system. Drivers then will drive autonomously for another 18 min. After 18 min, control group participants will experience another non-time-critical take-over request because of an exit ramp whereas experimental group participants will experience a time-critical or a non-time-critical request, depending on the criticality of the previously experienced request: Participants who already experienced a time-critical request will be faced with a non-time-critical request whereas participants who firstly experienced a non-time-critical request will be faced with a critical one. This counterbalance is done to avoid learning effects due to the order of the time-critical take-over situation. After the take-over, participants will be asked to drive manually until the next highway exit. Participants will drive to a specific parking space and will stop the car there. The procedure of the original experimental phase is graphically displayed in Fig. 4 separately for experimental and control group.

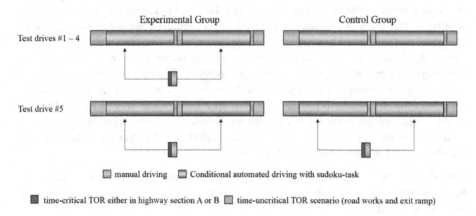

Fig. 4. Procedure

2.4 Participants

50 students with normal or corrected-to-normal sight and hearing and a valid driving license will be acquired. Participants will be paid 10 Euros per test trial and a bonus of 40 Euros as far as they will have taken part in all five repeated measurements. Participants will be randomly assigned to one of the two between subject conditions (experience with time-critical take-over situations vs. experience with non-time-critical takeover situations).

2.5 Measurements

Besides the simulation-based metrics, biometrics from the participant will be collected. The main biometrics here will be gaze-based interactions of the participant with the simulation and with objects aside the simulation (i.e. secondary task). For that purpose, a mobile eye tracking system will be used.

Eyetracking. For the assessment of the participants' glances behavior, the Dikablis Professional Glasses 3 (Ergoneers. Hard- und Software für Verhaltensforschung & Eye-Tracking n.d.) and the associated software D-Lab will be used. The binocular head-mounted eye-tracking system comprises an adjustable scene camera capturing the wearer's visual field with an aperture angle of about 90 and a resolution of 1920 × 1080 pixels. The two adjustable infra-red eye cameras, attached to a leg underneath the eyes, have a resolution of 648 × 488 pixels with a tracking frequency of 60 Hz. The cameras' housings contain small infra-red LEDs illuminating the eyes. The images of the illuminated eyes are digitally processed frame-wisely in order to detect the pupils with an accuracy of 0.05° visual angle. The glance direction is estimated with an accuracy of 0.1° to 0.3°.

The Dikablis glasses will be used wirelessly with the data being stored in a small transmission unit and sent via WiFi to the recording computer, a Dell Precision 7520 with the recording and analysis software D-Lab installed. In order to enable automatic analysis of glance behavior for different areas of interest, so-called markers, graphical black and white patterns of size 6 × 6 and 10 × 10 cm, will be arranged in the simulator. The markers are based on simple high contrast visuals (mostly similar to QR-code) and are placed in strategic positions on the main displays or on other objects. The analysis software D-Lab detects these reference points in the recording environment via image processing and determines the location of marker-coupled visual targets despite head movements. Three markers will be displayed directly in the middle screen corners of the simulation to tag the road. The speedometer unit will be coupled to one marker displayed left to the instrument cluster. Exterior mirror markers will be attached below the respective mirror screens and one marker will be displayed on the tablet which is used for the secondary task (Sudoku).

Behavior. To get insights into changes of behavioral adaptation over time, take-over behavior, reaction times and -correctness as well as driving behavior after take-over will be measured.

Stress/Demand. Subjective measurements will also been taken into account to get information concerning participants' stress and user experience over time. Stress and demand over time will be measured with the Rating Scale for Mental Effort (RSME, Eilers et al. 1986) whereas user experience will be measured with a pilot version of our newly developed UX questionnaire.

2.6   Analysis

Data will be analyzed with the statistic software SPSS and custom-designed software by Blickshift.

3   Results (Preview)

As the study is going to start in March 2019, the results cannot be presented at this point. It is planned to complete the collection of data by July, 2019. In parallel to the conduction of the study, data will be progressively preprocessed and analyzed so that at the HCII, first prospects of the change of driver behavior and attitudes over time will be presented.

4 Conclusion

The planning and execution of a longitudinal study with data collection over several weeks needs a high amount of investment. The advantage – compared to classical single data collection – is the knowledge about participants' behavior and experience development over time. In contrast to a very first contact to automated driving, this longitudinal research design will come somewhat closer to the reality of automated driving: When people own an automated driving system, they get used to it very quickly and change their usage behavior and experience within a few days (Arnon et al. 2014). Therefore, it is reasonable to assume that the planned experiment will reveal behavioral changes in drivers during the interaction with the automated system.

References

Arnon, M., Sonntag, E., Johannsen, U., Schmidt, C., Scherzinger, F., Engeln, A.: User experience für teilautomatische fahrfunktionen am beispiel des highway assist. In: VDI-Berichte. VDI-VW-Gemeinschaftstagung Fahrerassistenz und Integrierte Sicherheit, vol. 30. VDI, Düsseldorf (2014)

Böttcher, S., Nirschl, G., Schlag, B., Voigtländer, M., Weller, G.: INVENT-Teilprojekt Fahrerverhalten/MMI (FVM), Arbeitspaket 3100: Entwicklung eines Bewertungsverfahrens. Dresden: Fraunhofer Institut Verkehrs- und Infrastruktursysteme und Technische Universität, Verkehrspsychologie (2005)

Brooks, J.O., et al.: Simulator sickness during driving simulation studies. Accid. Anal. Prevent. **42**, 788–796 (2010)

DIN EN ISO 9241-210: Ergonomics of human-system interaction- Part 210: Human-centered design for interactive systems (2010)

Eilers, K., Nachreiner, F., Hänecke, K.: Entwicklung und Überprüfung einer Skala zur Erfassung subjektiv erlebter Anstrengung. Zeitschrift für Arbeitswissenschaft **40**(4), 215–224 (1986)

Engeln, A.: User Experience als Ansatz zur Gestaltung marktattraktiver Produkte. In: Tagungsband zum 4. HMID Symposium, Stuttgart, pp. 75–83 (2013)

Engeln, A., Engeln, C.: Customer Experience und kundenzentrierte Angebotsentwicklung. Was gehört dazu? In: Baetzgen, A. (eds.). Brand Experience. An jedem Touchpoint auf den Punkt begeistern, pp. 253–273. Schäffer Pöschel, Stuttgart (2015)

Ergoneers. Hard- und Software für Verhaltensforschung & Eye-Tracking (n. d.). http://www.ergoneers.com/. Accessed 21 Jan 2019

Gasser, T.M., Westhoff, D.: BASt-study: Definitions of automation and legal issues in Germany. In: Proceedings of the 2012 Road Vehicle Automation Workshop (2012)

Glaser, W.R., Waschulewski, H., Schmid, D.: I-TSA - a standardized procedure to assess the safety impact of driver information and driver assistance systems. In: "Der Fahrer im 21. Jahrhundert" (The driver in the 21st century), VDI-Berichte, Braunschweig, no. 1919, pp. 17–10 (2005)

Gold, C., Körber, M., Lechner, D., Bengler, K.: Taking over control from highly automated vehicles in complex traffic situations: the role of traffic density. Hum. Factors **58**(4), 642–652 (2016)

Itoh, M.: Toward overtrust-free advanced driver assistance systems. Cogn. Technol. Work **14**(1), 51–60 (2012)

Larsson, A.F.L., Kircher, K., Hultgren, J.A.: Learning from experience: familiarity with ACC and responding to a cut-in situation in automated driving. Transp. Res. Part F: Traffic Psychol. Behav. **27**, 229–237 (2014)

Manser, M., Creaser, J., Boyle, L.: Behavioural adaptation: methodological and measurement issues. In: Rudin-Brown, C.M., Jamsom, S. (eds.) Behavioural Adaptation and Road Safety, pp. 35–59. CRC Press, Boca Raton, FL (2013)

Mehler, B., Reimer, B., Lavalliere, M., Dobres, J., Coughlin, J.F.: Evaluating technologies relevant to the enhancement of driver safety. AAA Foundation for Traffic Safety, Washington, D.C. (2014)

National Transportation Safety Board: Factors that affect fatigue in heavy truck accidents. Safety Study NTSB/SS-95/01 and NTSB/SS-95/02. Washington, DC (1995)

NHTSA National Highway Traffic Safety Administration: Preliminary statement of policy concerning automated vehicles. National Highway Traffic Safety Administration (2013). www.nhtsa.gov/staticfiles/rulemaking/pdf/Automated_Vehicles_Policy.pdf

Nilsson, L.: Safety effects of adaptive cruise control in critical traffic situations. In: Proceedings of the Second World Congress on Intelligent Transport Systems, vol. 3, pp. 1254–1259 (1995)

projekt-tango-trucks: Project (2017). https://projekt-tango-trucks.c-om/en/. Accessed 30 Jan 2019

Rudin-Brown, C.M., Parker, H.A.: Behavioural adaptation to adaptive cruise control (ACC): implications for preventive strategies. Transp. Res. Part F-Traffic Psychol. Behav. **7**(2), 59–76 (2004)

SAE international: taxonomy and definition for terms related to driving automation systems for on-road motor vehicles (2018)

Sagberg, F., Jackson, P., Krüger, H.-P., Muzet, A., Williams, A.: Fatigue, sleepiness and reduced alertness as risk factors in driving. TOI report 739/2004. Institute of Transport Economics, Oslo (2004)

Stromberg, J.: This is the first licensed self-driving truck. There will be many more, 6 May 2015. https://www.vox.com/2014/6/3/5775482/why-trucks-will-drive-themselves-before-cars-do

Sullivan, J. M., Flannagan, M. J., Pradhan, A. K., Bao, S.: Behavioral adaptation to advanced driver assistance systems: a literature review. AAA Foundation for Traffic Safety (2016)

The Mercedes-Benz Intelligent Drive system offers compelling tech advances, 31 August 2015. http://www.loebermotors.com/blog/mercedes-benz-intelligent-drive-system/

WIVW (n.d.): Würzburg Institute of Traffic Science. https://wivw.de/en/

Young, K., Regan, M.: Driver distraction: a review of the literature. In: Faulks, I.J., Regan, M., Stevenson, M., Brown, J., Porter, A., Irwin, J.D. (eds.) Distracted Driving, pp. 379–405. Australasian College of Road Safety, Sydney (2007)

Your Autopilot has arrived, 14 October 2015. https://www.tesla.com/pt_PT/blog/your-autopilot-has-arrived

Exploratory Study into Designing Enhanced Commute Experiences in Autonomous Vehicles with Connected Sensors and Actuators

Zheng Sun[✉], Hugues Vigner, Shreyas Bhayana, Changhao Zheng, Ziwei Zhang, Euiyoung Kim, and Alice M. Agogino

University of California, Berkeley, CA 94720, USA
leceasun@berkeley.edu

Abstract. This paper discusses an exploratory study regarding user perceptions and behaviors associated with autonomous vehicles (AV). We conducted three research methods – interviews, observations, and surveys – to collect comprehensive user data for capturing useful insights. We found that comfort with light, vehicle safety, audio entertainment and itinerary transparency are the main four concerns that users in our study had in current transportation systems, especially in shared rides. Having these categories as our guidelines for design, we generated conceptsand built a low-fidelity prototype using connected sensors and actuators for user testing. Feedback from potential users and experts in the automobile industry were recorded to refine the proposed concept for further development.

Keywords: Automotive user interfaces · Traveler's behavior ·
User interfaces for (semi) autonomous driving · Women in transport ·
Car sharing in (semi) autonomous vehicles

1 Introduction

Breakthroughs in technology have allowed the opportunity to shift from full human-driven vehicles to autonomously driven vehicles (AVs), which partially or fully drive themselves and which may ultimately require no driver at all, based on different levels from 0 to 4. The current transition period of the vehicle assisting the decision-making for the human driver has introduced technologies such as crash warning systems, Adaptive Cruise Control (ACC), lane keeping systems, and self-parking technology [1]. The mobility sphere, which has undergone constant innovation in the past century, is going through a paradigm shift that seeks to transform the very core of the status quo [2]. The field of AVs has drastically revolutionized this mobility sphere, as dozens of startups and automobile manufacturers race to reach level 3, and 4 of autonomy [3].

With software controlling all the subsystems of the vehicles without human intervention, drivers will be able to use commute times more productively than ever before. Drivers (and passengers) will have the potential to use the travel time for a wide range of new tasks, such as preparing presentations and conducting mobile meetings or spending time on leisure activities, watching a video or reading a book during the

© Springer Nature Switzerland AG 2019
H. Krömker (Ed.): HCII 2019, LNCS 11596, pp. 115–127, 2019.
https://doi.org/10.1007/978-3-030-22666-4_9

commute time that wouldn't have otherwise been possible [4, 5]. While these are ideal mobility and travel experiences in future scenarios, previous research has shown that current commute experiences are typically negative with a wide range of onboard emotions, ranging from passengers feeling bored during a long-distance road trip to feeling extreme anxiety on the way to work [6]. The mobility experience is compounded while riding in a shared space. In a shared space, different users require different comfort levels; women, for example, revealed more concern than men in a constraint shared space [7].

In this paper, our research goal is to enhance the commuting experience in emerging mobility solutions (e.g., autonomous, shared, and connected vehicles) by employing connected sensors and actuators in the interior space of the vehicle. We employed a human-centered design methodology [8] to discover how users from various demographics interact with vehicles in daily life, and to identify pain points that need to be addressed in the context of emerging autonomous driving conditions for users. By analyzing the collected data, we aim to determine how to enhance the onboard experience in future mobility conditions. New concepts based on this user research were evaluated and a sample concept was selected to tackle identified user needs in various demographics. Finally, a reduced scale low-fidelity prototype was developed and tested. Feedback from potential users and domain experts were collected to refine the prototype for further research.

2 Methods and Analysis

Interviews, observations and a survey were triangulated to collect potential user information. We conducted in-person interviews focusing on potential sensorial experiences in automobiles and current sensors in automobiles; we distributed an online survey soliciting feedback before a prototype was developed; we also performed multiple observations of co-riders during rideshares trips to study their experience using the service. These different research methods allowed us to collect user behavior data. The findings from each research method were used to develop a prototype and served as the foundation for user testing. The sequence of research development and predicted outcomes is illustrated in Fig. 1.

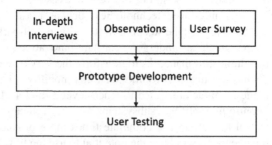

Fig. 1. Progression for research development and predicted outcomes.

2.1   Interviews

Interview Design. To obtain a broader view from the public, we conducted 19 in-depth interviews with potential users. We used snowball sampling—approaching potential interviewees via personal networks, co-riders during commercial shared rides, attendees in local tech conferences, etc. We selected interview participants based on their prior mobility experience in either shared or private transportation to gather their behavior, habits, and needs. We also targeted passengers with various dispositions. Through the interview method, we aimed to understand any specific issues or pain points with respect to the current transportation modes.

Interview Demographics. The demographic groups based on their age, gender, and occupation are detailed below.

Age. The age of interviewees ranged from 18 to 40. We mainly targeted young adults and working professionals as the recent study shows that more than 80% of rideshare users are aged between 18 and 49 [9]. In our interviews, the majority are either college students or young professionals.

Gender. In our study, we interviewed 19 rideshare uses – 12 males (65%) and seven females (35%). We focused on their ride-sharing experiences and their expectations in the development of autonomous vehicles. The extracted data from the seven female users shows that their perspective on autonomous vehicles are different than for male users, with a greater concern for security.

Profession. Thirty-six percent of the interviewees were college students who are attending either graduate or undergraduate programs. Among other interviewees, 26% of total interviewees work in technology fields.

Interview Analysis. With the data collected from the interviews, we applied grounded theory to analyze the gathered data and develop protocols and quantify the qualitative data. The analytic process takes the steps of coding data, developing, and refining theoretical categories [10]. Table 1 shows extracted sample scripts and resulting codes generated by two raters. To extract the key codes that are important to our research, two coders from our team recorded the categories of the selected code based on personal understanding. By comparing the similarity of our codes, we used the scored number to identify whether we successfully targeted the pain points of the interviewees. Two coders agreed on 65 out of the 74 scripts, which give us an 88% agreement in the data analysis.

According to the 2017 report on the demographics of Uber's user population in the U.S. [12], 48% of the Uber riders are female. In our research, we found female users tended to trust autonomous vehicles more in comparison to male users. Of all the interviewees that responded to the question of whether they would purchase/rent an autonomous vehicle for daily commutes, 83% females responded they would be looking forward to the bloom of autonomous vehicles whereas only 45% of males had a positive perception on the autonomous vehicle. Table 2 summarizes the breakdown of interests in using AV based on gender.

Table 1. Conduct grounded theory to quantify interview data (paragraph by paragraph) [11]

Script	Coder 1	Coder 2	Score
"I choose a rideshare service based on safety, convenience, cost. (It is) safer than bigger transportation. People taking pool are more mentally stable. There are crazy people on buses."	Mentally stable co-riders	Safety	0
"(I) don't care about the visual stuff, listening to music is nice. But I don't like staring at the screen. I am a software engineer, so I really value the time not staring at the screen."	Audio Entertainment	Audio Entertainment	1
"I want it (AV) to happen, I trust it more than a real person. Real people lose attention, people make stupid decisions. I personally really hate street lights. Bright light blinds me and distracts me from driving, especially at night!"	Comfortable light	Comfortable light	1
(I use ride share services) twice a week or more. I want to make sure that the drivers try to find where I am waiting for them... or at least make sure that they answer their phones. Some drivers might be politer."	Transparency	Transparency	1
"I hate driving in Indiana at night because it is so dark and people driving toward me [...] their high beams always blind me."	Comfortable light	Comfortable light	1
"My favorite thing in a car is listening to music. I want good sound in the car to have my relax time."	Audio Entertainment	Audio Entertainment	1
		Total inter-rater agreement score:	88%

Table 2. Number breakdown of whether users are interested in using AV based on gender

	Will use AV	Will not use AV
Males	5	6
Females	5	1

The extracted reasoning from the interviewees explain that most of the female drivers expressed concern about their driving skills under current road conditions. Female drivers predominantly perceived that autonomous vehicles would provide more coherent and agile driving experiences compared to human drivers [13, 14]. Table 3 shows example quotes from female interviewees: trust levels towards their own driving skillsets and autonomous vehicles.

Table 3. Example responses from female drivers on trust of autonomous vehicles

Interviewee initial	Response
DY	"I want it to happen as soon as possible. Since I am so bad at driving (laugh). Driving in China is insane!"
DU	"Would make tasks repetitive easier, trucks which transport (would be easier, concerned about safety but it should be fine once finished."
SK	"There will be a lot less accidents out there if I stop driving."
FA	"I want it to happen, trust it more than a real person... real people lose attention, people make stupid decisions. There is no fatigue."
LK	"Only two recorded cases of failing (for Tesla), safer than human driving."

2.2 Observations

The findings from the interviews indicate that users who are not informed about AV technologies, do not really trust the technology. However, users who are under the impression that they have a good level of understanding about AV technology, are much less wary of using the technology [15]. Other studies have identified trust as important factor in the adoption of autonomous vehicles [16]. Some of our interviews, highlighted opportunities where AV's could address users' need for improved safety, such as in understanding road conditions and witnessing the operations of the system.

To dig deeper into the findings from the interviews, we conducted five observations on rideshare trips. Ride-sharing simulates an 'autonomous' ride since the users do not need to drive. However, since there are actual drivers operating the shared rides, we also focused on their interactions with the passengers to see whether there are any pain points that we missed while interviewing the potential users.

The riders we observed were all under 30 and college students. The first observation was a 20 min shared ride around the campus of UC Berkeley. We observed the quality of driving, the personality of the driver, and the passengers' reactions when interacting with the driver. The driver was around 50 years old and was an enthusiastic environmentalist. During the ride, the driver constantly requested the riders to listen to her talk about her interests and upcoming projects riders could join in with. Even though the riders were mostly interested in the conversations, one of the riders seemed anxious by the impact of her distraction on her driving, since the driver was focused on her conversation and was distracted from driving. One of riders was constantly checking the route on her phone and looked outside of the window. She eventually asked to be dropped off before she arrived at her final destination. From this observation, it was clear that some riders were concerned about safety when they sensed the drivers were distracted. To seek security from a "dangerous" ride, the riders chose to constantly check their current route, as well as choosing to exit the vehicle before arriving at their destination.

In another similar observation of a rideshare, we observed a 30-min ride from downtown Oakland to downtown Berkeley on a rainy day. The shared ride driver was a heavy smoker and the vehicle had a pungent smell during the ride. One passenger asked the driver to roll down the window, but this was refused by the driver due to heavy rain. The lack of transparency on the route itinerary was also commented on. The driver also raised safety concerns of passenger due to using his phone and driving with a single-hand to the steering wheel. It was clear that the passengers felt that it was important to have fresh and breathable air in the vehicle, as well as the driver's attention to safe driving practices.

We also observed interesting patterns between passengers, luggage and the shared space that motivated new concepts in seat layout for a social setting, with the role of the driver being part of the social functions occurring [17]. We observed riders on a Friday night sharing a ride across downtown Berkeley. The vehicle was cramped, as there were already two people, a couple, in the vehicle. It was difficult to see where to sit in the car due to the darkness and tinted windows. It was also inconvenient for another two-person couple to arrange their eating to their satisfaction, when one person sat in the front and the second behind them, preventing them holding a conversation.

For the return trip one and a half hours later, one of the riders was using two of the seats for storage of their personal items, making it difficult for a 4th co-rider to use the remaining space. The space could not be rearranged, and the seat belt buckle had become inaccessible, comprising the safety of the 4th passenger. After dropping off two other riders, the drivers asked if the rider at the front wanted to sit in the back with her friend. Ridesharing in low comfort conditions like these underlines' possible improvements on the standardization of seat space when a high volume of different unique users occupy the space with low retainment.

Finally, we conducted two observations to consider the role that trip length played in passenger behavior. We observed that when riding with friends, for a short distance, most of the time individuals had conversations during the entire trip. For longer trips, it was typical for passengers to be energetic during the first hour. Later, when topics of conversation were depleted, and dialogues are difficult to maintain, users started checking their smartphones. Some riders would fall asleep during the second hour of the trip. The driver had to remain alert, listening to the music to stay awake. During most of the observed commutes, music was connected to the car infotainment system through the driver's smart device. However, in long trips, passengers could generally access the music controls by connecting their devices.

Onclusion, passengers showed concern for their safety, particularly if the driver appeared to be distracted or hid the itinerary during the trip. Passengers expected their ride to be in a hygienic and comfortable environment with access to their friends while travelling in groups. Entertainment is appreciated during relatively short trips, but riders also valued time for resting during long trips.

2.3 Online Survey

We conducted an online survey regarding user experiences in current vehicles, their expectations and their attitudes toward autonomous vehicles.

Respondent Demographics. Data from a total of 52 participants was collected through online surveys. Among the dataset, three incomplete surveys were excluded from analysis. For the remaining 49 participants (male: 27, female: 22), the age breakdown was as follows: the 18–25 range (22 participants), 25–35 range (13 participants), 35–45 range (6 participants), 45–55 range (2 participants), 55–65 range (2 participants). Among these 49 participants, there were 18 students, 16 engineers, 4 salespeople, 2 business managers, 2 teachers, 1 faculty member, 2 full-time house-wives, 1 entrepreneur, and 3 industrial workers.

Survey Design. The online survey used multiple choice questions, Likert scale, ordinal scale, categorical scale, and open-ended questions to collect participants' insights. The survey was composed of three sections and included a total of 20 questions (Part 1: 3 questions, Part 2: 11 questions, Part 3: 6 questions). The content and the design of the survey is shown in Table 4.

Table 4. Section breakdown for the designed survey

Section	Type of survey questions (quantity)	Functionality	Example question	Purpose of the survey design
Part (1)	Multiple Choice (3)	Gathers the demographic data of the participants including age, gender, and occupation	What's your current occupation?	Multiple choice questions enable us to categorize answers efficiently
Part (2)	Multiple Choice (6), Multiple Choice & Short Answer (2), Likert Scale (1), Ordinal Scale (1), Categorical Scale (1)	Inquiries into participants' habits in both ridesharing and private vehicles Asks participants' pain points when taking or driving vehicles	What are your favorite things to do while driving/being a passenger? What sensing experience do you value the most in a vehicle?	A combination of multiple choice and short answer question was used to assure that participants had a chance to justify their corresponding decisions thereby providing us with underlying insights The Ordinal Scale question asks the participants to compare and rank a range of core user needs[a]
Part (3)	Open-ended (5)	Gathers general opinions and expectations about the autonomous vehicle	How do you want to interact with your vehicle?	An efficient way to make respondents feel less constrained and express their deeper insights freely

[a]The core user needs included entertainment, privacy, comfort, safety, hygiene. It was rated by a 5-point scale, with (1st = 5 points, 2nd = 4 points, 3rd = 3 points, 4th = 2 points and 5th = 1 point).

Survey Analysis. We designed histograms to demonstrate the main reasons why users feel unsafe about a driver in a shared car-ride (Fig. 2). The result shows that the riders believe that the driver is the most distracted when he or her cannot see well. The riders also feel unsafe when the driver is in a physical health or emotional state that is unsuitable for driving, with concerns about health slightly above emotions.

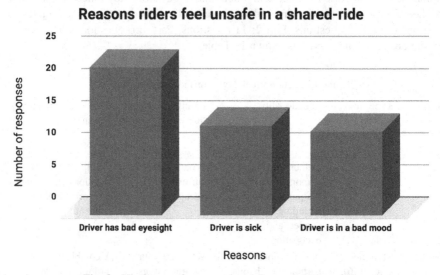

Fig. 2. Histogram of user needs scoring and demographic

Participants also valued sensing experience in autonomous vehicles. We manually processed the short answers and open-ended questions by categorizing participants' expectations, pain points, and demands by utilizing similar keywords. We found 35 participants showed their interest about how autonomous vehicle can improve their sensing experience including, lighting, sound system, visual entertainment, and comfort.

Among those 35, 12 outlined detailed scenarios that they wish to see in the future autonomous vehicle. For example, one participant envisions: "*I wish I can sleep, listen to music, play video games and work in autonomous vehicles. It would be like hiring a driver.*" Other quotes include: "*I am annoyed by the high intensity sunlight in California, sometimes I can't see anything outside or check my phone because it's too bright.*", and "*I enjoy the sightseeing and I want to see the views from all directions*".

2.4   Discussion

After analyzing the data collected from interviews, observations and surveys, we coded the four main pain points (ranked based on the number of times they are mentioned in our data) that concern most of our targeted population:

1. Comfort with light
2. Safety inside of the vehicle
3. Audio entertainment
4. Transparency of itinerary

Comfort with Light. Bright light, either sunlight or high beams from other vehicles was mentioned most frequently as a pain point. To solve this issue, we recommend future autonomous vehicles embed connected sensorial systems that could detect intense light that could cause discomfort. Actuators can potentially be used to adjust the user position to avoid the light. We performed a round of user testing for this concept, which is presented in the "User Testing" section.

Safety Inside of the Vehicle. Safety is a major concern with autonomous vehicles, even though statistically they appear to be relatively safe in test areas [18]. To help users feel safer in autonomous vehicles, we will simulate different concepts to improve real-life driving experiences inside a vehicle, to study whether these features will meet the safety demands of autonomous vehicle riders.

Audio Entertainment. Listening to music was an important feature while in automobiles since it helps relax both the drivers and the riders after a long day of work [19]. Having more engaging music on board can potentially enhance users' experience and fill the "chasms of silence" [20] that occur when traveling in groups. This motivation opens up many avenues for creative solutions to enhance the listening experience. For example, autonomous vehicles could potentially connect the music with other sensing and surroundings of the user. A sensor could detect the beat of the music and enhance the sensorial experience with actuators that could provide tactile vibrations, making the automobile acts like a sonic machine [19].

Transparency of Itinerary. Transparency of demographic data and GPS locations have an extensive effect on trust [21]. There are many options for improving the transparency of the itinerary or driving schedule, such as use of sharable displays next to each passenger. There is much to learn from current ridesharing applications allow riders to see the location of the car and the exact route the vehicle is taking to the rider's destination. Future improvements like this can be achieved by implementing real-time control of the vehicle, by the rider. For instance, users could choose to have the vehicle drop them off when they feel unsafe during the ride.

3   Prototype Development

Based on the analysis and identified pain points from the three methods we used, we found out that one of the major user concerns regarding comfort and safety is the high-intensity lights that project through the windows directly into passengers' eyes. After going through brainstorming, open card sorting, and a Pugh chart analysis [22], we selected the concept of a rotating mechanism for a chair that continuously monitors incident sunlight at eye level to help users avoid direct sun glare.

Subsequently, we built a low fidelity scaled prototype for the concept as a tool to test user feedback (Fig. 3). The prototype was built in the maker space in the Jacobs Hall at the University of California, Berkeley. The prototype contained a stepper motor that had a 3-D printed car seat mounted on it, along with a printed base and motor housing, and a photo-sensor on the seat. An Arduino Uno unit controlled the motor rotation based on input data from the sensor, so that any time when light passed an intensity threshold, the motor would rotate the seat a specified amount until the light intensity dropped below the defined threshold.

Fig. 3. Low-fidelity scaled prototype for user testing

4   User Testing

4.1   Prototype Demonstration

We demonstrated the prototype at a public exhibition at UC Berkeley campus to record the general feedback from the event participants. We collected feedback from 15 participants at the exhibition. Even though some believed that it was an innovative idea and should be developed further, many of the people we demonstrated the prototype to were concerned about its safety, as the chair rotation might limit the passenger's ability to see outside the front of the car. Some also suggested that it might take too much space inside the vehicle.

4.2   Expert Interviews

We presented the prototype and testing results to four industry experts in automotive, user experience design fields to evaluate the developed prototype. All the experts agreed that light is a common concern for passengers inside of a vehicle. However, they expressed concern on the safety of the mechanism when the vehicle has a sudden change of motion. One expert advised us to add a "sleep mode", or "manual option" that enables users to either lock the chair's movement or manually adjust the chair based on their personal preferences. Another expert also mentioned that the space arrangement inside of the vehicle needs to be more flexible to allow and reserve enough leg room 360° around the chair to achieve the higher extent of rotation. Another expert implored us to play the devil's advocate and view the problem from different angles and come up with more solutions that we can later judge as more or less feasible than the current one. One such example was to investigate the tradeoffs of tinted windows as a possible solution to the same problem or photochromatic lenses. *"Comparing two solutions often leads to more insight and possible generation of a third, integrated solution"*, recommended one of the experts.

5   Limitations

Although interviewing is an efficient method to have in-depth conversations with potential users in order to identify their pain points and user needs, the extensive time it takes to conduct an interview limits the total number of interviews that we were able to conduct given our time constraints. The observations provided qualitative comparisons to contrast differences in what people said and what they did. Although the number of respondents was larger, the survey may be less reliable than the interview or observations as the users participated remotely and individually without guidance from the moderator. As the surveys were anonymous, we opted out follow-up questions afterward.

6   Closing

To better understand how to improve passenger experiences in autonomous vehicles with embedded sensors and actuators, three user research methods we used: (i) user interviews for potential sensorial experiences on autonomous vehicles and current sensors in the transportation system, (ii) multiple observations of our co-riders during rideshare trips to study their experience with ride-sharing services, (iii) an online survey for feedback on our designed prototype. We identified four user pain points: comfort with light, safety concerns, audio entertainment and transparency in itinerary. Findings from the research gave us the necessary pointers to determine what sensors could be used to avoid external bright lights and glare from reaching the user during their transportation experience, to address the aspect of comfort with light. Other pain points that were identified, such as transparency of the itinerary, are less enabled by sensor integration but still help in providing guidance in designing displays for the future of shared rides, which would not be only visible to one user, but also benefit the whole group, considering that no single user would be operating the vehicles. The audio entertainment emphasis opens up creative opportunities that could integrate music with the surroundings of the user, using a fusion of both sensors, transducers and displays, to take advantage of new forms of interiors for AVs and their seating arrangements. Solutions might allow rotating chairs with a sleep mode option to control whether the chair will adjust automatically to the sunlight, as well as adding a screen in front of the user with a real-time display of the outside scenery.

7 Future Work

This exploratory study will be expanded to more subjects and used to create new designs and prototypes that address the pain points identified. We plan on designing a connected sensor system for detecting intense light that could cause discomfort. Actuators will be used to adjust the user position to avoid the light. Another future implementation will be an integrated common display to locate the car and the exact route the vehicle is taking to the rider's final destination, as well as to provide easy access to emergency intervention controls to the users. This could take the form of a window-embedded, in-wall, or projection form-factor adapted to autonomous vehicle shared spaces.

Acknowledgements. The authors would like to thank the members of the BRAVO lab and the Alliance (Renault-Nisan-Mitsubishi) for partially supporting this research.

References

1. Anderson, J.M., Nidhi, K., Stanley, K.D., Sorensen, P., Samaras, C., Oluwatola, O.A.: Autonomous Vehicle Technology: A Guide for Policymakers. Rand Corporation, Santa Monica (2014)
2. Chan, C.: Advancements, prospects, and impacts of automated driving systems. Int. J. Transp. Sci. Technol. **6**(3), 208–216 (2017). https://doi.org/10.1016/j.ijtst
3. Litman, T.: Autonomous vehicle implementation predictions implications for transport planning. Victoria Transport Policy Institute, pp. 3–4 (2018)
4. Theverge.com   Homepage.   https://www.theverge.com/2018/9/5/17822398/volvos-360c-concept-autonomous-car-electric-future. Accessed 11 Feb 2019, 15 Feb 2019
5. Cholsaipan P., et al.: Reimagining onboard experiences for autonomous vehicle in academic makerspaces. In: 3rd International Symposium for Academic Makerspace (ISAM 2018). Stanford University (2018)
6. Zhu, J., Fan, Y.: Daily travel behavior and emotional well-being: effects of trip mode, duration, purpose, and companionship. Transp. Res. Part A Policy Pract. **118**, 360–373 (2018)
7. Schoettle, B., Sivak, M.: A survey of public opinion about autonomous and self-driving vehicles in the U.S., The U.K., and Australia. The University of Michigan Transportation Research Institute Ann Arbor, Michigan 48109-2150 U.S.A. Report No. UMTRI-2014-21 (2014)

8. Beckman, S.L., Barry, M.: Innovation as a learning process: Embedding design thinking. Calif. Manag. Rev. **50**(1), 25–56 (2007)
9. GALLUP page. https://news.gallup.com/poll/237965/snapshot-uses-ride-sharing-services. aspx. Accessed 12 Feb 2018
10. Charmaz K., Belgrave L.L.: Grounded theory. Wiley Online Library (2015)
11. Charmaz, K.: Constructing Grounded Theory, 2nd edn. SAGE publications Ltd., Sonoma State University, Thousand Oaks (2014)
12. Globalwebindex Homepage. https://blog.globalwebindex.com/chart-of-the-day/uber-demographics/, last accessed 2019/02/13
13. Anderson, J.M., Nidhi, K., Stanley, K.D., Sorensen, P., Samaras, C., Oluwatola, O.A.: Autonomous Vehicle Technology: A Guide for Policymakers. Rand Corporation, Santa Monica (2014)
14. Ioannou, P.A., Chien, C.C.: Autonomous intelligent cruise control. IEEE Trans. Veh. Technol. **42**(4), 657–672 (1993)
15. Abraham, H., et al.: Autonomous vehicles, trust, and driving alternatives: a survey of consumer preferences. Massachusetts Institute of Technology, AgeLab (2016)
16. Choi, J., Ji, Y.: Investigating the Importance of Trust on Adopting an Autonomous Vehicle. Int. J. Hum.-Comput. Interact. **31**(10), 692–702 (2015). https://doi.org/10.1080/10447318. 2015.1070549
17. Merat, N., Madigan, R. and Nordhoff, S.: Human factors, user requirements, and user acceptance of ride-sharing in automated vehicles. International Transport Forum Discussion Papers, OECD Publishing, Paris (2017)
18. Kalra, N., Paddock, S.M.: Driving to safety: how many miles of driving would it take to demonstrate autonomous vehicle reliability? Transp. Res. Part A: Policy Pract. **94**, 182–193 (2016)
19. Brandon, L.: Pump up the bass—rhythm, cars, and auditory scaffolding. Senses Soc. **3**(2), 187–203 (2008). https://doi.org/10.2752/174589308X306420
20. Walsh, M.J.: Driving to the beat of one's own hum: automobility and musical listening, In: Denzin, N.K. (ed.) Studies in Symbolic Interaction, vol. 35, pp. 201–221 (2010)
21. Mittendorf, C.: The implications of trust in the sharing economy – an empirical analysis of uber (2017)
22. CITRIS Invention Lab. http://invent.citris-uc.org, last accessed 2019/02/11

Attentional Dynamics After Take-Over Requests: The Need for Handover Assistance Systems in Highly Automated Vehicles

Tobias Vogelpohl[(⊠)] and Mark Vollrath

Technische Universität Braunschweig, Gaußstraße 23,
38106 Brunswick, Germany
t.vogelpohl@tu-braunschweig.de

Abstract. Drivers in highly automated vehicles will frequently transition back to manual driving. Drivers performing Non-Driving Related Tasks (NDRTs) during automated driving are generally capable of deactivating automated systems approx. 3–6 s on average after a take-over request (e.g. Vogelpohl et al. 2018a). However, take-over time should not necessarily be considered a measure of take-over quality (e.g. Gold et al. 2016). In complex situations drivers may be prone to neglect lower priority sub-tasks (e.g. Richard et al. 2006). After take-over requests drivers may therefore be uncertain about the status of safety-relevant areas in the driving scene after they have deactivated the automation. This uncertainty can be characterized as a reduction in situation awareness (c.f. Johnson et al. 2017).

We present research which shows that drivers may be slow to rebuild situation awareness after take-over requests based on delayed visual attention to lower priority sub-tasks such as looking at the mirror and looking at the speedometer. We discuss why we believe drivers' attentional dynamics (how and when attention is shifted; c.f. Lee 2014) after take-over requests should be taken into account for the design of automated driving systems.

Future automated driving systems should consider not only how long a driver takes to deactivate the automation, but also take into account the process of how the driver transitions back to manual driving. If a driver neglects stages of the transition, a guided transition could ensure that uncertainty during the transition to manual driving is reduced and that situation awareness after the transition is regained as quickly and as efficiently as possible.

Keywords: Vehicle · Automation · Take over request · Situation awareness

1 Introduction

Taking back control from an automated driving system will be a common event for drivers in highly automated vehicles. A driver may decide that he or she wants to re-engage in the driving task or an automated driving system may ask the driver to resume control over certain parts of the driving task because of a degradation of sensor quality or other constraints imposed by the driving situation (e.g. entering an area which is not certified for automated driving). If a driver resumes control of the vehicle there is a shift

© Springer Nature Switzerland AG 2019
H. Krömker (Ed.): HCII 2019, LNCS 11596, pp. 128–142, 2019.
https://doi.org/10.1007/978-3-030-22666-4_10

of responsibility from the vehicle to the driver. The driver has to switch his or her attention back to the driving task, perceive, understand and interpret the driving situation and manually, tactically and strategically control the vehicle in the context of the driving situation. Unfortunately, in many cases the situation in which the driver is asked to resume control will not only be highly complex, but will also feature some sort of time-constraint imposed on the driver, which is determined by the car (c.f. Object and Event Detection Response as defined by NHTSA 2017). In other words, if the car asks the driver to resume control it will often do so because it can only guarantee safe automated driving for a certain amount of time after the take-over request.

Therefore, from an engineering perspective it appears crucial to determine how long a driver will take to resume control and thereby relieve the automated driving system of its responsibilities. Therefore, the amount of time necessary for the driver to take control of the vehicle determines the time that an automated vehicle has to provide in a take-over situation (c.f. Operational Design Domain as defined by NHTSA 2017). A number of studies have investigated the necessary reaction time and found that after a take-over request most drivers will deactivate an automated driving system after approx. 3–6 s on average (e.g. Gold et al. 2013, Vogelpohl et al. 2018a). Researchers have also found that the duration of this automation deactivation seems to depend on several factors: Engagement in a Non-driving related task (NDRT), the current take-over situation, previously experienced take-over situations and the time available to take-over seem to have the largest influence (see review by McDonald et al. 2019) on this "take-over time".

However, while it is crucial to know how long the average driver will take to take back control from the car, this perspective fails to consider what happens to the car and the driver after the driver has taken back control. It also ignores the drivers who fail to react within the average time frame. For example Jarosch et al. (2019) found that after prolonged driving with automation several participants in a driving simulator study completely failed to react to a take-over request. Similarly, other studies have shown that under certain conditions drivers will sometimes fail to react to take-over requests (c.f. De Waard et al. 1999, Stanton and Young 1998). In a previous study we (Vogelpohl et al. 2018) found that there were large individual differences between reaction times to take-over requests and that the average reaction time may not be representative for a safe transition to manual driving.

From a human factors perspective, as compared to an engineering perspective, it may therefore actually not be relevant *how long* a driver takes to deactivate the automation. What matters is *if* and *how well* the driver deactivates the automation and how well he or she performs the driving task after control has been transferred. If an automated driving system were able to measure in real time how long the driver is taking to deactivate the automation in a specific situation and how well he or she is performing during this transition, it could in turn determine when and how the automation can safely be deactivated. Such an Adaptive Automation (c.f. Corso and Moloney 1996, Kaber and Endsley 2004) would not rely on predefined predictions about how long an average driver takes, but would use the information available in the situation to determine how long a specific driver in a specific situation needs to safely transition to manual driving.

Kircher et al. (2014) speculate that "Drivers do not just seem to react to automation, but rather interact with automation […]" and that "[…] drivers integrate the behavior of the automation into their tactical planning of the whole situation instead of only reacting to the responses of the automation" (Kircher et al. 2014, p. 166). This statement can be also be taken to mean that in turn automation should incorporate the driver into its tactical planning and make the best use of the resources available to both the system and the driver in a take-over situation. If this perspective is applied, an automated driving system would need to answer the following questions:

- How long does a specific driver need to successfully switch back to the manual components of the driving task (instead of how long the average driver takes to switch back)?
- How does a specific driver switch from a NDRT to the driving task (regarding this as a dynamic process, rather than a moment in time)?
- How does a specific driver allocate his or her attention towards relevant objects in the driving scene (again, regarding this as a dynamic process)?
- How does a specific driver continue to drive after the responsibility for driving has been transferred?

If some or all of these variables are taken into account by the automated driving system, the system could adaptively decide which information to provide to a driver in a take-over situation. It could also predict if and when to hand control over to the driver and how to support the driver after the responsibility for driving has been shifted to the driver.

In summary, we believe that the narrow focus on reaction time in take-over situations falls short of what we actually need to understand to ensure a safe transition of control from the vehicle to the human driver. We may need to look at specific drivers and design the automated driving system to adaptively react to how individual drivers perform after a take-over request in a specific environment. Such a *handover assistance system* that provides a guided transition to manual driving could supervise the drivers' actions during a transition to manual driving and decide to adapt the information provided to the driver and the support provided by the automated driving system based on the drivers' performance in a given situation.

In the next sections we present what we believe to be evidence for a need to go beyond take-over times and to consider *attentional dynamics* of take-over situations. We discuss findings from our own research and from other researchers and explain why we think adaptive handover assistance systems in highly automated vehicles may improve the safety of automated driving systems.

2 Take-Over Times and Quality After Automated Driving

2.1 Take-Over Time

A number of studies have determined influencing factors on take-over times after automated driving (see McDonald et al. 2019 and Zhang et al. 2018, for recent reviews of take-over times). Such studies largely agree that average take-over times may lie anywhere between 2 and 6 s and that take-over time depends on a number of factors. The level of supervision of the automation, the NDRT before the take-over request, the time available for the transition and complexity of the situation are cited as possible influences on take-over time (McDonald et al. 2019, Vogelpohl et al. 2018a), as well as inner-subject factors such as experience with the automated driving, fatigue or age (e.g. McDonald et al. 2019, Vogelpohl et al. 2018b, Jarosch et al. 2019). Interestingly, one major factor also seems to be the time available for the transition. Specifically, longer take-over time budgets seem to correspond to slower take-over reactions (e.g. Gold et al. 2013, Gold et al. 2018, Payre et al. 2016, Zhang et al. 2018)

2.2 Take-Over Quality

Other studies have focused on the quality of driving after the transition to manual driving, rather than at the duration of the take-over (e.g. Gold et al. 2018, Happee et al. 2017). Studies focused on take-over quality have found that driving may be influenced for as long as 5 min after a transition to manual driving (Brandenburg and Skottke 2014). Among others, variables such as standard deviation of lane departure (e.g. Wandtner et al. 2018, Wiedemann et al. 2018, Zeeb et al. 2016), number of steering reversals (Merat et al. 2014), car following behavior or lane crossings (e.g. Zeeb et al. 2017, Strand et al. 2014, Wandtner et al. 2018) have been found to be influenced by transitions to manual driving.

Again, take-over time budgets seem to influence the quality of driving after a take-over request (c.f. Wan and Wu 2018, van den Beukel and van der Voort 2013, Mok et al. 2015, Gold et al. 2013, Gold et al. 2018): Specifically, lower time budgets will negatively influence lane and distance keeping quality (e.g. Wan and Wu 2018) and will also have an impact on higher level decision making processes (i.e. whether to brake, steer or both; c.f. Gold et al. 2013, Gold et al. 2018). Gold et al. (2016) argue that not only the available time budget may influence the quality of manual driving after a take-over request, but also the time drivers themselves allow for the transition.

The results from the studies on the take-over time and take-over quality have been interpreted to signify that drivers are physically and mentally taken out of the driving control-loop (Endsley and Kiris 1995) during automated mode and that they need time to resume the driving task. From research in other areas where human supervision of automated systems is required, we know that being out of the control-loop can have serious consequences for the human-machine system, such as a decrease in vigilance (Parasuraman and Davies 1977), a loss of situation awareness (Kaber et al. 2006) and mode confusion (Degani et al. 1999). This holds especially true if drivers have been performing immersive or motivating non-driving tasks during the automated ride (c.f. Vogelpohl et al. 2018a), which may make them vulnerable to task perseveration effects (Zeigarnik 1938, Fox and Hoffmann, 2002). Wandtner et al. (2018) observed task perseveration effects after automated driving with NDRTs. At the same time, it is likely that such motivating NDRTs will become more frequent as drivers become more confident with the safety of high automation levels and stop to supervise the automated driving system (e.g. Carsten et al. 2012, Jamson et al. 2013).

2.3 Situation Awareness After Driving with Automation

In addition to the immediate effects of take-over time on the transition to manual driving, effects on Situation Awareness after automated driving have been discussed in the literature. Situation Awareness is largely defined as "[a] the perception of elements in the environment within a volume of time and space, [b] the comprehension of their meaning, and [c] the projection of their status in the near future." (Endsley 1988, p. 97). Some studies have identified the complexity of a situation as a predictor for the time that drivers will take to transition to manual driving after a take-over request (e.g. Gold et al. 2016, Naujoks et al. 2014, Radlmayr et al. 2014). The complexity of a take-over situation can be linked to the level of situation awareness required to appropriately react to a take-over request after a drive where NDRTs have been performed (e.g. Merat and Jamson 2009, Merat et al. 2012).

In Vogelpohl et al. (2018a) we define the complexity of a take-over situation in congruence with the definition of Situation Awareness by Endsley (1988): The complexity of take-over situation could therefore be defined by "[a] the number of elements present during the transition to manual driving [b] the difficulty with which meaning can be attributed to the elements and [c] the predictableness of the future status of these elements" (Vogelpohl et al. 2018a, p. 466). This definition is in line with the findings from Gold et al. (2016) who found that participants in their study needed longer to take back manual control if the number of elements (the traffic density) was increased in a take-over situation. The authors argue that "longer takeover times are actually indicative of a better reaction, as participants took the necessary time to regain situation awareness before starting a maneuver" (Gold et al. 2016, p. 7). Vlakveld (2015) notes that we can only be sure that the driver has fully regained situation awareness when we know if driver is able to react to latent threats and hazards in the driving situation, during and after the transition to manual driving. Studies which focused on situation awareness after take-over request have found that eye-tracking measures may be promising to determine a drivers' readiness for a take-over (Braunagel et al. 2017) and that based on such measures drivers may take up to 7–12 s to obtain situation awareness after take-over requests (c.f. Lu et al. 2017, Vogelpohl et al. 2018a). Lu et al. (2017) suggest that it may take a driver even longer than this to infer predictive information about latent hazards and safety-relevant objects from a complex driving scene.

2.4 What Is a Safe Transition to Manual Driving?

It can therefore be argued, that neither the duration of the time which a driver takes to transition to manual driving after a take-over request nor the immediate quality of his or her driving is a sufficient measure of the ability to perform well in a complex driving situation. Instead of a measure of take-over quality, the take-over time and the driving quality after the take-over are only a function of the take-over situation and the individual driver. In other words, it would be insufficient to infer the readiness of the driver to react to latent threats after a take-over situation from the time provided to the driver to take over. In part this is because drivers will react within the time which is provided to them (as discussed above) and in part this is because drivers do not necessarily deactivate the automation when they are actually ready to drive (as argued in Vogel-pohl et al. 2018a). Instead, some drivers will deactivate the automation because the take-over request tells them to deactivate the automation (c.f. automation complacency, e.g. Parasuraman and Riley 1997) or because they fail to perceive the complexity of the situation.

If take-over time is not a sufficient predictor of the drivers' ability to perform after the responsibility of control has been shifted from the automated driving system to the driver, how can critical situations after take-over requests be avoided? An analysis of driving quality after the transition of control could be an indicator, but can only be measured after the fact. If a driver is found to badly follow his or her lane or is following very close behind a lead vehicle after the automation has been deactivated, this can be taken as an indicator that the transition to manual was unsuccessful or is indeed not yet completely finished. But it cannot be used to avoid such behavior in the first place. If a driver deactivates the automation very quickly after a take-over request this can be taken as an indicator that the driver has failed to carefully inspect the driving situation and will likely miss latent hazards or unexpected events. However, it could also mean that the driver has been monitoring the driving situation before the take-over request and is simply ready to take back control even before the take-over request is issued. To address these open issues, we discuss attentional dynamics as a more encompassing view of driver behavior in a take-over situation.

3 Attentional Dynamics After Take-Over Requests

Attentional dynamics are a concept first proposed for the context of driver distraction and partially automated driving systems by Lee (2014). Lee (2014) distinguishes attentional dynamics from attentional resources and argues that distraction during driving can also be described as a failure to appropriately allocate attentional resources. This includes aspects of task timing, task switching and task prioritization as well as task perseveration. Lee (2014) argues that understanding attentional dynamics, i.e. how, when and where drivers allocate attentional resources, will become increasingly important with higher levels of vehicle automation and that this concept may complement the perspective of attentional resources. We believe that this approach is not only valid in manual driving for frequent switches between the driving task and secondary tasks, but may also be applicable to the more detailed level of switching from a NDRT back to manual driving in higher levels of automated driving. As described by Lee (2014), interruption management becomes crucial to the safety of the driver, if a sudden task switch from the NDRT back to manual driving is required, as would be the case in a take-over situation.

If, as we argue, take-over time and take-over quality provide an incomplete picture of the transition to manual driving or are not sufficient to be used as a predictor for the outcome of a transition to manual, attentional dynamics could provide a new perspective to understand take-over situations. We propose regarding the transition to manual driving as a dynamic process with interdependent granular sub-tasks, which are indicative of the attentional dynamics and attentional resources of the driver. If sub-tasks are neglected or falsely prioritized, this may indicate insufficient situation awareness and a mismatch of attentional resources to the attentional demands of the take-over situation (c.f. Johnson et al. 2017, Vogelpohl et al. 2018a).

The attentional dynamics of a take-over situation could for example be described as follows: If a take-over request is issued while a driver is engaged in a NDRT, the driver is likely to redirect his or her attention towards the roadway in a certain amount of time (task switching), but will then often return visual attention to finish/interrupt the NDRT (task perseveration) or to find a place to store the task if it was performed on a handheld device. While, or after, the NDRT is stored (task prioritization), the driver will direct his or her visual attention towards areas which he or she believes to be relevant for the manual driving task (prioritization of lower priority sub-tasks, as discussed in Johnson et al. 2017). The gaze may be directed towards the forward roadway, it may be directed towards one of the rearview mirrors, the gauges of the car or some other visual area which the driver expects will help to understand the situation (expectancy). Visual attention will then be redirected multiple times in response to the first impression of the driving situation, the information which is gathered through other modalities (e.g. vibration from the road surface, acceleration of the car, approaching sirens of an ambulance) and the previous knowledge that is stored in the drivers' long term memory (prior knowledge and experience with take-over situations; c.f. Payre et al. 2016). While visual attention is spent on gathering information, simultaneously motor processes are engaged which move the drivers' hands and feet towards the controls of the car.

In a take-over situation, the visual, motor and other modalities will partially depend on one another and on the information gathered from the environment. If, for instance, visual attention is quickly directed towards the side mirror and a vehicle in the adjacent lane is perceived, the likelihood that the first steering reaction will be towards that lane may be reduced. At the same time, the likelihood that the driver will brake may increase if a previous visual check has detected a braking lead vehicle. In this way, every redirection of attention is partially linked to every previous redirection of attention. Thereby, the attentional dynamics of the driver are the sum of all observable reactions from the driver and their temporal relation to one another after a take-over request has been issued. The prioritization and allocation of attentional resources after a take-over request is moderated by prior experience with take-over situations. The allocation of attention before a take-over request is influenced by the trust placed in the automated driving system (which is in turn influenced by prior experiences with take-over situations). We have visualized an example for the process of disengaging from a NDRT and transitioning back to manual driving as related to the theory of attentional dynamics in Fig. 1.

How can we use the attentional dynamics during a take-over request to increase the safety of transitions from automated driving to manual driving? We argue that an automated driving system should be aware of the attentional dynamics and resources of the driver in a take-over situation. By observing and tracking the drivers' reactions and their temporal relation the automated driving system may be able to predict in real time how attentional resources of the driver are allocated and how this will influence the drivers' reactions to the take-over situation. In consequence, such a system would be able to react to what the driver is seeing and doing and improve the information and

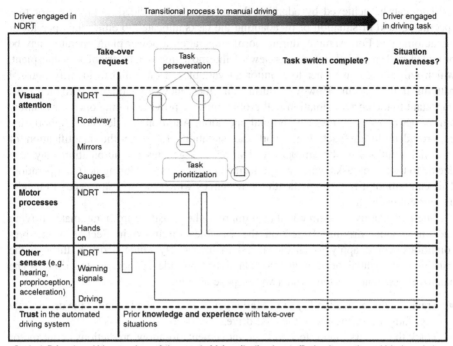

Fig. 1. Example model of attentional dynamics in a take-over situation.

support to the driver accordingly. This would make the automated driving system and "adaptive automation".

4   Adaptive Automation and a Handover-Assistance System

Adaptive automation has been widely discussed in contexts where humans supervise complex automated systems and where time critical decisions may be required from the human operators (e.g. power plants, aircraft; air control centers; c.f. Parasuraman et al. 1996, Parasuraman et al. 1999). In contrast to static automation, adaptive automation will react flexibly and context dependent to the requirements of a human operator. It will take into account what is known to the operator and provide information about the current system state in such a way that an optimized performance is achieved by the operator-machine team. Adaptive automation has been shown to reduce some of the negative effects of automation on human operators by improving situation awareness and avoiding overreliance and skill degradation (e.g. Hilburn et al. 1997, Scallen et al. 1995, Scerbo 1996). The goal of adaptive automation is to increase the performance of the human-automation system and to mitigate the effects of excessive workload as well as task induced fatigue.

This can be achieved by adapting the information provided to the operator to a specific context or situation or by adapting the tasks allocated to either the operator or the automation. For example, during normal operation a power plant operator may be provided with a scaled down overview of the most important variables of the plant, which will allow him or her to monitor all variables at once and to identify unusual patterns if they occur. However, during an emergency situation the variables most important to the current situation will prominently be provided to the operator to allow him or her to focus on restoring normal operation. Additionally, decision support may be provided during such high workload situations to guide the identification of potential solutions to the emergency. During emergencies the automation may also decide to take over tasks which are performed by the operator during normal operation. This can temporarily free up attentional resources which the operator needs to handle the current emergency.

Such an adaptive system could also improve the transition from automated driving back to manual driving. Based on the variables discussed in Sect. 2, some pre-conditions would apply to enable a system to actively guide a driver through the transition to manual driving and to adaptively provide support to the driver. The handover assistance system which we propose should:

- Track the driver availability (e.g. Marberger et al. 2017) and monitor the driver state (e.g. fatigue) during the automated drive.
- Track the drivers' hand, feet and eye movements to allow predictions about motor readiness and visual attention allocation.
- Include a sophisticated model of driver behavior which can quickly predict driver reactions based on the available variables (e.g. predict impending lane changes from glance behavior, see. Henning et al. 2008, Salvucci and Liu 2002).
- Include a sophisticated model of driver attention and driver workload which allows predictions about the current and future status of the drivers' attention allocation and workload.
- Include a sophisticated environmental model which interacts with the driver model (for example drivers may be less likely to use steering to avoid a crash if the road is slippery; drivers may be more likely to react by braking if the traffic density is high).
- Enable minimal risk maneuvers (e.g. Reschka and Maurer 2015) from complex driving situations as a fallback option in every take-over situation if the driver is judged unavailable or if the driver fails to appropriately react in a take-over situation.

Such a hypothetical system which is able to track the drivers' actions and reactions and his or her estimated allocation of attentional resources during a transition to manual driving could then use this information to:

- Direct the drivers' attention towards potentially relevant objects and controls.
- Improve the drivers' attention prioritization by providing guidance on "what to do next" and "where to look next".
- Avoid task perseveration effects by quickly removing NDRTs.
- Warn the driver about hazards which he or she may not have detected during the transition to manual driving.

- Decide whether to provide braking and/or steering support to the driver in a complex take-over situation.
- Decide if and when a shift of responsibility to the driver is safe or whether a minimal risk maneuver should be engaged.
- Decide when the transition to manual driving has been completed.
- Decide if a take-over request is likely to be beneficial to the safety of the driver, or whether a minimal risk maneuver should be engaged without trying to reengage the driver (e.g. if the driver is exhibiting strong signs of fatigue).

A handover assistance system based on attentional dynamics and attentional resource allocation has the potential to not only ensure that a transition to manual driving is safely performed. It could also render the transition more efficient by reducing the uncertainty of the driver about a situation (thereby improving situation awareness in complex situations, c.f. Johnson et al. 2017, Vogelpohl et al. 2018a) and by adapting to drivers who are experienced at take-over situations or perform well in a take-over situation to quickly transition to manual driving. On the other hand, inexperienced or slow drivers would be provided with the optimum level of assistance which could help to build trust in the automated driving system and to improve learnability of the system. We have summarized a first draft for an adaptive automation based on the prediction of attentional dynamics and attentional resources in Fig. 2.

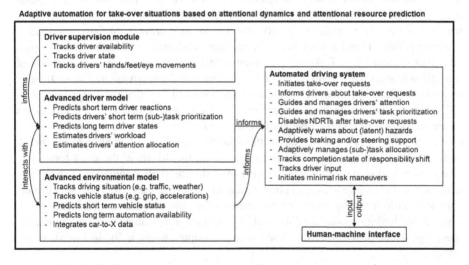

Fig. 2. Draft for an adaptive automation for a handover assistance system

5   Existing Research on Handover Assistance Systems

Some efforts have already been made to define and build handover assistance systems and human-machine interfaces which support the driver during transitions to manual driving after driving with automation. Walch et al. (2015) designed and tested a *handover assistant* in a driving simulator setting. However, the handover assistant in the study by Walch et al. (2015) did not meet the criteria of an adaptive automation in that the information provided to the drivers and the drivers' input after a take-over request was not adapted to the reactions of the drivers. The authors note that the majority of the participants "took over as soon as they were alerted, and did not report trying to perceive the situation before they intervened" (p. 17). Therefore, Walch et al. (2015) propose a *system monitored handover* during which the drivers' steering and braking inputs are monitored and may be adjusted by the automation to avoid impulsive reactions from the drivers. The handover assistance system proposed in Fig. 2 is different from such a system in that it could not only avoid inappropriate driver reactions, but actively improve the reactions of the driver to render the take-over more efficient. Also, the hypothetical system proposed in our paper goes beyond driver inputs such as steering and braking by using driver gaze analysis and hand/feet tracking to estimate visual attention allocation. Similarly, Braunagel et al. (2017) have proposed a measure to determine driver readiness for take-overs through the analysis of the drivers NDRT engagement, the current traffic situation and the drivers gazes at the roadway. The authors showed that the aggregation of this information could success-fully be used to issue pre-warnings to drivers who displayed a low driver readiness before a take-over request.

Clark et al. (2018) propose a speech based *handover assist* for planned, non-critical transitions of control. In situations where the car is able to rely not only on its sensors, but also on information from a navigation system or car-to-x data, take-over requests may be issued much earlier to allow the driver to transition to manual driving at his or her own speed (e.g. highway exits). Similar to time-critical take-over situations, researchers have found a wide range of reactions and reaction speeds to non-critical take-over requests (e.g. Eriksson and Stanton 2017). Therefore, for such non-critical transitions it would be equally important to guide drivers through the transition, thereby rendering their behavior more predictable and their driving more stable. Clark et al. (2018) based the design of their handover assist on checklists from other domains where handovers of control frequently occur, such as air traffic and medical care. The drivers in their study where guided through the transition of control by a speech assistant. Depending on the guidance strategy used, the drivers in the study took more or less time to transition to manual driving after driving with automation. The authors note, that in the experimental condition where no speech guidance was provided drivers paid very little attention to the environment before the automation was deactivated. This further highlights the need for a guided transition to manual driving. Again, the speech guidance provided in this study was not adapted to the individual needs of the driver.

Adaptive automation for driver assistance systems may be difficult to implement, but if warning signals and control allocation is adapted to the contextual needs of the driver such systems could be significantly improved. As an example, Kaß et al. (2018) proposed an adaptive warning strategy for collision avoidance systems which would be able to reduce unnecessary warnings to the driver. The authors discuss that by adap-tively taking into account the time to collision as well as the drivers' maneuver intentions system effectiveness as well as user acceptance could be increased.

6   Future Research Needs

Much research is needed to determine the correct variables to as input for the driver model, as well as to reliably measure these variables and to improve real-time predictions. The variables discussed in this paper (e.g. gazes to the mirrors, gazes to the gauges, hand/feet movements) can only serve as a first indication of what might actually be needed to reliably model and predict driver behavior in a take-over situation. Individual differences (such as age or driving experience) and driver states (such as fatigue or stress) may also need to feature prominently in the modeling of take-over situations. Finally, in the course of future studies it may become apparent that taking into account individual "live" variables during transitions to manual driving does not significantly increase the predictability of driver behavior in a take-over situation compared to what is already known about average take-over behavior. Empirical testing of adaptive handover assistance systems should determine whether and how they can improve the transition to manual driving.

Additionally, much work needs to go into the definition and technical improvement of minimal risk states for every conceivable driving situation. Due to the high variance in possible reactions after a take-over request, highly automated vehicles should always provide a fallback option for those situations in which the driver does not react to a take-over request or in which he or she reacts inappropriately. The models presented in this paper should only be taken as a starting point which can guide future research on the attentional dynamics of take-over situations and adaptive car to driver handover assistance systems. Future research will need to determine if the transition to manual driving can be improved by adaptive, human centered automation.

References

Brandenburg, S., Skottke, E.M.: Switching from manual to automated driving and reverse: are drivers behaving more risky after highly automated driving? In: IEEE 17th International Conference on Intelligent Transportation Systems (ITSC), pp. 2978–2983. IEEE, October 2014

Braunagel, C., Rosenstiel, W., Kasneci, E.: Ready for take-over? A new driver assistance system for an automated classification of driver take-over readiness. IEEE Intell. Transp. Syst. Mag. **9** (4), 10–22 (2017)

Carsten, O., Lai, F.C., Barnard, Y., Jamson, A.H., Merat, N.: Control task substitution in semiautomated driving: does it matter what aspects are automated? Hum. Factors **54**(5), 747–761 (2012)

Clark, J., Stanton, N., Revell, K.: Handover assist in highly automated vehicles: how vocal communication guides visual attention. In: Stanton, N. (ed.) AHFE 2018. AISC, vol. 786, pp. 295–306. Springer, Cham (2019). https://doi.org/10.1007/978-3-319-93885-1_27

Corso, G.M., Moloney, M.M.: Human performance, dynamic function allocation and transfer of training. In: Manufacturing Agility and Hybrid Automation, pp. 636–639 (1996)

de Waard, D., van der Hulst, M., Hoedemaeker, M., Brookhuis, K.A.: Driver behavior in an emergency situation in the automated highway system. Transp. Hum. Factors **1**(1), 67–82 (1999)

Degani, A., Shafto, M., Kirlik, A.: Modes in human-machine systems: constructs, representation, and classification. Int. J. Aviat. Psychol. 9(2), 125–138 (1999)

Endsley, M.R.: Design and evaluation for situation awareness enhancement. In: Proceedings of the Human Factors Society Annual Meeting, vol. 32, no. 2, pp. 97–101. SAGE Publications, Los Angeles, October 1988

Endsley, M.R., Kiris, E.O.: The out-of-the-loop performance problem and level of control in automation. Hum. Factors 37(2), 381–394 (1995)

Eriksson, A., Stanton, N.A.: Driving performance after self-regulated control transitions in highly automated vehicles. Hum. Factors 59(8), 1233–1248 (2017)

Fox, S., Hoffman, M.: Escalation behavior as a specific case of goal-directed activity: a persistence paradigm. Basic Appl. Soc. Psychol. 24(4), 273–285 (2002)

Gold, C., Damböck, D., Lorenz, L., Bengler, K.: "Take over!" how long does it take to get the driver back into the loop? In: Proceedings of the Human Factors and Ergonomics Society Annual Meeting, vol. 57, no. 1, pp. 1938–1942. SAGE Publications, Los Angeles, September 2013

Gold, C., Körber, M., Lechner, D., Bengler, K.: Taking over control from highly automated vehicles in complex traffic situations: the role of traffic density. Hum. Factors 58(4), 642–652 (2016)

Gold, C., Happee, R., Bengler, K.: Modeling take-over performance in level 3 conditionally automated vehicles. Accid. Anal. Prev. 116, 3–13 (2018)

Happee, R., Gold, C., Radlmayr, J., Hergeth, S., Bengler, K.: Take-over performance in evasive manoeuvres. Accid. Anal. Prev. 106, 211–222 (2017)

Henning, M.J., Georgeon, O., Wynn, T., Krems, J.F.: Modelling driver behaviour in order to infer the intention to change lanes. In: Proceedings of European Conference on Human Centred Design for Intelligent Transport Systems, vol. 113, April 2008

Hilburn, B., Jorna, P.G., Byrne, E.A., Parasuraman, R.: The effect of adaptive air traffic control (ATC) decision aiding on controller mental workload. In: Human-Automation Interaction: Research and Practice, pp. 84–91 (1997)

Jamson, A.H., Merat, N., Carsten, O.M., Lai, F.C.: Behavioural changes in drivers experiencing highly-automated vehicle control in varying traffic conditions. Transp. Res. Part C Emerg. Technol. 30, 116–125 (2013)

Jarosch, O., Bellem, H., Bengler, K.: Effects of task-induced fatigue in prolonged conditional automated driving. Hum. Fact. (2019). 0018720818816226

Johnson, A.W., Duda, K.R., Sheridan, T.B., Oman, C.M.: A closed-loop model of operator visual attention, situation awareness, and performance across automation mode transitions. Hum. Factors 59(2), 229–241 (2017)

Kaber, D.B., Endsley, M.R.: The effects of level of automation and adaptive automation on human performance, situation awareness and workload in a dynamic control task. Theor. Issues Ergon. Sci. 5(2), 113–153 (2004)

Kaber, D.B., Wright, M.C., Sheik-Nainar, M.A.: Investigation of multi-modal interface features for adaptive automation of a human–robot system. Int. J. Hum Comput Stud. 64(6), 527–540 (2006)

Kaß, C., Schmidt, G.J., Kunde, W.: Towards an assistance strategy that reduces unnecessary collision alarms: an examination of the driver's perceived need for assistance. J. Exp. Psychol. Appl. (2018)

Kircher, K., Larsson, A., Hultgren, J.A.: Tactical driving behavior with different levels of automation. IEEE Trans. Intell. Transp. Syst. 15(1), 158–167 (2014)

Lee, J.D.: Dynamics of driver distraction: the process of engaging and disengaging. Ann. Adv. Automot. Med. 58, 24–32 (2014)

Lu, Z., Coster, X., de Winter, J.: How much time do drivers need to obtain situation awareness? A laboratory-based study of automated driving. Appl. Ergon. **60**, 293–304 (2017)

Marberger, C., Mielenz, H., Naujoks, F., Radlmayr, J., Bengler, K., Wandtner, B.: Understanding and applying the concept of "driver availability" in automated driving. In: Stanton, N.A. (ed.) AHFE 2017. AISC, vol. 597, pp. 595–605. Springer, Cham (2018). https://doi.org/10.1007/978-3-319-60441-1_58

McDonald, A.D., et al.: Towards computational simulations of behavior during automated driving take-overs: a review of the empirical and modeling literatures. Hum. Factors (2019)

Merat, N., Jamson, A.H.: Is drivers' situation awareness influenced by a fully automated driving scenario? In: Human Factors, Security and Safety. Shaker Publishing (2009)

Merat, N., Jamson, A.H., Lai, F.C., Carsten, O.: Highly automated driving, secondary task performance, and driver state. Hum. Factors **54**(5), 762–771 (2012)

Merat, N., Jamson, A.H., Lai, F.C., Daly, M., Carsten, O.M.: Transition to manual: driver behaviour when resuming control from a highly automated vehicle. Transp. Res. Part F Traffic Psychol. Behav. **27**, 274–282 (2014)

Mok, B.K.J., Johns, M., Lee, K.J., Ive, H.P., Miller, D., Ju, W.: Timing of unstructured transitions of control in automated driving. In: 2015 IEEE Intelligent Vehicles Symposium (IV), pp. 1167–1172. IEEE, June 2015

Naujoks, F., Mai, C., Neukum, A.: The effect of urgency of take-over requests during highly automated driving under distraction conditions. Adv. Hum. Aspects Transp. **7**(Part I), 431 (2014)

NHTSA: National Highway Traffic Safety Administration. Automated Driving Systems (ADS): A Vision for Safety 2.0. NHTSA, DOT HS 812 442 (2017). https://www.nhtsa.gov/sites/nhtsa.dot.gov/files/documents/13069aads2.0_090617_v9a_tag.pdf

Parasuraman, R., Riley, V.: Humans and automation: use, misuse, disuse, abuse. Hum. Factors **39**(2), 230–253 (1997)

Parasuraman, R., Davies, D.R.: A taxonomic analysis of vigilance performance. In: Mackie, R.R. (ed.) vigilance, pp. 559–574. Springer, Boston (1977). https://doi.org/10.1007/978-1-4684-2529-1_26

Parasuraman, R., Mouloua, M., Hilburn, B.: Adaptive aiding and adaptive task allocation enhance human-machine interaction. In: Automation Technology and Human Performance: Current Research and Trends, pp. 119–123 (1999)

Parasuraman, R., Mouloua, M., Molloy, R.: Effects of adaptive task allocation on monitoring of automated systems. Hum. Factors **38**(4), 665–679 (1996)

Payre, W., Cestac, J., Delhomme, P.: Fully automated driving: impact of trust and practice on manual control recovery. Hum. Factors **58**(2), 229–241 (2016)

Radlmayr, J., Gold, C., Lorenz, L., Farid, M., Bengler, K.: How traffic situations and non-driving related tasks affect the take-over quality in highly automated driving. In: Proceedings of the Human Factors and Ergonomics Society Annual Meeting, vol. 58, no. 1, pp. 2063–2067. Sage Publications, Los Angeles, September 2014

Reschka, A., Maurer, M.: Conditions for a safe state of automated road vehicles. Inf. Technol. **57**(4), 215–222 (2015)

Richard, C.M., Campbell, J.L., Brown, J.L.: Task analysis of intersection driving scenarios: information processing bottlenecks (No. FHWA-HRT-06-033), Turner-Fairbank Highway Research Center (2006)

Salvucci, D.D., Liu, A.: The time course of a lane change: driver control and eye-movement behavior. Transp. Res. Part F Traffic Psychol. Behav. **5**(2), 123–132 (2002)

Scallen, S.F., Hancock, P.A., Duley, J.A.: Pilot performance and preference for short cycles of automation in adaptive function allocation. Appl. Ergon. **26**(6), 397–403 (1995)

Scerbo, M.W.: Theoretical perspectives on adaptive automation. In: Parasuraman, R., Mouloua, M. (eds.) Automation and Human Performance: Theory and Applications, pp. 37–63 (1996)

Stanton, N.A., Young, M.S.: Vehicle automation and driving performance. Ergonomics **41**(7), 1014–1028 (1998)

Strand, N., Nilsson, J., Karlsson, I.M., Nilsson, L.: Semi-automated versus highly automated driving in critical situations caused by automation failures. Transp. Res. Part F Traffic Psychol. Behav. **27**, 218–228 (2014)

van den Beukel, A.P., van der Voort, M.C.: The influence of time-criticality on situation awareness when retrieving human control after automated driving. In: 2013 16th International IEEE Conference on Intelligent Transportation Systems-(ITSC), pp. 2000–2005. IEEE, October 2013

Vlakveld, W.: Transition of control in highly automated vehicles. A literature review (Report No. R-2015-22). SWOV Institute for Road Safety Research (2015)

Vogelpohl, T., Kühn, M., Hummel, T., Gehlert, T., Vollrath, M.: Transitioning to manual driving requires additional time after automation deactivation. Transp. Res. Part F Traffic Psychol. Behav. **55**, 464–482 (2018a)

Vogelpohl, T., Kühn, M., Hummel, T., Vollrath, M.: Asleep at the automated wheel—sleepiness and fatigue during highly automated driving. Accid. Anal. Prev. (2018b)

Walch, M., Lange, K., Baumann, M., Weber, M.: Autonomous driving: investigating the feasibility of car-driver handover assistance. In: Proceedings of the 7th International Conference on Automotive User Interfaces and Interactive Vehicular Applications, pp. 11–18. ACM, September 2015

Wan, J., Wu, C.: The effects of vibration patterns of take-over request and non-driving tasks on taking-over control of automated vehicles. Int. J. Hum.-Comput. Interact. **34**(11), 987–998 (2018)

Wandtner, B., Schmidt, G., Schoemig, N., Kunde, W.: Non-driving related tasks in highly automated driving-effects of task modalities and cognitive workload on take-over performance. In: 9th GMM-Symposium on Automotive meets Electronics, AmE 2018, pp. 1–6. VDE, March 2018

Wandtner, B., Schömig, N., Schmidt, G.: Effects of non-driving related task modalities on takeover performance in highly automated driving. Hum. Factors (2018). 0018720818768199

Wiedemann, K., Naujoks, F., Wörle, J., Kenntner-Mabiala, R., Kaussner, Y., Neukum, A.: Effect of different alcohol levels on take-over performance in conditionally automated driving. Accid. Anal. Prev. **115**, 89–97 (2018)

Zeeb, K., Buchner, A., Schrauf, M.: Is take-over time all that matters? The impact of visual-cognitive load on driver take-over quality after conditionally automated driving. Accid. Anal. Prev. **92**, 230–239 (2016)

Zeeb, K., Härtel, M., Buchner, A., Schrauf, M.: Why is steering not the same as braking? The impact of non-driving related tasks on lateral and longitudinal driver interventions during conditionally automated driving. Transp. Res. Part F Traffic Psychol. Behav. **50**, 65–79 (2017)

Zeigarnik, B.: On finished and unfinished tasks. In: A Source Book of Gestalt Psychology, vol. 1, pp. 300–314 (1938)

Zhang, B., de Winter, J., Varotto, S., Happee, R., Martens, M.: Determinants of take-over time from automated driving: a meta-analysis of 93 studies, vol. 78, pp. 212–221 (2018). https://doi.org/10.13140/RG.2.2.33648.56326

Driving Experience

Combining Virtual Reality (VR) Technology with Physical Models – A New Way for Human-Vehicle Interaction Simulation and Usability Evaluation

Chao Ma and Ting Han[✉]

School of Design, Shanghai Jiao Tong University, Shanghai, China
{craigiemc,hanting}@sjtu.edu.cn

Abstract. With the popularity of virtual reality technology, it has already been wide applied in the current automotive industry. However, most of these applications are aimed at the consumer market, and remain to provide the consumers with a method to have an overall experience of the vehicle visually. Based on the development of the automotive industry and the virtual reality technology, this study attempts to utilize the current popular virtual reality technology to solve the contradiction between the increasingly complex technologies applied in the automotive and the gradual shortening design and development cycle of the automotive due to market pressure. This study innovatively combines virtual reality technology and physical model, produces a virtual model in the virtual reality environment with the 1:1 size in real world, and uses the car seat, the steering wheel and the physical air-conditioning button model with 3D printing to produce the preliminary simulation test bench for human-vehicle interaction. After analyzing the completion time and the operational validity of the participants' same interactive operations on the test bench and in the real car, the thesis preliminarily verifies the usability of the bench, and analyzes the advantages and disadvantages of it. The bench created by combining virtual reality technology and physical model can give the designers a better sense of immersion, and because of the addition of the physical model, the designers can evaluate their designs both visually and tactilely, with one more physical dimension. Besides, thanks to the advantages of virtual reality technology, the test bench of this study can achieve quick design update, provide the designers with more convenient solution comparison method, and can save a lot of time and money, which is a novel, concrete, and resource-saving design evaluation method with great application potentials.

Keywords: Human-vehicle interaction · Virtual reality · Evaluation

1 Background

Product development in the automotive industry is driven by a highly complex series of market requirements that stem from a wide range of product variants and functionalities. Stagnating sales volumes in traditional markets and increased competition are leading to both growing product diversification and reduced time-to-market processes

© Springer Nature Switzerland AG 2019
H. Krömker (Ed.): HCII 2019, LNCS 11596, pp. 145–160, 2019.
https://doi.org/10.1007/978-3-030-22666-4_11

[1]. Given the pressure on car companies to reduce time-to-market and to continually improve quality, original Equipment Manufacturer and suppliers have to develop more flexible assembly chains, manufacturing services and methods for job planning [2, 3]. In order to ensure the quality of cockpit design, the traditional method requires the production of a full-size physical prototype of cockpit, which is expensive and time-consuming. And it is hard to modify the physical prototype after it has been made, therefore a physical prototype is not suitable in the early phase of car development during which comparisons among various of versions and agile optimizations and iterations of design are involved. During the phase of car cockpit conceptual design, the responsible departments need to compare, evaluate, optimize and iterate different design plans. Design prototypes should be real enough in order to guarantee the quality of design evaluations, the prototype should also be easy to be modified to ensure the efficiency of the iterations. In respect of these needs during the early phrase of modern automobile development, traditional physical car cockpit prototypes are in a dilemma of increased technical complexity and decreased development period. Because of progressing digitization in the automotive industry, it is increasingly assessed using virtual 3D models [4]. The field of computer graphics is greatly increasing its overall performance enabling consequently the implementation of most of the product design process phases into virtual environments [5]. Virtual reality (VR) technology has the potential to make evaluations of computer-generated models possible in very early phases of the process, which could significantly reduce the number of required hardware mockups as well as the number of design iterations [6].

2   Current Research Status

2.1   Current Research of Automotive Design, Virtual Reality and Virtual Reality Applied in Automotive Design Industry

The academic background research of this paper will be carried out in three aspects, namely the traditional automobile design, virtual reality technology and its application in the automotive industry. The research on the traditional automobile industry is conducive to discovering the pain points and the blue ocean of virtual reality application. The research on virtual reality technology aims to figure out the current situation, functions and effects of virtual reality technology. Finally, the existing applications of virtual reality technology in the automotive industry will be learned about. A considerable number of scholars have conducted research in these three aspects (Table 1).

2.2   Virtual Reality

A complete virtual reality system consists of three modules, namely hardware system, interactive device and virtual content [11]. The hardware system supports the production and operation of virtual content. The interactive device is the bridge between the users and the virtual world. And the virtual content is the object directly perceived by the users. For the hardware system, firstly, there must be a high-performance

Table 1. Current research

Year	Author	Title	Main research content
2018	Winkelhake [7]	"Vision Digitised Automotive Industry 2030, in The Digital Transformation of the Automotive Industry: Catalysts, Roadmap, Practice"	This paper mainly discussed the tendency that established automotive companies are going to change into a comprehensive digitisation strategy and roadmap
2017	Stefania et al. [3]	"FCA Ergonomics Proactive Approach in Developing New Cars: Virtual Simulations and Physical Validation. in Advances in Applied Digital Human Modeling and Simulation"	In this paper, the authors have shown the approach used in FCA based on simulation methods and experimental facilities to analyse ergonomics aspects of future workcells
2017	de Clerk et al. [4]	"Interaction Techniques for Virtual Reality Based Automotive Design Reviews. in Virtual Reality and Augmented Reality"	In this paper, researchers investigated interaction techniques for the design assessment of automotive exteriors. The study results confirm that "Direct Touch" and "First Person" provide best overall quality with respect to these aspects
2017	Ihemedu-Steinke et al. [8]	"Virtual Reality Driving Simulator Based on Head-Mounted Displays, in Automotive User Interfaces: Creating Interactive Experiences in the Car"	As vehicles become more complex and connected, there is need to implement solutions that avoid and reduce any form of distraction to the driver. The application of DS is preferable to real-life experiments on the road because it is cost-efficient and enables rapid prototyping but most especially, it is a controlled environment and totally risk free for the users"
2014	Mihelj, Novak, and Begus [9]	"Introduction to Virtual Reality, in Virtual Reality Technology and Applications"	This article provides an overview of the history of virtual reality and briefly describes the main feedback loops used in virtual reality and the human biological systems used to interpret and act on information from the virtual world
2007	Guo, and Zhou [10]	"VR-Based Virtual Test Technology and Its Application in Instrument Development"	To supply a new approach for the solution of complex engineering test problems, this paper first gave a modified model of VR-based virtual test technology, and then discussed its application in instrument development

workstation system that can ensure the virtual world data can run smoothly in real time. Secondly, a software platform is necessary for building 3D virtual data, based on which, users can restore the real-world environment in the digital virtual world, or establish the world that is only imagined by people but does not exist in the real world. Furthermore, for a fully immersive experience in the virtual environment, extra control devices are necessary for an optimal interaction with virtual objects. These devices aid to navigate freely within the virtual world. For example, the Virtualizer (Cyberith 2014) aids to move freely (walk, jump, sit, and run) in the simulated world, tracking the position of head, hand, or the entire body [8]. For virtual content, there must be a virtual reality environment in the form of 3D data, whose content is what users can see visually. In the latest research, virtual reality is applied to test the efficiency of advanced driver assistance systems, verify the validity and assess the risk of autonomous vehicle algorithms, train simulation assembly in manufacturing industry and so on [12–14]. And it can be figured out in the research that the current virtual reality technology has reached a fairly high level of simulation, which is sufficient for a variety of simulation tasks.

Characteristics of Virtual Reality Technology in Automotive Design

There are five characteristics of the virtual reality technology applied in the automotive design industry, namely immersion, interactivity, imagination, flexibility and economy.

Immersion. It allows designers to immerse themselves in virtual cars and have almost the same experience in real cars. In the 3D car model created by the computer, everything inside and outside the car can be visible, audible, or touchable.

Interactivity. If virtual reality is to be realistic, it must respond to the user's actions; in other words, it must be interactive [9]. Designers could interact with cars in the virtual environment through the input and output of interactive devices to assess the rationality of the car's design plan.

Imagination. The production of physical models is often constrained by the real technologies. While, in the virtual environment, designers could get rid of the constraints of reality temporarily, use bolder ideas, and apply new, even immature, and still imaginary technologies to their own designs and show them.

Flexibility. The production of physical models requires a high time cost, and some designs may have been reversed during the production process. While the virtual models could be produced very fast, and the designers can select materials and set illumination according to different parts and feelings that need to be achieved. These modifications can be quickly reflected on the virtual models for comparison, analysis and improvement.

Economy. Although virtual design does not produce actual economic returns on it own, it will bring considerable economic benefits. In a usual 18-month molding cycle of a vehicle development project, it is necessary to produce the frames and interiors of several non-full-size or full-size data verification clay models, and color-texture verification models in all stages including option screening, modeling review, and data distribution [15]. By utilizing virtual design, the production cost of clay modeling in traditional development processes can be significantly reduced, and the time of modeling development can be saved.

Applications of Virtual Reality Technology

In *Simulation Sickness Evaluation While Using a Fully Autonomous Car in a Head Mounted Display Virtual Environment* by *Rangelova et al.* [16], the study used the virtual reality system to simulate the disease assessment in automated vehicles, and proposed the concept of HMD VR driving simulation (Fig. 1), using Unreal engine4 to create driving scenarios for the study (Fig. 2).

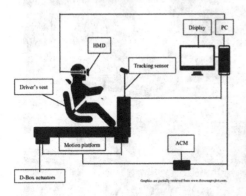

Fig. 1. Concept map of HMD VR driving

Fig. 2. Virtual scenario made by Unreal engine4

In 2015, EVOX Images launched a VR app "RelayCars" for consumers who want to buy a car, allowing users to experience almost all the latest cars on the market at home using their smartphones, tablets or VR helmets (Fig. 3), and the users can select the vehicles in the database of RelayCars, change the color, and view the stereo models from inside and outside.

Fig. 3. 3Relay Cars website show https://www.relaycars.com/

At the latest CES 2019 show, Unity demonstrated a real-time vehicle configuration solution (Fig. 4), in which users could select their favorite colors and interior style in 4S stores without having to communicate with salesmen by using mobile phones, VR devices or PCs.

Fig. 4. Real-time vehicle configuration function exemplified by Lexus LC 500 by Unity at CES 2019 show

As can be seen from the above examples, the virtual reality technology has been developed very mature, and it is possible for various research simulations, and the display of the models in the virtual environment has been able to achieve very realistic effects.

2.3   Advantages of Virtual Reality Technology in the Design and Evaluation Processes of Automotive

The application of Virtual Reality (VR) has been an important supporting method in human-vehicle interaction developments. During the design process of car cockpits, conducting human-vehicle interaction usability evaluations (usability means the validity of interaction processes and results, interaction efficiencies and user's satisfactions when a user is using a product to accomplish certain task in a certain environment) ahead of time can help designers to find perceived issues, reduce or even avoid revisions and realize the agile iteration cycle of "design-analyze-optimize".

2.4   Disadvantages of Virtual Reality Applied in the Design and Evaluation Processes of Automotive

In recent years there is a strong trend to leave the world of physical modeling as early as possible in the styling process and to create digital models even from the scratch. In consequence, industrial designers and deciders are losing the physical experience of the models [4].

Limited by invisibility of depth information and reduced size, common practice of visualization based on CAD 3D digital model is hard to be immersive enough to assess designer's and tester's real feelings and judgements. The applications of immersive VR techniques make the design evaluation more convenient by putting the designers and testers in a full-size 3D virtual car cockpit. However, purely virtual displays and interactions are not able to offer the reality of the real interactions with the product, VR peripheral devices will also distract the testers from the virtual car cockpit and thus interfere the reliability of car cockpit design evaluation.

Taking the perceived concerns above into consideration, a hybrid car cockpit simulation environment that combines VR display and physical models of operative parts is worth to be applied in human-car interaction usability evaluation. By touching and operating the physical models, testers can get real physical feedbacks. These physical models should:

1. Correspond with the size and shape of the certain operative parts that are displayed in the VR environment.
2. Be as modularized as possible in order to fulfill the need of design plan comparison, design iteration and saving budget.

Driven by computer hardware, advanced display and information technology, the applications of VR have been developed rapidly in recent years and thus provide decent technical condition for making car cockpit human-vehicle interaction simulation test benches. Such benches can reach a relatively high fidelity.

3   Methods

In order to verify the usability of the test bench built by combining virtual reality and physical model in the evaluation of the car design, that is, the spatial location and physical experience of the users of the bench can be close to the 1:1 cockpit physical model made with the traditional method, and even the real car, this study uses the user test method. The study uses a 2018 Mercedes-Benz E200L car as the experimental vehicle in control group, combines a virtual reality device and a physical model to build the test bench, selects the air-conditioning button as the example of this experiment, and demands the participants to perform the same operation in the actual vehicle and on the test bench. Finally, the measurers record and analyze the operation time of the participants, the operational validity, and the subjective feedback of the participants. The flow chart is shown in Fig. 5.

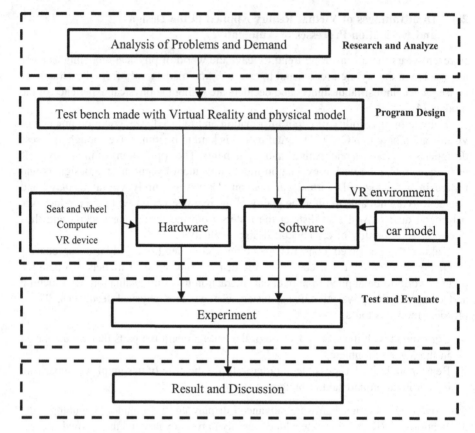

Fig. 5. Method and flow chart of the study

3.1 Establishment of Test Bench

A traditional PC and HTC vive virtual reality helmet are used as the hardware system for building virtual reality scenarios. First, the measurers build the virtual model with the 1:1 size of the frame and interiors of the real vehicle in 3d max and import it into Unity (Fig. 6).

The model is used to build a virtual reality scenario applicable for HTC vive in Unity (Fig. 7).

Meantime, in order to facilitate the subsequent research, the simulated driving bench consisting of Logitech G29 game steering wheel kit and real car seat is selected for the physical platform of the cockpit (Fig. 10).

Fig. 6. Full-size model of 2018 Mercedes-Benz E200L

Fig. 7. Driver's vision in HTC vive

Since the example of the experiment is adjusting the air-conditioning button, as well as the advantages brought by the virtual reality bench, it is only necessary to produce the physical model of the air-conditioning button. The measurers make it with 3D printing method, as is shown in Fig. 8. Number 1–6 is the air-conditioning buttons controlling the six air-conditioners near the driver's seat.

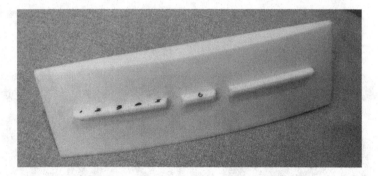

Fig. 8. Air-conditioning button model of Mercedes-Benz E200L by 3D printing

3.2 Participants

There are 12 participants in the study, aged between 21 and 27, 8 of whom are male, all having a driver's license, but none of them had ever driven this car.

3.3 Measures

The experiment takes the user test method. First, the participants are required to sit on the driver's seat in the real car and keep the hands on both sides of the steering wheel. The measurers first introduce the functions of each air-conditioning button to the participants, and after all the participants confirm that they understand the functions and keep them in mind, the measurers issue an instruction (the experiment demands the participants to touch the air-conditioner menu button Fig. 9), and the participants perform the corresponding operation, and then the measurers record the completion time of the operation with the stopwatch (touching the button is counted as the completion of the operation), and finally record the time and the operational validity (i.e. whether the participants touch the correct button) on the table.

Fig. 9. Demand participants to touch the menu button in the middle in the real car

Then, the participants are asked to sit on the test bench. Due to the limitations of the HTC vive hardware platform, the measurers replace the HTC vive handle with the hand model and fix it on the participants' hands. After putting on the device on the bench, each participant adjusts the relative position of the physical model of the air-conditioning button, and the hand model as well as the handle model in the virtual environment, to ensure that the position of the air-conditioning button seen by each participant in the virtual environment is consistent with that of the physical model in the real world. Then, the measurers issue the same instruction as in the control group, and record the operation time and the operational validity on the table. The experimental scenario is shown in Fig. 10.

Fig. 10. A participant carries out the experiment on the virtual reality test bench

Finally, the measurers interview and record the subjective feelings of the participants on the experiment.

4   Results

4.1   Data Analysis Results

In order to verify the usability of the bench combining virtual reality and physical model in the rational evaluation of human-vehicle interaction in automotive design, this study separately observes the time and the operational validity of the same operation performed by the participants in the control group and on the test bench, as is shown in Table 2.

Table 2. Completion time and operational validity of 12 participants' operations in the controlled experiment

Participant's number		1	2	3	4	5	6	7	8	9	10	11	12
Control group	Completion time	2″ 95	4″ 23	2″ 43	2″ 54	3″ 67	3″ 12	3″ 29	3″ 01	2″ 64	3″ 42	2″ 69	3″ 92
	Validity (Y/N)	Y	Y	Y	Y	Y	Y	Y	Y	N	Y	Y	Y
Test group	Completion time	3″ 42	2″ 98	5″ 52	2″ 79	2″ 62	3″ 74	2″ 82	3″ 21	2″ 71	2″ 98	4″ 41	2″ 97
	Validity (Y/N)	Y	Y	N	Y	N	N	Y	Y	Y	Y	Y	Y

It can be seen from the table that in the real car, 11 of the 12 participants have completed the operation, which means the operational validity ratio is about 91.7%. On the test bench, 3 of the 12 participants did not complete the test (they touched the wrong button), which means the operational validity ratio is 75%. For the 8 participants whose two operations are both valid, their average completion time in the control group is about 3.26 s, and that on the bench is about 3.07 s, as is shown in Table 3.

Table 3. Two groups of time data of 8 participants who have completed the tests

In order to verify whether there is significant difference between the two groups of data, this study carries out T-test (Tables 4 and 5), and obtains t value = 0.817, $p > 0.05$. In the circumstance that the degree of freedom is 14, and the significance level is greater than 0.05, the original hypothesis is supported. That is, there is no significant difference between the test bench and the control group in terms of the user performing the interactive operation.

Table 4. T-test group statistics

Group	Number of cases	Mean	Standard deviation	Standard error mean
Control group	8	3.2563	0.58612	0.20723
Test bench	8	3.0725	0.24633	0.08709

Table 5. T-test independent sample test

	Levene Variance Equivalence Test		Mean Equivalence T-test						
	F	Significance	t	Degree of Freedom	p	Average Error Value	Standard Error Value	95% Confidence Interval of the Difference	
								Lower Limit	Upper Limit
Equal Variances Assumed	4.442	0.054	0.817	14	0.427	0.18375	0.22478	-0.29836	0.66586
Equal Variances Not Assumed			0.817	9.398	0.434	0.18375	0.22478	-0.32148	0.68898

From the above results, limited by the positioning accuracy of the hardware platform, the users' operation accuracy in the control group is still higher than that on the virtual reality test bench, but for all the participants whose both operations are valid, their average completion times are very close. And it can be derived from the t-test that there is no significant difference between the test bench and the control group in terms of the user performing the interactive operation. Therefore, the test bench of this study can be used to evaluate the human-vehicle interaction.

In the post-experimental interviews with the participants, the measurers conclude that most participants believe that this is a very novel evaluation method by which they can see the interiors and structure of the car comprehensively and clearly on a virtual bench. Some of the participants who have previously contacted the VR device hold the view that it has one more dimension combined with the physical model compared with the pure VR, to help them feel the interiors, styling, etc. of the car. Besides, some participants have pointed out some shortcomings, such as the car in the virtual scenario cannot achieve the same texture as the real car, and sometimes the hand location offset would occur, and there is only one single hand model that leads to no feeling of the hand action.

In general, on the premise of accurate positioning, the interaction on the bench combining virtual reality and the physical model can get the experimental results close to those of the real car. Therefore, this experiment preliminarily verifies the usability of the bench.

5 Discussion and Conclusion

In order to provide automotive designers with a faster, concrete and resource-saving design test bench, this study makes use of the increasingly mature virtual reality technology, combined with the physical model to create a virtual test bench for human-vehicle interaction.

In order to figure out the rationality, achievability and usability of the bench application, this study starts from the research of the development of the traditional automotive design industry, the virtual reality technology and related fields, and then verifies the usability of the virtual test bench by experiments.

Traditional car design methods require a lot of time, money and manpower. While the application of virtual reality technology in automotive design can great help car designers to evaluate their designs in all aspects, and flexibly update their designs and compare multiple solutions. Some automakers have also used virtual reality technology to evaluate the styling and interior design in the early stage of car design. And due to the development of virtual reality technology, virtual reality devices and programs on the market can provide very high-precision images. These all verify the rationality and achievability of the bench.

However, most of the applications remain on the visual level, only allowing the users to observe the whole vehicle by rotating the device without interaction. This research is aimed at automotive design. It pioneers to combine physical models and virtual reality technology, allowing designers to not only "see" their designs visually, but also "touch" them tactilely, which provides the designers with simulated interaction solutions in the early stage of design. And because of the advantages of virtual reality technology, designers only need to produce the models that need to be tested, which greatly reduces the time and money costs, and is also conducive to the comparative evaluation of multiple solutions.

Finally, it is concluded from the experiment that the data and experience close to those in the real car can be obtained on the bench of this study.

Of course, there are still some shortcomings in this research. For example, the image in the virtual environment is not fine enough, and limited by the hardware devices, the movement of the participants' hands cannot be well simulated, and the positioning offset often occurs. If possible, the handle can be replaced by data gloves to simulate the movement of the hands.

Since it is only a preliminary verification of the usability of the bench, this research's experiment only includes the interactive test in the static process of the car, and there are still considerable potentials of the bench needing to be explored. For example, the internal interactive interface of the car can be tested and evaluated, and various processes and surface treatments can be added to the physical model to simulate a more realistic tactile sensation, and the real buttons can be added to the physical model to make it truly "interactive", and feedback is got in the virtual environment through circuits and programming.

Furthermore, this experiment is carried out when the car is stationary. In the further study, the simulation when the car is moving can be added to evaluate the impact of the driver's operation during the car's movement on the concentration, etc., and to assess the rationality of the design during driving, which is not possible with traditional full-size physical models.

All in all, the simulation test bench combining virtual reality and physical model has quite great application potentials.

References

1. Hirz, M., Dietrich, W., Gfrerrer, A., Lang, J.: Automotive development processes. In: Hirz, M. (ed.) Integrated Computer-Aided Design in Automotive Development, pp. 1–24. Springer, Heidelberg (2013). https://doi.org/10.1007/978-3-642-11940-8_1

2. Lawson, G., Salanitri, D., Waterfield, B.: Future directions for the development of virtual reality within an automotive manufacturer. Appl. Ergon. **53**, 323–330 (2016)

3. Stefania, S., et al.: FCA ergonomics proactive approach in developing new cars: virtual simulations and physical validation. In: Duffy, V. (ed.) Advances in Applied Digital Human Modeling and Simulation. Springer, Cham (2017). https://doi.org/10.1007/978-3-319-41627-4_6

4. de Clerk, M., Schmierer, G., Dangelmaier, M., Spath, D.: Interaction techniques for virtual reality based automotive design reviews. In: Barbic, J., D'Cruz, M., Latoschik, M.E., Slater, M., Bourdot, P. (eds.) EuroVR 2017. LNCS, vol. 10700, pp. 39–48. Springer, Cham (2017). https://doi.org/10.1007/978-3-319-72323-5_3

5. Bordegoni, M., Giraudo, U., Caruso, G., Ferrise, F.: Ergonomic interactive testing in a mixed-reality environment. In: Shumaker, R. (ed.) ICVR 2007. LNCS, vol. 4563, pp. 431–440. Springer, Heidelberg (2007). https://doi.org/10.1007/978-3-540-73335-5_47

6. Moehring, M., Froehlich, B.: Natural interaction metaphors for functional validations of virtual car models. IEEE Trans. Visual Comput. Graphics **17**(9), 1195–1208 (2011)

7. Winkelhake, U.: Vision digitised automotive industry 2030. The Digital Transformation of the Automotive Industry, pp. 77–126. Springer, Cham (2018). https://doi.org/10.1007/978-3-319-71610-7_5

8. Ihemedu-Steinke, Q.C., Erbach, R., Halady, P., Meixner, G., Weber, M.: Virtual reality driving simulator based on head-mounted displays. In: Meixner, G., Müller, C. (eds.) Automotive User Interfaces. HIS, pp. 401–428. Springer, Cham (2017). https://doi.org/10.1007/978-3-319-49448-7_15

9. Mihelj, M., Novak, D., Begus, S.: Introduction to virtual reality. In: Mihelj, M., Novak, D., Beguš, S. (eds.) Virtual Reality Technology and Applications, pp. 1–16. Springer, Dordrecht (2014). https://doi.org/10.1007/978-94-007-6910-6_1

10. Guo, T., Zhou, X.: VR-based virtual test technology and its application in instrument development. In: Shumaker, R. (ed.) ICVR 2007. LNCS, vol. 4563, pp. 468–477. Springer, Heidelberg (2007). https://doi.org/10.1007/978-3-540-73335-5_51

11. 邵景峰, 刘优中, and 喻志强, 基于汽车造型评审的虚拟现实环境设计. 上海汽车, pp. 3–9, April (2017)

12. Schuldt, F., Reschka, A., Maurer, M.: A method for an efficient, systematic test case generation for advanced driver assistance systems in virtual environments. In: Winner, H., Prokop, G., Maurer, M. (eds.) Automotive Systems Engineering II, pp. 147–175. Springer, Cham (2018). https://doi.org/10.1007/978-3-319-61607-0_7

13. Leudet, J., Mikkonen, T., Christophe, F., Männistö, T.: Virtual environment for training autonomous vehicles. In: Giuliani, M., Assaf, T., Giannaccini, M.E. (eds.) TAROS 2018. LNCS (LNAI), vol. 10965, pp. 159–169. Springer, Cham (2018). https://doi.org/10.1007/978-3-319-96728-8_14

14. Werrlich, S., Lorber, C., Nguyen, P.-A., Yanez, C.E.F., Notni, G.: Assembly training: comparing the effects of head-mounted displays and face-to-face training. In: Chen, J.Y.C., Fragomeni, G. (eds.) VAMR 2018. LNCS, vol. 10909, pp. 462–476. Springer, Cham (2018). https://doi.org/10.1007/978-3-319-91581-4_35

15. 唐明星, et al.: 全虚拟设计在 Vision-R 概念车研发中的应用. 机械制造 **55**(06), 7–10, 23 (2017)
16. Rangelova, S., Decker, D., Eckel, M., Andre, E.: Simulation sickness evaluation while using a fully autonomous car in a head mounted display virtual environment. In: Chen, J.Y.C., Fragomeni, G. (eds.) VAMR 2018. LNCS, vol. 10909, pp. 155–167. Springer, Cham (2018). https://doi.org/10.1007/978-3-319-91581-4_12

Designing Augmented Reality Navigation Visualizations for the Vehicle: A Question of Real World Object Coverage?

Nikolai Pärsch[1(✉)], Clemens Harnischmacher[1], Martin Baumann[2], Arnd Engeln[3], and Lutz Krauß[1]

[1] Dr. Ing. h.c. F. Porsche AG, Porschestr. 911, 71287 Weissach, Germany
nikolai.paersch@porsche.de
[2] Ulm University, Albert-Einstein-Allee 41, 89081 Ulm, Germany
[3] Stuttgart Media University, Nobelstr. 10, 70569 Stuttgart, Germany

Abstract. Augmented reality head-up displays are emerging in automotive industry. While they have obvious advantages by showing needed information directly within the field of vision and at the exact location, problems as overlaying real world objects could occur that might lead to driver inattention. We introduce relevant design factors for augmented reality navigation visualizations and present a driving simulator study, where different augmented reality navigation designs were evaluated and the degree of coverage was varied. Performance criteria as well as subjective measurements were recorded. We found differences between the designs concerning their subjective acceptance and their navigation performance, but none of those differences could solely be explained by their degree of coverage. Future work is needed to determine effects in terms of coverage.

Keywords: Augmented Reality · Navigation visualizations · Real world object coverage

1 Introduction

Augmented Reality (AR) display development in the automotive industry is proceeding and will probably be approaching series development in the next decade. So far, some use cases as distance indicators for parking are already implemented in the center display [1]. However, future concepts will address windshields as display location, where even more application scenarios are possible. Besides many advantages (e.g. higher usability of information, less visual distraction) there are also disadvantages. Kim [2] especially mentions driver attention issues. Because of their high salience [3], high rate of information switches [4] or visual clutter [5] interaction with these displays could result in decreased driver attention. Further problems concern the coverage of real world objects that also could result in driver inattention issues.

To avoid these possible disadvantages and fulfill future user requirements, designers and researchers are working on various solutions. This paper describes exemplary design factors for the application scenario "navigation with AR in the

© Springer Nature Switzerland AG 2019
H. Krömker (Ed.): HCII 2019, LNCS 11596, pp. 161–175, 2019.
https://doi.org/10.1007/978-3-030-22666-4_12

vehicle". First, we present a selection of related work in this sector. Second, our conducted experiment provides data about the effect of coverage by AR visualizations as well as about the comparison of different visualizations.

The contributions of our work are as follows:

(a) **Traffic safety:** Insights about the effect of real world object coverage by AR visualizations on driver performance.
(b) **Design:** Insights about acceptance of different design alternatives for AR navigation in the vehicle.

2   Related Work

2.1   AR in Vehicles

AR is a field of technology that is spreading over a huge variety of different application areas. For a good overview of areas and research conducted, please see Dey et al. [6] and their systematic review of AR usability studies. In our work, we focus on the application area "navigation and driving" and use the term AR visualization analogue to Bubb's definition of contact analog displays [7, p. 281]: *"a real representation of the environment, where artificial information is imbedded"*. This information is displayed at the same time and the exact location of real objects. First applications are dated from the mid-seventies [8] and were implemented in a real driving vehicle by Assmann [9].

The main goal of these applications is to reduce cognitive effort when interpreting information that is displayed by the in-vehicle information system and to increase situation and system awareness [10].

Until now, a lot more possible use cases have been proposed. For instance, Haeuslschmid et al. [11] present 96 different use cases that are suitable for windshield display applications and often involve some sort of augmentation. When introducing their design space, they present four different categories of application purposes: safety, vehicle monitoring, entertainment and navigation & geo information systems.

An example for safety applications is the marking of pedestrians and objects [12], whereas supporting conditionally automated driving [13] is an example for the category vehicle and monitoring. Social interaction with other drivers would belong to the entertainment category. However, a huge amount of application examples in the vehicle belongs to the category of navigation and geo information systems. Examples are intersection assistants [14] or navigation visualizations [15].

2.2   AR Navigation in Vehicles

The navigation task itself consists of wayfinding and motion as well as the linked cognitive and motor elements [16]. For more information about stages and processes, please see Düner et al. [16].

Narzt et al. [17] list some of the advantages, when displaying navigation information with AR in the vehicle: conventional navigation systems show only abstractions of navigation data (e.g. flat arrows indicating a turn), so the information presented is not clear and users have to abstract this information. AR eases this by showing clear information with no need for abstraction at the relevant location. Furthermore, it helps at situations with a higher level of ambiguity, for instance turning left when there are several junctions in a row and users have to count junctions with a conventional system. In addition, AR also could show information that is obstructed by other vehicles in the driver's view.

In the past years, researchers and designers presented various designs to show navigation information with AR. Those designs mostly could be divided into following categories:

a. arrow (e.g. Pfannmüller [10]): route painted in form of an arrow
b. tube (e.g. Narzt [17]): route painted in form of a tube or carpet
c. boomerang (e.g. Pfannmüller [10]): route painted in form of multiple boomerangs
d. virtual cable^{TM} (e.g. [18]): route painted in form of a line above the street-level activity of traffic
e. landmark indications (e.g. Bolton et al. [19]): route events (e.g. turns) marked by an arrow or a box that is highlighting a landmark
f. virtual follow-me car (e.g. Topliss et al. [20]): route presented by a virtual lead-car

When investigating these designs, we find several factors that are varied between those visualizations. In the following, we list some of those factors and introduce relevant research.

- **Metaphor.** Besides a traditional driving trajectory, there are other metaphoric approaches to guide drivers through the streets. For instance, Bolton et al. [19] evaluated trajectory arrows against landmark arrows and landmark boxes. The concept of landmark navigation resulted in improved decision times and success rates compared to arrow visualizations. They suggest that this is due to less ambiguity when navigating with landmarks. Topliss et al. [20] introduce another kind of navigation metaphor: the virtual lead vehicle. In their studies, it performed especially well when navigating at complex situations such as complex junctions.
- **Display concept.** Especially visualizations that are following the trajectory metaphor are varying regarding their specific design (e.g. arrows, tube or boomerang). Pfannmüller [15] tested arrows against boomerang navigation. Here, the arrow was rated worse in terms of clearness and was rated as more stressful than the boomerang visualization.
- **Color.** Several factors influence perception of AR colors, for instance different backgrounds and lighting conditions. Gabbard [1] describes these problems in detail. Solutions are for instance adaptive AR concepts, that consider those problems and adjust the AR graphics in real-time. In general, Merenda et al. [21] found out that blue, green, and yellow AR colors are robust to different background colors. Considering user preferences, Pfannmüller [15] found out, that users wish a simple, discreet design with less salient colors to avoid too much attention capturing by the visualization.

- **Presence.** Another important detail is the duration a visualization is visible (e.g. permanent or maneuver-based). Milicic [22] explains that a permanent presentation of these visualizations could lead to driver distraction in terms of cognitive capturing, a form of unconscious shift of attention from the road to the AR information, and therefore should be considered carefully.
- **Shapes.** Another design question is how to design the edges of a visualization. For instance, Tönnis et al. [23] evaluated three different arrow types: solid arrows with hard corners, flat arrows and solid arrows with rounded shapes in terms of their perception for larger distances. They found out, that especially rounded shapes reduce perception in large distances.
- **Tilt.** To achieve a higher salience the visualization could also be tilted along the longitudinal axis when driving towards a maneuver point. When evaluating this design against other design variants, Pfannmüller [10] found out, that a tilted version is ranked worst. Presumably, this is due to a higher perceived distraction because of its high salience or user problems when interpreting this concrete design ahead of maneuvers.
- **Area.** Another question is whether to cover a larger area with the visualization to increase salience and clearness of the navigation indication or reduce the covered area to avoid driver distraction issues by misleading attention and covering relevant traffic objects. Horrey et al. [24] speak of a possible degradation of the driver's ability to respond to truly unexpected traffic events when information covers the driver's line of sight. Pfannmüller [15] observed that in cases where the lead vehicle was covered by the augmented visualization the contact-analogue impression was destroyed in some cases.

3   Research Questions and Hypotheses

We learn from the previous section, that there is some research concerning the concrete design of AR navigation visualizations. However, some questions remain unanswered, especially when considering specific details of the design. One example is the before mentioned area of coverage of the navigation indication. Researchers stated that coverage per se is not preferable. The simplest solution to this problem would be to reduce the visualization area itself to lower the possibility of any coverage. On the other hand, this approach could lead to a salience that is too low, so the visualization will not be perceived anymore. Overall, our research tackles the question whether the degree of coverage will have any effects on driving performance or subjective perception. Our research hypotheses are as follows:

- *H1: Performance decreases with increasing coverage.*
- *H2: Subjective acceptance decreases with increasing coverage.*

4 Method

4.1 Study Design

The study follows a within-subject design, exposing all participants to following selected navigation visualizations and selected driving scenarios (independent variables).

In order to manipulate the degree of coverage, three different visualizations (AR tube max, AR tube min, AR two lines) are designed (also see Fig. 1). Concerning the metaphor we choose the most common visualization form of a trajectory path in front of the driver. We choose blue as one of the recommendations from Merenda et al. [21]. Visualizations are not shown permanent but are presented before and shortly after occurring maneuvers in order to avoid any negative effects reported by Milicic et al. [22]. Furthermore we select a flat design for the edges of the AR graphics, to allow good perception even in higher distances (in accordance with Tönnes et al. [23]). Our visualizations are not tilted, because we want to avoid any halo effects due to a large salience because of the tilt effect.

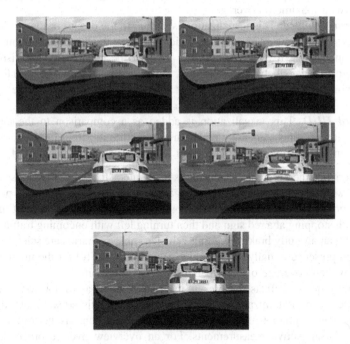

Fig. 1. Overview of evaluated navigation visualizations (from upper left to lower right: AR tube max, AR tube min, AR two lines, AR boomerang, head-up (HUD) conventional)

Besides the three visualizations relevant to our research hypotheses, we add two more visualizations to allow a comparison to other common visualizations present in literature (AR boomerang, HUD conventional). The final set of visualizations is presented below:

- **Tube with maximum coverage.** This visualization is selected because of its high degree of coverage. It consists of a tube-shaped blue layer that covers most of the upcoming route. This approach allows maximum visual attention for the route guidance with the disadvantage of covering many parts of underlying objects. Similar visualizations can be found in [17] and [25].
- **Tube with minimum coverage.** This visualization represents the counterpart to the tube with maximum coverage. It shows the same tube-shaped blue layer but does not cover other moving objects. Hereby the effect of coverage on inattentional blindness should be investigated.
- **Tube with sides.** This visualization is a mix of tube with and without coverage. It is reduced to two lines, each one on the left and right edge of the tube. These edges cover other moving objects. So there is coverage, but more reduced than in the version with maximum coverage.
- **Boomerang-shaped.** This visualization is similar to the design in Pfannmüller [15]. It consists of multiple arrows in-line, which represent the driving direction. This has the advantage that real objects are less likely to be covered by the visualization because of the gaps between the arrows. When driving towards curves, the arrows also tilt on the longitudinal axis in order to allow higher salience for the upcoming route.
- **Standard HUD.** This visualization is added as a comparison to the augmented designs. It represents standard navigation information in form of arrows, which can be seen in front of the driver but are not located at the exact position of the event.

To experience the different visualizations in an interactive way, we choose a driving simulation with different scenarios of navigation. Overall, three scenarios per visualization are driven by each participant: (a) a scenario that consists of a right turn, (b) a scenario with stopping at a red sign and then turning left with oncoming traffic and (c) a scenario with an abruptly braking leading vehicle. Those scenarios are selected because of their high presence in daily traffic, and especially (b) and (c) for the involved traffic and the provoked coverage of other cars by the visualizations.

To avoid sequence effects, we present the visualizations as well as the scenarios in random order. The randomization is carried out with the latin square method.

As dependent variables, we measure various dimensions as performance-based indicators and subjective measurements. For an overview over recorded dependent variables, please see Table 1.

To prove, if performance decreases with increasing coverage, brake reaction times are recorded. This measurement is implemented because of Horrey's et al. [24] research finding concerning decreased performance at unexpected traffic events. For general performance the navigation performance in form of finding the intended way of driving is used.

Table 1. Recorded dependent variables assigned to hypotheses

H1	Performance decreases with increasing coverage
	Brake reaction time
	Navigation errors
H2	Subjective acceptance decreases with increasing coverage
	User-experience-questionnaire
	Ranking of design
	Perception of coverage and disturbance rating

In order to investigate the effect on subjective perception, the designs are ranked and parts of the User Experience Questionnaire (UEQ) [26] are asked. Due to time constraints, we only use the dimensions perspicuity and attractiveness, as we suppose that differences between the visualizations would mostly express on these two dimensions. After experiencing all visualizations, we ask if participants noticed any form of coverage of objects in the driving simulation. In case a coverage is noticed, the degree of disturbance is ranked for each design.

4.2 Apparatus

The experiment takes place in the driving simulator of Porsche AG with motion dynamics. The simulator has a 6 degrees-of-freedom hexapod system. The platform is equipped with a high-fidelity mockup of a vehicle interior. The visualization system consists of a 4-sided cave with 11 projectors (120 Hz with WQXGA resolution) [27]. An image of the driving simulator can be found in Fig. 2.

Fig. 2. Driving simulator of Porsche AG with motion dynamics

As driving simulator software, we use VIRES Virtual Test Drive (VTD) 2.1. The AR navigation visualizations are integrated into the driving simulation as transparent layers on the objects of the driving simulation.

Driving data are recorded and directly transferred to a D-Lab 3.5 software suite. The sampling is carried out at a rate of 60 Hz.

4.3   Tasks and Procedure

After a first instruction of the experiment as well as the review and signing of a participant consent form, a short demographic questionnaire followed. Then participants entered the vehicle mockup and adjusted all relevant controls. Subsequently data recording started and subjects drove a first training phase to get used to the driving simulator. Participants drove about 700 m of country roads and then entered a city, took several turns and then drove another 700 m via country roads to the next city. There the first set of visualizations was presented.

The visualization of the routing started as soon as the upcoming intersection was visible and ended about 50 m after turning into the next road. From first to last appearance of a visualization type, the drive took approximately between 2 and 2.5 min. An exemplary sequence is shown in Fig. 3. Subsequent, participants stopped at the town exit to fill out a short questionnaire. After experiencing all visualizations, participants completed a post-scenario questionnaire and then finished the experiment. Overall, one experiment took about 45 min.

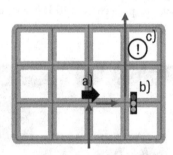

Fig. 3. Exemplary scenario sequence consisting of right turn, left turn with upcoming traffic and suddenly braking vehicle (events randomized)

4.4   Participants

Overall, 36 employees of Porsche AG participated in the study, that were recruited via newsletter at Porsche AG in Weissach, Germany. Seven participants dropped-out because of simulator sickness, which could be explained due to the moderate difficulty of the driving task with several turns and braking maneuvers. Another ten experiments were affected by data logging issues in one or more of the trials. These data logging issues were due to problems with our data recording software suite that stopped recording in some cases. This did not influence quality of any remaining cases.

In total, the final dataset consisted of 19 participants. 68.4% were male, 31.6% female. Five participants were between 18 and 24 years, eight between 25 and 39 years, three between 40 and 54 and three participants were older than 54 years. No participant was connected to HUD development and only one subject had a vehicle with HUD system. Three subjects already had driving simulator experience. The study design was approved by a committee consisting of experts from human resources, data privacy and members of the work council.

5   Results

To examine differences between the visualizations, repeated measurements general linear models (GLM) were used (SPSS Version 25). Before analysis, data was checked for any inconsistencies.

5.1   H1 "Performance Decreases with Increasing Coverage"

Figure 4 presents the values of the brake reaction times during the different visualizations in form of boxplots. Means range from 0.92 s (conventional HUD) to 1.09 s (two lines AR HUD). There are no significant differences between the various visualizations (F value: 0.936; sig.: p = 0.449; partial eta sq.: 0.053).

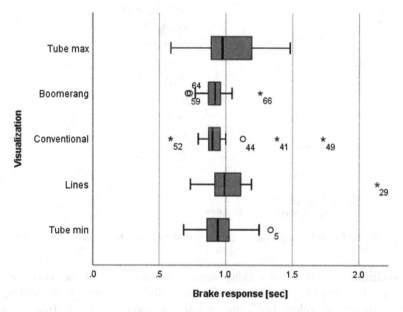

Fig. 4. Boxplots of brake response times

Overall performance was observed by counting navigation errors. Statistics are presented in Table 2. With the "tube max" and "two lines" visualization, five participants took the wrong way. With "tube min", only one participant took the wrong turn. Boomerang and conventional were error-free. In total, there are significant differences concerning navigation errors between the visualizations (F value: 3.816, sig.: p = 0.007; partial eta sq.: 0.145).

Table 2. Overview of navigation errors

	Tube max	Boomerang	Conventional	Lines	Tube min
Navigation errors	5	0	0	5	1

5.2   H2 "Subjective Acceptance Decreases with Increasing Coverage"

Due to time constraints, only selected items of the UEQ were asked: one item "good-bad" for the dimension attractiveness and all items for the dimension perspicuity. Results are shown in Fig. 5. Although mean ratings differ, those results are not statistical significant, neither on attractiveness (F value: 2.257, sig.: p = 0.069; partial eta sq.: 0.091) nor on perspicuity (F value: 2.219, sig.: p = 0.073; partial eta sq.: 0.090).

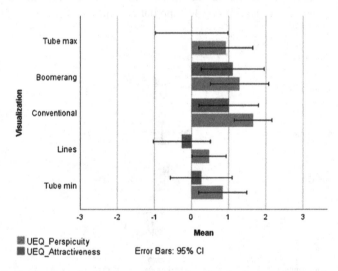

Fig. 5. UEQ results (scale ranges from −3 (horribly bad) to +3 (extremely good))

Participants carried out a ranking after experiencing all of the visualizations. Results can be seen in Fig. 6. There are significant differences between the rankings of the visualizations (F value: 5.680, sig.: p = 0.000; partial eta sq.: 0.202). The conventional head-up visualization is ranked best with a mean rank of 2.05, followed by the boomerang concept with a mean rank of 2.31. Tube with minimum coverage follows with a mean of 2.63 and tube with maximum coverage with a mean of 2.95. Least ranked visualization is "lines" with a mean rank of 3.63.

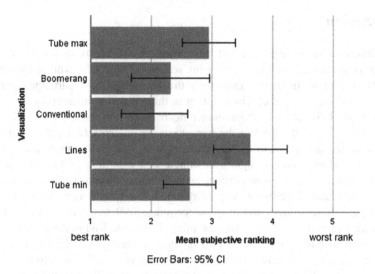

Fig. 6. Mean Subjective ranking (1 = ranked best, 5 = ranked worst)

In the end, participants were asked if they noticed any form of coverage. 32% (6 participants) stated, that they perceived the coverage of simulation objects by navigation visualizations. Those participants rated the degree of disturbance resulting from this coverage for each design. Results are presented in Fig. 7. Perceived disturbance is rated highest for tube with maximum coverage and the boomerang design. Disturbance is rated lower and similar for the remaining designs.

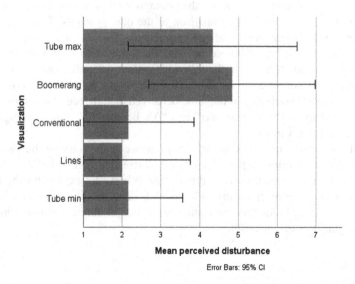

Fig. 7. Mean perceived disturbance of coverage (1 = not disturbing, 7 = very disturbing)

6 Discussion

In the described study, different AR navigation visualizations in the vehicle were compared regarding their degree of covering real world objects as well as their general subjective perception. In order to experience the different designs, participants drove in a driving simulation with each visualization in three different maneuvers.

Concerning hypothesis 1 *"Performance decreases with increasing coverage."*, we found ambiguous results. To test the hypothesis we selected two different measurements: brake reaction times to unexpected events as well as the observation of navigation errors. There were no significant differences regarding brake reaction times. However, navigation errors when driving with each visualization differed significantly between the variants. There were more navigation errors when driving with the variants with highest (AR tube max) and moderate coverage (AR two lines) compared to the variant with minimum coverage (AR tube min). This could lead to the assumption that coverage has an effect on the ability to navigate correctly. On the other hand, the comparison variants "AR boomerang" and "HUD conventional" had also some sort of moderate coverage but performed very well when navigating through the streets. So we tend to reject the hypothesis and assume that differences concerning the navigation performance arise from another factor then coverage. The variant "AR two lines" had a very low salience, because of its minimal area of visualization; this could have lead to perception problems in this case. Further explanations might consider usability of the specific designs that contributed to the navigation performance while using.

To evaluate hypothesis 2 *"Subjective acceptance decreases with increasing coverage."* we selected three different measurements: the UEQ scales attractiveness and perspicuity, a ranking of the visualizations and a rating that asked for the disturbance that resulted from coverage. Concerning the UEQ scales, we found no significant differences although attractiveness of the variants "HUD conventional" and "AR boomerang" was rated higher in comparison to the other variants. When ranking the variants there were larger and significant differences: "HUD conventional" and "AR boomerang" were rated best followed by "AR Tube min" and "AR Tube max", last-ranked was the variant "Lines". In the following questions, participants were asked, if they recognized any sort of coverage. About one third of the participants recognized the coverage of real world objects and rated the perceived disturbance of this coverage. By trend, the disturbance of "AR tube max" and "AR boomerang" was rated higher than that of the remaining variants.

Overall we could state, that rating differs between the variants but we cannot explain this by the varying degree of coverage. Interesting is, that besides all advantages of AR the conventional visualization is ranked best. We explain this by the fact, that users in many cases react negatively to novel devices and do not see the advantages at first glance [e.g. 28]. Another interesting result is that only about one third of the sample recognizes coverage at all.

7 Conclusion

Overlaying real world objects by head-up display visualizations could result in negative effects, which has been reported by Horrey et al. [24] and Pfannmüller [15]. In our experiment, we found differences between the five tested visualizations that were varied by their degree of overlaying real world information. Those differences concerned navigation performance as well as the subjective ranking of the designs. We cannot explain these differences solely by their degree of coverage. However, practitioners could use our insights on subjective preferences of different designs to develop their visualizations in a user-centric way.

Limitations to this work concern the realization of the visualizations. They were integrated into the driving simulation as additional layers, which is not precisely augmented but virtual reality. The setting of a driving simulator was chosen on purpose in order to better realize large visualizations without technical restraints as well as minimizing any interfering factors. Future work should transfer those results in a real driving scenario.

Furthermore, our experiment was realized as a practical setting with several design differences between the visualizations that are interesting but maybe did not solely vary in terms of coverage. In our future work, we want to eliminate any confounding variables and focus on coverage itself. One possible approach would be to vary the degree of transparency of the visualization instead of varying the area of coverage that resulted in quite large conceptual differences.

References

1. Gabbard, J.L., Fitch, G.M., Kim, H.: Behind the glass: driver challenges and opportunities for AR automotive applications. Proc. IEEE **102**(2), 124–136 (2014)
2. Kim, H., Gabbard, J.L.: Quantifying distraction potential of augmented reality head-up displays for vehicle drivers. In: Proceedings of the Human Factors and Ergonomics Society Annual Meeting, vol. 62, no. 1, p. 1923. SAGE Publications, Los Angeles, September 2018
3. Sharfi, T., Shinar, D.: Enhancement of road delineation can reduce safety. J. Saf. Res. **49**, 61-e1 (2014)
4. Wolffsohn, J.S., McBrien, N.A., Edgar, G.K., Stout, T.: The influence of cognition and age on accommodation, detection rate and response times when using a car head-up display (HUD). Ophthalmic Physiol. Opt. **18**(3), 243–253 (1998)
5. Donkor, G.E.: Evaluating the impact of head-up display complexity on peripheral detection performance: a driving simulator study. Adv. Transp. Stud. (28) (2012)
6. Dey, A., Billinghurst, M., Lindeman, R.W., Swan, J.: A systematic review of 10 years of augmented reality usability studies: 2005 to 2014. Front. Robot. AI **5**, 37 (2018)
7. Bubb, H.: Automobilergonomie. Springer, Wiesbaden (2015). https://doi.org/10.1007/978-3-8348-2297-0
8. Bubb, H.: Untersuchung über die Anzeige des Bremsweges im Kraftfahrzeug. Technische Universität München (1975)
9. Assmann, E.: Untersuchung über den Einfluss einer Bremsweganzeige auf das Fahrverhalten. Dissertation an der Technischen Universität München. München: Technische Universität München (1985)

10. Pfannmueller, L., Kramer, M., Senner, B., Bengler, K.: A comparison of display concepts for a navigation system in an automotive contact analog head-up display. Proc. Manuf. **3**, 2722–2729 (2015)
11. Haeuslschmid, R., Pfleging, B., Alt, F.: A design space to support the development of windshield applications for the car. In: Proceedings of the 2016 CHI Conference on Human Factors in Computing Systems, pp. 5076–5091. ACM, May 2016
12. Bergmeier, U.: Kontaktanalog markierendes Nachtsichtsystem - Entwicklung und experimentelle Absicherung. Dissertation an der Technischen Universität München. München: Technische Universität München, 26 October 2009
13. Schoemig, N., Wiedemann, K., Naujoks, F., Neukum, A., Leuchtenberg, B., Vöhringer-Kuhnt, T.: An augmented reality display for conditionally automated driving, pp. 137–141 (2018). https://doi.org/10.1145/3239092.3265956
14. Blavšic, M., Bubb, H., Duschl, M., Tönnis, M., Klinker, G.: Ergonomic design and evaluation of augmented reality based cautionarywarnings for driving assistance in urban environments. In: 17th World Congress on Ergonomics (International Ergonomics Association, IEA), Beijing. Technische Universität München, München (2009)
15. Pfannmüller, L.: Anzeigekonzepte für ein kontaktanaloges Head-up-Display. Dissertation an der Technischen Universität München. München: Technische Universität München, 01 February 2017
16. Dünser, A., Billinghurst, M., Wen, J., Lehtinen, V., Nurminen, A.: Exploring the use of handheld AR for outdoor navigation. Comput. Graph. **36**(8), 1084–1095 (2012)
17. Narzt, W., et al.: Augmented reality navigation systems. Univ. Access Inf. Soc. **4**(3), 177–187 (2006)
18. MVS - California, LLC Homepage. http://www.mvs.net/. Accessed 16 Jan 2019
19. Bolton, A., Burnett, G., Large, D.R.: An investigation of augmented reality presentations of landmark-based navigation using a head-up display. In: Proceedings of the 7th International Conference on Automotive User Interfaces and Interactive Vehicular Applications, p. 56–63. ACM, September 2015
20. Topliss, B.H., Pampel, S.M., Burnett, G., Skrypchuk, L., Hare, C.: Establishing the role of a virtual lead vehicle as a novel augmented reality navigational aid. In: Proceedings of the 10th International Conference on Automotive User Interfaces and Interactive Vehicular Applications, pp. 137–145. ACM, September 2018
21. Merenda, C., Smith, M., Gabbard, J., Burnett, G., Large, D.: Effects of real-world backgrounds on user interface color naming and matching in automotive AR HUDs. In: IEEE VR Workshop on Perceptual and Cognitive Issues in AR (PERCAR), pp. 1–6. IEEE, March 2016
22. Milicic, N.: Sichere und ergonomische Nutzung von Head-Up Displays im Fahrzeug. Doctoral dissertation, Technische Universität München (2010)
23. Tonnis, M., Klein, L., Klinker, G.: Perception thresholds for augmented reality navigation schemes in large distances. In: Proceedings of the 7th IEEE/ACM International Symposium on Mixed and Augmented Reality, pp. 189–190. IEEE Computer Society, September 2008
24. Horrey, W.J., Wickens, C.D., Alexander, A.L.: The effects of head-up display clutter and in-vehicle display separation on concurrent driving performance. In: Proceedings of the Human Factors and Ergonomics Society Annual Meeting, vol. 47, no. 16, pp. 1880–1884. SAGE Publications, Los Angeles, October 2003
25. Israel, B.: Potenziale eines kontaktanalogen Head-up-Displays für den Serieneinsatz. Dissertation an der Technischen Universität München. Technische Universität München, München (2012)

26. Laugwitz, B., Held, T., Schrepp, M.: Construction and evaluation of a user experience questionnaire. In: Holzinger, A. (ed.) USAB 2008. LNCS, vol. 5298, pp. 63–76. Springer, Heidelberg (2008). https://doi.org/10.1007/978-3-540-89350-9_6
27. Baumgartner, E., Ronellenfitsch, A., Reuss, H.C., Schramm, D.: Using a dynamic driving simulator for perception-based powertrain development. Transportation Research Part F: Traffic Psychology and Behaviour (2017)
28. Dillon, A., Morris, M.G.: User acceptance of new information technology: theories and models. In: Annual Review of Information Science and Technology. Information Today, Medford (1996)

A Method to Assess the Effect of Vertical Dynamics on Driving Performance in Driving Simulators: A Behavioural Validation Study

Arben Parduzi[1], Joost Venrooij[1(✉)], and Stefanie Marker[2(✉)]

[1] BMW Group, Research, New Technologies, Innovations, 80788 Munich, Germany
{arben.parduzi,joost.venrooij}@bmw.de
[2] Technical University of Berlin, Chair of Naturalistic Driving Observation for Energetic Optimisation, Gustav-Meyer-Allee 25, 13355 Berlin, Germany
stefanie.marker@tu-berlin.de

Abstract. This paper reports on a study which investigated the effects of adding vehicle vibrations on the driving simulator's validity in terms of the stabilization performance. For this reason a motion-based simulator with three degrees of freedom was upgraded to present a realistic perception of engine and road surface induced vehicle vibrations which improves the overall immersion of the simulator. The study showed that representing vertical dynamics to the participants improves the lateral stabilization performance at nearly constant longitudinal stabilization performance and self-reported simulation sickness scores. This improvement leads partially to absolute validity for the stabilization performance. However the study also shows that the presentation of vehicle vibrations leads to an increase in self-reported audible and temporal mental demand. To improve the self-reported mental workload further examinations are necessary.

Keywords: Virtual environments · Driving simulation · Validity · Vehicle vibrations · On-road study

1 Introduction

A multitude of virtual methods, amongst which driving simulators, are increasingly used in the automotive product development process. An important driver for this is the demand for shorter and more cost-efficient development processes [1,2]. Driving simulators offer the opportunity to carry out driving tests in a safe, reproducible and controlled environment [3]. The testing scope for driving simulators includes human-machine interactions, such as driver distraction studies for operating concepts, assessment of assistance functions and automated driving functions [4] while also involving aspects like suspension set-ups utilizing hardware-in-the-loop simulations [5,6].

© Springer Nature Switzerland AG 2019
H. Krömker (Ed.): HCII 2019, LNCS 11596, pp. 176–189, 2019.
https://doi.org/10.1007/978-3-030-22666-4_13

Despite the continuous improvement, driving in a simulator has been and always will be merely a simulated reproduction of reality. Due to an increasing number of optimizations and a strong dependency between individual components of the driving simulation, a holistic evaluation of further developments in the driving simulation is necessary. With a larger number of applications and an ever advancing technical development, the need to prove the validity of driving simulator tests has been increasing. This requires methods to analyse the differences between simulators and to quantify the effects of improvements in simulator configuration. With the method reported in this paper the influence of the reproduced vertical dynamics on behavioural validity will be examined. The National Highway Traffic Safety Administration (NHTSA) recommends testing on driving simulators especially for driver distraction related studies [7]. Resulting from the need to evaluate, for example, new touch screen control elements, a suitable method for presenting vertical dynamics is necessary. For the implementation of vehicle vibrations, the simulation software was upgraded and a static simulator was equipped with four linear actuators (Fig. 2). The four actuators allowing a simulator motion with three degrees of freedom (3-DOF: *heave, pitch and roll*).

2   Driving Simulator Validity

Driving simulator validity is an active research topic [1,8,9] and an important requirement for transferring development tasks from the road into driving simulator laboratories [10]. Depending on the desired focus of the research, different types of simulator validity are used. According to Blana et al. [11,12] the term of driving simulator validity can be divided into:

- Behavioural validity describes the correspondence between driving behaviour in the simulator and on-road.
- Physical validity measures the extent to which a driving simulator can reproduce the visual and vestibular cues while driving an actual vehicle.
- Absolute validity is defined as a quantitative criterion which is achieved if the numerical values of these performance differences are about equal or the same.
- Relative validity is achieved if the difference is in the same direction and relative size of the effect of the measure as in reality.

To examine the validity of the simulator in this study the definition of behavioural validity is used. In this case the behavioural concept refers to the driving performance and the self-reported mental workload.

3   Background and Research Question

The research in [13] provided some evidence that adding a 3-DOF motion system to a static simulator can improve the lane keeping performance and decrease the self-chosen vehicle speed on country roads. Another study showed that adding vibrations to a 3-DOF motion system can alleviate the symptoms of motion sickness [14]. Neither study, however, investigated the influence on behavioural validity. Another approach focused on improving the realism of a driving simulator by presenting vehicle vibrations to the participants [15] and validating the influences on the self-chosen speed [16], but the findings were based on a advanced driving simulator with 6-DOF.

The current paper aims to add to existing knowledge by providing a detailed analysis of the influence of vertical dynamics on behavioural validity on highway conditions in a cost-efficient mid-size 3-DOF simulator.

Reproducing vertical dynamics leads to vehicle vibrations, which are perceived by the driver. On the one hand the presence of vehicle vibrations can be interpreted as an additional cue for the driver and results in an increased immersion. On the other hand the resulting vehicle movements can be interpreted as a disturbance input for the driver-vehicle-environment control loop. To determine the influence of the vehicle vibrations on the driving performance the following research question will be answered:

How does the reproduction of vehicle vibrations on a driving simulator influence the stabilization performance, the-self reported mental driver workload and the self-reported simulation sickness?

4   Methodology

To examine the research questions two empirical studies were performed. In May 2018 a real driving study on an highway near Munich was conducted with 30 participants. This study will be referred to here as the *On-Road Study*. Furthermore, a driving simulator study in one of the BMW Group's driving simulation facilities with 41 participants was conducted in August 2018.

4.1   Apparatus

On-Road Study. Figure 1 shows the experimental vehicle used for the on-road study. The used car was a 4-door sedan BMW 5-Series with 248 horse power. For data acquisition the internal vehicle bus was recorded with 20 Hz using a data logging system. At the bottom left on Fig. 1 the experimenter interface is shown. The interface was displayed on a 7″ display which can be positioned near the passenger or rear seat. The right-hand side of Fig. 1 shows the driver position. The experimental car was equipped with a car radio to perform the selected driving tasks (see Sect. 4.4). The modifications of the experimental vehicle were implemented such that the general appearance for the participants, apart from the radio, is not different from a series-production vehicle.

Simulator. The driving simulator used in this validation study is shown on Fig. 2. The vehicle mockup consisted of the front section of a BMW 5-Series chassis and 5-Series cockpit elements, which was mounted on a electromotive 3-DOF

Fig. 1. Experimental vehicle used for the on-road validation study

platform. The platform was actuated using 4 D-Box actuators (D-BOX TECH-NOLOGIES INC., Gen II Actuators 6″). The vehicle dynamic model described the same vehicle type which was used for the on-road study. All typical control elements such as direction-indicator control, brake and acceleration pedal and gear selector lever were fully functional. The steering torque feedbacks came from an electric actuator (Sensodrive GmbH, SENSO-Wheel SD-LC) that delivers a nominal torque of 7.5 Nm. For the visualisation of the virtual environment five video projectors were used which enable a 220° horizontal and 45° vertical field of view. The views seen in the exterior and interior rear view mirrors were provided by three LCD screens.

Fig. 2. 3-DOF driving simulator system used in the validation study

4.2   Vehicle Vibrations

The main aspect of this study was to examine the influence of vehicle vibrations on driving simulator validity. Vehicle vibrations can be classified according to Heissing et al. [17] in three categories. Vibrations with frequencies from 0–25 Hz are perceived by tactile sense and are categorized as *tangible*. Frequencies higher than 100 Hz are mainly noticed as audible noise and hence categorized as *audible*. Between these frequency bands exists a so called *transition zone*. Phenomena in this transition zone can be perceived by both, the acoustic and tactile senses and are therefore both *audible* and *tangible*. To present these effects to the participants three types of actuators were used:

- 3-DOF motion-system (D-BOX TECHNOLOGIES INC., Gen II Actuators 6″) for frequencies up to 30 Hz, i.e. the *tangible* range
- Structure-borne sound converter (SONIC IMMERSION TECHNOLOGIES, IBEAM VT-200 Transducer) for frequencies from 30–125 Hz, i.e. reproducing roughly the *transition zone*
- Four channel sound system mainly for frequencies over 100 Hz up to 20 kHz

The structure-borne sound converter was located under the driver seat to represent phenomena like road roughness, road bumps and engine vibrations. Vehicle vibrations with frequencies up to 100 Hz are mainly caused by tyre-road interactions [18]. For an adequate representation of these effects a realistic simulation of road surfaces is necessary.

Therefore the road surface for the simulated road was derived from laser-measured data of the highway on which *on-road study* was performed. The measured road excitations are provided to the driving simulation by the OpenCRG file format [19] with a resolution of 10×10 mm. To obtain the resulting vehicle accelerations and angular rates, a non-linear double-track model is used for the vehicle dynamics simulation. To present the vehicle vibrations to the participants the calculated movements are transferred to the motion-system and the structure-borne converter.

It should be noted that this implementation also allows for the reproduction of steering torque effects caused by road excitations from road damages, such as grooves, potholes and cracks. Furthermore, it has to be taken into account that in the real world road damages are not only perceived as vestibular or tactile cues but also by visually cues. Therefore the existing road damages were also reproduced graphically (see Fig. 3).

4.3   Participants

In the *on-road study* 30 BMW employees (28 men and 2 women), with at least six years of driving experience and a minimal annual mileage of 5000 km, took part. The age of the participants ranged from 25 to 50 (mean (μ) = 36.13 years, standard deviation (σ) = 8.28).

The *simulator study* was conducted with 41 BMW employees (34 men and 7 women), with at least four years of driving experience and an minimal

Fig. 3. Virtual highway with road damages

annual mileage of 5000 km. The age of the participants ranged from 20 to 59 ($\mu = 37.12$ years, $\sigma = 12.08$). 31 of 41 participants had prior experience in a driving simulator. The remaining 10 participants took part for the first time in a driving simulator study.

All participants were included on a voluntary basis. The participants received written instructions and provided their informed consent before the experiment commenced. They were informed about that they could abort the experiment at any time in case of discomfort.

4.4   Study Design

Independent Variables. The driving simulation study was designed as a within-subject design with the two independent variables *'Flat'* and *'Ride'* combined with the real driving study *'Real'*.

- Real: On-road study which serves as benchmark for a realistic stabilization performance.
- Flat: Only roll- and pitch angles induced by vehicle accelerations were presented to the participants by the motion system.
- Ride: This condition extends the *'Flat'* condition by representing vehicle vibrations due to road-excitations by the motion system and a structure-borne sound converter. See also Sect. 4.2 for more information.

The driving task consisted of a highway ride in which the so-called radio tuning task was carried out according to the specifications of AAM [20]. In the *on-road* study participants were instructed to follow a lead vehicle at a distance of 50 m. The lead vehicle, driven by on of the experimenters, drove in the right lane at a speed of 92 km/h. The written instruction for the participants was to focus on lane keeping and following the lead vehicle with a constant distance. This task will be referred to here as the baseline condition. In order to allow an analysis of relative effects (i.e., relative validity) the above described driving task was also conducted while the driver was instructed to simultaneously perform a secondary task. The secondary task was the so-called radio tuning task (radio). In the radio tuning task, the participants were asked to select a given frequency on a radio, which was installed according to the AAM [20]. The radio which was used for the radio tuning task is shown on the right-hand side in Fig. 1.

The drive to the highway took about 20 min. During this time the participants could familiarize themselves with the car. At the beginning of the highway drive the participants practiced the baseline task for about 5–10 min. The radio task was not practiced before.

After the on-road study the radio was installed into the simulator mock-up. The test procedure in the simulator study was the same as in the real study. Before the test cases have been started the participants drove along urban, rural and highway roads for around 10 min to get familiar with the environment. The lead vehicle on the highway was controlled automatically by the simulation software. The highway section where the *on-road* study was conducted, was reproduced for the simulation including the laser-measured road surface.

To calculate the performance indicators six repetitions of the radio tuning task and the baseline task with a duration of 60 s were performed. The order in which the participants performed the two tasks was randomized.

Figure 4 summarizes the experimental design and provides an overview of the independent and dependent variables of the experiment. In the next section the selected dependent variables will be described.

Dependent Variables. The above described scenario was used for the examination of the stabilization performance, whereby the stabilization task can be divided into lateral and longitudinal stabilization [21]. The lateral stabilization performance was measured with the standard deviation of lane position (SDLP). To calculate the SDLP the distance between the right-hand front wheel and the

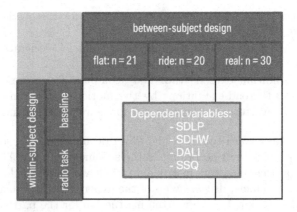

Fig. 4. Schematic overview of the study design

side stripes was chosen. The longitudinal stabilization performance was measured with the standard deviation of head-way (SDHW) [22,23], in this case the head-way was defined as the distance between the front and rear bumper of the experimental and the lead vehicle.

In addition the self-reported driver mental workload was recorded with the driver activity load index (DALI).

The DALI is a revised version of the NASA-TLX, adapted to the driving task [24] and consist seven workload dimensions:

- effort of attention
- visual demand
- auditory demand
- temporal demand
- interference
- situational stress
- total score (consisting of the average of the other dimensions)

Each dimension has been valuated by the participants with a numeric rating scale from 0 (low demand) to 5 (high demand) after each of the three repetitions. Afterwards the average of both submitted ratings was calculated.

Simulator sickness is a common issue in the context of driving simulation and a major obstacle to the use of driving simulators for research, training and driver assessment purposes [25,26]. Therefore the self-reported simulation sickness was quantified by Kennedy's SSQ Simulator Sickness Questionnaire (SSQ) [27] to analyse the effect of vehicle vibrations on the participant's sense of well-being. After data acquisition it is possible to provide simulator sickness scores classified in three symptoms referred to as *nausea* (N), *oculomotor disturbance* (O), *disorientation* (D). Additionally a *total score* (TS) can be derived from the three symptoms sub scales.

5   Results

To test whether the data meet the requirements for a parametric independent analysis of variance (ANOVA), which are that the residuals are normally distributed and have homogeneity of variances [28], the Lilliefors and Levene test were performed [29]. If the conditions are not fulfilled the significance test was performed using the non-parametric Kruskal-Wallis-Test [30]. For the post-hoc tests the p-values were corrected with the least significant difference method [31].

5.1   Lateral Stabilization Performance

Figure 5 shows the results (mean and 95% confidence interval) for the lateral stabilization performance quantified by the SDLP. On the *'real'* condition the participants produced the lowest mean for the SDLP followed by the *'ride'* and *'flat'* condition. The statistical tests reveals significant differences for the SDLP during the baseline and radio task (baseline: $F(2) = 6.30$, $p < .01$, radio: $F(2) = 4.83$, $p = .01$). Regarding to the post-hoc tests the results for the condition *'ride'* on both tasks do not differ significantly from the on-road study (baseline: $p > .30$, radio: $p > .08$). These results indicates that the addition of vehicle vibrations are required to obtain absolute validity, i.e., no significant difference with the *real* condition.

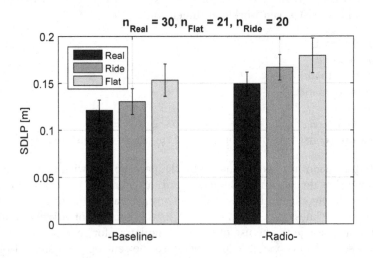

Fig. 5. Bar chart: results for standard deviation of lane position (SDLP) based on 71 participants. $n_{Real} = 30, n_{Flat} = 21, n_{Ride} = 20$

5.2 Longitudinal Stabilization Performance

Figure 6 illustrates the results for the longitudinal stabilization performance quantified by the SDHW. For the baseline task the means of all three conditions are similar. For the radio task the mean for the *'real'* condition is the lowest followed by the *'ride'* and *'flat'* condition. The ANOVA for the SDHW on baseline condition shows no significant differences ($H(2) = 1.16$, $p > .55$). The AMOVA test for the radio task shows significant differences between the three test conditions ($H(2) = 8.08$, $p = .018$). The post-hoc contrasts, however, show no significant differences between the simulator and on-road conditions (*real-flat*: $p = .058$, *real-ride*: $p = .804$).

5.3 Mental Driver Workload

The mental workload quantified by the DALI total score is shown in Fig. 7). A statistical analysis showed a significant effect of task but no significant difference between the conditions *real*, *ride* and *flat*. For closer investigation the six DALI sub scales were also analysed (see Fig. 8) which showed that the auditory demand

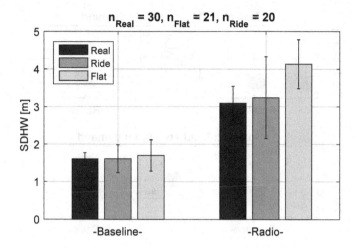

Fig. 6. Bar chart: results for standard deviation of head-way (SDHW) based on 71 participants. $n_{Real} = 30, n_{Flat} = 21, n_{Ride} = 20$

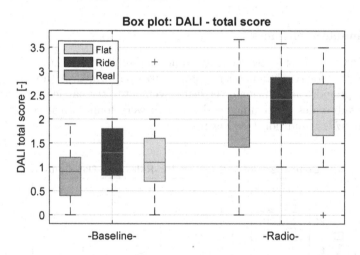

Fig. 7. Box plot: results for DALI total score sub-scale based on 71 participants. $n_{Real} = 30, n_{Flat} = 21, n_{Ride} = 20$

for the baseline task for the *'ride'* condition is significantly higher ($H(2) = 6.31$, $p = .043$). Furthermore the temporal demand is perceived higher for the *'ride'* condition during the baseline task ($H(2) = 6.15$, $p = .046$). Regarding the radio task, the descriptive analysis shows also higher values for the auditory and temporal demand but the differences are not statistically significant.

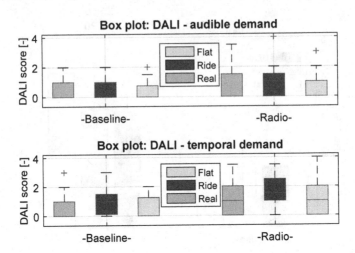

Fig. 8. Box plot: results for DALI auditory and temporal demand sub-scale based on 71 participants. $n_{Real} = 30, n_{Flat} = 21, n_{Ride} = 20$

5.4 Simulator Sickness

To analyse the effect of vertical dynamics on simulator sickness a Wilcoxon rank sum test for two independent samples was performed. Although the *'ride'* condition provides an additional vertical movement to the participants, regarding the SSQ sub-scales no significant differences were found when comparing the *'flat'* and *'ride'* condition (see Fig. 9).

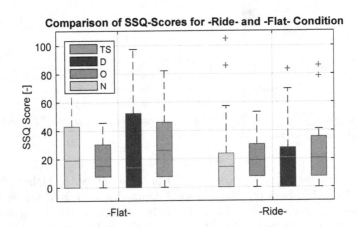

Fig. 9. Box plot: SSQ-Scores sorted by the four SSQ-Subscales N: Nausea, O: Oculomotor, D: Disorientation and TS: Total Score [27]

6 Discussion and Outlook

The results of this approach underline the importance of vehicle vibrations on lateral stabilization performance. Absolute validity could be achieved after presenting vehicle vibrations for the SDLP in both conditions. These results show that the presented additional cues provide useful information for the lateral stabilization task. For the head-way performance (SDHW) absolute validity is reached for all simulator conditions. This means that presenting road excitations to the subject has no negative effect on longitudinal stabilization performance. It will be assumed that the perceived visual cues are sufficient to perform the longitudinal stabilization task.

Although the DALI total score shows no significant differences between the two test conditions - *flat* and *ride* - the descriptive analysis indicates an increasing overall mental workload for the *ride* condition. This increase seems to be caused by a higher temporal and audible demand.

Possible explanations for the higher audible demand could be that the cues, provided by the vehicle vibration system in the *ride* condition, did not present the participants a realistic experience or caused additional noise that did not aid the drivers in their task. Another explanation may be that the study was conducted with experienced driving simulator drivers which were not familiar with the newly presented cues in this experiment. The possible correlation between the increased audible demand and temporal demand is to be investigated.

A 3-DOF motion system in combination with vehicle vibrations delivers a cost-efficient solution to improve the lateral stabilization performance without influencing the longitudinal stabilization performance. An negative effect regarding to the SSQ could not be found. A possible next step is to investigate the specific reasons for the higher demanded mental workload while driving with simulated vehicle vibrations.

To analyse the effect of vehicle vibrations on motion simulators with more than 3-DOF the study should be repeated in classical motion simulators with a hexapod motion base. Furthermore this approach will be extended by a country road scenario to evaluate the implementation of vehicle vibrations, with more behavioural aspects. In this future study additional behavioural aspects, such as a subjective evaluation of the realism of the driving scenario and perceived driving speed, will be included.

References

1. Zöller, I.M.: Analyse des Einflusses ausgewälter Gestaltungsparameter einer Fahrsimulation auf die Fahrerverhaltensvalidität. Dissertation, Technische Universtät Darmstadt, Darmstadt (2015)
2. Hassan, B., Gausemeier, J.: A design framework for developing a reconfigurable driving simulator. IARIA Int. J. Adv. Syst. Meas. **8**, 1–17 (2015)
3. de Winter, J.C., van Leeuwen, P.M., Happee, R.: Advantages and disadvantages of driving simulators: a discussion. In: Spink, A., Grieco, F., Krips, O., Loijens, L., Noldus, L., Zimmerman, P. (eds.) Proceedings of Measuring Behavior 2012, pp. 47–50 (2012)

4. Winner, H., ed.: Handbuch Fahrerassistenzsysteme: Grundlagen, Komponenten und Systeme für aktive Sicherheit und Komfort; mit 45 Tabellen. 1. aufl. edn. ATZ-MTZ-Fachbuch. Vieweg + Teubner, Wiesbaden (2009)
5. Fioccardi, A., Dusini, L.: How to reduce development time: Winter session at the driving simulator (2017)
6. Moscatelli, S.: Virtual simulation hil & sil: Using driving simulator and hardware in the loop in calibration of magneto-rheological damper controller (2017)
7. NHTSA: Visual-manual NHTSA driver distraction guidelines for in-vehicle electronic devices (2012)
8. Klüver, M.: Can we trust driving simulator studies? the behavioral validity of the daimler AG driving simulators. Dissertation, Johannes Gutenberg Universität Mainz, Mainz (2016)
9. Reich, D., et al.: Gestaltungsaspekte immersiver Fahrsimulationsumgebungen. Dissertation, Fraunhofer-Institut für Produktionsanlagen und Konstruktionstechnik and Fraunhofer IRB-Verlag (2017)
10. Harms, L.: Driving performance on a real road and in a driving simulator: results of a validation study. VTI särtryck **1996**(267), 19–26 (1996)
11. Blana, E.: Driving simulator validation studies: A literature review (1997)
12. Hoskins, A.H., El-Gindy, M.: Technical report: literature survey on driving simulator validation studies. Int. Journey Heavy Veh. Syst. **13**(3), 241–252 (2006)
13. Parduzi, A., Bezikofer, F., Comulada-Simpson, E., Marker, S.: Comparable evaluation of a 3DOF mid-size simulator concept. In: Kemeny, A., Colombet, F., Mérienne, F., Espié, S. (eds.) Proceedings of the DSC 2018 Europe Conference, pp. 207–208 (2018)
14. Guillaume, L., Kemeny, A., Paillot, D., Colombet, F.: A simulation sickness study on a driving simulator equipped with a vibration platform. In: Kemeny, A., Colombet, F., Mérienne, F., Espié, S. (eds.) Proceedings of the DSC 2018 Europe Conference, pp. 19–22 (2018)
15. Bolling, A., et al.: Improving the realism in the VTI driving simulator. Shake final report (2012)
16. Ahlström, C., Bolling, A., Sörensen, G., Eriksson, O., Andersson, A.: Validating speed and road surface realism in VTI driving simulator III (2012)
17. Heißing, B., Ersoy, M., Gies, S.: Fahrwerkhandbuch: Grundlagen, Fahrdynamik, Komponenten, Systeme, Mechatronik, Perspektiven. Vieweg+Teubner Verlag/Springer Fachmedien Wiesbaden GmbH Wiesbaden, Wiesbaden (2011). https://doi.org/10.1007/978-3-8348-8168-7. 3., überarbeitete und erweiterter auflage edn
18. Thomaier, M.: Optimierung der NVH-Eigenschaften von Pkw-Fahrwerkstrukturen mittels Active-Vibration-Control. Dissertation, Technische Universtät Darmstadt, Darmstadt, 16 June 2008
19. Dupuis, M.: OpenCRG: managing the road surface ahead. http://www.opencrg.org/
20. Alliance of Automobil Manufactures: Statement of principles, criteria and verification procedures on driver interactions with advanced in-vehicle information and communication systems: Including 2006 updated sections (2006)
21. Bubb, H., Bengler, K., Grünen, R.E., Vollrath, M.: Automobilergonomie. A. Springer, Wiesbaden (2015). https://doi.org/10.1007/978-3-8348-2297-0
22. Östlund, J., Engström Johan, Keinath, A., Horst Dorit et. al.: Driving performance assessment: methods and metrics: aide adaptive integrated driver-vehicle interface (2005)

23. Papantoniou, P., Papadimitriou, E., Yannis, G.: Review of driving performance parameters critical for distracted driving research. Transp. Res. Procedia **25**, 1796–1805 (2017). World Conference on Transport Research - WCTR 2016 Shanghai. 10–15 July 2016

24. Pauzié, A.: A method to assess the driver mental workload: the driving activity load index (DALI). IET Intell. Transp. Syst. **2**(4), 315 (2008)

25. Helland, A., Lydersen, S., Lervag, L.E., Jenssen, G.D.: Driving simulator sickness: impact on driving performance, influence of blood alcohol concentration, and effect of repeated simulator exposures. Accid. Anal. Prev. **94**, 180–187 (2016)

26. Neukum, A., Heidi, G.: Kinetose in der fahrsimulation: Simulation von einsatz-fahrten im auftrag des präsidiums der bayerischen bereitschaftspolizei, abschluss-bericht - teil ii

27. Kennedy, R.S., Stanney, K.M., Harm, D.L.: Configural scoring of simulator sickness, cybersickness and space: Adaptation syndrome: Similarities and differences? https://ntrs.nasa.gov/archive/nasa/casi.ntrs.nasa.gov/20100033371.pdf

28. Rasch, B., Friese, M., Hofmann, W., Naumann, E.: Quantitative Methoden 2. Springer-Lehrbuch. Springer, Heidelberg (2014). https://doi.org/10.1007/978-3-662-43548-9

29. Bortz, J., Schuster, C.: Statistik für Human- und Sozialwissenschaftler. Springer-Lehrbuch. Springer, Heidelberg, 201 (2010). https://doi.org/10.1007/978-3-642-12770-0. 7, vollständig überarbeitete und erweiterte auflage edn

30. Du Prel, J.B., Röhrig, B., Hommel, G., Blettner, M.: Choosing statistical tests: part 12 of a series on evaluation of scientific publications. Deutsches Arzteblatt international **107**(19), 343–348 (2010)

31. Salkind, N.J.: Encyclopedia of Research Design. Sage, Thousand Oaks (2010)

Trust Is Good, Control Is Better? – The Influence of Head-Up Display on Customer Experience of Automated Lateral Vehicle Control

Seda Aydogdu[1]([✉]), Corinna Seidler[2]([✉]), and Bernhard Schick[2]([✉])

[1] MdynamiX AG, Heßstraße 89, 80797 Munich, Germany
Seda.Aydogdu@mdynamix.de
[2] University of Applied Sciences Kempten,
Bahnhofstraße 61, 87435 Kempten, Germany
{corinna.seidler,bernhard.schick}@hs-kempten.de

Abstract. Advanced driver assistance systems (ADAS) support the driver in certain traffic situations and can increase road safety. For this appropriate interaction, concepts between the driver and the assistance systems are required, which focuses on driver's needs. In a user-centered study with N = 48 subjects, interviews and questionnaires were conducted during a test drive in real road traffic in order to test and evaluate the lane keeping assistant system (LKAS) with head-up display (HUD). In addition, two current premium vehicles from various manufacturers were used to investigate the influence of the HUD on the user experience and to derive optimization potential for current and future automatic driving functions. In comparison to the test rides with LKAS in combination without HUD (with head-down display), it can be determined that there is a positive influence of HUD on the experience with LKAS.

Keywords: User experience · Head-up display ·
Lane keeping assistance system · Customer acceptance · User-centered design

1 Introduction

Mobility is constantly changing worldwide. The automobile has always been associated with attributes such as passion, commitment and high emotionality. At the same time, new developments such as Advanced Driver Assistance Systems (ADAS) and Automated Driving (AD) have introduced both solutions and challenges. To achieve high user acceptance and an emotional driving experience, a system transparency and sufficient feedback - as in conventional driving mode - are necessary. In particular, the driver has to understand the system so that they can intervene at any time, which is also currently required for level 2 automation [1, 2]. An appropriate human-machine interface (HMI), which focuses on the needs of the driver, is essential here [3–5]. The person who hands over control to a machine wants to feel well informed about the status of the system, which plays a major role in his or her wellbeing and for the user experience (UX) [6]. Man and machine should not be considered separately, but as a whole, whose

© Springer Nature Switzerland AG 2019
H. Krömker (Ed.): HCII 2019, LNCS 11596, pp. 190–207, 2019.
https://doi.org/10.1007/978-3-030-22666-4_14

strengths complement each other and weaknesses balance each other out [7]. Thus, the goal of the development of human-computer interaction (HCI) is a seamless interaction of devices, so that the user must do as little as possible. Because the best interaction with a device is when the device suspects what the user wants [8]. It is about non-verbal communication in which both parties "understand" each other to manage the tradeoff between over-trust and under-trust. Over-trust leads to higher risk, whereas under-trust leads to non-usage of the system [9]. Only systems that offer added value are used by people [10], and only used systems can generate their benefits.

A pilot study with N = 50 subjects, conducted by MdynamiX and University of Applied Sciences Kempten on automated lateral vehicle control respectively lane keeping assistance systems (LKAS), has shown that customer acceptance is poor for all three vehicles tested. In the case of one of the three vehicles, a better evaluation and a more positive UX by the customers compared to the experts/engineers were found. Our hypothesis: The Head-Up Display (HUD), this one vehicle was equipped with additionally, has a substantial influence on the LKAS evaluation [1].

2 User Experience with LKAS

Firstly, for a better understanding, the following section deals with the subjects walk-thru, based on the method customer walkthrough according UX [10] and related driving experience during the pilot study with LKAS in combination with a head-down display (HDD) [1, 3].

- The test person gets into the car, has a certain expectation of the system and comes into contact with the HMI of the LKAS the first time by switching it on. The first hurdles appear here, as some car manufacturer use complicated non-intuitive operating functions, such as lever functions or switching on buttons via the board menu. The respondent is overstrained and needs support. Only in a car model with a one-button operation, the respondent is able to switch on the system without support.
- System is switched on, test person drives with LKAS, which becomes active from a speed of approx. 65 km/h. They can follow the status of the system via the LKAS symbol in the HDD. A driver-vehicle interaction is created. There are differences in display and design between the manufacturers. The size of the LKAS symbol can thus vary between 1.5–3 cm depending on the manufacturer. If the system is active and the camera detects the lane, the symbol is green. In order to monitor the system status, the test person must look away from the road every few seconds to read the status in the HDD. The test person experiences a certain quality of lane guidance. The system does not follow a precise lane, so an oscillating between the lane markings is noticeable.
- The vehicle approaches the lane markings and the test person feels unsafe. The system drops for an inexplicable reason. At best, a warning comes exactly at the moment of the system drop-off (LKAS symbol in the HDD changes from green to grey, an audible warning signal is sometimes issued). In this short time, the test person cannot even watch the HDD and perceive the system status. There may also be no warning at all. The subject is surprised and asks why the LKAS dropped-off even though the conditions were perfect.

- The drop-offs are not reproducible, so that a familiarization effect can develop little or not at all. Thus the expected availability of the system is not given. This also affects the driver's safety feeling and the desired relief is not noticeable. The driver is disappointed, stressed and feels unsafe.

In summary, while driving with LKAS the drivers do not only have the task of keeping the lane:

- They also must monitor the system via the HDD all the times.
- They also must be ready to intervene at any time due to the sudden drop-offs.
- They also must adjust to the fact that transparency with regard to the system status is not given always and all the time.

2.1   Results Overview of the Pilot Study

In the preliminary customer study with $N = 50$ subjects, the customer wishes/acceptance of the LKAS were/is tested, evaluated and compared on public roads. In addition, three latest premium vehicles from various manufacturers were used to derive optimization potential for current and future automated driving functions. About 50% of the subjects were already familiar with the LKAS and the reference vehicle (vehicle 2). To avoid the inaccuracy caused by tiredness, the subjects were randomly divided into two groups, each comparing two of the three vehicles. In total, 100 test drives over 4000 km were conducted. The average age of the people was 35 years. One third of the participants (30%) were women [1].

The three vehicles were equipped as follows:

- Vehicle 1 with LKAS (center guidance) in combination with HDD
- Vehicle 2 with LKAS (center guidance) in combination with HDD
- Vehicle 3 with LKAS (center guidance) in combination with HUD

The results show that for all LKAS criteria, such as "human-machine-interface" (HMI), "lane tracking quality", "edge guidance", "vehicle reaction", "driver-vehicle-interaction", "availability", "safety feeling" and "degree of relief" [3], each manufacturer has a "GAP" of importance to the degree of fulfillment. However, it is also shown that the LKAS of one vehicle performs better than the competitor vehicles in seven of the eight criteria. In summary, it can be seen that the importance of the criteria is not fulfilled in the implementation. On the basis on one example result of the Market Opportunity Map (MOM) the main findings of the study should be illustrated. This map shows serious deficits of individual manufacturers in a different way. The MOM is an illustrative opportunity for customer management to gain an overview of the areas in which action is needed to optimize the performance. It is based on the customer satisfaction portfolio and the strengths and weaknesses portfolio [11]. It represents the mean values for each LKAS criterion for each test vehicle.

It becomes clear that the better the criterion HMI (circled with dotted stitches) is evaluated, the better the criterion degree of relief (circled with continuous stitches) is evaluated. The statement that the LKAS of vehicle 3 performed best in this benchmark study was confirmed by other questions and is illustrated by the following figure.

It should be noted that vehicle 3 was additionally equipped with a HUD compared to the other two test vehicles. This led to the assumption whether the HUD was the reason for a better assessment of the LKAS.

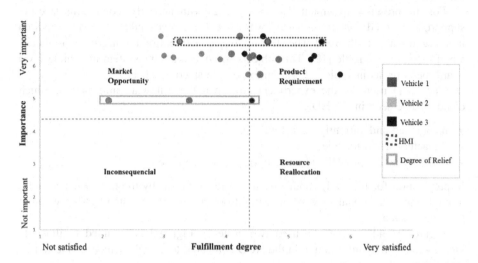

Fig. 1. Market Opportunity Map (MOM) for the LKAS criteria of all tested three vehicles of pilot study

3 Experimental Design

To verify our hypothesis, the customer experience and evaluation of the LKAS with and without HUD were tested on public roads over more than 5900 km. Two current premium vehicles from different manufacturers (equipped with LKAS and HUD) were used to examine the influence of the HUD on a test drive with LKAS as benchmark.

3.1 Participants

In total, N = 48 subjects (M_Age = 37.67; SD = 15.69, Range: 19–70), 11 females (23%) and 37 males (77%), participated in the test drives. The employment status of the sample was composed as follows: 31% of the subjects were college students, 30% employees, 15% professors. Participants were not paid for completing the experiment. They were recruited through a call for tenders at the University of Applied Sciences in Kempten. The remaining 24% of the participants were three pensioners, one journalist, one engineer, an office clerk and one person was engaged in purchasing. 41 test persons (85%) had an annual mileage more than 6000 km.

38% of the test persons had already participated in at least one study with the LKAS of MdynamiX and the University of Applied Sciences Kempten. All subjects had normal visual acuity or visual acuity correction.

3.2   Vehicles and Equipment

In order to test the two ADAS in the benchmark, two premium vehicles from different manufacturer were used. Both vehicles are equipped with HUD that can be switched on and off via board menu while driving with LKAS.

For the present experiment, LKAS (type 2) with an early, continuous steering support, a so called "center guidance", was used. The center guidance is also supported in the central area of the road. The steering torque intervention is comparable to that of a half pipe or a V-profile [12]. The subjects could activate the system in vehicle 1 via board menu or, as in vehicle 2, via button on the steering wheel.

The HUD used for the experiment contained the following information, which could also be seen in the HDD:

- recognized and currently valid traffic sign
- the actual speed travelled
- currently active ADAS and its state (in this case LKAS)

In preparation for the study, both systems were examined by the test manager and a more detailed functional survey was carried out, e.g. whether a step-by-step warning had been issued.

Figures 2 and 3 show the used two systems. Figure 3 is intended to illustrate visually the display information in the HUD compared to a HDD, which was tested by participants while driving.

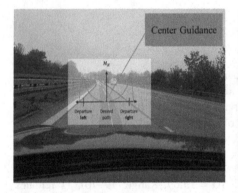 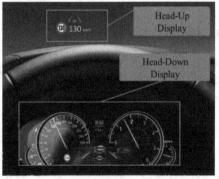

Fig. 2. Lane keeping assistant system with early support (center guidance) [3]   **Fig. 3.** Head-up display and head-down display (picture bmw.de)

3.3   Test Route

The test route should reflect a "real and customer-oriented car journey" and enable the test persons to recognize the potentials and advantages of the two assistance systems as well as weak points and disadvantages. This enables an overall evaluation of the display concept. It had a length of 61.1 km (per vehicle) and the average driving time was 45 min (including an acclimatization period). It contained 16 cornering events and consisted of two sections (federal highway B12 and B19) that were 90% identical.

Only the length of the test drive on B19 with LKAS combined with HUD was extended by one exit. This should give the subject a sufficient time slot to test the two assistance systems. Each test person drove the two vehicles immediately one after the other on the defined test route, with defined maneuvers and a speed of 100 km/h–130 km/h during their test participation.

3.4   Methods

In order to place the human being at the center of the development, methods such as Quality Function Deployment (QFD) [11] and KANO [13] as well as the Technology Acceptance Model (TAM) [14] and Rodgers Diffusion of Innovation Model were applied.

In addition product specific criteria were developed. Therefore, the LKAS evaluation matrix, so called "Level Model", was extended by HUD evaluation criteria for a structured HUD assessment on customer and expert level. Main customer criteria of the LKAS-Level Model are "human-machine-interface" (HMI), "lane tracking quality", "edge guidance", "vehicle reaction", "driver-vehicle-interaction", "availability", "safety feeling" and "degree of relief" [3]. The main criterion Human-Machine-Interface (HMI) has been replaced by the extended user-oriented HUD Level Model, which has been developed under consideration of different situations the driver is confronted with when driving with LKAS. For each evaluation criterion, the influences of the driver, the vehicle and the environment must be taken into account.

Following use cases were identified during a test ride with LKAS, which the driver experienced during the journey (Table 1).

Table 1. Method of the use case-oriented development (pictures bmw.de)

		Use case 1	Use case 2	Use case 3	Use case 4	Use case 5
Use cases						
What is happening?		LKAS off: Driver steers	LKAS on: No line detected; Driver steers	LKAS on: One line detected; Driver steers	LKAS on: Two lines detected; Vehicle steers	LKAS on: Hands-off time reached; Vehicle steers Intervention necessary
Driver Vehicle Environment		What must appear in the display?	What must appear in the display?	What must appear in the display?	What must appear in the display?	What must appear in the display?
		What kind of warning must be given?	What kind of warning must be given?	What kind of warning must be given?	What kind of warning must be given?	What kind of warning must be given?
		What must be taken into account during monitoring?	What must be taken into account during monitoring?	What must be taken into account during monitoring?	What must be taken into account during monitoring?	What must be taken into account during monitoring?

Furthermore, a detailed description of the criteria "operation", "display", "design", "monitoring" and "warning" has been developed, that have been included in the Level

Model (Fig. 4) with sub criterions. This matrix enables a holistic evaluation of the HUD in combination with an LKAS. To illustrate the methodology, the following figure shows a section example of the Level Model with regard to the criterion "warning".

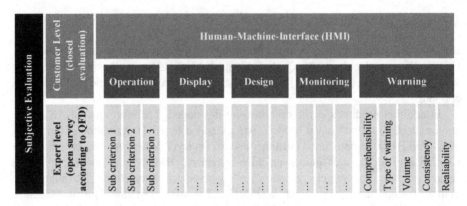

Fig. 4. Level Model to evaluate head-up display or head-down display

3.5   Procedure

The philosophy of the study was based on gaining knowledge in driving situations that were as real as possible. Such studies are not yet known for the LKAS. In most cases, comparable studies are carried out only in driving simulators. The environmental situations and vehicle reactions as well as the risk potential and the resulting feeling of safety for subjects are sometimes perceived completely differently than in the real world test ride. The driving study should gain valuable insights in this respect. The higher implementation effort, the increased risk as well as the comparability of the results represent a special challenge for real test rides. Traffic and weather conditions, driving maneuvers, special events, number of driver interventions and the resulting LKAS system behavior are more difficult to control and plan. This randomness, however, offers valuable insights. In order to achieve comparability even of real test rides, the route selection was fixed, a uniform time window under consideration of the traffic conditions, driving instructions such as speed limit, lane selection, lane excitations and the framework for comparable weather conditions were defined. All of these aspects were summarized in a so-called road book (guideline) based on test routes. All instructors and interviewers were trained in this context.

Since the LKAS system is essentially stimulated by the road (road markings, topology, etc.) and the vehicle speed, the system behavior and the events of system drops were reproducible in this respect. Due to the clear position reference, the events for the selected route sections could be planned very well. For all sections, highly accurate Ground Truth maps were also available.

The adaptation of a product does not happen instinctively, but it rather is a process that lasts over a period and contains certain actions. People only speak of an innovation when they themselves acquire knowledge and know-how about a product. Everett Rodgers' Diffusion of Innovation Model divides the adaptation process into five phases, "Knowledge", "Persuasion", "Decision", "Implementation" and "Confirmation", which this study has taken into account [15].

The design of the study was based on the driver-vehicle-environment closed loop principle as well as on specific driving scenarios, respectively on driving maneuvers, which were exactly described in a guideline. At the beginning of the study, the test persons were welcomed, their driving license checked and a declaration of liability issued. An explanation of the procedure was given before the test persons received the first part of the questionnaire. As an introduction, two video clips was shown which demonstrated and explains the HUD and LKAS. The following test drive was carried out with two premium vehicles in which both systems were installed. Additionally the test persons were instructed in the test vehicles. After the subjects have adjusted their driver's seat, the side mirrors, the rear-view mirror, width, height and inclination angle/rotation of HUD, the test drive began after a short familiarization period. The HUD standard setting, the time of intervention of LKAS (early) and the volume of the warning (middle) could not be changed in order to ensure equal weighting.

The test persons were divided into 2 groups, so that 24 test persons first started with LKAS in combination with HUD, the other 50% started with LKAS in combination with HDD. Each test person drove both vehicles immediately one after the other as part of his test participation, so that the defined test route was driven four times. To support the test persons, the interviewers explained various possibilities for testing the LKAS and the HUD while driving. Crossing the lane without indicating and the resulting warning was one maneuver. The interviewers guaranteed the proper use of both systems for the complete test drive.

The defined test drives were conducted randomized on the test route as follows:

- Lap 1: with LKAS in combination with HDD
- Lap 2: with LKAS in combination with HUD
- and vice versa

3.6   Questionnaire and Interview

For the type of survey, the personal interview (called paper and pencil interview or face-to-face interview [11]) beside several fix question sheets before and after the test ride was chosen. With fixed questions before the trip it was essential to learn a lot about the test persons and their attitude to the technology. During the ride, the main aim was to reproduce the test persons' driving experience and to collect requirements and wishes based on interview, and after the ride to compare the systems and evaluate the overall impressions based on a fixed evaluation sheet of the LKAS level model [3]. The interview method while real test ride was chosen in order to obtain as much information as possible from the real experiment. Sudden events, which were predictable for the instructor due to the route stimulation and driving instructions, generated very different reactions and vocabulary of the test persons. These were processed by questions and re-questions in the following rout sections so that the customer's needs and suggestions can be recorded better and can be compared. The conversation can be directed in certain directions and possible misunderstandings are dealt with directly. A refusal or omission of certain questions is excluded and ambiguities in content can be clarified [2]. The questionnaire/interview was designed in such a way that the customer has the possibility to formulate his thoughts and ideas in detail according to the QFD principle through many open questions [11]. No fixed psychological questionnaires were used, because the study was conducted on real routes and the questions were triggered by external influences. It was important to observe the reactions of the driver, their gestures and facial expressions just in time as an interviewer and to formulate the questions specifically. Existing user experience questionnaire sheets are too general and hardly applicable for LKAS during a real test rides, because of their special function and the clarity of the terms for the subjects. Furthermore, the procedure should provide insights into how a user experience questionnaire could be designed in the future in the context of the ADAS/AD driving functions.

The questionnaire/interview for the study was divided into three parts (before, during, after test ride) in order to obtain a comprehensive picture of the usability of HUD and LKAS and to capture its user experience (UX). Because UX is not only limited to the time of use, but also the time before and after the use of the product [10]. UX thus focuses on the holistic view of subjectively experienced product quality [15].

The evaluation criteria of the LKAS (lane tracking quality, edge guidance, vehicle reaction, driver-vehicle interaction, availability, safety feeling, degree of relief, which have been extended with HUD criteria (operation, display, design, monitoring, warning), form the basis of the interview. In the expert level of the level model, an interview was conducted with the test persons according to the survey method of the QFD procedure. Like that, all customer wishes and requirements could be recorded directly. All interviews - both spoken and written - were conducted in native German.

The following research questions should be answered within the scope of the study:

Research question: How does the HUD influence the driving experience and the rating of the different evaluation criteria?

- Will LKAS be better experienced and evaluated by customers due to HUD usage?
- What will the human-machine interaction of the subjects look like when driving with and without HUD?

4   Results

53% of subjects said they had a mediocre (4) to no (1) level of knowledge. A Likert scale from 1 (no knowledge) to 7 (excellent knowledge) was used.

43 participants (90%) of the benchmark want to be supported by a HUD in any situation. Considering the sense and purpose of this technical supplement, not to distract the driver from the actual driving situation, this result is pleasing. Very few drivers will switch off the HUD when entering city traffic in order to use the head-down display again. Thus, the HUD can be used as complementary support to many other ADAS such as Adaptive Cruise Control (ACC) or LKAS. For this reason, the development of the HUD requires a holistic use case-oriented approach together with other

assistance systems. A completely separate development of the system would not be target-oriented and would not be tailored to the needs of the customers.

The subjects expect the main benefit of a HUD to be that all important information is displayed on the street (42%) and thus the view remains focused on the street (42%). 24% also hope for more safety and 17% less distraction by no longer having to rely on the display behind the steering wheel. In order to be able to assess the importance of different ADAS from the test persons' point of view, ten assistance systems are to be evaluated by awarding points. For this purpose, a total of 100 points are available, which are allocated proportionally in steps of ten. With an average of 15.9 points, Adaptive Cruise Control is the ADAS with the highest importance. The Emergency Brake Assistant (15.2 points) and the Blind Spot Assistant (15 points) occupy the places behind them. The Head-up Display with 11.7 points and the Lane Keeping Assistant with 10.4 points are less important, but are still in the upper midfield. Fatigue monitoring (5.9 points) and the night vision system (4.4 points) are not very important for the test persons and are in the last two places.

In order to assess a warning concept that makes sense from the customer's point of view, the test persons are asked how they would like to be warned if the LKAS is unavailable (LKAS switches suddenly off). 43% of the participants stated that they had missed a warning at least once when driving without HUD (in vehicle 2) and 37% with HUD (in vehicle 2), which did not occur despite the system being dropped.

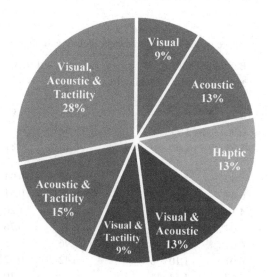

Fig. 5. Preferred warning concept for unavailability of lane keeping assistant system

As shown in Fig. 5 the triple warning is most frequently selected (visually & acoustically & tactile) with 28%. In 69% of the answers, at least the acoustic warning is desired, in 65% the haptic and in 59% the visual. This result is very interesting in view of the fact that people perceive 83% through sight, 11% through hearing and only 1.5% through touch [16] even if the haptic channel is the fastest to perceive [17]. The reason for the warning over three sensory channels could be the dropping of LKAS during the journey without any warning. Because the more senses are addressed, the greater is the effect of the recording [18]. 43% of the participants stated that they had missed a warning at least once when driving without HUD (in vehicle 2) and 37% with HUD (in vehicle 2), which did not occur despite the system dropped. The fear of being distracted and therefore not perceiving the warning visually probably explains this result. Nevertheless, it is important to reduce the amount of information to a minimum in order not to overload the short-term memory.

The questioning after QFD provides for example for the criterion "warning" following results:

According to the test persons, an intensive and clear warning must be given before the system switches off. A warning at the same time as LKAS has dropped is unacceptable. A written warning that the system has shut down is not effective, as the time for the driver to comprehend is too long. In addition, open windows or motorway journeys, for example, must not drown out the warning. Important for the participants is the triple warning in visual & acoustic & tactile form, which is standardized.

The questioning of the importance of the criteria for LKAS and HUD was carried out before and after the test drive on the given Likert scale (1 = not important to 7 = very important). Thus, a feeling for a sensitization of the test person during the test drive could be detected. The following Table 2 shows the mean values and the standard deviations of the importance before and after the test drive.

Table 2. Importance of the evaluation criteria before and after the test ride

	Before test ride			After test ride		
	Mean value	SD	Ranking by rating	Mean value	SD	Ranking by rating
Lane tracking quality	6.31	0.97	3	6.58	0.92	4
Edge guidance	6.00	1.21	8	6.62	0.75	3
Driver-vehicle interaction	5.58	1.42	10	6.40	0.81	9
Availability	5.80	1.39	9	6.47	0.94	6
Safety feeling	6.49	1.10	2	6.87	0.51	1
Degree of relief	4.89	1.68	12	5.56	1.20	12
Vehicle reaction	6.07	0.96	5	6.47	0.76	6
Operation	6.07	1.12	5	5.96	0.95	10
Display	6.07	1.10	5	6.42	0.62	8
Design	5.07	1.54	11	5.73	1.10	11
Monitoring	6.29	0.92	4	6.58	0.54	4
Warning	6.69	0.63	1	6.82	0.44	2

The results show that the "feeling of safety" and the "warning" before and after the test drive are in the top two places. The "design" and the "degree of relief" prove the least important criteria. "Driver-vehicle interaction" experienced the largest increase with 13% after the test drive. The subjects have become much more interested in communicating the functional intention of the system. The interaction between ADAS and the driver is not only of great importance with regard to acceptance and trust, but also in order to avoid possible operating errors, regular, comprehensible feedback is of immense importance.

4.1   Results of the Evaluation for Vehicle 1

The rating of the LKAS in combination with HUD or with HDD was rated on the given Likert Scale from 1 = "very bad" to 7 = "very good".

A positive influence of the HUD is recognizable for all evaluation criteria except the edge guidance and lane tracking quality. A significant positive influence of the HUD in the sense of a better evaluation of the criterion with versus without HUD was proven with a t-test for dependent samples. The following criteria were evaluated significantly better: "Monitoring effort" (+31%; $p < .001$), "warning" (+17%; $p < .001$), "display" (+17%; $p < .001$), "feeling of safety" (+12%; $p < .001$), "driver-vehicle interaction" (+11%; $p < .001$), "degree of relief" (+10%; $p < .001$) and "design" (+8%; $p = .007$). Figure 6 shows the evaluation for vehicle 1 with both conditions (with HUD/without HUD = with HDD).

With over 31% increase, the point of monitoring experiences the largest increase by switching on the HUD. With 5.86 points, it achieves the highest rating of all criteria. This makes it clear how important the monitoring of the system is for the subject.

The other criteria, which are increasing by a double-digit percentage, are also related to monitoring. The HUD makes the visual warning that the LKAS has switched

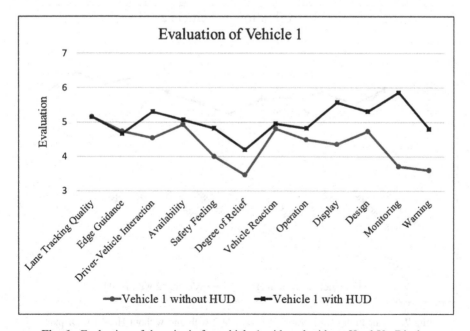

Fig. 6. Evaluation of the criteria for vehicle 1 with and without Head-Up Display

off much easier to perceive. The degree of relief and the driver-vehicle interaction also increase significantly, which is accompanied by a better feeling of safety. (Focusing of the eye on HDD not required can perceive everything more quickly).

The display of LKAS, which is directly in the test person's field of vision when the HUD is switched on and is no longer significantly more inconspicuous in the HDD, can achieve the second highest increase of 17% together with the warning.

4.2 Results of the Evaluation for Vehicle 2

The following section deals with the evaluations for vehicle 2.

A positive influence of the HUD is recognizable for all evaluation criteria except the lane tracking quality and operation. A significant positive influence of the HUD in the sense of a better evaluation of the criterion with HUD versus without was proven with a t-test for dependent samples. The following criteria were evaluated significantly better: "Monitoring effort" ($+22\%$; $p < .001$), "warning" ($+8\%$; $p < .001$), "display" ($+16\%$; $p < .001$), "feeling of safety" ($+10\%$; $p < .001$), "driver-vehicle interaction" ($+12\%$; $p < .001$), "degree of relief" ($+11\%$; $p < .001$) and "design" ($+4\%$; $p = .017$).

The evaluation of the operation without HUD refers to the operation of the LKAS and with HUD to the operation or setting of the HUD (Fig. 7).

Significantly more negative evaluation has been found for the criterion "operation" (-13%; $p < .001$). This is because switching LKAS on/off with a button directly on the steering wheel was easier. The HUD, on the other hand, was operated via the board menu.

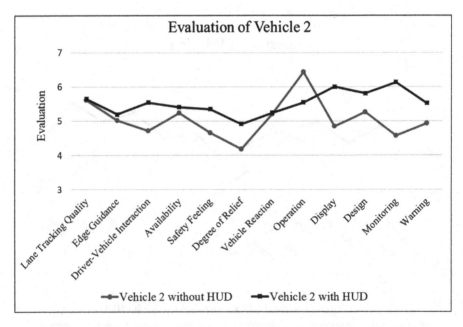

Fig. 7. Evaluation of the criteria for vehicle 2 with and without Head-Up Display

4.3   Results for Vehicle 1 and Vehicle 2

To verify the assumption that the order of the test conditions (1st round with HUD/2nd round with HDD or 1st round with HDD/2nd round with HUD) has an influence on the evaluation of the test persons, the evaluations of the 48 test persons were considered separately. In the figures below, the mean values were calculated for all criteria except lane tracking quality, edge guidance, availability and vehicle reaction. With these LKAS criteria, the difference in evaluation proves to be small (see Sects. 4.1 and 4.2).

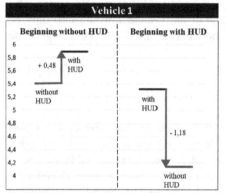

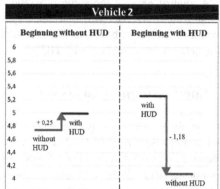

Fig. 8. Comparison of evaluation according to initial condition for vehicle 1

Fig. 9. Comparison of evaluation according to initial condition for vehicle 2

For vehicle 1 as well as for vehicle 2 the difference of the evaluation becomes visible especially when starting with HUD. With over 1.18 points, both vehicles are rated worse. At the beginning without HUD (with HDD), vehicle 1 scores on average 0.48 points better, vehicle 2 0.25 points better. There are also differences in the entry-level scores (Figs. 8 and 9).

Switching off the HUD takes away the easier monitoring of the system from the test persons, which is accompanied by worse evaluation of the criteria. The HDD is located - depending on the vehicle type - at a distance of approx. 0.5 to 0.8 m, the projected image of the HUD at a distance of approx. 2.20 m. A rapid change of view between the HDD and the road thus has a maximum possible accommodation time (focusing the eye on an object), as this means a change between infinite and near focus. The human eye needs for this process up to 0.5 s. [18]. This represents an increased potential of danger, as the driver is distracted from the actual driving task. Added to this is the strain caused by the constant adaptation of the eyes between road traffic (distance) and HDD (proximity). During the test drives, it was also possible to record and observe how difficult it was for the test persons to continue the test drive without HUD.

After Davis TAM a technology to be used by users, must be perceived by them as useful and simple to use [14]. The subjects were therefore asked whether they would recommend the systems to their friends, family or acquaintances because they fulfilled the TAM criteria. Table 3 shows the results.

Table 3. Recommendation of ADAS for vehicle 1 and vehicle 2 in [%]

	LKAS in combination with HDD	LKAS in combination with HUD
Number of recommendations in vehicle 1 [%]	12%	88%
Number of recommendations in vehicle 2 [%]	17%	83%

A total of 88% of the subjects would recommend LKAS of vehicle 1 in combination with HUD. The two assistance systems of the vehicle 2 would also only be recommended in combination. These results show once again that LKAS can generate its benefits in combination with HUD.

5   Summary and Discussion

As a result, a clearly positive influence of the HUD on the driving experience, comfort feeling and assessment of the LKAS can be determined across almost all evaluation criteria. In particular, a positive influence of the HUD could be observed in the LKAS monitoring effort. With an improvement of the rating by 31% (vehicle 1) and 22% (vehicle 2), the criterion "monitoring" had the largest impact. Simple and intuitive monitoring is very important, especially for ADAS that switches off regularly due to the condition of the road, weather influences or system boundaries that a non-expert cannot understand. It was discovered that a HUD can compensate the weaknesses of LKAS. On average, the evaluation criteria of vehicle 1 are almost 10% better and those of vehicle 2 with 7% better the HUD being switched on. In particular, the HUD's highly rated criteria of "feeling of safety" (6.87 out of 7 points after driving) and "monitoring" (6.58 out of 7 points after driving) in the importance survey increased the degree of fulfillment. With Rogers' "Diffusion of Innovations", the participants of the benchmarking study were made aware of the potential of a HUD.

The study results make it clear that a better user experience is possible by using both assistance systems. Meanwhile, 52% of the subjects were the opinion that LKAS and HUD should only be sold in combination. On average, the test persons stated that they had 35% more system transparency when driving with LKAS in combination with HUD than when driving with LKAS in combination with HDD. The resulting transparency and feedback made it easier for subjects to recognize the status of LKAS. Thereby the driver can focus more intensively to the environmental and traffic scenarios. Other studies have also revealed the importance of system transparency during a journey with LKAS [19]. Furthermore, such positive evaluation and user experience would lead to an increasing use of LKAS to reach initial goal of ADAS for more safety and comfort.

The following illustration shows the most important and most frequently mentioned product characteristics during the pilot study with regard to the LKAS. These were structured again according to Maslow's pyramid of needs (Fig. 10).

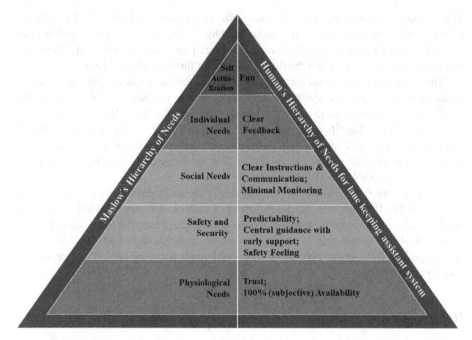

Fig. 10. Pyramid of needs for lane keeping assistant system [1] according to Maslow [20]

Eight out of 10 of the test persons' requirements above could be met with a HUD. The subject wants a predictability of the system, so that they can feel safe. Clear instructions and continuous communication whether the system is actually active or not are necessary. They need minimal monitoring so that he can concentrate on the driving situation and can immediately perceive and respond in case of failure. Subjects want to be able to trust the system, which only works if all requirements are met. Trust can only be built if the system behaves in the same way, the driver has transparency and receives feedback at all time. Without confidence in the system, no acceptance and driving pleasure can arise [1]. Arndt (2011) also points to the importance of the emotional experience for the acceptance of the systems [21].

In summary, it can be concluded from the available results that a HUD is a meaningful addition in the area of ADAS/AD. However, an ergonomically designed display concept that takes into account the functionality and capacity of human information processing is essential and may help to compensate for limitations of the technology to a certain extent in the future. In spite of increasing levels of automation in the vehicle, people must continue to be able to understand the processes and action steps of the vehicle at all times. System understanding, transparency, trust and the feeling of "indirect" control are essential for acceptance, well-being and the associated positive user experience.

The "Ironies of Automation" according to Bainbridge (1983) must also be considered here. The more automated a system is and the smoother it works, the less often humans have to intervene themselves. However, it is precisely this loss of practice and potential loss of attention that makes it all the more difficult for them to do so [22]. In addition, drivers can distract themselves if the load caused by the driving task is low. Accidents can therefore occur more frequently if the demands placed on drivers are low [19]. Therefore, it is recommended to maintain the driver's attention through adapted HCI.

The study is based only on subjectively collected data and was carried out on public roads in order to achieve the most realistic customer journey possible. During the interview, it was possible to collect essential requirements for a HUD when driving with LKAS. For objectification and higher reproducibility of the results, a driving simulator study in combination with gaze recording and recording of physiological data is recommended. The study clearly shows that the subjective perception of ADAS plays an indispensable role for humans.

In order to record the subjective experience of driver in dealing with ADAS and to incorporate the findings into the development of ADAS, use case-oriented acceptance studies with potential end user are of central importance. Here, the focus should be on the needs of the human and a holistic view of the systems should also be carried out with regard to AD. At the end of the day, the human being alone will decide on the success of the systems.

References

1. Aydogdu, S., Schick, B., Wolf, M.: Claim and reality? Lane keeping assistant - the conflict between expectation and customer experience. In: Fahrzeug- und Motorentechnik, Aachen (2018)
2. Seidler, C., Schick, B.: Stress and workload when using the lane keeping assistant: driving experience with advanced driver assistance systems. In: Fahrzeug- und Motorentechnik, Aachen (2018)
3. Schick, B., Seidler, C., Aydogdu, S., Kuo, Y.-J.: Fahrerlebnis versus mentaler Stress bei der assistierten Querführung ATZ, vol. 121, no. 2, pp. 74–79 (2019)
4. Turner, P.: A Psychology of User Experience. Springer, Cham (2017). https://doi.org/10.1007/978-3-319-70653-5
5. Aydogdu, S., Schick, B., Wolf, M.: Current lane keeping assistance systems in benchmarking-accepted or rejected by customers? In: VDI/VW Gemeinschaftstagung Fahrerassistenz und automatisiertes Fahren, vol. 34, pp. 33–45
6. Altendorf, E.: Assistenz versus Kontrolle beim hochautomatisierten Fahren - eine Akzeptanzanalyse (2016)
7. Thaller, G.E.: Interface Design: die Mensch-Maschine-Schnittstelle gestalten; Konzepte für Programm- und Web-Oberflächen
8. Reuter, C.: Sicherheitskritische Mensch-Computer-Interaktion: Interaktive Technologien und Soziale Medien im Krisen-und Sicherheitsmanagement. Springer Vieweg, Wiesbaden. https://doi.org/10.1007/978-3-658-19523-6
9. Gold, C., Körber, M., Hohenberger, C., Lechner, D., Bengler, K.: Trust in automation – before and after the experience of take-over scenarios in a highly automated vehicle. Procedia Manufact. 3, 3025–3032 (2015)
10. Robier, J.: Das einfache und emotionale Kauferlebnis: Mit Usability, User Experience und Customer Experience anspruchsvolle Kunden gewinnen. Springer Fachmedien Wiesbaden, Wiesbaden (2015). https://doi.org/10.1007/978-3-658-10130-5

11. Saatweber, J.: Kundenorientierung durch Quality Function Deployment: Systematisches Entwickeln von Produkten und Dienstleistungen, 3rd ed. Symposion Publishing GmbH, Düsseldorf

12. Winner, H., Hakuli, S., Lotz, F., Singer, C. (eds.): Handbuch Fahrerassistenzsysteme: Grundlagen, Komponenten und Systeme für aktive Sicherheit und Komfort, 3rd edn. Springer Vieweg, Wiesbaden (2015). https://doi.org/10.1007/978-3-658-05734-3

13. Hölzing, J.A.: Die Kano-Theorie der Kundenzufriedenheitsmessung: Eine theoretische und empirische Überprüfung, 1st edn. Gabler, Wiesbaden (2007)

14. Davis, F.D.: A technology acceptance model for empirically testing new end-unser information systems: Theory and results. Dissertation, Massachusetts Institute of Technology (1985)

15. Messner, T.: Von Usability zu User Experience: Auswirkungen auf die Praxis, Fachhochschule Nordwestschweiz (2016)

16. Braem, H.: Die Macht der Farben, 7th edn. Wirtschaftsverl: Langén Müller/Herbig, München (2004)

17. Scott, J.J., Gray, R.: A comparison of tactile, visual, and auditory warnings for rear-end collision prevention in simulated driving. Hum. Factors **50**, 264–275 (2008)

18. Biesemeister, B.B., (Ed.): Die Neuro-Perspektive: Neurowissenschaftliche Antworten auf die wichtigsten Marketingfragen, 1st ed. Haufe-Lexware, Freiburg im Breisgau (2016)

19. Höfer, M.: Fahrerzustandsadaptive Assistenzfunktionen, Universität Stuttgart (2015)

20. Maslow, A.H.: Motivation and Personality. Harpers, Oxford (1954)

21. Arndt, S.: Evaluierung der Akzeptanz von Fahrerassistenzsystemen: Modell zum Kaufverhalten von Endkunden. Zugl.: Dresden, Technical University, Dissertion 2010, 1st ed. Wiesbaden: VS Verl. für Sozialwiss (2011)

22. Bainbridge, L.: Ironies of Automation, pp. 775–779 (1983)

Do Drivers Prefer Female Voice for Guidance? An Interaction Design About Information Type and Speaker Gender for Autonomous Driving Car

Wang Ji[1](✉), Ruijie Liu[2], and SeungHee Lee[1]

[1] University of Tsukuba, Tsukuba 305-8577, Japan
wangjige@foxmail.com
[2] Shaanxi Normal University, Xi'an 710062, China

Abstract. This research is to get a better understanding of what kind of intention is needed to communicate between drives and autonomous car. What types of communication methods should be used to offer a better user experience to drivers and passengers. In this study, we examined the effects of information type ("why" message, "how" message) and speaker gender (female voice, male voice) voice messages that provide communication about the car operations in advance before the autonomous car's action. 30 participants took part in this experiment. Results showed that female voice was more trustworthy, acceptable and pleasure than the male voice, because participants were more familiar with female voice as it was widely used in smart device and navigation system. Also, the female voice is easier to hear than the male voice in the noisy condition. More than half of the participants changed their preference for the speaker gender, which reminds researchers to be cautious about the results of an online survey of drivers' attitude to speaker gender. How message is preferred by participants. It is considered to be more intelligent and interesting than why message as it described what automated activity it is undertaking directly, participants didn't need to take part in the decision process which may add their cognitive load. Additionally, how message uses the first person word like "I" "we" or "us" to communicate with the drivers increase participants' trust and satisfaction.

Keywords: Voice interaction design · Autonomous car · Speaker gender · Information type · Usability evaluation · User experience

1 Introduction

1.1 Background

The autonomous car has been predicted in the science fiction and discussed in popular science media for many years. Recently, with the development of new technology and the need for the new transportation revolution, autonomous cars have been an essential strategy of the motor corporations. Most of them and some internet companies have announced their plans to begin selling such cars in a few years. In the new century, it is

© Springer Nature Switzerland AG 2019
H. Krömker (Ed.): HCII 2019, LNCS 11596, pp. 208–224, 2019.
https://doi.org/10.1007/978-3-030-22666-4_15

widely believed that technology will continue to increase the efficiency and safety of transportation. One of the major breakthroughs will be the autonomous car. It is expected to alleviate or completely solve all these serious problems the traditional car faces. By automating the vehicles, human error can be significantly reduced since the intelligent system will never get drunk or be distracted, and it can be designed to execute often appropriate maneuvers that avert the crash entirely. Also, smaller headway and better driving behaviors can be achieved, which will bring out less traffic congestion. Autonomous car will also benefit specific groups of people who are not able to drive by themselves for any reason, they will enjoy the mobility facilitated by the new technology. In summary, it is probable that the autonomous car will likely have a profound impact on society.

1.2   Levels of Automation

Fully autonomous car and semi-autonomous car are the two main type of intelligent cars. Fully autonomous cars are completely controlled by their automated systems such as artificial intelligence, machine learning, GPS, deep learning and land changing technology. Semi-autonomous cars need drivers control together with some automatic functions such as automated parking, cruise control and so on. Multiple classification systems are published to understand better and direct the development of autonomous car technology. Recently, the NHTSA updated its standardization to aid clarity and consistency in the federal automated vehicles policy6 [1]. It adopts the SAE International (Society of Automotive Engineers) definitions for levels of automation. SAE definitions divide vehicles into levels based on "who does what and when" [2].

1.3   Manufactures

This market for autonomous vehicles is expected to be quite large in the coming decades, which attracts many companies invest a lot of money and human resources to start their research in this sunrise industry.

Automakers are the main driving engines behind the research and development of autonomous car technology in the major car-producing countries. These manufacturers are comprised of two main subgroups, traditional car manufacturers, and new technology innovators. Those tech giants play the role of innovators to develop new strategies and models for automation and single occupancy cars. But they still need to cooperate with the traditional automakers as they do not have the equipment and factory to produce automobiles. According to a report from the Brookings Institution, several key challenges will be faced as intelligent cars emerge in order to make all the great dreams come true [3]. These crucial challenges include technical challenges or societal action. Each of these matters poses problems for autonomous vehicles and their success in the market.

1.4   Public Acceptance

Whatever the future translation will be, the core part will never change—humans. No matter how autonomous the car is and what kinds of automaker the company is, all those new technology and products are developed for supporting a better life for humans, and the company needs to make a benefit from their customers. Automakers and researchers must consider the psychological dimensions of technological design. Even the greatest technology, such as vehicles that drive themselves, is of little benefit if consumers are unwilling to use it.

Ultimately, the public must feel comfortable with the autonomous car for this market to develop. As with any emerging technologies, it takes a long time for individuals to accept new models of travel. According to a survey from AAA [4], More than 75% American drivers feeling "afraid" to ride in a self-driving car. In new emerging market countries like China, people appear more open to vehicular experimentation. The view was echoed in a separate survey undertaken by the Roland Berger consulting firm. It found that "96% of Chinese would consider an autonomous vehicle for almost all everyday driving, compared with 58% of American and Germans" [5]. Because Chinese people do not have a long history to own a private car, their emotional relationship related to the car is not so positive as the Americans, Germans or Japanese who have their unique profound car culture, so they are more amenable to autonomous vehicles. Before drivers are ready to switch their control to an autonomous car, the automaker and new technology companies have to take this question into considering: how to design the car, especially the interaction system between a car and a driver, what can help to establish and increase drivers' trust in their autonomous car.

According to a public opinion survey in the US, many people prefer to use voice or sound to interact with their autonomous car. When asked about their preferences, 36.2% of Americans said they prefer to tell their self-driving vehicles their route or destination by voice command. More than 90% of Americans prefer to be notified by a combined warning mode that includes sound (sound + vibration + visual 59.4%, sound + visual 19.4%, sound + vibration 9.7%) or only sound warning (7.9%) when they need to take control of their partially self-driving vehicle [6]. The voice has a long history in Human-Computer Interaction, often as a supporting function in the user interface as a way to notify the user about their actions. In the traditional human-vehicle interactions, the main function of the voice message is to alert the driver to the actions of the car and the changes in the surrounding environment. The researcher found that professional drivers preferred auditory signals since it could provide a warning as well as arouse the drivers [7]. Thus, in a traditional in-vehicle warning system, auditory signals are often used to notify drivers about impending danger, operational feedback, and malfunction warnings. With existing high demands on a driver's visual attention, many on-board computer systems utilize speech technologies. While speech-based interfaces may seem to be a safer option than graphical user interfaces in cars, there remains a concern that the cognitive demands associated with richer communication will damage driving safety [8]. Natural language dialog may become the preferred mode to inform drivers of control issues, as well as changing road and environmental conditions, to help give an explanation for the decision that the vehicle makes as it is in a natural way people communicate with each other.

Speaker Gender. Speaker gender is the first impression a user gets about the products and company [9]. A combination of real world mishaps and controlled experimental studies has shown that several factors significantly affect driver responses to voice interfaces in cars, including perceived speaker gender, emotion and even age. The BMW5-series released in Germany included a voice-based navigation system, featuring a computer-generated voice with female characteristics [10]. Although these drivers were well-aware that the voice was computer-generated, they reacted with gender-stereotyped responses, ultimately rejecting the female voice and demanding a product recall.

From our life experience, we can find that most of the in-vehicle information systems, such as navigation systems are using a female voice. Also, most internet companies develop the voice assistant as a female, such as Apple Siri, Microsoft Cortana, Google and Amazon Alexa. Voice is also a brand and copyright of the company. Using voice can reinforce the company's brand identity. The study found that listeners can encode speakers specific voice attributes in memory along with information about the specific word [11]. Also, another study reported that if participants' familiarity with a talker's voice will lead to higher intelligibility of speech enunciated by that speaker, suggesting a memory link between listener sensitivity to speaker-specific voice and overall speech intelligibility [12]. So voice branding can be a carrier for conveying a memorable message to targeted consumers, taking advantage of the powerful memory sense of sound. Sound or voice used in the in-vehicle system can help the car companies to increase their brand awareness, distinguish themselves from their competitors and connect with their users on a deeper level. Taking Apple Siri as an example, Siri is available in all the ecosystems of Apple products. Users can talk with their Siri on their iPhone, iPad, Apple Watch or Macbook, they don't need to learn how to interact with their "voice assistant" in a new device as they are so familiar with the voice, just like talking with an old friend. Gender is one of the most psychologically powerful social categories.

The Intelligibility also affect people's impression of voice. Intelligibility of the female and male voice is equivalent in most normal conditions. Because of the small differences between female and male acoustic speech signals, the intelligibility of the voice one can be different in certain situations such as in the condition with high levels of noise. A study of aircraft cockpits suggested that the lessened interference between cockpit noise and female voices made female voices easier to be heard in the cockpit than male voices [13].

Little is known about the effects of the in-vehicle information system of the autonomous vehicle on the driver, as it is a sunrise industry. Autopilot is a new function for the automobile industry, but in the aviation industry, pilots have worked with autopilot from the 1930s. The autopilot in aircraft is a system used to control the trajectory, direction and altitude without a pilot handling the wheel. Autopilot assists the pilot in controlling the aircraft, allowing them to pay attention to broader aspects of the operation and can significantly reduce workload during critical phases of flights [14]. So the methods of studies on the cockpit information system are very valuable.

More recent researches, however, have indicated that the original popular hypothesis may be unreliable since more females have been employed as pilots and air traffic controllers. General human factors now indicate largely that, either due to current culture or changing attitudes, an automated female voice is no more or less effective than a male voice [15].

One important characteristic used to distinguish between gender is the pitch. The pitch is associated with the physical sensation of frequency. In humans, the auditory frequency is positively correlated with the perceived pitch. Higher voice pitches are usually associated with the female voice, and they also tend to elicit higher perceived urgency [16].

A study found that both acoustic and no acoustic differences between male and female speakers are negligible [17]. Therefore, they recommended, the choice of the speaker should depend on the overlap of noise and speech spectra. The female voice did, however, appear to have an advantage in that they could portray a greater range of urgencies because of their usually higher pitch and pitch range. They reported an experiment showing that knowledge about the gender of a speaker has no effect on judgments of perceived urgency, with acoustic variables accounting for such differences. A study from Defense Research and Development Canada in Toronto [18], found that with simulated cockpit background radio traffic, a male voice rather than a female voice, in a monotone or urgent annunciation style, resulted in the largest proportion of correct and fastest identification response times to verbal warnings, regardless of the gender of the listeners.

The study of social psychology has demonstrated that gender makes a difference on a lot of human-human interaction [19], this gender stereotype also appeared in human-machine interaction as computers are social actors. A computer-generated speech study showed that the male-voiced computer exerted greater influence on the user's decision than the female-voiced computer and was perceived to be more socially attractive and trustworthy [20]. More strikingly, gendered synthesized speech triggered social identification processes, such that female subjects conformed more to the female-voiced computer, while males conformed more to the male-voiced. Similar identification effects were found on social attractiveness and trustworthiness of the computer. As gender can influence behavior, it is important to choose a proper voice for the in-vehicle information system of the autonomous car.

Voice Content. One of the essential factors for the success of the autonomous car is providing feedback to the driver. Norman pointed out that the problem with automation is that its condition is inadequately communicated to the drivers [21], even this kind of communication can assist the drivers into the control loop. He noted that the inappropriate feedback had made to some serious incidents in the aviation domain. The results of some previous studies [22] also suggested that providing feedback to make the driver understand the car's operations and the surrounding situation is a key feature of autonomous driver assistance systems. Moreover, the context, integration and abstraction should be proper. Otherwise, the driver may not understand the feedback [23].

If we consider the driving task as teamwork, driver and the car are teammates who collaborate to ensure the driving activity will be safely and efficiently. How to get a better understanding of what kinds of voices need to be used and what types of messages should be sent. Hence, the study and design of an appropriate feedback model is a foundation for increasing the drivers' trust in the autonomous car.

A study from Koo [24] in the USA claimed that the nature of feedback is to tell drivers the outcome of the system's operation, and deliver the operation information after the event. However, in the condition of autonomous driving, the information about the operations and situations should be provided to drivers ahead of the event, he named this kind of information: "feed-forward" information. As drivers transmit their control power to the car, how drivers perceive and accept the autonomous function of the car become increasingly. For instance, when there is a red light ahead and the car is about to stop automatically. The automated system could provide the information about how the car is going to act or supply message regarding why the car is going to conduct that activity. That's the two main types of information the drivers are interested in. In Koo's study. They explored those two types of feed forward information: How message (what automated activity it is undertaking) and Why message (reason the car is acting that way) in the study to explore which types of information can allow the drivers to respond appropriately to the situation and gain trust the autonomous car is taking control for good reason. In their study, they tested three different design settings. This study found that the driver preferred receiving the message only with why information, which created the least anxiety and high trust. This study took place in the US and explored how the context of semi-autonomous affect driver's attitude. Is there any difference in the condition of a fully autonomous car as the driver does not need to operate? Whether there is a cross-cultural difference between the western and eastern drivers will be an important issue, as the previous study showed that East Asians are holistic thinkers [25]. Holistic thinking encourages people to have a worldview in which all kinds of events and phenomena are interrelated and perpetually changing. This implies that East Asians are more attentive to the behavior or flow of objects. In contrast, Americans are more analytic. They pay attention primarily to the object and the categories to which it belongs and using rules, including formal logic, to understand its behavior. So the East Asian drivers may have a different preference on the type of feedforward information. The current auto industry is a global business, offering the product suited to local users will bring great benefit to the automakers.

1.5   Present Study

An autonomous vehicle is a new model of future transport; few studies were done on the in-vehicle information system. For vehicles the most important part is safety, not only at the technology level but also giving the driver a feeling of safety and trust, this is also a very significant field that the researcher should pay attention to. The present study examines speaker gender as well as voice content; the former study only focused on voice content or speaker gender. Our study explores two main questions about the in-vehicle information system: (a) whether participants have a preference for speaker gender; (b) whether participants have a preference for voice content.

2   Method

2.1   Study Design

A 2 (male voice, female voice) × 2 (Why message, How message) factorial design was used.

2.2 Participants

30 participants (15 males and 15 females) aged between 21years and 35 years (mean = 24.6 years, standard deviation = 2.58) took part in this experiment. 24 participants are Chinese (12 males and 12 females), and the other 6 participants are Japanese (3 male and 3 females). All of the 30 participants have a good English ability, and they can understand all the English prompts used in the experiment. They all had a driver license for more than one year, and none had any obvious hearing abnormalities. Before the experiment, each participant was briefly introduced about the purpose of the experiment, and a short interview was conducted to get the information about their driving experience, personal preference about speaker gender of voice assistant (Including the voice assistant of their smart device and in-vehicle information system such as navigation system). Each participant was paid 500 Japanese yen (500 yen = $4.59) for the total test duration of around 35 min.

2.3 Apparatus

In this experiment, we use TTS (Text to Speech) technology to make the voice message. Text-to-Speech, abbreviated as TTS, is a technology that converts digital text into the spoken voice output. Text-to-Speech systems were first developed to aid the visually impaired. They are nowadays ubiquitous, having an extremely broad field of application ranging from voices giving directions on navigation devices to voices for public announcement systems and virtual assistants [26]. We use the TTS provided by Ivona (https://www.ivona.com/us/) and chose the British English Brian as the speaker of the male voice message and American English Kimberly as the female voice. Both of the voice with no particular inflections or moods.

Table 1 Voice prompts used in the experiment. The same line is the same car action. For example, when there is a red light in front of the car and the car is going to brake to stop. In why message experiment setting, the voice content is "Red light ahead", in how message experiment setting, the voice content is "Car is braking."

Table 1. Voice prompts used in the experiment.

Why message	How message
"Autopilot start"	"I will take the control"
"Red light ahead"	"Car is braking"
"Road clear"	"Speed up"
"Vehicle in front is too slow"	"We will pass the vehicle in front of us"
"Intersection ahead"	"Car is barking"
"Roadworks ahead"	"Turning around"
"Autopilot ending"	"Car is stopping"

The video used in the experiment is from an autopilot demonstration of a Tesla Model SP85D. The 4 min 6 s video includes highway, urban, and suburban road situations with stop signs, road work sign, and traffic signals. The car in the video conducts the following operations: start the autopilot model, brake, speed up, overtake, cross intersection, steering, turn around, and ending the autopilot model. In the experimental conditions, whenever the car changed its operation, the voice message was generated. All the voice messages were sent $1 \sim 2$ s before the car conducts the operation. Such a voice message ahead of the operation allows the participants to understand the current situation. To minimize effects caused by the lack of sound feedback that exists in the real driving situation, we added the background sound of driving in the video, an engine sound was simultaneously provided as a cue of an accelerating action every time the car was speeding up or starting. All the voice prompts used in the video are in English as participants are from different countries, but all of them have a good English ability.

The experiment videos was run on lab computer—A Macbook Air 13.3 2015 was used to play the videos of all experiment setting. The videos were projected using an ASUS DLP S1 projector, the biggest brightness is 200 lm. Speaker in the projector was used to output the audio for all videos.

2.4 Questionnaire

Attitudinal measures were based on self-reported data on adjective items in the post-drive questionnaire. The questionnaire was adapted from a published model from the CHIMe Lab at Stanford University that is used to measure driver experience [27, 28]. Participants were asked to rank each item on a nine-point Likert scale ranging from "Describes very poorly (= 1)" to "Describes very well (= 9)" The questionnaire combined an English and Japanese version on one paper. Four same questionnaire papers were given to the participants, and required them to finish one questionnaire just after they watched the video, the time to finish was no limited.

2.5 Procedure

This experiment was conducted in a small quiet room without light and all the participants were tested individually. After they had came to the room, they received a packet consisting of an approved human subject consent form, four questionnaires and a pencil. The experimenter asked them to read the consent form and sign the form if they agree to participate in the experiment (See Appendix A and B). Then, the experimenter gave a brief introduction of the whole experiment. An interview was conducted with the participants to know their preferences regarding the speaker gender of their smart device voice assistant and in-vehicle navigation system. Also they were asked to explain the reason of their preferences. Three questions were be asked as follows:

Q1 Do you have a driver license? How many years have you driven a car?

Q2 In the future, if the self-driving car is available what kind of voice system do you hope it to be, in terns of kind of information or voice type? And why would you like this?

Q3 As we know most navigation system sin the car now are female, which do you prefer in the future self-driving car? Could you tell me the reason why you like......

If they preferred a female voice, the order was Female-Why—Female-How—Male-Why—Male-How, otherwise the order was Male-Why—Male-How—Female-Why—Female-How. For the convenience of data collected, we use "WF" for Why message Female voice, "HF" for How message Female voice, "WM" for Why message Male voice and "HM" for How message Male voice. The details about participants' order and preference are shown in Appendix D.

Participants were asked to sit in the middle of the room, about 2 m away from the screen, they can take the seat and adjust their gesture. The videos were projected onto a wall directly in front of the participants. The size of projected area was about 1.5 m diagonal; the participants had about a 60-degree field of view. The brightness of the display is 80% of the biggest brightness, about 160 lm. The intensity levels of test voice ranged between 55–75 dB measured by sound level meter depending on the comfortable threshold of participants. When they were ready to begin the experiment, they should motion to the experimenter by hand. Then experimenter turns off the light in the room and played the video and projected the frame on the screen. After watching the video, participants filled out the questionnaire that assessed their overall experience. They filled in the first questionnaire consisting of general information such as gender and age in addition to nationality. The same as the second, third and fourth trial.

After all the four trials had been finished, a short post-experiment interview was conducted to explore whether participants have a new attitude towards the speaker gender, voice content and explain the reason they change their preference. Three questions were asked:

Q1 After you watched the four videos, you still prefer your choice at the beginning or did anything change?
Q2 About the information types which do you prefer? Why?
Q3 Is there any place you think the voice system can be improved?

Afterwards, the experimenter thanked participants for their participant and gave 500 Japanese yen (500 yen = $4.59) as the gift. All the participants were debriefed at the end of the experiment.

This experiment had received the permission from the research ethics review committee of School of Art and Design, University of Tsukuba, research ethics number 150422038.

3   Results

3.1   Descriptive Statistics

Figures 1, 2, 3 show the distribution of 30 participants' preference for speaker gender and voice content. Interesting data is that 57% participants changed their speaker gender preference after the experiment. For the female participants, 2/3 of them changed from male voice to female voice, and 1/3 from female voice to male voice. Most of the male participants (7 out of 8) made a change on the speaker gender, from female voice to male voice.

■ male voice■ female voice

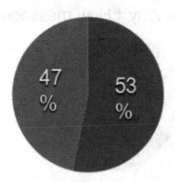

Fig. 1. The distribution of 30 participants' preference for speaker gender before experiment. 57% participants said they prefer female voice as the voice used in the future in-vehicle information system autonomous car while 43% participant would like to choose male voice.

■ male voice■ female voice

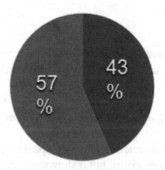

Fig. 2. The distribution of 30 participants' preference for speaker gender after experiment. 53% participants thought male voice is a better choice for the autonomous car voice system and 47% participant viewed the female voice is more suitable for voice assistant in autonomous car.

3.2 Chi-square Test

A Chi-square test was performed to examine the relation between a participant's gender and their preference for speaker gender before the experiment. The relation between these variables was significant, $X^2 = 6.652, p = 0.01 < 0.05$. Female participants were more likely to choose male voice than the male participants, and the male participants ware more likely to choose female voice than the female participant.

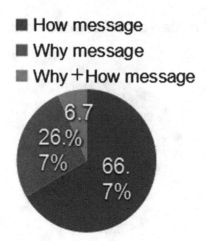

- How message
- Why message
- Why + How message

Fig. 3. The distribution of 30 participants' preference for voice content. 66.7% participants said they like How message, 26.7% prefer Why message, other 2 participants thought Why + How message is the best choice for in-vehicle information system.

A Chi-square was conducted to examine the relation between participant's gender and their preference for speaker gender after the experiment. There is no significant difference between these variables, as $X^2 = 0.536$, $p = 0.464 > 0.05$.

3.3 ANOVA

A speaker gender (female voice vs. male voice) x voice content (why message vs. how in message) repeated—measures ANOVA was conducted with the ratings of all 15 adjective words as dependent variable. The purpose of the analysis was to compare the effect of speaker gender and voice content on ratings of adjective words which were used to evaluate the participants' experience. The summary results of ANOVA are shown in Appendix E.

There was no significant interactive effect on speaker gender and voice content in the rating of all the 15 adjective words, but it shown significant main effects on speaker effect gender and voice content as follow.

There was a significant main effect on speaker gender in the rating for "trustworthy", the rating for female voice ($M = 6.133$, $SE = 0.221$) is higher than the rating for male voice ($M = 5.650$, $SE = 0.276$), $F (1,29) = 5.608$, $p = .025 < .05$. These results suggest that speaker gender really does have a effect on participants' trust in the voice system. Specifically, the result suggests that a female voice is more "trustworthy" than a male voice.

A significant main effect of voice content is shown in the rating for "uninterested", the rating for why message ($M = 4.183$, $SE = 0.310$) is higher than the rating for how message ($M = 3.583$, $SE = 0.256$), $F (1,29) = 8.534$, $p = .007 < .01$. These results suggest that voice content really does have an effect on participants' interest in the voice system. The results show that the why message is less is "interesting" than the how message.

There was a significant main effect of voice content in the rating for "intelligent", the rating for how message ($M = 6.550$, $SE = 0.260$) is higher than the rating for why message ($M = 5.983$, $SE = 0.215$), F $(1,29) = 6.227$, $p = .019 < .05$. These results suggest that voice content really does have an effect on participants' view about the intelligent level of voice system. Specifically, the results suggest that the how message is more "intelligent" than the why message.

The ANOVA reveals that participants felt the how message is more "stimulating" than the why message. The rating of "stimulating" for how message ($M = 4.600$, $SE = 0.316$) is higher than the rating for why message ($M = 4.000$, $SE = 0.292$), F $(1,29) = 6.178$, $p = .019 < .05$. These results suggest that voice content really does have an effect on participants' attitude to the stimulated level of voice system. The results suggest that the how message is more "stimulated" than the why message.

What's more, there was a marginal significant main effect of speaker gender in the rating for "acceptable", F $(1,29) = 3.955$, $p = .056 < 0.1$, female voice ($M = 6.350$, $SE = 0.212$) is higher than the male voice ($M = 5.950$, $SE = 0.263$), which means the female voice is more "acceptable" than male voice. Another marginal significant main effect of speaker gender is observed in the rating of "pleasure", F $(1,29) = 3.754$, $p = .064 < 0.1$, female voice ($M = 5.450$, $SE = 0.248$) is higher than the male voice ($M = 4.983$, $SE = 0.271$).

4 Discussion

The major goal of this study is to find our whether drivers have a specific preference on the speaker gender and voice content in the future autonomous car driving. The results showed that participants did not have a preference on speaker gender, but the female voice had a significantly higher rating than the male voice at some adjective words. In the voice content, both results of the descriptive statistics and ANOVA showed that how information is more popular and welcomed by the participants. Those points will be discussed more thoroughly below.

According to the research from Cozby [29], in the repeated measure design, the order of presenting the treatments may have an effect on the dependent variable. The ANOVA found that the order of the experiment has no significant effect on participants' preference for speaker gender and voice content in this study.

The result showed that the number of participants who prefers female voice is reduced after the experiment. In the pro-experiment gender voice choice, male participants preferred to choose female voice, 8 out of 12 reported that their voice assistant they used such as Siri, Google now or the navigation system in their vehicle are female voice, they hope the autonomous car can keep this design. A study of warning signals also suggested that when introducing a new signal or voice, it should not be too different from the existing ones [30]. Rather, Female participants prefer male voice, 7 out of 10 reported that the reason they like male voice was because their early memory about car and driving were always together with their male family members who drove them everywhere, taught and guided them the experience and skills about driving, so they reported that it would make them feel relaxed and trustworthy if the autonomous car uses a male voice to interact with them.

More than half of the participants changed their choice. Most of them claimed the voice was not what they expected. The autonomous car is not yet available to the public so the participants can only image they are driving an autonomous car, but there can be a huge gap between the real situation and their imagination. Even though in this study, we tried to provide a reliable method for manipulating the experiment; there is a fair amount of difference in fidelity between the experiment setting and real-life autonomous driving.

7 male participants who changed their preference reported that the female voice used in this experiment was a little annoying and shrill because of female high voice pitch. Meanwhile, 8 female participants switched to choose female voice after the experiment; they said the male voice was too vague to hear clearly. Edworthy and Waring's study also found that a quiet, low-pitched sound or a loud, high-pitch sound are less pleasurable than the sounds between these two extremes [31].

This finding can be explicated by the fact that auditory frequency is positively correlated with perceived pitch. Lower voice pitches are usually associated with the male voice, and the female voice is contacted with higher. Studies from neuroscience showed male and female have a significant difference in hearing sensitivity, that female show a greater sensitivity to sound than the male [32]. The results were consistent with the study of human-robot interaction [33], that male and female subjects both anthropomorphized the robot with the same-gender human voice more strongly compared to the robot with opposite gender human voice, they showed greater acceptance and felt psychologically closer to the robot which shared the same gender with them. Considering all of those facts, it can explain why the males and females changed their preference to the voice of the same gender with them.

The Chi-square test result showed no significant difference on participants' preference for the speaker gender after experiencing all four experiment stimuli. To investigate further, we conducted a repeated-measures ANOVA with the ratings of all 15 adjective words as the dependent variables to compare the effect of speaker gender.

Participants rated enunciated by the female as more trustworthy, acceptable and pleasurable than message enunciated by the male. In the theory of "Computer As Social Actors", it proposes that people actually engage in the same kinds of social response to the computer as they do with humans [34]. Nass's studies found that users treat the voice of a computer as the social response but not an unseen human "behind" the computer, usually a programmer [35]. While this theory approaches to human-vehicle interaction, it is a new way to look at the autonomous car as the source of information. The participants do not define the voice as a driver for the autonomous car who is taking control of their car, but as the voice assistant in the car. Almost all were the same as the navigation system they used in the car were female voice. Participants were wholly aware that the voice was a narrator for the autonomous car system (as opposed to some human "behind" the car). From their experience, they would rather trust a voice they are familiar with, that can be the explanation of this result. Also, 8 participants reported that the male voice is on a low pitch, the message is not clear, this result supported the study that female voices are easier to hear in the noisy condition than male voice [11], as we add the background sound in the experimental videos. The lower Intelligibility of the male voice in the noisy condition may also contribute to the result that participants rated male voice lower on "trustworthy", "acceptable" and "pleasurable".

Even though, studies about the cockpit warning voice show a male voice resulted in a larger proportion of correct and faster response times to verbal warning than a female voice [15], the role and function of speech systems in the cockpit and autonomous car are completely different. Warning in the cockpit is asking the pilot to take control of the aircraft and deal with the emergency situation. However, in the autonomous car, the only thing that drivers need to do is switching their control to the car. Moreover, vehicle drivers were not got the professional training for years to acquire a license as the aircraft pilots. So the lessons from aviation have limited application to autonomous car design due to the difference in operating methods for the different machines in different "road" condition [36].

Less than 1/3 of participants like the why message more than the how message and rated the why message as "uninterested", not "intelligent" and not "stimulating". In Endsley's model of situation awareness (SA) [37], he explained the concept of SA. He defined the SA as the ability to perceive the related elements of the surrounding, to understand the current situation. However, firstly the participants need to be aware that the autonomous system already took control of the car and is driving. We can imagine one scenario: an autonomous car gives the driver a message "Stop sign ahead!". At this moment, the driver will look for a stop sign, after he notices the stop sign the car is braking to stop, so the driver must process two types of information: the situational status and the car's status. Then the driver can know what the car is doing and the reason it performs such an action. As a consequence, the driver's cognitive resource is overloaded, as an autonomous car owner, he may think the car is not "Intelligent", he also needs to take part in the decision process together with car. Moreover, sometimes the stop sign is not easy to find, which will easily cause the driver lose their patience to enjoy the interaction with the car. The next time, the voice message comes again, the driver will show no interest to listen carefully. In addition, the why message is not considerate to provide information that can help the driver know what to do next. Lee's study showed that the content used in displaying a system's operational status play a key role in building driver trust in the vehicle's automation capabilities [38]. As how message described what automated activity it is undertaking directly, but why message informed the divers the reason the car is acting that way, the result is consistent with Lee's finding.

The finding that participants prefer the how message can also be explained by the concept of anthropomorphism. A previous study showed the user who drove the anthropomorphized vehicle with enhanced humanlike features (name, gender, voice) reported trusting their vehicle even more [39]. Technology appears better able to perform its intended design when it seems to have a humanlike mind. How information uses the first person word like I, we or us ("I will take the control" "we will pass the car in front of us") to interact with the driver, which make them like they are combined with the car, what the car does is just like what they do. Anthropomorphism of in-vehicle information system predicts trust in that car.

5 Conclusion

The female voice was found be more trustworthy, acceptable and pleasure than the male voice, because participants are more familiar with female voice as it is widely used in smart device and navigation system. In addition, the female voice is easier to hear than the male voice in the noisy condition. More than half of the participants changed their preference for the speaker gender, which reminds researchers to be cautious about the results of an online survey of drivers' attitude to speaker gender of the autonomous car, as there is a fair amount of difference in fidelity between real autonomous driving and participants' subjective imagination. Additionally, researchers should take the difference of autonomous car and aircraft into consideration when they apply the methods and lessons from aviation studies.

How message is preferred by participants. They thought the how message is intelligent as it described what automated activity it is undertaking directly, they didn't need to take part in the decision process which may add their cognitive load. In addition, how message uses the first person word like "I" "we" or "us" to communicate with the drivers. Those anthropomorphized features made participants trust it more.

5.1 Limitation

One limitation of our study is that the sample group is limited demographically: participants are university students, most of them are less than 30 years old. A wider range of participants-including newer drivers as well as elderly drivers with longer experience but possibly slower perception and reaction times-could yield an opportunity to more broadly generalize our results or to produce different findings.

As the technology and time were limited, we only used two speaker voice, a wider range of speaker voice—including young as well as the middle age also with different accents and tumble could yield an opportunity to produce different results. Therefore more study is needed to see is the speaker gender, or just the speaker's timbre affecting the participants' choice. Meanwhile, study already showed that people like a nature voice more than a synthetic voice, this part also need to be studied further.

5.2 Future Work

In this study, we used text-to-speech (TTS) voice, rather than pre-recorded human voice. Previous research has focused on speech quality, human voice versus TTS, and how it affects user's attitude change. Future studies should examine how speech quality interacts with speaker gender in the study of the in-vehicle information system.

Some participants reported that they did not like the why information, because the time between the voice and car action is too short to catch the point of what the car is going to do. We are going to set an experiment to test what timing is better for the interaction between the autonomous car and the driver, and whether the best timing is correlated with the on-going car speed.

References

1. Federal Automated Vehicles Policy—Accelerating the Next Revolution in Roadway Safety. https://www.transportation.gov/AV. Accessed 2016
2. Automated Driving Levels of Driving Automation Are Defined in New Sae International Standard J3016. https://www.sae.org/misc/pdfs/automated_driving.pdf. Accessed 2016
3. West, D.M.: Moving forward: self-driving vehicles in China, Europe, Japan, Korea, and the United States. Brooking (2016)
4. Three-Quarters of Americans Afraid to Ride in a Self-Driving Vehicle, 1 March 2016. http://newsroom.aaa.com/2016/03/three-quarters-of-americans-afraid-to-ride-in-a-self-driving-vehicle/. Accessed 2016
5. Schoettle, B., Sivak, M., Transportation, S.W.: Motorists' Preferences for Different Levels of Vehicle Automaton: University of Michigan Sustainable Worldwide Transportation (2016)
6. Chi, C.F., Dewi, R.S., Huang, M.H.: Psychophysical evaluation of auditory signals in passenger vehicles. Appl. Ergon. **59**, 153–164 (2017)
7. Mittal, R.: In-vehicle Multimodal Interaction. Doctoral dissertation, Arizona State University (2015)
8. Bem, S.L.: Gender schema theory: a cognitive account of sex typing. Psychol. Rev. **88**(4), 354 (1981)
9. Nass, C., Brave, S.: Wired for speech: how voice activates and enhances the human-computer relationship (2005)
10. Palmeri, T.J., Goldinger, S.D., Pisoni, D.B.: Episodic encoding of voice attributes and recognition memory for spoken words. J. Exp. Psychol. Learn. Mem. Cogn. **19**(2), 309 (1993)
11. Nygaard, L.C., Sommers, M.S., Pisoni, D.B.: Speech perception as a talker-contingent process. Psychol. Sci. **5**(1), 42–46 (1994)
12. Nixon, C., et al.: Female voice communications in high level aircraft cockpit noises–part II: vocoder and automatic speech recognition systems. Aviat. Space Environ. Med. **69**(11), 1087–1094 (1998)
13. "Automated Flight Controls" (PDF). faa.gov. Federal Aviation Administration. Accessed 20 Feb 2014
14. Samuel Gibbs: China's Baidu could beat Google to self-driving car with BMW. 10 June 2015. https://www.theguardian.com/technology/2015/jun/10/baidu-could-beat-google-self-driving-car-bmw. Accessed 2016
15. Hellier, E., Edworthy, J., Weedon, B., Walters, K., Adams, A.: The perceived urgency of speech warnings: semantics versus acoustics. Hum. Factors J. Hum. Factors Ergon. Soc. **44**(1), 1–17 (2002)
16. Edworthy, J., Hellier, E., Rivers, J.: The use of male or female voices in warnings systems: a question of acoustics. Noise Health **6**(21), 39–50 (2003)
17. Arrabito, G.R.: Effects of talker sex and voice style of verbal cockpit warnings on performance. Hum. Factors **51**(1), 3–20 (2009)
18. Eagly, A.H.: Gender and social influence: a social psychological analysis. Am. Psychol. **38**, 971–981 (1983)
19. Lee, E.J., Nass, C., Brave, S.: Can computer-generated speech have gender? An experimental test of gender stereotype. In: Extended Abstracts on Human Factors in Computing Systems, CHI 2000, pp. 289–290. ACM, April 2000
20. Norman, D.A.: The "problem" with automation: inappropriate feedback and interaction, not "over-automation". Philos. Trans. Roy. Soc. London B Biol. Sci. **327**(1241), 585–593 (1990)

21. Stanton, N.A., Young, M.S.: Vehicle automation and driving performance. Ergonomics **41**(7), 1014–1028 (1998)
22. Walker, G.H., Stanton, N.A., Young, M.S.: The ironies of vehicle feedback in car design. Ergonomics **49**(2), 161–179 (2006)
23. Lee, J.D., Seppelt, B.D.: Human factors in automation design. In: Handbook of Automation, pp. 417–436 (2009)
24. Koo, J., Kwac, J., Ju, W., Steinert, M., Leifer, L., Nass, C.: Why did my car just do that? Explaining semi-autonomous driving actions to improve driver understanding, trust, and performance. Int. J. Interact. Des. Manuf (IJIDeM) **9**(4), 269–275 (2015)
25. Nisbett, R., Peng, K., Choi, I., Norenzayan, A.: Culture and systems of thought: holistic versus analytic cognition. Psychol. Rev. **108**(2), 291–310 (2001)
26. What is Text-to-Speech? https://www.ivona.com/us/about-us/text-to-speech/. Accessed 2016
27. Takayama, L., Nass, C.: Assessing the effectiveness of interactive media in improving drowsy driver safety. Hum. Factors J. Hum. Factors Ergon. Soc. **50**(5), 772–781 (2008)
28. Takayama, L., Nass, C.: Driver safety and information from afar: an experimental driving simulator study of wireless vs. in-car information services. Int. J. Hum.-Comput. Stud. **66**(3), 173–184 (2008)
29. Mullennix, J.W., Stern, S.E., Wilson, S.J., Dyson, C.L.: Social perception of male and female computer synthesized speech. Comput. Hum. Behav. **19**(4), 407–424 (2003)
30. Lemaitre, G., Susini, P., Winsberg, S., McAdams, S., Letinturier, B.: The sound quality of car horns: designing new representative sounds. Acta Acustica united with Acustica **95**(2), 356–372 (2009)
31. Edworthy, J., Waring, H.: The effects of music tempo and loudness level on treadmill exercise. Ergonomics **49**(15), 1597–1610 (2006)
32. McFadden, D.: A speculation about the parallel ear asymmetries and sex differences in hearing sensitivity and otoacoustic emissions. Hear. Res. **68**(2), 143–151 (1993)
33. Eyssel, F., Kuchenbrandt, D., Bobinger, S., de Ruiter, L., Hegel, F.: 'If you sound like me, you must be more human': on the interplay of robot and user features on human-robot acceptance and anthropomorphism. In: Proceedings of the Seventh Annual ACM/IEEE International Conference on Human-Robot Interaction, pp. 125–126. ACM, March 2012
34. Reeves, B., Nass, C.: How People Treat Computers, Television, and New Media Like Real People and Places. CSLI Publications and Cambridge (1996)
35. Nass, C., Moon, Y.: Machines and mindlessness: social responses to computers. J. Soc. Issues **56**(1), 81–103 (2000)
36. Koo, J., Shin, D., Steinert, M., Leifer, L.: Understanding driver responses to voice alerts of autonomous car operations. Int. J. Veh. Des. **70**(4), 377–392 (2016)
37. Endsley, M.R.: Toward a theory of situation awareness in dynamic systems. Hum. Factors J. Hum. Factors Ergon. Soc. **37**(1), 32–64 (1995)
38. Lee, J.D., See, K.A.: Trust in automation: designing for appropriate reliance. Hum. Factors J. Hum. Factors Ergon. Soc. **46**(1), 50–80 (2004)
39. Waytz, A., Heafner, J., Epley, N.: The mind in the machine: anthropomorphism increases trust in an autonomous vehicle. J. Exp. Soc. Psychol. **52**, 113–117 (2014)

Development of Immersive Vehicle Simulator for Aircraft Ground Support Equipment Training as a Vocational Training Program

Yongjae Park[✉], Yonghyun Park, and Hyungsook Kim

Department of Human Arts and Technology, Inha University, Incheon, Korea
{mayamind,yhpark81,khsook12}@inha.ac.kr

Abstract. Due to the increase in the number of flights and passengers at Korea's airports, there is an urgent need for airport personnel. A demand survey showed that the number of airport ramp services personnel was insufficient for the increasing amount of air traffic. Free access to various parts of the airport is difficult due to the security system, which limits on-the-job training. The purpose of this study was to develop an immersive vehicle simulator that can be used as a vocational training system to assist airport ramp agents in training to understand the process of performing the job and to master ground handling services before using the actual equipment at the airport. To aid in the development of this simulation, focus group interviews were conducted with ramp service personnel engaged in ground handling services, and the requirements for the position were identified. The simulator is largely divided into three parts: a hardware platform, a simulator software, and vocational training educational simulation content. To enhance the immersion of the experience, a display device was installed on the front and rear of the vehicle simulator. The software environment replicates the actual design of the Incheon airport, incorporates a 3D model of an actual vehicle and is simulated with the UNITY program. The simulator was used for vocational training in concert with the developed training program. Additionally, we collected feedback from trainees who participated in the vocational training program and identified necessary improvements to the hardware and software.

Keywords: Virtual reality systems for learning · Immersive vehicle simulator · Aircraft ground support equipment · Vocational training program

1 Introduction

Over the past several years, an increasing number of immersive virtual environment experiences have become available for both educational and entertainment purposes.

Participants in entertainment experiences now number in the hundreds of millions, yet adoption in educational settings remains limited [1].

In this paper, we will discuss the development of immersive simulation for training airport ramp agents. International airports are complex systems that require efficient operation and the coordination of all departments. Related research include the development of scheduling solutions to efficiently manage the work of airport ground

© Springer Nature Switzerland AG 2019
H. Krömker (Ed.): HCII 2019, LNCS 11596, pp. 225–234, 2019.
https://doi.org/10.1007/978-3-030-22666-4_16

crews [2], the study of aircraft coordination simulation and education to improve student understanding of aircraft coordination [9], and Cessna 172 Aircraft Simulation studies on virtual training [10].

In the past few years, the common direction of research on educational simulation development for airport-related work has been the construction of immersive environments. Focused research is being conducted to construct environments in which users can be immersed in various situations through the combination of hardware and software according to the characteristics of each field, along with a training environment configured using virtual reality (VR) and augmented reality. However, it is difficult to find simulations developed to replicate the work of airport ramp agents for educational purposes related to vocational training. The demand for such research is high for the personnel at Incheon International Airport.

Since its opening in 2001, Incheon International Airport has grown steadily to become an international hub airport and has developed into the world's seventh largest international airport and the world's third largest international freight airport. The Incheon airport has been ranked No. 1 in the International Airport Service Quality (ASQ) for 12 consecutive years and the number of passengers at the airport has surpassed 62 million per year. The third phase of the airport construction project was recently completed and Incheon International Airport successfully opened a second passenger terminal on January 18, 2018. With the addition of the new terminal, Incheon International Airport covers a total land area of 22,397,000 m^2, with three runways, two passenger terminals, 3,085,000 m^2 of passenger moorings and 1,155,000 m^2 of cargo moorings [3].

According to the 2017 aeronautical statistics for airport use in Korea, 850,214 flights, 143,331,106 passengers and 4,611,766 tons of cargo pass through Korean airports annually [4], with the number of passengers increasing by about seven million every year. In 2017, Incheon Airport had the highest amount of passenger traffic in Korea, accounting for 44% of the total number of passengers at all airports nationwide. According to a statistical survey from Korea's national statistics portal, the number of employees required for the airport passenger industry is increasing annually. From 2017 onward, it is estimated that between 2,000 and 2,500 additional aviation crews, comprising four to five percent of existing employees, will be needed [5]. Thus, it is more urgent than ever to train agents capable of airport ramp service.

At airports, a variety of ground handling services are required to load and unload passenger cargo as it enters and leaves the aircraft. The functions of ground handling agents have a very significant impact on timely execution of the flight network by the air carrier. The main tasks of ground handling agents consist of taking care of the aircraft before and after the flight, specifically, handling passengers, baggage and the aircraft [6]. According to the results of a survey of company personnel demands regarding airport ground handling services [7], the personnel for the ramp service is insufficient for completing the aforementioned services. It has since been confirmed that few staff members are capable of performing the duties of aircraft marshaling, towing with pushback tractors, and luggage handling with Belt Loaders and baggage carts.

Ground handling is one of the most important processes at the airport and is directly related to the operation of the aircraft. Efficient management and the competitiveness of ground handling services are both business assets and a survival strategy for airlines. In particular, ground handling accounts for a large portion of cargo transportation services and these services should be differentiated. According to the current guidelines, operating companies are selected based on the criteria set out in the International Air Transport Association (IATA) Ground Handling Agreement regardless of the nature of the airport or the size of the airline. There is no standardization of the method for selecting a ground operation company to complement the work of the airline. Consequently, these companies are selected according to the judgment of each airline. Different selection strategies are required taking into consideration the business area of the airline, the passengers and cargo, and the airport situation of the ground operator [8].

Based on the results of a demand survey, the Inha University Institute of Advanced Human Resource Development (IAHRD) developed a 160-h training program to educate airside ramp service agents who can immediately apply the training to their work. Due to the nature of the job, many occupations require the use of specialized equipment. Therefore, it is necessary to practice rather than learn theoretical concepts in order to develop the skills to use the equipment. However, the security system of the airport and the characteristics of specialized equipment make practice difficult. Not only is the airport's airside ramp difficult to access, but the equipment required for practice is specialized and can only be used at the airport. Therefore, a virtual simulator is needed to help trainees indirectly experience the work of an airside ramp services agent by facilitating practical training for the use of specialized equipment.

The purpose of this study is to develop immersive equipment for airport ramp agent trainees to help them understand the process of performing the job and to learn the ground handling services tasks before using the actual equipment at the airport. This equipment will be used for vocational training to train the ground handling services personnel needed at Incheon International Airport and was developed to be adaptable to other airports of similar size. Therefore, this study contributes to a practical environment that can be used universally in vocational training education so that airport ramp agents can utilize the equipment necessary for their job in a virtual environment.

2 Related Work

Research related to the development of ground handling services equipment was difficult to find. Studies on the development of simulation equipment in the aeronautical field that combine hardware and software similar to this study are described below.

2.1 Aircraft Handling Simulator

The report titled "Creating an Aircraft Handling Qualities Simulator for the USAF Test Pilot School" [9] contains a description of aircraft coordination simulation equipment and training programs designed to improve the understanding of aircraft handling for students at the USAF Test Pilot School. The equipment used in the study included hardware from the actual equipment located in a separate laboratory where users could practice. The hardware configuration consisted of a Cockpit & Cockpit Inceptor Control (stick, rudder pedals, throttles), five displays, Image Generator, Image Driver Generator, Control Room, Master Sim, and Backup. For the purpose of general purpose aircraft coordination, the software included basic controls or a display with general purpose GUI, and the simulation software was configured to select and practice the operation of various aircrafts. The goal of the simulator is to enable instructors and students to easily create lesson plans that include the characteristics of various aircrafts.

2.2   Cessna 172 Aircraft Simulator

The study "Development of a Cessna 172 Aircraft Simulator with a Glassless Open-type VR Screen for Virtual Training" [10] was conducted for pilot flight simulations based on the Cessna 172 model, which is widely used for the training of aircraft pilots around the world. The simulator provides virtual training for piloting and aircraft instrument operation procedures. The content used in this simulator was developed using the PREPAR3D engine program provided by Lockheed Martin, which includes a physical interface to express the physical aspects of the aircraft. The equipment is largely divided into three parts: a simulator interface, a simulator operating software, and pilot training educational simulation content. The display utilizes a Glassless Open-type VR Screen instead of a regular monitor, which is a positive factor for practitioners to improve immersive experiences during pilot exercises. This simulator has been certified by the Ministry of Land for Grade A training for flight simulation equipment, which is the same as the Flight Simulator Level C of the Federal Aviation Administration (FAA). The efficacy of the aircraft simulator was verified for Cessna 172 aircraft virtual simulation training through two field trials with a group of experts in VR and aviation related fields.

3   Simulator Requirements (Development Process)

In order to develop virtual simulation training equipment, focus group interviews were conducted 3 times with ramp services agents currently performing ground handling services (Table 1).

Table 1. Development requirements for ground handling services

Necessary task	Marshaling	Tug Car	Loader	Belt Loader	Aircraft Tractor
Job description	Guide the aircraft in and out of the parking position	Move cargo from the warehouse to the airplane or the plane to the warehouse	Move cargo from the aircraft to the ground or from the ground to the aircraft	Move cargo from the aircraft to the ground or from the ground to the aircraft	Tow aircraft
Requirements	• Hardware production based on various vehicle operating environments and user interfaces. • Marshaling: Implementing a situation in which the aircraft reacts in real time to movement signals from the marshal to induce parking. • Tug Car – Implementation of Tug Car backward interface for connection with Tug Car and Dolly – Implementation of a display device that allows trainees to check the situation behind simulation training – Implementation of a hardware experience that allows the user to feel the movement of the vehicle when driving in the airport • Loader – Implementation of a realistic user interface (with up and down movement of the chair) reflecting the characteristics of the equipment – Describing in detail the moment of contact between the aircraft and the Loader • Aircraft Tractor: Representation of the connection between the tractor and the Tow-bar • Simulation of the Tug Car in which the process is similar to that of work at the airport • Development of the working environment by simulating day, night, and rainy weather • Implementation of unexpected situations due to obstruction or collision with the progress route of other vehicles during work • Assessment of the students' practice and ranking announcement system				

4 Configuration

4.1 Hardware

According to the results of the demand survey, most of the simulation work required by the company was related to the operation of special equipment. In order to implement the simulation in accordance with the requirements of the company, a universal hardware structure capable of presenting a common interface for different vehicles was required. The hardware platform largely consists of display devices, a controller, a frame, and a six-axis motor. The displays consist of two 42-in. monitors used to represent the front and rear views. The controller consists of a cockpit chair, steering wheel, transmission, brake pedal, accelerator pedal and option buttons. The frame is custom made using steel. In order to increase the immersion feeling by conveying physical movements in various situations to the driver, an electric motor composed of six axes is used as the base of the whole equipment (Fig. 1).

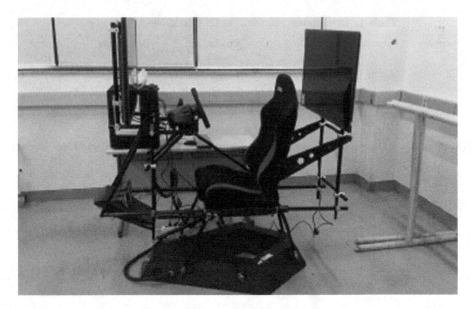

Fig. 1. Simulation hardware architecture

4.2 Computer

High polygon modeling data was used for realistic representation of the airport's overall virtual space and 3D object configuration. The following computer hardware specifications are required for real-time rendering including physical environments: Intel i7-4790 3.6 GHz quad-core CPU, 32 GB memory, Geforce GTX 980 4 GB graphics card, 500 GB SSD drive, and Windows 7 64 bit operating system. However, since the software was designed to arbitrarily adjust the 3D graphics resolution considering various platforms, there should be no problems using it on a computer with lower specifications than those proposed above.

4.3 Software

We created 3D objects for the terrain, buildings, structures, vehicles, cargo, and aircrafts that can be seen in the ramp area of the airport. We simulated real-time operations that may occur for inbound and outbound freight at airports using the Unity program. In detail, the character representing the ramp services agent was dressed in an outfit with the same design as the actual work outfit, and each vehicle was modeled in 3D based on the actual vehicle operating in the field. In addition, the physical environment such as slippery road surfaces, changes in braking distance, changes in illumination, etc., which can occur at nighttime or in rainy weather, was also represented to better reflect the characteristic and differences in the work depending on weather changes due to the nature of work performed outside. Six electric motor shafts were used to replicate a realistic driving experience by moving the driver's seat in real time in response to the driver's manipulation of the direction of the vehicle, collision of the vehicle, and up and down movement of the equipment. In order to evaluate the students, an evaluation system was developed to show individual points and rankings taking into consideration the operating time and the delay time due to equipment malfunction. The displayed view can be selectively changed between the first and third person perspectives in the middle of the practice session according to the situation, and the basic resolution for the display is set to 1920 × 1080 (Fig. 2).

Fig. 2. Ramp services agent, Tug Car, and Loader driving screen shot

4.4 Ground Handling Equipment

For all vehicles that appear in the simulation, users can operate the transmission, accelerator pedal, and brake simultaneously, which are necessary for traveling (forward and backward). The crash response caused by external shock is transmitted to the cockpit. The main function of the Tug Car is driving and cargo transportation, which can be achieved by connecting up to three cargo containers. The exterior of the vehicle is modeled after a 3D object in the same form as the Toyota TD25. Due to the design of the vehicle, new ramp services agents experience difficulties connecting a Dolly to the Tug Car. Therefore, the simulation for connecting the Tug Car to a Dolly is presented through a driving interface that shows a backward view, similar to the situation in the field. The Loader is a device that comes into direct contact with the aircraft. The loader simulation has a function to learn the procedure and method of the Bridge Seesaw and the Elevator

Seesaw which are necessary to carry out the controlling practice in contact with the aircraft and the process of loading and unloading the cargo. Belt loader is a device that is mainly used to move small packages and it shows the situation to practice the procedures of operation of the equipment. Aircraft tractor is a device that pulls and moves the aircraft directly. The user can practice connecting the aircraft wheel and Aircraft Tractor to the Tow-bar, and also practice dragging and moving the aircraft (Figs. 3 and 4).

Fig. 3. Tug Car and Loader equipment screen shot

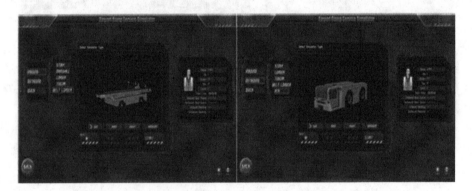

Fig. 4. Belt Loader and Aircraft Tractor equipment screen shot

5   Vocational Training Program

The Inha University Institute of Advanced Human Resource Development (IAHRD) developed a 160-h, 20 - day vocational training offline training program for 15 job seekers who wish to perform aircraft operations at airports such as Incheon Airport and Gimpo Airport. The main contents of the training program are utilization of the logistics information system, air transportation management, storage and unloading management, land transportation management, ramp operation safety and security education, ground moving support equipment operation training, air cargo handling and handling equipment operation training, training on aircraft towing and communication. The equipment training on campus is carried out using simulation equipment, and the last 6 days of the training program consist of practice using actual equipment at the Incheon Airport work site.

The practice sequence of this equipment is divided into inbound, outbound, and elective practice. Inbound practice is performed in the order of Marshalling, Loader, Belt Loader, Tug Car operation, and Cargo terminal movement starting from the time the aircraft lands, while the outbound process proceeds in the reverse order. If sudden and ad-hoc decisions need to be made while practicing with the equipment, the instructor will demonstrate the process and interact with the students to provide information that enhances the completeness of the education.

6 Results and Discussion

The Ramp Services Simulator developed according to these requirements was put into practice for education, and 49 students participated in the training program. After training with the simulator, the students completed a questionnaire. The results of the survey on the merits of simulator training are as follows. First, the 3D graphics environment created to be similar to the actual airport environment was realistic, which made the educational experience more immersive. Secondly, the vibrations and movements of the vehicle during the training were delivered to the cockpit and the steering wheel realistically. Third, the structure of the airport in the educational content and the moving line for the car were similar to those at the actual airport, so the users were able to adapt quickly to the actual work.

The disadvantages of simulation education according to the students are as follows. Only some aspects of airport operations were simulated; therefore the students could not practice with all the equipment they needed for the work. Additionally, it was inconvenient that the equipment was so bulky that practice was only possible in a specific room. Lastly, it was not possible for the students to practice for the desired amount of time.

Evaluation of the students' participation in the practice revealed two areas that required improvements. First, in order to better represent the comprehensive work of a ramp services agent, it is necessary to reorganize the existing simulation content to express all aspects of the actual work. Second, the hardware platform requires various improvements for convenient use of the equipment. For example, if the display method is implemented in a VR environment using a head mounted display instead of a monitor, the user will be able to experience various tasks in a narrower space. This will make it possible to create a simplified program that enables the user to experience the simulation with no time or space restrictions. We expect this to be possible.

After the development was completed, the students provided feedback on their experiences for technical improvement. First, when there was a collision between vehicles during operation and sudden decision making was required, the implementation of collision considering various physical environments was found to be insufficient. This situation was attributed to the lack of exception processing information that can be generated during driving. Second, in one section of the simulation, animation of the linking action between mechanical parts of the equipment was not expressed accurately. This was because mechanical drawings of the equipment were unavailable and the simulation was created using photographs of the equipment as a reference. Consequently, simulation of the connection of parts of the equipment was inaccurate. Third, the expression of reactions such as the speed of movement according to the cargo weight and the physical force experienced when the cargo is moved or loaded were insufficient. As the airport's security system provides limited information, data on how to handle various exceptions was unavailable.

7 Conclusion

The proposed simulator equipment is displayed in the equipment room at the Inha University Institute of Advanced Human Resource Development. Individuals undergoing vocational training and students in the airport ramp training program are using the equipment and providing feedback to help improve the product. We also continue research on UX/UI improvements based on ongoing user experience. We plan to complement the software and add content to provide similar educational content for a variety of industries. To improve the quality of vocational training in various fields that require experiential training and to improve the academic achievement of students, it is necessary to increase the universal utilization of these types of simulation equipment. In order to do this, it is necessary to continue research to expand the platform environment to multi-platform environments such as VR or online and to develop new content.

References

1. Dawley, L., Dede, C.: Situated learning in virtual worlds and immersive simulations. In: Spector, J.M., Merrill, M.D., Elen, J., Bishop, M.J. (eds.) Handbook of Research on Educational Communications and Technology, pp. 723–734. Springer, New York (2014). https://doi.org/10.1007/978-1-4614-3185-5_58
2. Rodič, B., Baggia, A.: Airport ground crew scheduling using heuristics and simulation. In: Mujica Mota, M., Flores De La Mota, I. (eds.) Applied Simulation and Optimization 2, pp. 131–160. Springer, Cham (2017). https://doi.org/10.1007/978-3-319-55810-3_5
3. Namhyun, J., Jinyoung, K., Seryoung, K.: Incheon international airport pavement management system. Korean Soc. Road Eng. **20**(1), 5–14 (2018)
4. https://www.airport.kr/co/ko/cpr/statisticOfLocalAirport.do
5. http://kosis.kr
6. Artur, K., Tomasz, K.: Conception of logistic support model for the functioning of a ground handling agent at the airport. In: Probabilistic Safety Assessment and Management 2014 (2014)
7. Swissport Korea Ltd.: Report of survey of company personnel demands regarding INCHEON airport ground handling services (2014)
8. Insu, K., Jinwoo, P.: A study on the relative importance of air cargo ground handling service. J. Distrib. Manage. Res **21**(2), 119–126 (2018)
9. Gray, W.R., Kemper, J.C.: creating an aircraft handling qualities simulator for the usaf test Pilot School. In: U.S. Air Force T and E Days 2009 (2009)
10. Sung, B., Kim, B., Yi, S.: Development of a Cessna 172 aircraft simulator with a glassless open-type VR screen for virtual training. Int. J. Pure Appl. Math. **118**(19), 1963–1973 (2018)

Proposal of Driving Support Agent Which Speak Based on Politeness Theory

Tomoki Miyamoto[1(✉)], Daisuke Katagami[2], Yuka Shigemitsu[2],
Mayumi Usami[3], Takahiro Tanaka[4], Hitoshi Kanamori[4],
Yuki Yoshihara[4], and Kazuhiro Fujikake[4]

[1] Graduate School of Tokyo Polytechnic University,
Atsugi-shi 243-0297, Japan
d1985001@st.t-kougei.ac.jp
[2] Tokyo Polytechnic University, Atsugi-shi 243-0297, Japan
[3] National Institute for Japanese Language and Linguistics,
Tachikawa-shi 190-0014, Japan
[4] Institutes of Innovation for Future Society, Nagoya University,
Nagoya-shi 464-8603, Japan

Abstract. In this research, we propose a driving support agent which speak based on politeness theory. Driving support agent is drawing attention as a new form of driving assistance for driving, but knowledge on effective utterance for driving support has not been established. The proposed agent selects an utterance based on the politeness theory in consideration of the age of driver, gender and driving characteristics. In the previous research, it has been pointed out that the driving support agent needs to be designed according to the driver's age and ability. Experimental results suggest that agent using PPS is effective for improving familiarity between agent and participants. Since agent using NPS gave impression that it was carefully supporting, it was suggested that driver feel that they convey accurate information.

Keywords: Smart vehicle interaction · Driving support agent ·
Politeness theory

1 Introduction

In this research, we propose a driving support agent which speak based on politeness theory [1]. Driving support agent is drawing attention as a new form of driving assistance for driving, but knowledge on effective utterance for driving support has not been established. The proposed agent selects an utterance based on the politeness theory in consideration of the age of driver, gender and driving characteristics. In the previous research, it has been pointed out that the driving support agent needs to be designed according to the driver's age and ability [2].

Politeness theory is proposed by Brown and Levinson is employed in this research. This research focuses on positive politeness strategies (PPS) and negative politeness strategies (NPS). PPS and NPS are representative utterance strategies in the five strategies defined in politeness theory. PPS is a strategy for actively reducing

© Springer Nature Switzerland AG 2019
H. Krömker (Ed.): HCII 2019, LNCS 11596, pp. 235–244, 2019.
https://doi.org/10.1007/978-3-030-22666-4_17

psychological distance to the opponent. They are categorized into fifteen sub-strategies. NPS is strategy for keeping psychological distance to the opponent and keeping it, there are ten sub-strategies.

In the research field of Human-Robot Interaction, politeness theory has been used for robot's utterance design in the situation of cooking, but it does not correspond to languages and cultures other than American English [3]. So this research we are uses Japanese language and Japanese native speakers.

2　Politeness Theory

Of the two individuals interacting with one another, we define the speaker as S and the listener as H [1]. According to Brown and Levinson, S and H both desire to form an interpersonal relationship with one another. This desire is called face [1]. In general, in dialog, S wishes to hold H's face; however, depending on the action, the result may be to threaten H's face. Such an action is called a face-threatening act (FTA).

When S needs to perform an FTA to H, S estimates the weight possessed by the FTA. Here, the weight of the FTA is calculated as

$$Wx = D\,(S,\ H) + P\,(H,\ S) + Rx \tag{1}$$

where D is a value indicating the social distance of S and H, P is the amount of force H exerts on S, and Rx is a value indicating how much the FTA is considered a burden in the given culture and society. More specifically, weight Wx of the FTA is the sum of D, P, and R. Since P and R fluctuate in different cultures and societies, the resulting weight of the FTA varies depending on the given culture and society, even in the same speech act. Here, if the degree of the FTA is relatively high, NPS is selected; however, if the degree of the FTA is relatively low, PPS is selected. Furthermore, PPS comprises 15 strategies, whereas NP comprises 10 [1].

As noted above, PPS consists of 15 strategies, including praising the other individual, sympathizing with the other individual, giving a gift to the other individual, and so on.

3　Proposed System

An overview of the proposed system is shown in Fig. 1. The proposed system supports the driver with the following flow. The system gains information concerning the driver. Each driver inputs their gender, age, driving characteristics and so on before driving. In this research, we use RoBoHoN [4] as an agent. Also, driver's individual characteristics is used to evaluate driving characteristics [5].

1. During driving, in the situation when a driver needs assistance, the system sends the acquired driving situation to stochastic model and estimate politeness strategy according to the driver and the driving. The construction of the stochastic model is based on sample data collected from the result of the questionnaire on the driver's information, and the questionnaire asking impressions of agent which uses PPS and/or NPS.

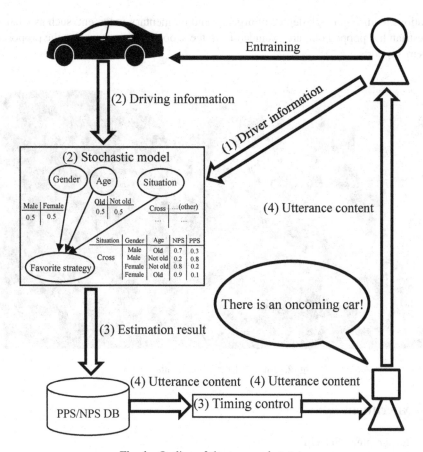

Fig. 1. Outline of the proposed system.

2. An instruction sends to extract the utterance to the utterance DB based on the estimation result by the stochastic model that can estimate the politeness strategy according to the information of the user. The utterances, utterances according to the driving situation are categorized into each politeness strategy and stored in DB. The utterance DB is constructed by converting instructor's utterances to NPS and PPS. They are accumulated them in the DB, utilizing the utterance data from the driver's instructor. The utterances extracted from the utterance DB is sent to the utterance timing control mechanism.

3. The agent utters sentences sent from utterance timing control mechanism to the driver by speech synthesis. It should be assumed that the utterance timing control can be performed sufficiently.

As described above, the proposed system is thought to select a utterance according to the attributes and circumstances of the driver, and to make it possible to support driving with high acceptance for a wide range of users. Moreover, in this paper, situations requiring a high degree of urgency and momentary judgment, such as when a pedestrian has popped out, are excluded from the scope of the system that the proposed system supports.

Fig. 2. Image of the experimental video.

4   Experiment

4.1   Experiment Set Up

The purpose of this experiment is to investigate the impression on the utterance strategy used by the agent in the driving support scene as a preliminary experiment for developing the proposed system. In this experiment, we also focus on the difference between the end-of-sentence style (honorific words - non-honorific words), which is a representation method of distinctive psychological distances in Japanese conversation. Sample data collection for constructing a probabilistic model of the proposed system will be conducted for participants with a wide range of attributes based on this experiment.

The experiment participants watch movies in which the authors manually converted RoBoHoN's utterance in a part of the reference movie [6] to NPS and PPS in the participant plan. Figure 2 is the image of the experimental video. In this experiment, NPS is "to show respect" (use of honorifics), "to question/to obscurate", "to apologize", PPS, in addition to non-honorific words, Incorporate, "exaggerate sympathy" are used. Take counter balance on the order of the videos you watch. The length of the video is about 2 min for both NPS and PPS. Table 1 examples of utterance contents of agents. In the utterance example of the first NPS in Table 1, we show respect by using honorific expressions and it is said that "Do you mind if I enter behind a white car?" We are doing support. Language behaviors that do not compel actions against opponents correspond to NPS's "Question and ambiguity" [7]. In the first utterance example of PPS, the proximity of psychological distance is expressed by using non-polite words. In the second example of NPS, for the utterance that the driver "avoids pedestrians," it is dangerous because you do not know when to jump out "While using the ambiguous utterance" I do not think so "while agreeing, I use NPS's" "Question and ambiguity". In the example of PPS, we use "to emphasize empathy" by PPS by not only empathizing but also emphasizing expressions such as "it is true!" Participants will answer questionnaires by question paper for agents of NPS and PPS after watching movies.

Table 1. Example of agent's utterance.

NPS (Formal)	PPS (Informal)
合流の調停が完了いたしました.白い車の後ろに入っていただいてもよろしいですか? (Conciliation mediation is completed. May you enter the back of a white car?)	合流の調停が完了したよ.白い車の後ろに入って! (The mediation of the confluence was completed. Enter behind a white car!)

The experiment participants were 46 university students/graduate students (20 men, 26 women, average age 21.7 years old, driving history: 9 years or less).

As the evaluation criterion, we use the characteristic adjective scale [8] (Table 2) and the impression evaluation items created by the authors (Table 3). The characteristic adjective scale is a seven-step SD method, the impression evaluation item is a seven-step Likert scale method (1: absolutely not applicable, 2: hardly applicable, 3: it does not apply, 4: neither, 5: Somewhat true, 6: fairly applicable, 7: perfectly applicable). The following 15 items are used for the evaluation items.

Table 2. Characteristic adjective scale [8].

No.	Item
1	Aggressive-Passive
2	Of bad character-Of good character
3	Sassy-Cheeky
4	Friendly-Severe
5	Cheerful-Pretty
6	A wide mind-narrow minded
7	Unsociable-Sociable
8	Responsible-No sense of responsibility
9	Impolite-Prudent
10	Shameless-Shy
11	Heavy-Frivolous
12	Sunk-Excited
13	Dignified-Subservient

(continued)

Table 2. *(continued)*

No.	Item
14	Bad feeling-Feeling good
15	Down to earth-Indiscriminate
16	Approachable-Friendless
17	Lethargic-Ambitious
18	Not confident-Confident
19	Patient-Short temper
20	Unkind-kind

Table 3. Evaluation items.

No.	Item
Q1	I felt robot can be trusted.
Q2	I felt secure by having the robot in the car
Q3	I felt that I could accept robot advice and instructions
Q4	I felt the robot uttered intentionally
Q5	I am irritated against robot
Q6	I can feel good with robot
Q7	I felt that I could get along with the robot
Q8	I think that the robot gets tired quickly
Q9	I want the robot to be like a family member or best friend
Q10	I felt the robot is moving as designed by someone
Q11	I think that the existence and advice of a robot leads to safe driving
Q12	If there is a robot, it seems to be more fun than driving by alone
Q13	If the robot is in place, I'm afraid that I feel less anxious than driving by alone
Q14	I want to use a robot
Q15	I'd like my robot to use it for my family

Of the 15 items above, Q5, Q8, and Q10 are reversal items. For example, in "Q5: Irritated against a robot", as the evaluation value approaches seven points, the evaluation is "not irritated".

4.2 Experimental Result

Figures 3 and 4 shows the results of impression evaluation. In the figure, the numerals attached to the graph represent the average point of the evaluation with one digit after the decimal point. This also applies to subsequent graphs. Also, "#" is added to the item number as a mark in the reversal item. In order to compare the evaluation values of NPS and PPS in each item, significance judgment was made by using Wil-coxon's signed rank order test with significance levels of 1% and 5%. A significant difference in the result of the test, "Q 7: I felt that I could get along with the robot" ($p < 0.01$), "Q 8: I think that the robot gets tired quickly", "Q 9: I want the robot to be like a family member or best friend", "Q12: If there is a robot, it seems to be more fun than driving

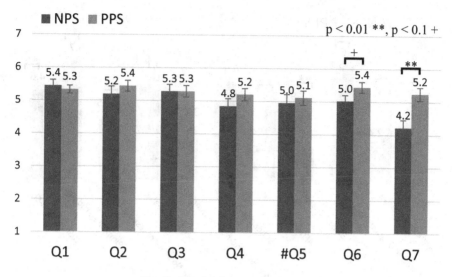

Fig. 3. Result of the evaluation (a).

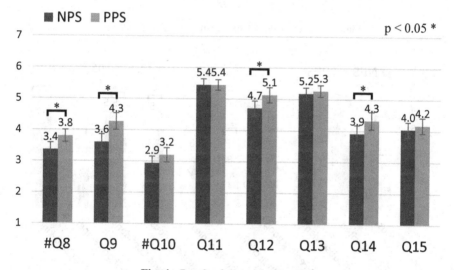

Fig. 4. Result of the evaluation (b).

by alone", "Q14: I want to use a robot" ($p < 0.05$). In addition, a marginally significant ($p < 0.1$) was observed in "Q 6: I can feel good with robot". In the other items, no statistical difference was found.

Figures 5 and 6 shows the results of the characteristic adjective scale. In this figure, for example, in the item of "1 Aggressive-Passive", one point is "Aggressive" and seven points are "Passive". As a result of Wilcoxon's signed ranking test, significant difference in "1 aggressive - Passive", "3 Sassy-Cheeky", "4 Friendly-Severe", "8 Responsible-No sense of responsibility", "9 Impolite-Prudent", "10 Shameless-Shy",

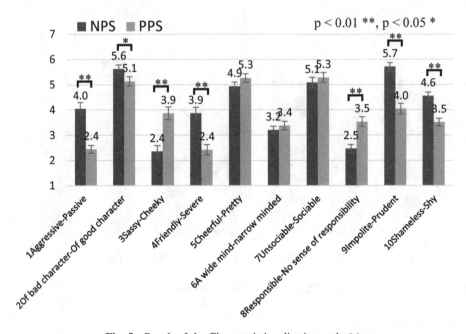

Fig. 5. Result of the Characteristic adjective scale (a).

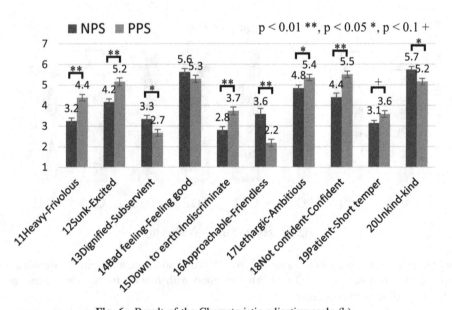

Fig. 6. Result of the Characteristic adjective scale (b).

"11 Heavy-Frivolous", "12 Sunk-Excited", "15 Down to earth-Indiscriminate", "16 Approachable-Friendless", "18 Not confident-Confident" ($p < 0.01$), "2 Of bad character-Of good character", "13 Dignified-Subservient", "17 Lethargic-Ambitious", "20 Unkind-kind" ($p < 0.05$). In addition, a significant trend ($p < 0.1$) was observed in "19 long-short temper". In the other items, no statistical difference was found.

5 Discussion

In impression evaluation, PPS overall got high evaluation. Especially in "Q7: I felt that I could get along with the robot", the average PPS was 5.2 points compared to the average NPS score of 4.2, and a score difference of about 1.2 times was seen. From this, it is suggested that driving support agent using PPS support is more effective for improving familiarity than NPS. On the other hand, NPS did not have any items that received a significantly high evaluation. However, since the experiment participants are all students in their twenties in this experiment, different results may be seen when widening the age range of participants. For example, when conducting experiments with relatively high age group participants, it is considered that PPS may be evaluated poorly by agents. In previous research, it has been reported that acceptance of driving support agents differs for each age of participants [2].

According to the results of the characteristic adjective measure, PPS has an average score of 4.0 (neither) on the average of 5.7 (cautious) of NPS in "9 Impolite-Prudent", about 1.4 times the score difference Was observed. From this, it is suggested that by using NPS, careful support is provided and the user may feel that it is conveying accurate information. On the other hand, it also gives the impression that it is "inaccessible" and "less friendly" compared with PPS, so suggesting that continuing support by NPS for too long may not be expected to improve familiarity.

Experimental results suggest that agent using PPS is effective for improving familiarity between agent and participants. Since agent using NPS gave impression that it was carefully supporting, it was suggested that driver feel that they convey accurate information. Our study approach will help design the driving support agent. The experiment participants are limited to students in this research. Future research requires to conduct experiments on a wider range of users.

6 Conclusion

In this paper, we selected agents that support driving by selecting a politeness strategy according to driver's attributes and operating conditions. In addition, we focused on the difference in style of end of sentence, and conducted an impression evaluation experiment of agents. Experimental results suggested that PPS could be more effective in improving familiarity than NPS to university students and graduate students. While NPS gave the impression of careful support, it gave an impenetrable impression compared to PPS. By verifying the utterance effect of the driving support agent for each utterance strategy and targeting participants with a wide range of age and driving characteristics as in this paper, we established a support method with a language corresponding to a wider attribute It is thought that it will be possible to go.

In the future, we plan to develop a proposal system and examine interaction experiments considering utterance strategy effective other than the end-of-sentence style.

References

1. Brown, P., Levinson, S.C.: Politeness: Some Universals in Language Usage. Cambridge University Press, Cambridge (1987)
2. Tanaka, T., et al.: Effect of difference in form of driving support agent to driver's acceptability-driver agent for encouraging safe driving behavior (2). J. Transp. Technol. **8**(3), 194–208 (2018)
3. Torrey, C., Fussell, S.R., Kiesler, S.: How a robot should give advice. In: Proceedings of the 2013 International Conference on Human-Robot Interaction. IEEE Press (2013)
4. SHARP: RoBoHoN. https://robohon.com/. Accessed 20 Sept 2018
5. Ishibashi, M., Okuwa, M., Akamatsu, M.: Development of metrics for driver's individual characteristics. Mazda Tech. Rev. **22**, 155–160 (2004)
6. Nagoya University COI vision. https://www.youtube.com/watch?v=m6m8u1lX5ao. Accessed 21 Sept 2018
7. Takiura, M.: Poraitonesu nyumon (An introduction to politeness), Kenkyusha (2008)
8. Hori, H., Yoshida, F.: Psychological Scale II - to understand the connection between human beings and society <Social Relations/Values>, Saiensu-Sha Co., Ltd. (2001)

Do You Shift or Not? Influence of Trajectory Behaviour on Perceived Safety During Automated Driving on Rural Roads

Patrick Rossner[✉] and Angelika C. Bullinger

Chair for Ergonomics and Innovation, Chemnitz University of Technology,
Chemnitz, Germany
patrick.rossner@mb.tu-chemnitz.de

Abstract. There is not yet sufficient knowledge on how people want to be driven in a highly automated vehicle. Currently, trajectory behaviour as one part of the driving style is mostly implemented as a lane-centric position of the vehicle in the lane, but drivers show quite different preferences, especially with oncoming traffic. A driving simulator study was conducted to investigate naturally-looking reactive driving trajectories on rural roads in an oncoming traffic scenario to better understand people's preferences regarding driving styles. 30 subjects experienced a static and a reactive (based on manual driving) trajectory behaviour on the most common lane widths in Germany: 2.75 m and 3.00 m. There were twelve oncoming traffic scenarios with vehicle variations in type (trucks or cars), quantity (one or two in a row) and position (with or without lateral offset to the road centre) in balanced order. Results show that the reactive trajectory behaviour leads to higher perceived safety in comparison to the static trajectory behaviour. Type of the oncoming traffic as well as lane width do have an influence on perceived safety. A small lane width (2.75 m) and oncoming trucks (type) result in lower perceived safety. There is an effect of quantity and position of oncoming traffic, too. Vehicles with a lateral offset to the road centre lead to lower safety ratings as well as more approaching vehicles. The results of the study help to design an accepted, preferred and trustfully trajectory behaviour for highly automated vehicles.

Keywords: Automated driving · Trajectory behaviour · Perceived safety · Driving experience

1 State of Literature and Knowledge

Sensory and algorithmic developments enable an increasing implementation of automation in the automotive sector. Ergonomic studies on highly-automated driving constitute essential aspects for a later acceptance and use of highly automated vehicles [1, 2]. In addition to studies on driving task transfer or out-of-the-loop issues, there is not yet sufficient knowledge on how people want to be driven in a highly automated vehicle [3–5]. First insights show that preferences regarding the perception and rating of driving styles are widely spread. Many subjects prefer their own or a very similar driving style and reject other driving styles that include e.g. very high acceleration and

© Springer Nature Switzerland AG 2019
H. Krömker (Ed.): HCII 2019, LNCS 11596, pp. 245–254, 2019.
https://doi.org/10.1007/978-3-030-22666-4_18

deceleration rates or small longitudinal and lateral distances to other road users [6, 7]. Studies show that swift, anticipatory, safe and naturally-looking driving styles are prioritized [8, 9]. In existing literature, trajectory behaviour as one part of the driving style is mostly implemented as a lane-centric position of the vehicle in the lane. From a technical point of view this is a justifiable and logical conclusion, but drivers show quite different preferences, especially in curves and in case of oncoming traffic [10, 11]. In manual driving, subjects cut left and right curves and react on oncoming traffic by moving to the right edge of the lane. When meeting heavy traffic, subjects' reactions are even greater [12–14]. The implementation of this behaviour into an automated driving style has not been realized and investigated so far, but includes high potential to increase perceived safety during highly-automated driving. In the present study, this natural trajectory behaviour is transferred into an automated driving behaviour to investigate whether a natural-looking reactive trajectory is preferred over a static trajectory behaviour. First insights show that the concept is positively experienced and leads to less discomfort during highly-automated driving on rural roads [15].

2 Method and Variables

The aim of the study was to investigate naturally-looking reactive driving trajectories on rural roads in an oncoming traffic scenario to better understand people's preferences regarding driving styles. A fixed-based driving simulator (Fig. 1) with an adjustable automated driving function was used to conduct a within-subject design experiment. 30 subjects experienced a static and a reactive trajectory behaviour on the most common lane widths in Germany: 2.75 m and 3.00 m. This resulted in four experimental conditions that were presented in randomized order to minimize potential systematic biases. All subjects were at least 25 years old and had a minimum driving experience of 2.000 km last year and 10.000 km over the last five years (see Table 1 for details). The static trajectory behaviour kept the car in the centre of the lane throughout the whole experiment whereas the reactive trajectory behaviour moved to the right edge of the lane when meeting oncoming traffic. There were twelve oncoming traffic scenarios that varied in type (trucks and cars), quantity (one or two in a row) and position (cars in the middle of the oncoming lane and cars with lateral offset to the road centre) in balanced order – see Fig. 2. The participants were required to observe the driving as a passenger of an automated car.

During the drive subjects' main feedback tool was an online handset control to measure perceived safety as shown in Fig. 3. This tool provides information about the occurrence of safety concerns in each location of the track and could be recorded in sync with video, eye-tracking, physiological or driving data. After each experimental condition subjects filled in questionnaires regarding acceptance [16], trust in automation [17] and subjectively experienced driving performance [18] and were interviewed at the end of the study. The questionnaire also required single item ratings regarding perceived safety, driving comfort, driving joy and driving style on a 11-point Likert scale with values from 0 (very low) to 10 (very high).

Fig. 1. Driving simulator with instructor centre (left) and an exemplary subject (right)

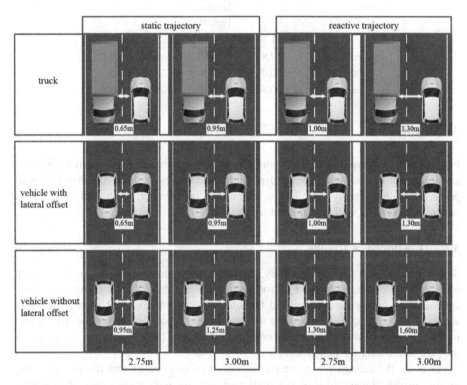

Fig. 2. Variations of oncoming traffic, resultant lateral distances to the ego-vehicle on two different lane widths (2.75 m, 3.00 m) in two different trajectory behaviour models (static, reactive)

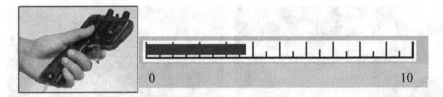

Fig. 3. Handset control (left) and visual feedback (right) for the online-measurement of perceived safety while driving highly automated. Higher values indicate higher perceived safety.

Table 1. Subjects characteristics

	Number	Age		Driver's license holding [years]		Mileage last five years [km]	
		M	SD	M	SD	M	SD
Female	12	29.8	7.9	10.6	4.2	40,083	32,745
Male	18	30.9	6.8	11.9	6.1	68,333	43,661
Total	30	30.4	7.1	11.3	5.3	54,208	41,501

3   Results

First, results of the questionnaire's single item ratings for each experimental condition are presented. In a second step, handset control data analysis shows the distribution of perceived safety during each drive depending on number, type and position of oncoming traffic as well as lane width.

3.1   Questionnaire Results

Single item ratings of perceived safety, driving comfort, driving joy and the assessment of driving style were compared performing two-factor ANOVAs with repeated measurements including lane width and trajectory behaviour. Figure 4 shows the mean values of the dependent variables for all four drives, Table 2 describes the overall and interaction effects of the two independent variables and Table 3 gives an overview of the post hoc t-tests for paired samples.

Overall, the ratings for perceived safety, driving comfort and driving style are located in the highest third of the scale and increase with wider lanes and reactive trajectories. The data show a left-skewed distribution for all experimental conditions, which is caused by many high and few very low ratings (perceived safety: 5 statistical outliers (SO) over $\pm$ 1 SD; driving comfort: 7 SO over $\pm$ 1 SD, 1 SO over $\pm$ 2 SD; driving style: 4 SO over $\pm$ 1 SD, 1 SO over $\pm$ 2 SD). That effect is higher for reactive than for static trajectory behaviour. Values for driving joy are situated in the middle third of the scale and increase with wider lanes and reactive trajectories. The data shows a symmetric distribution with no statistical outliers for all experimental conditions.

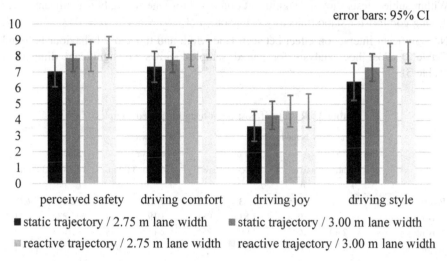

Fig. 4. Mean values of single item ratings from 0 (very low) to 10 (very high)

Table 2. Results of two-factor ANOVAs with repeated measurements including lane width and trajectory behaviour

Dep. variables	Independent variables	F	p	η_p^2
Perceived safety	Trajectory behaviour	2.67	.111	.09
	Lane width	*11.81*	*.002*	*.29*
	Trajectory behaviour × lane width	.36	.553	.01
Driving comfort	*Trajectory behaviour*	*4.47*	*.043*	*.13*
	Lane width	6.69	.066	.11
	Trajectory behaviour × lane width	.08	.778	<.01
Driving joy	*Trajectory behaviour*	*5.54*	*.026*	*.16*
	Lane width	2.30	.141	.07
	Trajectory behaviour × lane width	2.38	.134	.08
Driving style	*Trajectory behaviour*	*8.41*	*.007*	*.23*
	Lane width	*4.59*	*.041*	*.14*
	Trajectory × lane width	1.76	.195	.06

Perceived Safety

Within-subject tests show no significant difference for trajectory behaviour, but significant lower perceived safety for lane width of 2.75 m, $F(1, 29) = 11.808$, $p = .002$, $\eta_p^2 = .289$. Post hoc t-tests for paired samples confirm these results with significant differences between same trajectory behaviours on different lane widths (Table 3). No significant interaction effect between lane width and trajectory behaviour is found as seen in Table 2.

Driving Comfort

Within-subject tests show no significant difference for lane width, but significant lower driving comfort for static trajectory behaviour, $F(1, 29) = 4.471$, $p = .043$, $\eta_p^2 = .134$. No significant interaction effect between lane width and trajectory behaviour is found (Table 2). Post hoc analysis for paired samples shows no significant differences (Table 3).

Table 3. Results of post hoc t-tests for paired samples

	Trajectory behaviour				Lane width			
	2.75 m		3.00 m		Reactive		Static	
	Static–reactive		Static–reactive		2.75 m–3.00 m		2.75 m–3.00 m	
	t	p	t	p	t	p	t	p
Perceived safety	−1.73	.095	−1.33	.195	−2.82	*.009*	−2.19	*.037*
Driving comfort	−1.74	.088	−1.81	.081	−1.19	.244	−1.33	.194
Driving joy	−2.40	*.023*	−1.09	.286	−2.17	*.038*	−.10	.919
Driving style	−2.96	*.006*	−1.92	.065	−2.18	*.038*	−.512	.612

Driving Joy

Within-subject tests show no significant difference for lane width, but significant lower driving joy for static trajectory behaviour, $F(1, 29) = 5.542$, $p = .026$, $\eta_p^2 = .160$. Post hoc t-tests for paired samples approve these results and additionally find significant differences between different lane widths for reactive trajectory behaviour (Table 3). No significant interaction effect between lane width and trajectory behaviour is found as seen in Table 2.

Driving Style

Within-subject tests show significant difference for trajectory behaviour, $F(1, 29) = 8.411$, $p = .007$, $\eta_p^2 = .225$, as well as for lane width, $F(1, 29) = 4.586$, $p = .041$, $\eta_p^2 = .137$. Post hoc t-tests for paired samples deliver significant differences between trajectory behaviours on 2.75 m lane width and for lane width combined with reactive trajectory behaviour (Table 3). No significant interaction effect between lane width and trajectory behaviour is found as seen in Table 2.

3.2   Handset Control Results

For a detailed analysis, the handset control data was reversed and cumulated for all subjects to identify clusters that represent low perceived safety. Figure 5 gives an overview of the whole test route with its different types of oncoming traffic and shows highlights for the absence of high perceived safety – hereafter stated as perceived safety concerns. The graphs show the static and the reactive trajectory behaviour in comparison on 2.75 m (upper section) and 3.00 m (bottom section) lane width each.

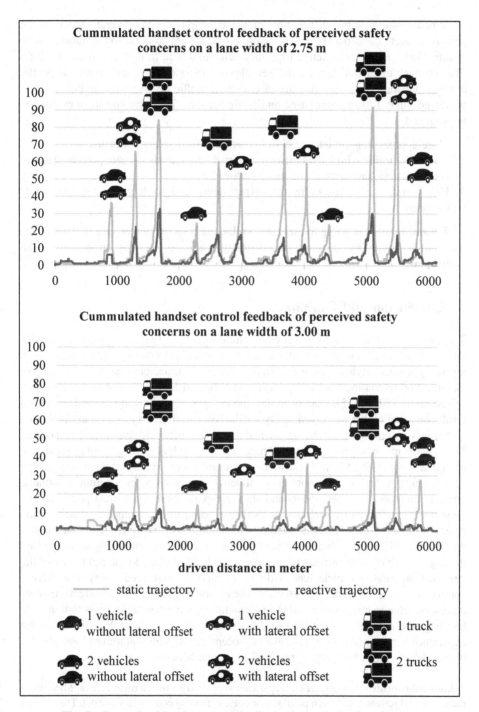

Fig. 5. Cummulated handset control feedback of perceived safety concerns

When looking at the distribution of the descriptive data, several tendencies of perceived safety concerns are able to be observed that are conform to the questionnaire results. Wider lanes and reactive trajectory behaviour lead to higher perceived safety. The feedback of the handset control set allows a more detailed and situation-specific analysis. Position, type and quantity of oncoming traffic do also have an influence on perceived safety (assumption based on descriptive data, inference statistical evaluation in progress).

1. More approaching vehicles lead to higher perceived safety concerns.
2. Oncoming traffic with lateral offset to the road centre leads to more perceived safety concerns than lane-centric oncoming traffic.
3. Heavy traffic (e.g. trucks in this experiment) lead to higher perceived safety concerns.

Further analysis is going to include correlations between perceived safety concerns and number, type and positon of oncoming traffic as well as cross lane width evaluations.

4 Conclusion and Outlook

The aim of the study was to investigate naturally-looking reactive driving trajectories on rural roads in an oncoming traffic scenario to better understand people's preferences regarding driving styles. The use of manual drivers' trajectories as basis for implementing highly-automated driving trajectories showed high potential to increase perceived safety [10, 11, 15]. Data from the studies' questionnaire and the handset control results revealed higher perceived safety for the reactive trajectory as well as significant higher driving comfort and driving joy and significant better driving style assessments. The effect is much larger for the tested narrower lane width of 2.75 m, but also existent for the lane width of 3.00 m. We also identified traffic density, lateral position and type of oncoming vehicles as factors that influence perceived safety during automated driving. In order to better understand the impact of these different aspects, further inference statistical and correlation analysis are going to be conducted. Based on the results so far, we conclude that factors which influence perceived safety in manual driving [11–14] are also factors influencing perceived safety during highly-automated driving. As drivers cannot react to oncoming traffic by shifting to the right edge of the lane, the automated vehicle has to do so to increase perceived safety and driving comfort of the passenger. Therefore, it seems most relevant to investigate manual trajectory behaviour in more detail to implement better reactive trajectories that include less negative side effects and lead to a better driving experience. A positive driving experience has the potential to improve the acceptance of highly-automated vehicles [5, 9] and therefore has both ergonomic and economic benefits.

Acknowledgements. This research was partially supported by the German Federal Ministry of Education and Research (research project: KomfoPilot, funding code: 16SV7690K). The sponsor had no role in the study design, the collection, analysis and interpretation of data, the writing of the report, or the submission of the paper for publication. We are very grateful to Konstantin Felbel and Marty Friedrich for their assistance with data collection and analysis.

References

1. Banks, V.A., Stanton, N.A.: Keep the driver in control: automating automobiles of the future. Appl. Ergon. (2015)
2. Elbanhawi, M., Simic, M., Jazar, R.: In the passenger seat: investigating ride comfort measures in autonomous cars. IEEE Intell. Transp. Syst. Mag. **7**(3), 4–17 (2015). https://doi.org/10.1109/mits.2015.2405571
3. Gasser, T.M.: Herausforderung automatischen Fahrens und Forschungsschwerpunkte, 6, Tagung Fahrerassistenz, München (2013)
4. Radlmayr, J., Bengler, K.: Literaturanalyse und Methodenauswahl zur Gestaltung von Systemen zum hochautomatisierten Fahren. In: FAT-Schriftenreihe, vol. 276. VDA, Berlin (2015)
5. Siebert, F.W., Oehl, M., Höger, R., Pfister, H.-R.: Discomfort in automated driving – the disco-scale. In: Stephanidis, C. (ed.) HCI 2013. CCIS, vol. 374, pp. 337–341. Springer, Heidelberg (2013). https://doi.org/10.1007/978-3-642-39476-8_69
6. Festner, M., Baumann, H., Schramm, D.: Der Einfluss fahrfremder Tätigkeiten und Manöverlängsdynamik auf die Komfort-und Sicherheitswahrnehmung beim hochautomatisierten Fahren. 32nd VDI/VW-Gemeinschaftstagung Fahrerassistenz und automatisiertes Fahren, Wolfsburg (2016)
7. Griesche, S., Nicolay, E., Assmann, D., Dotzauer, M., Käthner, D.: Should my car drive as I do? What kind of driving style do drivers prefer for the design of automated driving functions? In: Contribution to 17th Braunschweiger Symposium Automatisierungssysteme, Assistenzsysteme und eingebettete Systeme für Transportmittel (AAET), ITS automotive nord e.V., pp. 185–204 (2016). ISBN 978-3-937655-37-6
8. Bellem, H., Schönenberg, T., Krems, J.F., Schrauf, M.: Objective metrics of comfort: developing a driving style for highly automated vehicles. Transp. Res. Part F Traffic Psychol. Behav. **41**, 45–54 (2016)
9. Hartwich, F., Beggiato, M., Dettmann, A., Krems, J.F.: Drive me comfortable: customized automated driving styles for younger and older drivers. 8. VDI-Tagung "Der Fahrer im 21. Jahrhundert" (2015)
10. Bellem, H., Klüver, M., Schrauf, M., Schöner, H.-P., Hecht, H., Krems, J.F.: Can we study autonomous driving comfort in moving-base driving simulators? A Validation Study. Hum. Factors **59**(3), 442–456 (2017). https://doi.org/10.1177/0018720816682647
11. Lex, C., et al.: Objektive erfassung und subjektive bewertung menschlicher trajektoriewahl in einer naturalistic driving study. VDI-Berichte Nr. **2311**, 177–192 (2017)
12. Dijksterhuis, C., Stuiver, A., Mulder, B., Brookhuis, K.A., de Waard, D.: An adaptive driver support system: user experiences and driving performance in a simulator. Hum. Factors **54**(5), 772–785 (2012). https://doi.org/10.1177/0018720811430502
13. Mecheri, S., Rosey, F., Lobjois, R.: The effects of lane width, shoulder width, and road cross-sectional reallocation on drivers' behavioral adaptations. Accid. Anal. Prev. **104**, 65–73 (2017). https://doi.org/10.1016/j.aap.2017.04.019
14. Schlag, B., Voigt, J.: Auswirkungen von Querschnittsgestaltung und laengsgerichtet Markierungen auf das Fahrverhalten auf Landstrassen. Berichte der Bundesanstalt fuer Strassenwesen. Unterreihe Verkehrstechnik (249) (2015)

15. Roßner, P., Bullinger, A.C.: Drive me naturally: design and evaluation of trajectories for highly automated driving manoeuvres on rural roads. In: Technology for an Ageing Society, Postersession Human Factors and Ergonomics Society Europe Chapter 2018 Annual Conference, Berlin (2018)
16. Van der Laan, J. D., Heino, A., Waard, D.D.: A simple procedure for the assessment of acceptance of advanced transport telematics. Transp. Res. Part C Emerg. Technol. 5(1), 1–10 (1997)
17. Jian, J.Y., Bisantz, A.M., Drury, C.G.: Foundations for an empirically determined scale of trust in automated systems. Int. J. Cogn. Ergon. 4(1), 53–71 (2000)
18. Voß, G., Schwalm, M.: Bedeutung kompensativer Fahrerstrategien im Kontext automatisierter Fahrfunktionen. Berichte der Bundesanstalt für Straßenwesen, Fahrzeugtechnik Heft F 118 (2017). ISBN 978-3-95606-327-5

Automobile Auditory Experience:
A Pilot Study

Yang Zhao[✉] [ID] and Hua Dong[✉] [ID]

Tongji University, Shanghai, China
zhaoyangisvip@163.com, h.dong@lboro.ac.uk

Abstract. The automobile industry is in the transformation period: from traditional combustion vehicles to new energy vehicles. It is of great significance for automobile companies and researchers to understand users' concerns during this period, as well as the changes in automobile sound brought by the transformation. This paper reports on a mixed-method study of a survey of typical users and interviews with ten experts from a large automobile company. The aim was to identify current concerns of automobile and identify methods to improve the auditory experience. The key findings are: (1) automobile's power was the most concerned factor; the second group of user concerns relates to appearance, price, and driving experience, followed by configuration, accelerating performance, and fuel consumption. Environmental friendliness were the least concerned factors. (2) Human evolution, human needs, driving contexts, the human five senses, and natural sounds consistent with the particular frequency band were current industrial understanding of the design of automobile auditory experience; and consistency, anti-interference mechanism, and redundancy design were considered as the design principles of auditory experience by automobile developers; and they use comparison and users' subjective evaluation for auditory experience evaluation; a holistic framework of auditory experience was expected by developers to enhance their understanding of it. This paper contributes to a deeper understanding of automobile auditory experience by bringing together the viewpoints of automobile users and developers.

Keywords: Auditory experience · Electric vehicle sound · User concerns

1 Introduction

Global car sales reached 79 million in 2018 [1]. The development of automobiles is taking new directions: electrification, intelligent networking, sharing and internationalization [2]. Among them, the strategy of electrification has become the direction vigorously developed by the automobile industry and the government [3], and the electric car sales worldwide rose to 2 million sales in 2018 [4].

During the transformation from traditional combustion energy vehicles to electric vehicles, the sound system of vehicles has changed greatly. To a large extent, the change has impacts on the automobile stakeholders: car users and buyers. Their perceptions and purchase decisions as well as expected auditory experience in driving were affected [5, 6]; for automobile companies, the difficulties and challenges of design

© Springer Nature Switzerland AG 2019
H. Krömker (Ed.): HCII 2019, LNCS 11596, pp. 255–269, 2019.
https://doi.org/10.1007/978-3-030-22666-4_19

have increased [7] and users' concerns often affect the popularity of vehicles in the market; for governments and industrial agencies, the scope of supervision and management of automobile has enlarged [8]. For example, the change of sound systems bring a series of safety issues for drivers, passengers, and pedestrians [9], which would require more detailed rules and regulations.

To date, most studies in the field of automobile research focus on mechanical and electronic technologies. However, the ultimate beneficiaries of automobiles are human beings, there are few studies that have investigated from automobile users' perspectives.

This paper focuses on automobile auditory experience from the users' and car developers' perspectives. Two studies were conducted: (1) An analysis of 96 comments from automobile users, relating to 32 vehicles; (2) interviews with ten participants to understand their insights into automobile sound. The participants are car developers, i.e. designers, engineers, researchers, and managers from an automobile company.

This paper contributes to a deeper understanding of automobile auditory experience by bringing together automobile users' and developers' viewpoints.

2 Understanding Auditory Experience

Hearing is critical to human beings. All kinds of sounds and noises come into ears all the time. Hearing is one of the important information acquisition channels for individuals to perceive the physical world [10]. Individuals experience sounds mainly through the auditory system which is composed of the ear and other biological tissues [11].

For a long time, acoustics and philosophy research suggests that auditory experience was closely related to the instinct of the body and is dominated by subjective uncertainty factors and it is difficult to measure or describe [10, 12, 13].

However, people's auditory experience does exist in every day's perceptual experience. Individuals can understand how the aesthetic experience of sound occurs and its related characteristics from three aspects: (1) the human's auditory system; (2) the sound and (3) user experience. These three aspects can help to understand how the sound that is used as the norm arises from an intuitive unconscious to a conscious experiential behavior [14].

The human auditory system is divided into three parts: the outer ear, the middle ear and the inner ear. The outer ear is an auricle composed of external skin. The function of the outer ear is to help the human ear locate the sound source and enhance the sound at certain frequencies. The function of the middle ear is to transmit periosteal vibration to the lymph of the cochlea and to protect the auditory system when the external sound source is too loud. The main component of the inner ear is the cochlea which is to convert mechanical vibration into electrical nerve impulse signals for the utility of processing by the brain [11].

Different researchers have adopted different definitions of sound, with similar connotations. According to Everest and Pohlmann, the sound is a kind of wave motion in the air or other elastic media, it can be regarded as an excitation [15]. Sound can also be thought of as a stimulus to the auditory system, which makes human beings feel the sound. In this case, a sound is a sensation [15]. Michele Chion argued that there are two kinds of definitions on sound: physical and philosophical. Physics defines sound as 'a motion of molecules, through a medium such as air, water, or rock, caused by a vibrating body'; whilst philosophical definition stresses that 'sound is a sense, an experience of the sensory organs' [13].

Before going into automobile auditory experience, let us first look into user experience. There are many publications on the concept, composition, measurement, evaluation, and application of user experience. The most widely accepted definition of user experience comes from ISO 9241-210:2010: 'a person's perceptions and responses resulting from the use and/or anticipated use of a product, system or service' [16]. The definition points out that user experience is generated in the process of user-product interaction, including the user's psychological and physical feelings. The main result of user experience is the user's perception and reaction, including emotional and physiological reaction [17].

Auditory experience is based on the theory of user experience. The object of auditory experience is the physical sound, and the sensory organ. From the definition of user experience, and the definition of auditory system and sound, it is possible to preliminarily define the auditory experience as a person's perceptions and responses of sound, a physical stimulus, by the auditory system. Here the sound usually comes from products, systems or services used.

The automobile auditory experience is the perceptions and responses of automobile users (i.e. drivers or passengers) to the sound within and surrounding the automobile. Specifically, automobile auditory experience refers to the perceptions and reactions of the driver or passenger to the sound made by the automotive when driving or riding the car. In this paper, the automobile auditory experience is from the perspective of the driver, and the sound is limited to the sound produced by the automobile, e.g. the combustion sound, and functional warning sounds.

3 Methods and Analysis

Mixed research methods were used in this paper, using a combination of quantitative and qualitative research approaches [18]. Two preliminary studies were conducted, and quantitative and qualitative data collected. The first study was to analyze the qualitative data of the users' concerns of automobiles. The second study was a set of semi-structured interviews with experts in the fields of sound design and research of electric vehicles. The interviews were conducted with ten participants from a large automobile company in China.

3.1 User Survey

In this study, we focused on the automobile evaluation section of the Autohome website (www.autohome.com.cn), a large community website in China where users can freely comment on vehicles and choose their favorite ones. The main reasons for choosing Autohome as the source of data collection were as follows: (1) the accessibility of users' feedback data. Autohome provides professionally produced and user-generated content, a comprehensive automobile library, and extensive automobile information to potential consumers, covering the entire car purchase and ownership cycle [19], so it is of great convenience to get the data we want; (2) Autohome Incorporation is the leading online destination for automobile consumers in China [20]. It is listed on the New York Stock Exchange, NYSE: ATHM, and have been publishing regularly audited operation of the company, providing reliable sources of data.

With the comprehensive user-generated content library, we extracted 96 comments of automobile users on 32 vehicles; three users' comments were selected for each vehicle. These users can be regarded as typical expert users; they tend to have a profound understanding of vehicles; their comments have covered 32 vehicles selected in which seven are new energy ones. The 32 vehicles sample has exceeded the sample size of 30 suggested by Robson for single-group observations [21].

The strategies of sample selection were as follows: (1) 32 vehicles were selected, each vehicle was evaluated by three users, and a total of 96 evaluation comments were obtained; (2) five dimensions, i.e. type, country of making, power, release time, and price were used as the selection criteria, and this helps cover global vehicles sold in China to some extent.

As shown in Table 1, the selected 32 vehicles came from eight countries: China, Germany, the United Kingdom, the United States of America, France, Japan, South Korea and Sweden. We selected newly released vehicle models, about 72% were released in 2017 and 19% in 2018. These models to some extent represent the latest and most advanced design and manufacturing capabilities of automobile companies. Their price ranged from 66,800 CNY to 1,091,000 CNY, which covers most of the prices of vehicles sold on the market. Future more, vehicle types were diverse, covering sedan, Sport Utility Vehicle (SUV) and Mini Passenger Van (MPV) so as to include different comments and evaluations.

Table 1. 32 selected vehicles

Brand	Series	Type	Country	Engine	Time	Price*
DONGFENG	AX4	SUV	China	Gas	2017	6.68–10.18
ZOTYE	T600	SUV	China	Gas	2017	7.98–14.28
MG	AZS	SUV	China	Gas	2017	7.38–11.58
ROEWE	RX3	SUV	China	Gas	2018	8.98–13.58
GAC	GS7	SUV	China	Gas	2017	14.98–20.98
SAIC MAXUS	D90	SUV	China	Gas	2017	15.67–26.69
BORGWARD	BX5	SUV	Germany	Gas	2017	12.38–21.98
LUXGEN	U5	SUV	China	Gas	2017	7.58–9.98
LAND ROVER	Range Rover	SUV	UK	Gas	2017	51.74–104.80
PEUGEOT	5008	SUV	French	Gas	2017	18.77–27.97
BUICK	Regal	Sedan	USA	Gas	2017	13.18–22.58
VOLKSWAGEN	Golf	Sedan	Germany	Gas	2018	7.92–15.59
HONDA	Civic	Sedan	Japan	Gas	2016	11.59–16.99
VOLKSWAGEN	Gran Lavida	Sedan	Germany	Gas	2017	11.29–16.29
AUDI	A6	Sedan	Germany	Gas	2017	31.21–41.98
SKODA	Octavia	Sedan	Germany	Gas	2018	7.79–14.19

(*continued*)

Table 1. (*continued*)

Brand	Series	Type	Country	Engine	Time	Price*
HYUNDAI	Sonata 9	Sedan	South Korea	Gas	2018	13.98–24.98
MAZDA	MX-5	Sedan	Japan	Gas	2018	29.98–34.00
BMW	4 series	Sedan	Germany	Gas	2017	36.98–83.00
JAGUAR	XFL	Sedan	UK	Gas	2018	26.62–59.38
ACURA	NSX	Sedan	Japan	Gas	2016	289
VOLVO	V60	Sedan	Swedish	Gas	2017	22.29–52.29
CHEROLET	Malibu XL	Sedan	USA	Gas	2017	17.99–26.99
BENZ	C Class	Sedan	Germany	Gas	2017	31.08–48.68
BYD	Song MAX	MPV	China	Gas	2017	7.99–12.99
LINCOLN	MKZ	Sedan	USA	Hybrid Electric	2017	23.68–38.38
TESLA	Model S	Sedan	USA	Electric	2015	74.23–109.10
ROEWE	e i6	Sedan	China	Hybrid Electric	2017	20.18–22.28
BYD	Song EV300	SUV	China	Electric	2017	26.59–27.59
ROEWE	e RX5	SUV	China	Hybrid Electric	2017	23.19–29.68
AUDI	Q7 e-tron	SUV	Germany	Hybrid Electric	2017	80.88
BYD	Song MAX	MPV	China	Hybrid Electric	2017	7.99–12.99

Note: *Price: ten thousand CNY; SUV = Sport Utility Vehicle; MPV = Mini Passenger Van.

Firstly, users' evaluation comments on the 32 vehicles were extracted from the evaluation module of the Autohome website, and a total of 96 categories of comments were extracted. Context analysis method was used to code and recognize the key words, which can be viewed as the key features of the vehicle from the perspective of the user. For instance, in the following user's comment, the key features mentioned by the user were: **price**, **space**, **sound**, **power**, **steering control**, **fuel consumption**, and **configuration**.

> "... *the most satisfactory of the car should be the **price**, it was the most cost-effective among the same level cars... there's absolutely enough **space** for me to ride... the **sound** of speed-up is very good and the **power** is strong and smooth... the car is in an accurate **steering control**... **fuel consumption** is in line with my expectations... very rich in **configuration**... [22]"*

These features reflected the user's concerns of the vehicle. Then the cluster analysis method was used to group key features where typical user concerns were identified. Cluster analysis is a method for estimating similarities, and can be used to find out which object in a set are similar [23]. In this study, cluster analysis was based on (1) semantic similarity of terms and (2) the taxonomic similarity of items, i.e. words with similar meanings or connotations were clustered into one group and expressed in the paper with one term. For instance, (1) these terms, 'modeling', 'appearance', 'good-looking', 'exterior form' and 'shape' all described similar meaning, i.e. the appearance of the vehicle, so they were uniformly clustered into 'appearance'; (2) these terms, 'panoramic sunroof', 'leather seat', 'driving image assist system', and 'Head Up Display', all reflected the 'configurations' of the vehicle.

3.2   Interviews

A series of semi-structured interviews with a convenient sample drawn from one of the largest Chinese automobile company were undertaken, including ten participants: (1) experts representing a diverse range of knowledge from different sections relating to automobile auditory experience design; (2) experts who know the procedures and details of automobile sound design or research; (3) experts with at least three years' working on automobile sound design or research. These experts came from: (1) Vehicle Integration Section; (2) Noise, Vibration and Harshness Section; (3) Perceived Quality Section; (4) Human-Machine Interaction Section; (5) Market Research and Analysis Section.

Interviews were conducted with each of the ten participants and took place in the expert's working space. Interview time ranged in duration from 45 min to one hour. The interview started with general questions about the participant's working related characteristics: (1) the nature of daily work activities; (2) their relevant work relating to auditory experience, followed by specific questions relating to auditory experience design research. The interview's protocol is shown in Fig. 1.

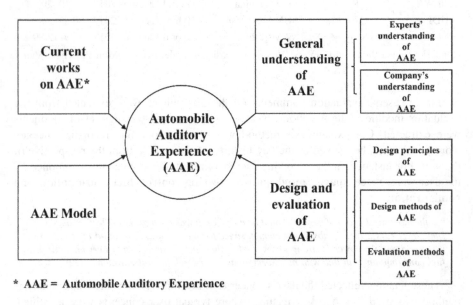

* AAE = Automobile Auditory Experience

Fig. 1. The protocol of the interview

For the interviews, the questions were predetermined, however, the question order and wording were flexible and where necessary, further explanations were given [24]. The interviews were transcribed into words of text in full. The transcribed interviews contain rich and descriptive data which were analyzed by content analysis, a systematic, replicable technique for compressing many words of text into fewer content categories based on explicit rules of coding [25, 26]. Content analysis enables researchers to sift through large volumes of data with relative ease in a systematic fashion [26]. It is a useful technique for discovering and describing the focus of individual, group, institutional, or social attention [26, 27]. Coding was made according to the interview questions. Experts' insights with the same or similar meaning were classified and combined after key insights were extracted from the original words of the interview transcriptions.

4 Results

4.1 Key Features of Automobile

In total 266 features were identified, which reflected the key concerns of users. They were extracted from the users' comments using content analysis method. Features were shown in Table 2.

Table 2. Features of each selected vehicle

Brand	Features evaluated by users
DONGFENG	modeling, anti-collision, power, gearbox, chassis, sound insulation, appearance, driving, configuration
ZOTYE	material, configuration, shock absorption, seat, power, gearbox, appearance, texture, voice interaction, noise, suspension
MG	appearance, interior, space, gearbox, fuel consumption, price, cost, driving, power, performance, configuration, feel
ROEWE	chassis, vibration, price, sound, gearbox, appearance, space, human-computer interaction, performance, power, vibration, ride, comfort, tranquility
GAC	modeling, appearance, interior, driving feeling, performance, user experience, grade sense, price
SAIC MAXUS	appearance, interior, fuel consumption, acceleration, driving, details, seats, air conditioning, driving mode
BORGWARD	appearance, price, material, quality, power, gearbox, cost performance
LUXGEN	power, fuel consumption, driving, configuration, appearance, price
LAND ROVER	cost performance, price, space, driving, comfort, maneuverability, stability, design, appearance, interior, engine, acceleration
PEUGEOT	good-looking, interior, sense of technology, suspension, space, driving feeling, power, appearance, texture
BUICK	gearbox, appearance, cost-effective, modeling, interior decoration, materials, configuration, engine
VOLKSWAGEN	performance, configuration, gearbox, driving experience, shock absorption
HONDA	power, configuration, fuel consumption, price, acceleration, performance, engine
VOLKSWAGEN	power, fuel consumption, configuration, comfort, space
AUDI	chair, engine, price, space
SKODA	energy saving, power, fuel consumption, road noise, shock absorption, driving, interior decoration, configuration, materials, appearance, space
HYUNDAI	cost-effective, engine, chassis, comfort, driving, quality

(*continued*)

Table 2. (*continued*)

Brand	Features evaluated by users
MAZDA	fun, engine, feel, pleasure
BMW	acceleration, fuel consumption, price, power, fun, exterior form
JAGUAR	ride, power, sound, comfort, seat, details
ACURA	power, personality, comfort, vibration, space, manipulation
VOLVO	shock absorption, ride, comfort, experience, power, sound, acceleration, feeling, appearance, price, performance, suspension, seat
CHEROLET	performance, configuration, price, gearbox, driving experience, comfort, fuel consumption, acceleration, power
BENZ	space, power, price, driving feeling, acceleration
BYD	acceleration, interior, seat, material, endurance, power, comfort, driving experience, configuration
LINCOLN	price, power, fuel consumption, maneuverability, comfort, space, driving feeling, endurance
TESLA	acceleration, silence, quietly, driving feeling, driving, stability, continuity, appearance, design, details, configuration, power, suspension
ROEWE	appearance, interior, shape, technology, power, suspension, gearbox, driving, space, practicability, comfort, acceleration, price
BYD	acceleration, interior, seat, material, duration, power, comfort, driving experience, configuration, acceleration
ROEWE	driving experience, sound insulation, power, vibration, appearance, quality, fuel consumption
AUDI	fuel consumption, engine, charging, weight, environmental protection
BYD	price, shock absorption, acceleration, cost, durability, fuel consumption, interior decoration, materials

The 266 features were further processed by the principle of similar semantics, and the frequency of occurrence was counted. A total of 24 key features were identified: Power/Engine, Appearance, Price, Driving Experience, Configuration, Acceleration, Fuel Consumption, Interior Space, Interior Material, Sound, Gearbox, Performance, Vibration, Seat, Chassis, Details, Quality, Texture, Voice Interaction, Stability, Air Conditioning, Charging, Weight, and Environmental Protection.

The frequency of occurrence is shown in Fig. 2.

As Fig. 2 shows, (1) the power/engine was users' most concerned factor; the appearance, price, and driving experience were in the second place; the configuration, acceleration, fuel consumption, interior space, sound, gearbox, performance, vibration, and seat were the third concerned factors; air conditioning, charging, weight, and environmental friendliness were the least concerned factors. (2) Users' concerns on the driving experience exceeded their concerns on configuration, accelerating, fuel consumption of vehicles.

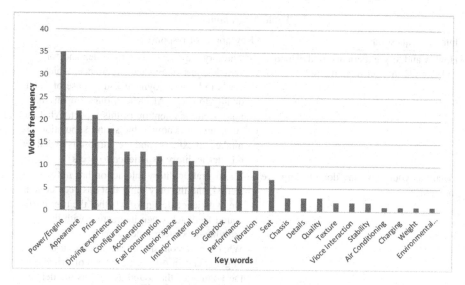

Fig. 2. Distribution of user comments

4.2   Automobile Auditory Experience Design and Research in Industry

Participants' responses to interview questions were analyzed and the results are shown in Table 3. This highlights the key insights obtained from the participants.

Table 3. Interview questions and the key points of responses

Interview questions	Key points of responses
What is your understanding of Automobile Auditory Experience (AAE)?	– The word 'experience' itself means that an object has experienced something, that is, the auditory system has experienced the sound event, so it is more accurate to say that 'auditory experience' rather than 'sound experience'
What kind of AAE do you think is good?	– When driving, only the ears are free. AAE is based on interaction. Good AAE comes from the perspective of 'pleasure' – Vehicles with good auditory experience must do well in the areas where users are mostly concerned. AAE needs to be able to perceive the context of use – Natural sounds are the best sounds, which are human preferences engraved in genes over a long evolutionary process. For many vehicles with good auditory experience, their sound must be consistent with frequency bands and in line with the natural sound

(*continued*)

Table 3. (*continued*)

Interview questions	Key points of responses
How is auditory experience understood and defined in your department?	– Auditory experience is a rigid demand for vehicles, just like visual experience, which needs to bring enjoyment and a sense of good quality to users. At present, our company has not raised the attention of this aspect. What the company does now is the sound simulation of intake and exhaust, and the purpose is to deliver a better auditory experience to users
What has your company done to improve the AAE?	– The overall auditory interaction has not yet started, and the only thing that has been done is the experience design of whether the user needs sound feedback when interacting with the multimedia system touch screen – What we are considering are whether users want to get sound in different contexts of driving and the volume of the sound as well as the delivery priority of the sound
What do you think are the principles of designing for AAE?	– Consistency: does the auditory experience conforms to user's auditory cognition? e.g. whether all auditory experience relating to the alarm is in the same dimension of user's past experiences – Anti-interference mechanism: the whole system must clearly determine the priority of sound delivery – Redundancy design: to ensure that the sound can be heard after it is sent out, and take measures in advance to make up for the failure of the sound equipment
What do you think are the design methods of AAE?	– Confucianism as AAE design methods. The Chinese character "儒(Confucianist)" is composed of "人 (Individual)" and "需 (Need)". So AAE can be designed according to the real needs of users, rather than to make a luxury and expensive car – Consider providing directional and customized AAE to each user
What methods are used to evaluate AAE in your department?	– Compatibility with functionality: whether expected functionality can be achieved – Compared with other vehicles. By measuring and collecting the data of other competitive vehicles, and then making a horizontal comparison with our own vehicles – Subjective scoring method of evaluation committee: usually using 10 points system, 1 'totally unacceptable', 10 'excellent'

(*continued*)

Table 3. (*continued*)

Interview questions	Key points of responses
What are the company's expectations by enhancing AAE?	– Hope to create a sound image and help promote the brand of automobiles of enhancing AAE
If there is a theoretical framework/model of AAE, what do you expect from it?	– To be able to quantify all subjective feelings, because the engineering emphasis is implementable and measurable
	– The model is expected to be a set of indicators, including all evaluation indicators related to sound and the functional relationship between indicators
	– The model must be subdivided. There are different strategies for different levels of AAE evaluation, such as reporting to the general manager, simplicity is the most important. But for designers and engineers, it must be very specific
	– The model is expected to include both objective evaluation and the optimization method

The participants' daily work were all relevant to automobile auditory experience. There were five participants with expertise in auditory experience design, two in auditory experience research, two in the evaluation of auditory experience and one in the charge of auditory experience design and research. The auditory working experiences of the ten participants were all more than three years.

As shown in Table 3, (1) automobile auditory experience should be improved from the perspectives of human evolution, human needs, driving contexts, the human's five senses, and natural sounds consistent with the particular frequency band; (2) consistency, anti-interference mechanism and redundancy design were considered as the design principles of auditory experience; (3) comparison and users' subjective evaluation were the main means of auditory experience evaluation; (4) a framework or model of auditory experience design was expected by developers to enhance holistic experience of automobile.

5   Discussion

5.1   Key Features as the Concerns of Users on Automobile

This study was designed to determine the users' concerns of automobile from actual users' comments, which can be summarized in two aspects (Fig. 2): (1) vehicles' basic functional properties as a means of transportation, e.g. the power, appearance, price, configuration, acceleration performance, and fuel consumption, which account for more than 52% of the comments. (2) driving experience related factors, e.g. sound, voice interaction, vibration. So, the vehicle's basic functional properties were most important to the users. The results show that, in general, products such as automobiles must have sound basic functions, and all other auxiliary functions including user experience should be based on sound basic function services.

Figure 2 shows users pay more attention to the driving experience than the technical configuration, acceleration performance and fuel consumption; and among the factors related to driving experience, sound, vibration, internal space, and voice interaction account for the majority of the comments, so sound, vibration, and space appear to be the key factors affecting driving experience. The study also revealed that users pay the least attention (only 2%) to the environmental factors, weight, charging and the air conditioning system of the vehicle.

Several previous research articles have shown the drivers' needs from the perspectives of automobile safety systems and user experience by using mixed methods, e.g. Future Workshops, The Repertory Grid Technique and In-depth Accident study [28–30]. These prior studies mainly focused on the expected future needs/concerns of drivers, rather than the actual current drivers' concerns. At the same time, the conclusions drawn by these prior studies were more macro, without details, e.g. Control, Versatility, Safety, Driving Please, Comfort and Convenience [28]. This paper tried to uncover detailed and specific micro-suggestions, e.g. Power, Appearance, Price, Gearbox, Voice Interaction. In spite of the differences, the conclusions tended to be consistent, i.e. trying to achieve better user experience based on the satisfaction of such factors as safety [29] and convenient transportation.

5.2   Insights of Automobile Auditory Experience from Industry

The key points summarized in Table 3, to some extent, represented the typical work, understanding, design and research methods and strategies toward auditory experience in the automobile industry.

The current works on automobile auditory experience were interdepartmental collaboration, involving many departments, i.e. the design department, the vehicle integration department, the powertrain department and marketing. This kind of multi-sectoral cooperative work mode can solve the problems quickly and accurately. However, it is also for this reason that the current automobile auditory experience lacks a unified standard. Each department had internal guidelines, which hindered the improvement of automobile auditory experience from a holistic perspective.

In terms of the understanding of automobile auditory experience, designers, engineers, researchers in the company have begun to realize the importance. Automobile developers have made it clear that auditory experience should be improved from the perspectives of human evolution, human concerns, and human needs, as well as from human's five senses. However, there were no specific strategies on auditory experience from the company's new product development process. By April 2018, very detailed auditory experience improvements had taken place, e.g. improving the auditory experience by adjusting the intake and exhaust sounds of the car engine; improving the experience by focusing on the driving context.

The sound problems of automobiles have always existed, which were more prominent due to the particularity of the electric vehicle's sound system. The design principles and methods as well as evaluation methods of automobile auditory experience were largely rely on the practices of traditional vehicles. Consistency, anti-interference mechanism and redundancy design were considered as design principles; human needs centered design and customized design methods were considered. Comparison methods and user subjective evaluation were the main means of auditory experience evaluation.

Importantly, participants' insights and expectations on the framework/model of automobile auditory experience were relatively specific and clear (Table 3). This indicates, to a certain extent, that a framework or model is needed as references for the design or research of automobile auditory experience.

Interviews were focused on a unique and holistic perspective of automobile sound research: auditory experience. Most previous studies, however, focused on the optimization and evaluation of automobile sound quality, e.g. the relationship between psychoacoustic parameters and perceived sound quality of automobile users [30], the development of assessing methods for external sound of electric vehicles [31], and the evaluation of noise quality of automobile doors [32, 33], etc. Although these prior studies have made contributions understanding to the automobile sound, they lack system and holistic thinking or suggestions on the automobile auditory experience.

The paper opened a window for the researchers engaged in design research to get an insight into the needs and requirements of automobile companies for auditory experience design and research, as well as the current industry measures adopted to the design of automobile sound during the period of automobile transformation.

5.3 Limitations

There are limitations regarding data collection and analysis.

In terms of data collection: (1) the number of electric vehicles accounted for 22% of the total sample size, which was relatively low. This is mainly due to the relatively small number of electric vehicles on the market; (2) the comments were from Chinese users, with no comments from other countries. It is difficult to judge whether these 96 comments were objective, impartial and representative; (3) for the interview, participants were all from one company.

In terms of the analysis methods used: when using content analysis to encode users' comments, the coding categories may risk losing a few key features. Involving more researchers in the coding process might have helped.

6 Conclusion

By collecting and analyzing the data of automobile users and industrial automobile developers, this paper has identified the users' concerns and developers' insights of automobile.

To conclude, in the period of automobile transformation, users are still most concerned with vehicle's functional attributes, i.e. automobile's power/engine was the most concerned factor. The second group of user concerns relates to appearance, price, and driving experience. The third group of concerns include configuration, accelerating performance, and fuel consumption. Environmental friendliness were the least concerned factors. These findings help uncover the real concerns of users.

This article concludes with the implications of the research findings for the design of automobile auditory experience, i.e. although designers, engineers and managers have realized the importance of automobile auditory experience, but the automobile company has not raised the attention of this aspect, and the design and research of automobile auditory experience needs to be integrated across departments. However, cross-departmental cooperation is the current mode of work, which often meet great obstacles. Automobile developers, under these circumstances, have their own means, understanding and expectations of dealing with specific issues relating to the auditory experience, i.e. (1) human evolution, human needs, driving contexts, the human's five senses, and natural sounds consistent with the particular frequency band were their understanding of the design of automobile auditory experience; (2) consistency, anti-interference mechanism, and redundancy design were considered as the design principles of auditory experience by automobile developers; (3) comparison and users' subjective evaluation are current evaluation methods for auditory experience; (4) a framework of auditory experience design was expected by developers.

This research is timely, given the users' concerns of vehicles and the automobile industrial strategies, understanding and expectations to enhance auditory experience, it would enable automobile market researchers, designers, engineers as well as decision makers have a better understanding of the factors contributing to users' auditory experience.

Acknowledgements. This research was supported by the China Scholarship Council and Design Research Society. The authors thank all the participants involved in the research.

References

1. Scotiabank: World auto sales growth stalls in 2018 (2018)
2. Chen, Z.: Focusing on the "New Four Modernizations" of the automobile industry and leading quality strategic transformation and upgrading through innovation. Shanghai Quality, pp. 20–22 (2018)
3. Yang, B.: Focus on innovation to accelerate the transformation of development - new energy vehicles development trends and prospects. Explore Sci. **3**, 66 (2016)
4. Kane, M.: Global October sales: new record for plug-in electric cars (2018). https://insideevs.com/global-october-sales-new-record-electric-cars/
5. Egbue, O., Long, S.: Barriers to widespread adoption of electric vehicles: an analysis of consumer attitudes and perceptions. Energy Policy **48**, 717–729 (2012)
6. Ziefle, M., Beul-Leusmann, S., Kai, K., Schwalm, M.: Public perception and acceptance of electric vehicles: exploring users' perceived benefits and drawbacks. In: International Conference of Design (2014). http://dx.doi.org/10.1007/978-3-319-07635-5_60
7. Nakai, Y.: Electric vehicle (EV) manufacturers' challenge: R&D strategy of battery safety units seen in WIPO data. In: Innovation Conference (2013). https://doi.org/10.1109/SIIC.2013.6624165
8. Ying, L., Zhan, C., Jong, M.D., Lukszo, Z.: Business innovation and government regulation for the promotion of electric vehicle use: lessons from Shenzhen. China J. Cleaner Prod. **134**, 371–383 (2016). https://doi.org/10.1016/j.jclepro.2015.10.013

9. Gillibrand, A., Suffield, I., Vinamata, X., Williams, R., Brückmann, A.: An initial study to develop appropriate warning sound for a luxury vehicle using an exterior sound simulator. SAE Technical Paper (2011). https://doi.org/10.4271/2011-01-1727

10. Li, Q.: Soundscape study and soundscape design. Doctoral thesis. Tsinghua University (2004)

11. Howard, D.M., Angus, J.A.S.: Acoustics and Psychoacoustics, 4th edn. Posts & Telecom Press, Beijing (2014)

12. Birkin, G.: Aesthetic complexity: practice and perception in art & design. Nottingham Trent University (2010)

13. Chion, M.: Sound. Peking University Press, Beijing (2013)

14. Lu, Y.: Acoustic experience: the study on sound consciousness and auditory aesthetics in design. Doctoral thesis, China Central Academy of Fine Arts (2017)

15. Alton Everest, F., Pohlmann, K.C.: Master Handbook of Acoustics, 5th edn. Post & Building Press, Beijing (2016)

16. Standardization (2010) ISO 9241-210: 2010 Ergonomics of human-system interaction Part 210 Human-centred design for interactive systems (2010)

17. Ding, Y., Guo, F., Hu, M., Sun, F.: A review of user experience. Ind. Eng. Manage. **19**, 92–97 (2014). https://doi.org/10.3969/j.issn.1007-5429.2014.04.013

18. Johnson, R.B., Onwuegbuzie, A.J., Turner, L.A.: Toward a definition of mixed methods research. J. Mixed Methods Res. **1**, 112–133 (2007). https://doi.org/10.1177/1558689806 298224

19. Home A. http://ir.autohome.com.cn/about-us

20. About Autohome. http://ir.autohome.com.cn/

21. Robson, C., McCartan, K.: Real World Research. Wiley, Chichester (2016)

22. Autohome. https://k.autohome.com.cn/detail/view_01bzkvve3n64w32d1m6wrg0000.html?st=0&piap=#pvareaid=3454575

23. Romesburg, C.: Cluster analysis for researchers (2004). Lulu.com

24. Haines, V., Kyriakopoulou, K., Lawton, C.: End user engagement with domestic hot water heating systems: design implications for future thermal storage technologies. Energy Res. Soc. Sci. **49**, 74–81 (2019). https://doi.org/10.1016/j.erss.2018.10.009

25. Berelson, B.: Content analysis in communication research. Am. Polit. Sci. Assoc. **46**(3), 869 (1952)

26. Stemler, S.: An overview of content analysis. Pract. Assess. Res. Eval. **7**, 137–146 (2001)

27. Weber, R.P.: Basic content analysis, vol. 49. Sage (1990)

28. Gkouskos, D., Normark, C.J., Lundgren, S.: What drivers really want: investigating dimensions in automobile user needs. Int. J. Design **8**, 59–71 (2014)

29. Elslande, P.V., Fouquet, K.: Drivers' needs and safety systems. In: European Conference on Cognitive Ergonomics: the Ergonomics of Cool Interaction (2008). https://doi.org/10.1145/1473018.1473029

30. Lee, Y.J., Shin, T.J., Lee, S.K.: Sound quality analysis of a passenger car based on electroencephalography. J. Mech. Sci. Technol. **27**, 319–325 (2013). https://doi.org/10.1007/s12206-012-1248-z

31. Singh, S., Payne, S.R., Jennings, P.A.: Toward a methodology for assessing electric vehicle exterior sounds. IEEE Trans. Intell. Transp. Syst. **15**, 1790–1800 (2014). https://doi.org/10.1109/TITS.2014.2327062

32. Parizet, E., Guyader, E., Nosulenko, V.: Analysis of car door closing sound quality. Appl. Acoust. **69**, 12–22 (2008). https://doi.org/10.1016/j.apacoust.2006.09.004

33. Kuwano, S., Fastl, H., Namba, S., Nakamura, S., Uchida, H.: Quality of door sounds of passenger cars. Acoust. Sci. Technol. **27**, 309–312 (2006). https://doi.org/10.1250/ast.27.309

User Experience in Real Test Drives with a Camera Based Mirror – Influence of New Technologies on Equipping Rate for Future Vehicles

Corinna Seidler[1], Seda Aydogdu[2(✉)], and Bernhard Schick[1(✉)]

[1] University of Applied Sciences Kempten, Bahnhofstraße 61,
87435 Kempten, Germany
{corinna.seidler,bernhard.schick}@hs-kempten.de
[2] MdynamiX AG, Heßstraße 89, 80797 Munich, Germany
seda.aydogdu@mdynamix.de

Abstract. The variety of products on the market that are offered with advanced driver assistance systems is huge. Besides systems like the lane keeping assistant also camera based systems are more and more in the foreground. The Full Display Mirror (FDM), a camera-based hybrid mirror, is also being installed in vehicles. User experience in the sense of acceptance and peace of mind as well as benefit of the product are essential not only to know the systems are installed, but also to know how to use them. In a customer study with N = 60 persons, interviews and questionnaires were conducted during a test drive in real road traffic with various scenarios in order to test and evaluate the FDM. In comparison to the normal mirror, the FDM scored very well - especially in the field of view, the feedback was extremely positive. The increased safety with the use of the system, among other things due to the considerably larger field of view, is an absolute plus point of this camera-based mirror version. The fast familiarization with the system as well as the high user friendliness make the FDM a meaningful invention.

Keywords: User experience · Camera based mirror · Field of view ·
Customer acceptance · User friendlieness · Trust

1 Introduction

Technological innovations are increasing more and more - what once seemed unthinkable, today is the most normal thing in the world (Stappenbeck 2017). A good usability of systems, task-appropriate functions as well as emotionally appealing user experiences are the goals of good human-computer interaction and thus also the basis for a successful innovation. Aspects such as attractiveness, user intention, and simplicity play a major role in capturing user experience aspects such as usefulness and user-friendliness (Reiterer and Geyer 2013).

In the ISO definition, usability is described as the extent to which products can be used in a particular context to achieve specific objectives effectively, efficiently and

© Springer Nature Switzerland AG 2019
H. Krömker (Ed.): HCII 2019, LNCS 11596, pp. 270–281, 2019.
https://doi.org/10.1007/978-3-030-22666-4_20

satisfactorily (DIN EN ISO D8241-11 1999; DIN EN ISO 9241-210 2010; Mentler 2018; Robier 2015). An evaluation of the system is therefore only possible after use. The user experience, on the other hand, is defined as "unique, subjective and individual experience that is evoked in the interaction with a system" (Körber et al. 2013, p. 14). Consequently, it is always about the interaction of a user with a system and can therefore be captured at several points in time, whether before, during or after use (Mentler 2018). The focus is therefore on the well-being of the user and not on the performance, as with the usability concept (DIN EN ISO 9241-210 2010; Mentler 2018). Due to this distinction of terms, in the following sections the focus will be on the user experience when using the product and less on the usability. User experience in this context also affects the equipment of vehicles with driver assistance systems and the development of automated vehicles. To ensure that the in the vehicle installed system is actually used, its development must be user-oriented. This means that in addition to cost, technical feasibility, standards, human capabilities and desires must also be included (Aydogdu et al. 2018; Held and Schrepp 2018; König 2015; Seidler and Schick 2018).

An interaction with different products and systems in vehicles is unavoidable due to the increasing digitalization, whereby the intuitive interaction has moved further and further into the foreground of research. This describes an interaction that is fast, unconscious and automatic (Blackler and Hurtienne 2007; Blackler et al. 2010; Macaranas et al. 2015; Ullrich and Diefenbach 2010a). However, the intuitive use of the system is also countered by its acceptance. This must be given, so that a use of the product comes at all. The technology acceptance model, an adaption of theory of reasoned action, served as basis to record the use of the product. This model is concerned that perceived usefulness and perceived ease of use influence the attitude towards use as well as the actual product use (Davis 1985; Davis et al. 1989; Venkatesh and Davis 2000). If, consequently, the comparison of actual performance in terms of experience in product use and the expectation, i.e. the target performance is compatible, customer satisfaction results. This in turn leads to confirmation and consequently to satisfaction (Nerdinger et al. 2015). The aim is therefore to keep the investigated product users satisfied by ensuring that the actual and target data of the product match. A supplement to this model is the KANO model, which examines the respective product requirements. These requirements are subdivided into enthusiasm, performance and basic requirements and classified according to the strength of the influence on customer satisfaction. It is also examined whether the characteristic is indifferent or questionable, in the sense of spending money on it. Thus, the KANO model can be used to analyze which problems exist with the product and which needs customers have in this respect (Hölzing 2007; Nerdinger et al. 2015). The market opportunity map also enables the structuring of customer wishes and requirements. In this representation, products can be classified according to importance and satisfaction. Among other things, this helps to classify the current acceptance stage of the product and how it is perceived by customers (Aydogdu et al. 2018).

The development of automation in the automotive industry is constantly progressing. Despite this, the rear-view mirror is still an integral part of vehicle equipment. The driver is accustomed to searching for information covering rear traffic at these points. However, the timely development away from a mirror towards a display is to be expected (Lee 2012). The development of a camera based mirror, following named as Full Display Mirros (FDM) starts here. The system will be installed in vehicles and may replace the mirror in the future, at least as a supplement or extension of it. The FDM operates as a standard mirror as well as display mode with streaming a video. Subjects can toggle the version manual by themselves. One of the advantages of the FDM is the mirror intergrade LCD for optimal rear vision. Thus, this paper presents the results of the evaluation and feedback of the users' experience by using the FDM. The results are to be evaluated statistically and presented with regard to the KANO model, the technology acceptance model and the market opportunity map. The goal is the investigation of human-computer interaction. The human being should be considered as the focal point in the use of the FDM in order to generate the best possible evaluation of the system.

2 Methods

2.1 Participants

In order to develop the Full Display Mirror (FDM) as a user-friendly system, it was tested in three studies with N = 60 to have a holistic view. The following criteria were taken into account when planning the experimental study:

- Different daytimes
- Different routes
- Different manoeuvres
- Interview
- Questionnaire
- Different methods
- Eye-Tracking.

The three studies included the following number of subjects

- Daylight study with n = 40
- Night ride study with n = 5
- Eye-tracking study with n = 15

In the following results, the contents of the test drive by daylight are described in more detail.

In total, n = 40 test drivers ($M_{Age} = 30.87; SD = 12.38$, $Range : 18 - 66$), 12 females (30%) and 28 males (70%), participated in the test drives. On average, the annual mileage was given as $M = 17,779$ km/$year$ ($SD = 8,378$). The technology affinity (from 1 = technology affinity very high to 7 = no technology affinity) was rather high in the sample ($M = 2.10$; $SD = 1.32$). The employment status of the sample was composed as follows: 19 subjects were college students, 13 full-time and three part-time employees, two officials, and in each case one pupil, apprentice and pensioner. Participants were not paid for completing the experiment. They were won through a call for tenders at the University of Applied Sciences in Kempten. Due to the central limit value theorem, the normal distribution is used as a basis for the following calculations (Döring and Bortz 2016).

2.2 Materials

Testing Equipment

The FDM installed in the test vehicle is a camera based mirror in a hybrid version. This means that mirror mode and display mode are combined, allowing the driver to switch between the two modes. The mirror mode operates as a standard mirror, whereas the display mode is streaming a video. The reason for this hybrid mode is to be able to distinguish between different driving scenarios, weather conditions and driver preferences. The test persons should be able to switch between the versions according to their preferences. This should facilitate a comparison of both modes. The following Fig. 1 shows the difference in looking behind in mirror mode versus display mode.

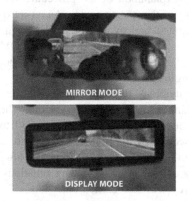 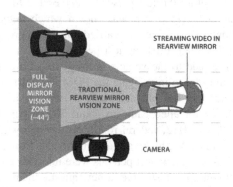

Fig. 1. Mirror mode versus display mode (see "Introducing the New Gentex Aftermarket Full Display Mirror (FDM)")

Interview and Questionnaire

Before and after the test drive on a public road, the subjects received a questionnaire. In addition to get information about user experience, the confidence in the system as well as the anticipated control and intended use was examined. The intuitive interaction questionnaire (INTUI) was used for this purpose. The questionnaire describes the factors "effortless" (5 items), "gut feeling" (4 items), "magical experience" (4 items) and "intuitive interaction" (1 item) (Ullrich 2014; Ullrich and Diefenbach 2010a, 2010b). In addition, questions based on the questionnaire of Gold et al. (2015) were used, especially the factors "trust in automation" with 12 Items and "safety" with four Items. The question about trust in automation included how reliably the system is evaluated, whether it offers security and whether one is confident in the system or whether there is distrust of it. The evaluation was made on a scale of 1 = absolute vote to 7 = absolute vote not to. The "safety" factor could also be answered on this scale and included, among other things, questions on support for hazards and road safety.

The product-specific criteria ("display", "degree of relief", "monitoring", "feeling of safety", "availability", and "field of view"), which the subjects should evaluate, were developed in a preliminary workshop. These criteria were evaluated by the participants before the test drive according to importance on a scale of 1 = not important to 7 = very important. After the test drive these criterias were evaluated.

Furthermore, an interview with the subjects was carried out during the journey in order to obtain further information on specific driving manoeuvres, such as the overtaking process and lane change. Further questions were also asked, including the topics of focusing, positive and negative aspects of the FDM, its ratings and importance, comparison of FDM and mirror, purchase decisions and so on. All interviews both spoken and written were conducted in German.

2.3 Procedure

At the beginning of the study, the test persons were welcomed, their driving licence checked and a declaration of liability issued. An explanation of the procedure was given before the test persons received the first part of the questionnaire. As an introduction, a video clip was shown which demonstrated and explains the FDM. The following test drive was carried out with a premium vehicle in which the Gentex FDM was installed. First the test persons were instructed in the test vehicle, then they got accustomed to the vehicle in the sense of a running-in period.

In order to investigate the FDM in all driving situations, three driving speeds were included in the planned route as well as a parking situation was carried out. The distance travelled came to about 70 km per subject. The speeds varied from 80 to 100 km/h (B12, federal road), on the following section of the highway from 140 to 160 km/h (A7, highway) and in the city from 30 to 60 km/ h (City). During and after the test drive, the test persons were interviewed about the respective driving situations. Following the test drive, a new questionnaire survey was also carried out.

3   Results

The intuitive interaction questionnaire (INTUI) revealed that the product is equated to/with effortlessness, as well as with ease, simplicity ($M = 5.85$; $SD = 0.97$) - more than with effort, disorientation or high attention. Furthermore, the use of the FDM in terms of good feeling was neither thoughtless, unconscious or unreasonable, nor deliberate, conscious or justified ($M = 4.16$; $SD = 1.30$). The product is equated more to/with an enthusiastic, magical experience or stirring product ($M = 4.73$; $SD = 1.14$) – more than as insignificant or irrelevant. As well, the use of the FDM is rated as intuitive ($M = 5.90$; $SD = 1.01$). The reliability, which describes the reliability of the measurement method, indicates the degree of measurement accuracy (Döring and Bortz 2016). To measure the internal consistency, Cronbach's Alpha was calculated. Acceptable to good values were achieved in all factors (effortlessness $\alpha = .86$; gut feeling $\alpha = .75$; magical experience $\alpha = .87$) (Field 2009).

The questions used on trust in automation included whether the system can be trusted in principle, to what extent its reliability is given or whether the system is

deceptive. The twelve items of the factor trust in automation resulted in a high reliability (Cronbach's Alpha $\alpha = .90$). The values recorded on a scale from 1 = "high
confidence" to 7 = "no confidence" resulted in an average high confidence in the FDM
($M = 2.59$; $SD = 1.07$). The experienced feeling of safety, which was recorded with
four items, was also rated rather highly by the test persons using the FDM
($M = 3.32$; $SD = 1.35$). As in the trust in automation factor, good reliability could be
recorded (Cronbach's Alpha $\alpha = .80$).

A comparison of the evaluation criteria of rear view mirror and FDM shows, that
the FDM was rated better than the normal mirror in all criteria except "availability".
The subjects stated that the field of view of FDM is 95% better or much better than the
mirror. In all three route sections the category "field of view" of the FDM is rated
significantly better than the normal mirror (B12: $t(39) = -9.21; p < .001$; A7:
$t(39) = -5.42; p < .001$; City: $t(39) = -8.02; p < .001$). Like this, also the criterion
"monitoring" was rated better in the FDM than in the normal mirror (B12: $t(39) = -3.31; p = .002$; A7: $t(39) = -2.87; p = .007$; City: $t(39) = -4.01; p < .001$).
There were also significant differences in the sections of the route in terms of "display",
"feeling of safety" and "degree of relief". If one now considers the differences in the
information on the importance of the criteria versus the later evaluation, it shows that
some factors, even if only minor, are rated significantly worse afterwards. Thus, the
feeling of safety ($t39) = -1.73$; $p = .091$) and the display ($t(39) = -2.98$; $0 = .005$) as well as the overall factor ($t(38) = -1.79$; $p = .081$) of all evaluation criteria are rated worse after the test drive than its importance is indicated beforehand. The
other criteria such as operation, field of view, monitoring, availability and degree of
relief were not evaluated significantly differently after the test drive than the importance
before the test drive. The following graphic illustrates this result (Fig. 2).

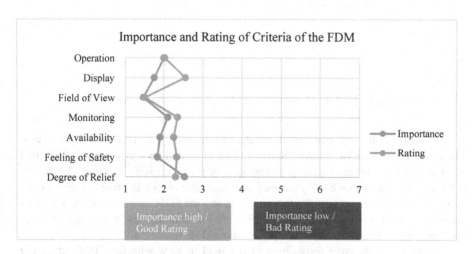

Fig. 2. Importance and rating criteria of the FDM

In addition, interesting customer statements could be recorded for the respective criteria, which enable further development with regard to increased customer requirement, acceptance and improvement of the actual use of the system. Looking at these results in the market opportunity map, the following graph illustrates this. The results of customer requests are divided into four quadrants with the axis cross importance versus fulfillment degree. The quadrant "Market Opportunity" defines requirements that are high in importance but low in satisfaction. The quadrant "Product Requirements", in which almost all evaluations can be mapped, contains product criteria that are very important and also record a high level of satisfaction. The graphic shows/demonstrates/reveals, that the mirror is rated worse than the FDM in all categories, no matter how important the criterion appears to be. The least important criterion is the degree of relief. The most important criterion is the field of view. The largest difference in the evaluation of mirrors and FDM is also reflected in this criterion (Fig. 3).

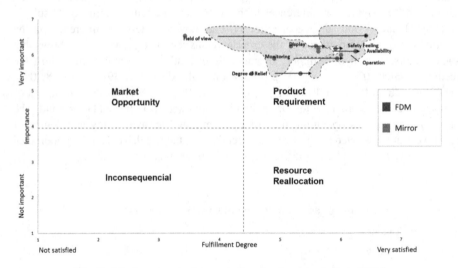

Fig. 3. Market opportunity map of evaluation criteria from FDM

Furthermore, the test persons were asked to describe the FDM with three words. The following entries were made: Safety (23%), clear and good overview (43%), field of view (13%), innovative (23%), picture quality positive (37%), comfort (13%), fun (13%), useful and intuitive (37%) as well as improvable (47%). In order to gain an even more precise insight into how the test drivers assess the FDM, it was recorded what was particularly positive or negative. The following evaluation gives an exemplary overview of which entries the field of view received (Fig. 4).

To obtain a subjective recognition of the field of view with the FDM versus the rear-view mirror, the test persons were asked to make an entry in a drawing. Here they could indicate how broad the field of view is perceived in the respective versions. The following figure shows the result. A total of 85% more field of view with the FDM

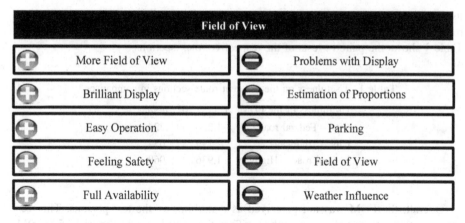

Fig. 4. Positve and negative aspects of field of view with FDM

instead of the rearview mirror was given. The subjects also stated that on average, the hidden area when using the mirror was given as 35.3% (Range: 7% to 60%).

In order to dispel the concerns that problems of focussing might occur, the question was asked whether the subjects had problems with this. A total of 75% of the respondents stated that they had no problems to focus on the FDM. A total of 10% reported focusing problems, 15% of the participants had partly problems. Further wishes regarding further development of the FDM towards design as well as additional features, could also be derived (Fig. 5).

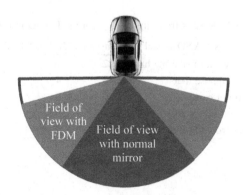

Fig. 5. Field of view with normal mirror versus FDM

In order to capture user-friendliness and usefulness, the questions were based on the technology acceptance model. The users rated the FDM on a scale of 1 = very user friendly with 7 = not user-friendly on average very positive ($M = 2.08$; $SD = 1.56$). Feedback on user-friendliness such as idiot proof, super easy, intuitive to fold etc. supported this result. Also the usefulness was rated very good on a scale from 1 = very useful to 7 = not useful with $M = 2.60$ ($SD = 0.55$). Here, however, it was shown that usefulness is rated most useful in the city ($M = 2.13$; $SD = 1.64$) compared to federal road ($M = 2.48$; $SD = 1.24$) or even highway ($M = 2.98$; $SD = 1.98$). The following Table 1 shows the paired t-tests of the respective track sections.

Table 1. Usefulness on the different route sections in comparison

Usefulness of the FDM	t	df	p-value
City – Federal road	−1.236	39	.224
City – Highway	−2.199	39	.034
Federal road – Highway	−1.936	39	.060

Overall, the FDM was not perceived as disturbing by the test persons. The recommendation rate was also very high at 75% in the hybrid version versus 12.5% FDM without the possibility of folding into the mirror versus no recommendation 12.5%. The purchase decision for the tested hybrid version was 57.5% (pure FDM 15% and no version, i.e. normal mirror 27.5%). The assessment of the degree of maturity also revealed that this was rated relatively high by the test persons ($M = 71.5\%$; $SD = 15.56$; range 0–100%), which illustrates a high product quality. The evaluation of the questions on the KANO model showed that the FDM was an enthusiastic feature. 22 of the test persons stated that it would make them happy if an FDM was installed in their next vehicle. The following Table 2 gives an overview of the respective response categories. The question was once asked what would happen if the FDM was installed and once if it was not installed.

Table 2. Results from questions from the KANO model

Answers KANO model	Installed	Not installed
That would make me happy	22	1
I presuppose that	0	4
I do not care	10	18
I could accept that	7	16
That would bother me a lot	1	1

4 Discussion

The aim of this study was to gather customer feedback on the FDM. End users should be involved as they will and should use the system in their vehicles in the future. Therefore, this large-scale customer study was carried out to involve users and get feedback on how the use of the system is perceived. The evaluation of the interview using theoretical models and existing questionnaires is intended to show how the customer's requirements and wishes affect opinions regarding the product and evaluations of them.

Results showed that there are significant differences in the evaluation of FDM and normal mirror. Based on this, it can be determined, that the FDM as a new technology is already positively evaluated and accepted after a short test drive. A comparison of the FDM and the mirror clearly shows that the broader field of view of the FDM, in particular, receives many plus points in the evaluation. Automation in the sense of using an FDM offers great potential for improving safety, since the field of view is much larger and thus offers more detection potential for hazards. The possibility to switch manually from FDM to mirror by means of a folding mode, even increases the confidence in the system. The occurrence of failures has a strong impact on confidence in automation, but the hybrid version can compensate for malfunctions or technical failures (Feldhütter et al. 2016; Gold et al. 2015).

The low proportion of subjects with focusing problems shows that this affects only a few. Possibly this problem could be excluded with long-term use. If the subject becomes accustomed to the system, it is easier for the test persons to focus, which could reduce the problems. In order to generate further information in this respect, it would be possible to record the accommodation moderation and adaptation. The data of the eye-tracking study will be used for this purpose.

Further results from the study show similar tendencies in user experience. This suggests the conclusion that FDM as an extension of the mirror is an important system. The user experience was rated very positively. The usability and usefulness of the FDM also show, how well the system already works. The test persons' desire to expand the system shows that there is a desire for digitisation. With a further development of the system, the rear traffic could be detected by vehicle detection. A possible extension would be to add a warning system, including information on the time to collision or a colour marking in relation to the speed of the vehicle (Sun et al. 2012). However, this question can be examined further and in more detail. Further research is suggested, to analyze whether added information in the display could be helpful for the monitoring and to increase the benefits for the driver.

In the future, the installation of driver assistance systems in new cars will become more and more important. The trend is towards supporting the driver with these systems. More and more vehicles will be equipped to increase safety in traffic (Köllner 2018). However, the basis for this must be the customer acceptance of the systems, which are only then also used (Schick et al. 2019). In the future, driver assistance systems will certainly be available, but the implementation and use of these systems by customers is currently doubtful. For the actual use of the customer-oriented application, however, a further development of the systems with the involvement of the end users is indispensable.

References

Aydogdu, S., Schick, B., Wolf, M.: Claim and reality? lane keeping assistant - the conflict between expectation and customer experience. In: 27 Aachener Kolloquium Fahrzeug- und Motorentechnik (2018)

Blackler, A., Hurtienne, J.: Towards a unified view of intuitive interaction: definitions, models and tools across the world. MMI-Interaktiv **13**, 36–54 (2007)

Blackler, A., Popovic, V., Mahar, D.: Investigating users' intuitive interaction with complex artefacts. Appl. Ergon. **41**, 72–92 (2010). https://doi.org/10.1016/j.apergo.2009.04.010

Davis, F.D.: A technology acceptance model for empirically testing new end-unser information systems: Theory and results (Dissertation). Massachusetts Institute of Technology (1985)

Davis, F.D., Bagozzi, R.P., Warshaw, P.R.: User acceptance of computer technology: a comparison of two theoretical models. Manage. Sci. **35**, 982–1003 (1989)

Döring, N., Bortz, J. (eds.): Springer-Lehrbuch. Forschungsmethoden und Evaluation in den Sozial- und Humanwissenschaften. Springer, Heidelberg (2016). https://doi.org/10.1007/978-3-642-41089-5

DIN EN ISO 9241-21: Ergonomie der Mensch-System-Interaktion Teil 210: Prozess zur Gestaltung gebrauchstauglicher interaktier Systeme. Beuth, Berlin (2010)

DIN EN ISO D8241-11: Ergonomische Anforderungen für Bürotätigkeiten mit Bildschirmgeräten Teil 11: Anforderungen an die Gebrauchstauglichkeit. Beuth, Berlin (1999)

Feldhütter, A., Gold, C., Hüger, A., Bengler, K.: Trust in automation as a matter of media influence and experience of automated vehicles. Proc. Hum. Factors Ergon. Soc. Ann. Meet. **60**, 2024–2028 (2016). https://doi.org/10.1177/1541931213601460

Field, A.: Discovering Statistics using SPSS: and Sex and Drugs and Rock 'n' Roll. SAGE Publications Ltd, London (2009)

Gold, C., Körber, M., Hohenberger, C., Lechner, D., Bengler, K.: Trust in automation – before and after the experience of take-over scenarios in a highly automated vehicle. Procedia Manufact. **3**, 3025–3032 (2015). https://doi.org/10.1016/j.promfg.2015.07.847

Held, T., Schrepp, M.: UX-Professionals im Spannungsfeld zwischen Nutzern und Entscheidern. In: Hess, S., Fischer, H. (eds.) Mensch und Computer 2018 - Usability Professionals, pp. 427–436. Gesellschaft für Informatik e.V. und German UPA e.V, Bonn (2018)

Hölzing, J.A.: Die Kano-Theorie der Kundenzufriedenheitsmessung: Eine theoretische und empirische Überprüfung, 1st edn. Gabler, Wiesbaden (2007)

Introducing the New Gentex Aftermarket Full Display Mirror (FDM). http://www.gentex.com/media-kit-files/Aftermarket-FDM/Aftermarket-FDM-Flyer.pdf

Köllner, C.: Immer mehr Assistenz für Fahrer (2018). https://www.springerprofessional.de/fahrerassistenz/automatisiertes-fahren/immer-mehr-assistenz-fuer-den-fahrer/15544504

König, W.: Nutzergerechte Entwicklung der Mensch-Maschine-Interaktion von Fahrerassistenzsystemen. In: Winner, H., Hakuli, S., Lotz, F., Singer, C. (Eds.) ATZ/MTZ-Fachbuch. Handbuch Fahrerassistenzsysteme: Grundlagen, Komponenten und Systeme für aktive Sicherheit und Komfort, 3rd ed., pp. 621–632. Springer Vieweg, Wiesbaden (2015). https://doi.org/10.1007/978-3-658-05734-3_33

Körber, M., Eichinger, A., Bengler, K., Olaverri-Monreal, C.: User experience evaluation in an automotive context. In: 2013 IEEE Intelligent Vehicles Symposium Workshops, pp. 13–18 (2013). https://doi.org/10.1109/IVS.2013.6629440

Lee, E.: Der Rückspiegel als Display. ATZ **7**, 108–113 (2012)

Macaranas, A., Antle, A.N., Riecke, B.E.: What is intuitive interaction? balancing users' performance and satisfaction with natural user interfaces. Interact. Comput. **27**, 357–370 (2015). https://doi.org/10.1093/iwc/iwv003

Mentler, T.: Usability engineering and user experience design sicherheitskritischer systeme. In: Reuter, C. (ed.) Sicherheitskritische Mensch-Computer-Interaktion: Interaktive Technologien und Soziale Medien im Krisen- und Sicherheitsmanagement, pp. 41–60. Springer Vieweg, Wiesbaden (2018). https://doi.org/10.1007/978-3-658-19523-6_3

Nerdinger, F.W., Neumann, C., Curth, S.: Kundenzufriedenheit und Kundenbindung. In: Moser, K. (ed.) Wirtschaftspsychologie, 2nd edn, pp. 119–137. Springer, Heidelberg (2015). https://doi.org/10.1007/978-3-540-71637-2_8

Reiterer, H., Geyer, F.: Mensch-computer-interaktion. In: Kuhlen, R., Semar, W., Strauch, D. (eds.) Grundlagen Information und Dokumentation: Handbuch zur Einführung in die Informationswissenschaft und –praxis, pp. 431–440. De Gruyter Saur, Berlin (2013)

Robier, J.: Das einfache und emotionale Kauferlebnis: Mit Usability, User Experience und Customer Experience anspruchsvolle Kunden gewinnen (Online-ausg). EBL-Schweitzer. Springer Fachmedien Wiesbaden, Wiesbaden (2015). https://doi.org/10.1007/978-3-658-10130-5

Schick, B., Seidler, C., Aydogdu, S., Kuo, Y.-J.: Fahrerlebnis versus mentaler Stress bei der assistierten Querführung. ATZ **121**, 74–79 (2019)

Seidler, C., Schick, B.: Stress and workload when using the lane keeping assistant: driving experience with advanced driver assistance systems. In: 27 Aachener Kolloquium, Fahrzeug- und Motorentechnik (2018)

Stappenbeck, B.: Technologische Innovation wird die Weltwirtschaft umwälzen (2017). https://www.bertelsmann-stiftung.de/de/themen/aktuelle-meldungen/2017/april/technologische-innovation-wird-die-weltwirtschaft-umwaelzen/

Sun, C., Ma, G., Dwivedi, M., Zeng, Y., Song, X.: A rear view camera based tailgate warning system. in society of automotive engineers of China & international federation of automotive engineering societies. In: Proceedings of the FISITA 2012 World Automotive Congress: Volume 9: Automotive Safety Technology, pp. 449–460. Springer, Heidelberg (2012). https://doi.org/10.1007/978-3-642-33805-2

Ullrich, D.: Intuitive Interaktion: Eine Exploration von Komponenten, Einflussfaktoren und Gestaltungsansätzen aus der Perspektive des Nutzererlebens (Dissertation). Technische Universität Darmstadt, Darmstadt (2014)

Ullrich, D., Diefenbach, S.: From magical experience to effortlessness: an exploration of the components of intuitive interaction. In: Hvannberg, E.T. (Ed.) ACM Digital Library, Proceedings of the 6th Nordic Conference on Human-Computer Interaction Extending Boundaries, pp. 801–804. ACM, New York (2010a)

Ullrich, D., Diefenbach, S.: INTUI exploring the facets of intuitive interaction. In: Ziegler, J., Schmidt, A. (Eds.) Mensch & Computer 2010, pp. 251–260. Oldenbourg Verlag, München (2010b)

Venkatesh, V., Davis, F.D.: A theoretical extension of the technology acceptance model: four longitudinal field studies. Manage. Sci. **46**, 186–204 (2000)

Mobility and Transport

Improving Mobility in University Communities Using a Collaborative Prototype

Erick López-Ornelas[✉]

Information Technology Department, Universidad Autónoma Metropolitana,
Mexico City, Mexico
elopez@correo.cua.uam.mx

Abstract. Mexico City is one of the cities with the highest vehicular traffic in the world. Santa Fe is a commercial and business zone located in the west side of this City. Also, three of the most important Universities in the country are located in this area (IBERO, UAM-C and ITESM). Santa Fe stands out as a particularly conflictive place in terms of vehicular traffic due to the high number of floating population, few access roads and non-connection with the massive public transportation of the City. This investigation analyzes the problem of urban mobility with an interdisciplinary approach. Some UX research methods were applied in order to identify the context and identify some interesting findings related with urban mobility. The main idea of this paper is the design of a digital tool only available for the University Community. It will be filled with information uploaded by the same Community. Through an APP, they will be able to visualize different options of routes and kinds of transport, also alerts and location. Everything is going to be modified and visualized in real time. This project will be a tool that will empower the University Community.

Keywords: Mobility · Traffic · Collaborative prototype

1 Introduction

Road congestion is one of the most common problems suffered by the major cities of the world, this problem has worsened in different parts of the world and has reached extreme levels that seriously affect the quality of life of the citizens. Aspects such as environmental, economic, health or insecurity are affected because of mobility problems in these cities, which involve private cars and the transportation system alike.

Studies such as the published by INRIX Global Traffic Scorecard in 2016 [1] show the complexity of the problem of mobility and traffic in certain cities, for example, it indicates that the inhabitants of the city of Los Angeles, in USA, spend 104.1 h a year stuck in traffic.

Other cities within the count made by INRIX are Moscow, with 94.1 h in traffic, New York (89 h), San Francisco (82.6 h) and Bogota (79.8 h), Mexico City has a calculation of 61.5 annual hours that its citizens pass in traffic.

On the other hand, TOM TOM TRAFFIC INDEX 2016 [2] places Mexico City first in its ranking where it calculates the "extra time" that is consumed in a route due to

© Springer Nature Switzerland AG 2019
H. Krömker (Ed.): HCII 2019, LNCS 11596, pp. 285–294, 2019.
https://doi.org/10.1007/978-3-030-22666-4_21

road congestion compared to the same route with clear roads, and in Mexico City the time on the road increases by 66% extra due to congestion.

Different actions have been taken to try to solve a problem that has gone out of control since a couple of decades, these initiatives range from imposing taxes on fuels, recovering public space to give priority to pedestrians and alternative ways of transportation or expanding the public transport offer.

While some of these actions have had a positive effect on reducing the mobility problem and the effects it entails, there is still a long way to go to achieve efficient mobility, especially in some cities where government policies and infrastructure are not enough. For this reason, we can see that Universities, Organizations and the private sector are developing new plans to improve mobility.

Gradually people begin to take actions in their hands to solve the problem of mobility and begin to organize with other citizens, therefore, it is necessary to expand the range of platforms and tools that give greater strength to these actions.

This paper presents the process of research, methodology and proposal of a digital tool that finds its main strength in the collaboration between users to solve the problem of mobility in one of the most conflictive points for mobility in Mexico City, Santa Fe.

2 Mexico City: A Big City

Santa Fe is the biggest commercial and business zone in Mexico City. It is located in the West side of this city. Also, three of the most important Universities in the country are located in Santa Fe (Fig. 1). Due to the high rental costs, the majority of the population that works in the area doesn't live there. A big percentage of people move daily to reach and leave the area after work, school or social hours (floating population).

In 2012, it was estimated that floating population in Santa Fe was of 233,000 people each day. Divided into 78,000 with permanent jobs, 40,000 with temporal jobs,

Fig. 1. Location of the area of interest.

100,000 visitors and 15,000 students [3]. Also, students, professors and workers of the three Universities have difficulties to reach their destinations.

Santa Fe is isolated from downtown and public transportation that cover the area is slow, insecure, inefficient and have few options. Also, it isn't connected with the main mass transport lines of the city such as the subway, bicycle network and metrobus (a bus network that has a special lane, similar to a Tranway but with buses).

When Santa Fe was planned, the automobile was the priority. Therefore, the streets and roads are not made for pedestrians or bicycles. This situation leaves few and complicated options for people who have to reach and leave the area.

From the central area of Mexico City to Santa Fe, there are only three main roads. They are constantly saturated because of the number of cars and public transports.

We found out that the community conformed by the three of Santa Fe universities (Universidad Autónoma Metropolitana, Universidad Iberoamericana and Tecnológico de Monterrey) are really affected by these mobility problems. But we also know there is a big potential in solutions made by this community itself.

3   Background

Mobility and transportation are global issues. There are plans around the world that seek to reduce the mobility problems in all sectors of the population. For example, the University of Valladolid in Spain integrated into its institutional development plan a strategy of urban integration and mobility. The idea is to have an "integrated campus" because their buildings are all around the city.

With this plan, they managed to connect the different areas through pedestrian routes, bicycle lanes and open spaces for the community. They connected roads and areas, opened public spaces and made one big University zone [4].

Cambridge is also an example where they tried to give a solution to this problem, but the approach was different. They were looking "to create places where people want to live and work". Their vision is that Cambridgeshire will be a place with strong, growing, prosperous and inclusive communities supported by excellent public services. There, people can fulfill their potential and live healthier lifestyles [5].

On the other hand, Mexico City, also have initiatives in which civil associations have worked together with the government. For example, with the Ecobici program [6]. It is a transportation system for shared bicycles. Registered users can use the bicycles and return them in any bike station. Although it is an initiative that helps many people, the system is only in some areas of the city.

Cities like Amsterdam and Copenhagen have shown public bicycles as a sustainable solution to mobility problems with programs like "smart bike" and "City Bike" [7].

Also, there are companies that promote carpool such as Bla Bla Car [8], worldwide, and Aventones [9], in Mexico, they help to reduce costs and the number of cars on the streets.

We also found projects that are based on digital platforms to analyze and/or solve this problem. For example, the "Twitter Jam" a University of Porto´s project. They know which areas have more traffic based on the Twitts analysis [10].

Another interesting project is the one of the University of Óbuda in Budapest. They gather information from people attending events in Facebook. Also, download information such as weather and the current state of transit. This way, they have enough information to know the movement of the city and to be able to predict congestions or even to prevent accidents [11].

4 Collaboration as an Important Element

To give a better definition of what we understand by collaboration, we return to the proposal by Crook who defines it as: "A coordinated and synchronized activity, the result of a sustained attempt to build and maintain a shared conception of a problem" [12].

From this concept we build the core of the digital tool, that is, to coordinate and synchronize collective actions that generate a shared vision of this problem and create a common channel between the users, so they can work on the solution.

We have the hypothesis that through collaboration and collective action, civil society can solve structural problems. That is why the design of the plan we propose is based in collaboration at different levels (between members of the same or different Universities). The strategy has an elaboration, communication and dissemination strategy seen from the communicative perspective and collaborative design.

"People in their daily lives, intent on their daily struggle with problems, opportunities, and ultimately the meaning of life. We observe how, more and more often, these people (re)discover the power of collaboration to increase their capabilities, and how this (re)discovery gives rise to new forms of organization (collaborative organization) and new artifacts on which they base enabling solutions" [13].

On the other hand, we took a more socially-oriented point of view and built the communication strategy based on Paramio's theory of collective action [14].

Based on this theory, we know that to motivate the use of the platform and collaboration, will be offering some kind of rewards.

5 Methodology

For the development of this project, we will rely on the methodology proposed in the "Developing and Implementing a Sustainable Mobility Plan" manual made by the European Union in 2013 [5]. This manual is divided in 4 phases: Research (or preparation), definition of objectives, elaborating the plan and implementation. Each stage includes a series of flexible and adaptable steps.

To complement the phases, we include tools to collect information in qualitative and quantitative forms. We took this tools from the Communication for Development model developed by the Swiss Agency for Development and Cooperation (SDC) [15]. SDC promotes development and actions for social change through community participation. Also, this document has practically the same phases previously indicated in the European Union manual.

In order to learn more about how the problem affected the University Community, we began our research with the community of the Universidad Autónoma

Metropolitana, Cuajimalpa. This way we understood how the mobility situation affected the community. The approach was made from a quantitative and a qualitative perspective.

For the research phase, we use different tools such as surveys for students, professors and workers; interviews, statistical data analysis, media monitoring, and 5 focus groups. That way we defined the first steps we needed to follow for the construction of the solution.

According to the surveys we made in the University, 17% of the students travel from 0 to 29 min each way, 25% from 30 to 59 min, 16% from 60 to 89 min, 14% from 90 to 119 min, 16% from 120 to 149 min and 12% 150 min or more (Fig. 2).

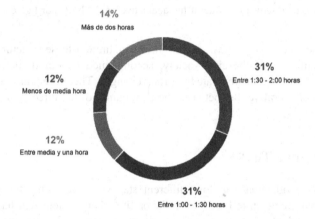

Fig. 2. Some discovering issue of the quantitative research.

In the qualitative e approach, we made focus groups with some students. This helped us to complete the quantitative information previously collected. First we carried out an activity with a physical map where the students could mark their daily route with threads and pins. The color of the threads represented the average ratings they had so far. This dynamic served both for us and for the students to achieve visualize the routes and have more clear the distance they must travel every day. Our discovering's are shown in Fig. 3.

This first stage helps us reach two important conclusions for the plan's construction.

The first one is that the mobility problem has a lot of different factors. This means that it's impossible to think that there is just one solution. We learned that the problem should be approached in different ways, that's why our plan is divided in phases.

In second place, we can conclude that in order to accomplish our objectives and the mobility plan, the solution has to come from the collaboration and collective actions of the community and not depending on the government.

We recognize that it is not viable from our field of study to make structural changes in roads and the transportation system which are controlled by the government and concessionaires.

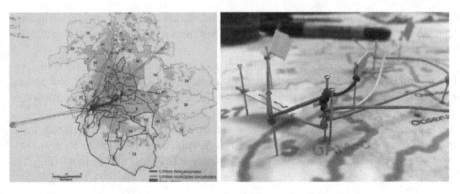

Fig. 3. Some discovering issue of the qualitative research. (Color figure online)

On the other hand, we are aware that, however limited these structural problems may be, the initiative of the civil society, accompanied by our digital tool, and a communication strategy can generate important changes. Therefore, our main objective is to design a collaborative tool between the community of the three main universities of Santa Fe.

6   Collaborative Tool

We decided to divide phase in four different stages: 1. Designing the information system's general architecture based on collaboration. This system will have information on routes, schedules and availability in the transport. 2. The creation of the prototype of the system. 3. Design an institutional collaboration strategy for the launch of the system and for monitoring its use. 4. Plan the communication and dissemination strategy focused on the University Community of the three institutions.

Our surveys showed that almost all members of the University Community have a smartphone and mobile internet plans, so it is feasible that the main communication tool for the community is a mobile app. This way they can share their location and know the status of the transport system. Also, the app will be used while the user is in motion, therefore, it must be intuitive and simple.

In this stage we implement the prototype and start the usability tests. Paper prototype help us to know what information should be included on each screen and to define the system architecture. This process helped us to complement the general functions that were missing and to be clearer how each screen should be. The designs of all the hand-drawn were made and allows us to make a first usability test, where the navigation and the distribution of the buttons were mainly reviewed (Fig. 4).

The information system will be a mobile application where the University Community can view possible routes from point x to the University and the other way around. This way, users can make the most appropriate decision according to their time and preferences cost. If people collaborate to arrive and leave the University together, they will do it faster by sharing a car or a taxi.

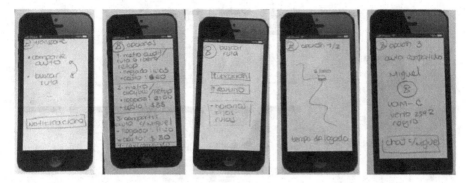

Fig. 4. Paper prototype of the tool.

The app will use an Api from Google Maps [16], and it will be powered by:

- Routes, stations and schedules of subway, *metrobus* and bike line. We decided to use only these transportation systems because they have the most established schedules.
- User profile data: academic institution to which they belong, type of transport used, registration and schedules (schedules may change). Also, users will fill if they own a car.
- Routes taken and kind of transport so that the system can define meeting points either by car or without car.
- Alerts regarding eventualities that arise on the road to Santa Fe such as: collisions, strikes, road congestion and transportation saturation.
- Internal chat so that, once you find a match, the users can agree on the meeting point and time.
- Each University will have a specific server with encrypted information. Therefore, access to the databases will be exclusive to each University and the administrators of each University will not be able to access data from the other institutions.
- The application is responsible for displaying data, not storing information.
- To register in the system will require the user to have a University number to guarantee that they belong to one of the educational institutions.

In order to achieve the project objectives and as part of the final delivery of results, we will develop the prototype of the system. However, we will keep the evidence and results of the work to make improvements.

The next Fig. 5 represents how the system works: first, the information that the system needs from the official public transportation data. Second, recollect the data from the community (University, schedule, routes, Facebook). Finally, the information generated by the interaction between the system and users will create collaborative schedules.

To deal with the economic aspect, we will expect the institutional support of the three universities in economic resources and infrastructure. Also, the plan considers alliances with the private sector and civil associations.

Fig. 5. Collaborative prototype

Database is an important element in the administration of the information. At the moment when users start using the application, a database will be generated and it will give us important information about the different stops and times that have the most affluence. If in any stop the influx is very high, we could talk to the authorities to request a new route of school bus. Also, the community may make requests for new stops and schedules.

These data will give us valuable information about the routes and times of the community that may serve for future projects or updating it. This database shows name, gender, occupation, if the user has a car or not, last stop and time used, score, qualification and if you have requested any additional stop to those that currently exist. Figure 6 shows a schema of the database implemented.

	A	B	C	D	E	F	G	H	I
1	Nombre	Sexo	Ocupación	Auto	Parada	Horario	Puntos	Calificación	Solicitud de parada
2	Juan Pérez	Masculino	Alumno	No	Las Águilas	11:00	40	5	San Ángel
3	Mariana Fernández	Femenino	Alumno	No	Tacubaya	10:30	20	5	
4	Patricio Juárez	Masculino	Profesor	Sí	Balderas	07:00	18	4	
5	Adela García	Femenino	Trabajador	No	Satélite	07:00	7	4	
6	Maria Teresa Gómez	Femenino	Alumno	No	Las Águilas	13:00	33	5	
7	Fernando López	Masculino	Profesor	No	CC Santa Fe	12:00	-3	2	
8	Gerardo Vazquez	Masculino	Trabajador	Sí	Huxquilucan	07:00	10	5	
9	Daniela Fernández	Femenino	Alumno	Sí	El Yaqui	09:00	12	4	
10	Erick Hernández	Masculino	Profesor	No	El Yaqui	08:00	24	4	
11	Tania Mendoza	Femenino	Trabajador	No	Satélite	08:30	20	5	Ecatepec
12	Cecilia Laguardia	Femenino	Alumno	No	Balderas	07:00	19	4	
13	Paulina Loera	Femenino	Profesor	No	Balderas	06:00	35	5	
14	Sergio Suárez	Masculino	Alumno	No	Observatorio	13:00	55	4	
15	Mariana Pérez	Femenino	Alumno	Sí	Ibero	10:30	30	4	Metepec
16	Rodrigo Velázquez	Masculino	Alumno	Sí	Las Águilas	11:00	-1	4	
17	Mauricio Vega	Masculino	Profesor	No	El Yaqui	08:00	15	4	
18	Rebeca Juárez	Femenino	Alumno	Sí	Observatorio	09:00	9	5	
19	Sofía Lara	Femenino	Profesor	No	UAM-L	07:30	13	5	
20	Fernanda Martínez	Femenino	Alumno	No	Balderas	07:00	23	5	San Ángel
21	Alejandro Martínez	Masculino	Profesor	No	Observatorio	08:00	12	4	

Fig. 6. Database design

7 Conclusions

It is important to mention that this digital tool is an element that is part of a larger scale mobility master plan for the three Universities located in Santa Fe, which will include an exclusive transportation system with real time location for the University Community, the diversification of internal and external routes and a more extensive inter-university collaborative work.

With this collaboration tool, we seek to lay the foundations for citizens to have new tools that allow them to face the daily problems that seem to have no solution. We believe that if society begins to collaborate, it will be more feasible to generate an important social change based on small actions taken together.

To the extent that the University Community makes use of this collaboration tool, we see a potential for improvement in the economic aspect of the users; in the environmental aspect, since a contribution is made to reduce the emission of CO_2 with the reduction of the traffic and a more rational use of the car, besides improving the efficiency of the routes and the performance of the members in the daily activities.

The plan is designed for a social sector that has demonstrated its willingness to work together. They look through different ways to improve mobility in the Santa Fe area and are aware of the potential to do something by themselves, and not only depend on government actions that have not been effective.

We consider that this project has a solid theoretical, methodological and contextual base. This means that it has an important potential to make a noticeable change in the way the University Community relates with the mobility in Santa Fe.

From the academy, it is necessary to continue with the study of the city and sustainable mobility from different areas of knowledge. Mobility problems have very different causes, and this allows for theoretical and practical approaches that contribute to the solution of them, like this project which was made by the interdisciplinary approach of communication, design and information technologies.

References

1. INRIX Global Traffic Scorecard (2016). http://inrix.com/scorecard/. Accessed 13 Feb 2018
2. TOM TOM TRAFFIC INDEX (2016). https://www.tomtom.com/en_gb/trafficindex/. Accessed 13 Feb 2018
3. Órgano de difusión del Gobierno del Distrito Federal y SEDUVI (2012) Programa parcial de desarrollo urbano de la zona de Santa Fe. http://www.data.seduvi.cdmx.gob.mx/portal/docs/transparencia/articulo15/fraccionxi/PPDU/PPDU_AO/PPDU_ZONA-SANTA-FE_AO-CM.pdf. Accessed 18 Dec 2017
4. De la Rivas, J.: Campus Universitario de Valladolid integración urbana y movilidad. Bitáctora Urbano Territorial **18**(1), 139–156 (2011)
5. Wefering, F., Rupprecht, S., Bührmann, S., Böhler-Baedeker, S.: Devoloping and implementing a sustainable urban mobility plan. European Commission (2013)
6. Ecobici Homepage. https://www.ecobici.cdmx.gob.mx/es/informacion-del-servicio/que-es-ecobici. Accessed 13 Feb 2018
7. De Maio, P.: Smart bikes: public transportation for the 21st century. Metrobike 9–12 (2013)
8. Bla Bla Car Homepage. https://www.blablacar.es/. Accessed 13 Feb 2018

9. Aventones Homepage. https://www.aventones.com/. Accessed 13 Feb 2018
10. Rebelo, F.: Twitter jam: identification of mobility patterns in urban centers based on Twitts. In: IEEE Smart Cities Conference (2015)
11. Mezei, M.: Urban mobility by Facebook events. by: INEES. In: 20th Jubilee IEEE International Conference on Intelligent Engineering Systems (2016)
12. Crook, C.: Computers and the Collaborative Experience of Learning. Routledge, London (1994)
13. Manzini, E.: Design When Everybody Designs. The MIT Press, London (2015)
14. Paramio, L.: Teorías de la decision racional y de la acción colectiva. Sociológica **19**(57), 13–34 (2005)
15. Jenatsch, T., Baue, R.: Comunicación para el desarrollo. Una guía práctica. Suiza. SDC (2014)
16. Google API Homepage. https://developers.google.com/. Accessed 02 Feb 2018

A Comprehensive Persona Template to Understand Citizens' Mobility Needs

Svenja Polst[✉] and Phil Stüpfert

Fraunhofer IESE, Fraunhofer Platz 1, 67663 Kaiserslautern, Germany
{svenja.polst,phil.stuepfert}@iese.fraunhofer.de

Abstract. In the city of Kaiserslautern, an urban district is being built, that aims to be climate-neutral. To achieve this aim, innovative concepts and technologies for energy generation and mobility services are to be applied. For the mobility concept, we needed personas as inhabitants and stakeholders are not living there yet. Since we could not find a persona template for the documentation of travelers' needs, we designed a persona template that provides a comprehensive picture of travel needs. We took a template based on Alan Cooper's idea and customized it to fit the topic of mobility, adding personality traits that are the basis of concrete mobility requirements. We also added characteristics of a person according to their stage of life, because personality traits hardly change over a person's lifetime. In addition, we included attitudes towards sustainability and technology. A description of the life situation together with the usual travel destinations and trip companions is also part of the persona template. The design of our template allows it to be used in other projects as well. We created personas based on our template and used them in two different workshops. The participants were able to use the personas without much explanation. A plan for further evaluation is presented in the discussion.

Keywords: Travelers' needs · Persona template · Urban mobility

1 Introduction

In the city of Kaiserslautern, Germany, an urban district is being built that aims to be climate-neutral by the year 2029. To achieve this aim, innovative concepts for energy supply and mobility services are being developed, supported by digital services. The concepts describe technologies, services, and measures that are necessary to achieve climate neutrality while assuring high quality of life in the district. A group of researchers from the areas of software engineering, energy supply, and mobility are cooperating with practitioners from the city's department of urban development to develop a mobility concept. The concept describes mainly three areas of action. First, it describes which sustainable means of transportation and which smart services regarding mobility are recommended for this district in the year 2029. The aim is for the combination of these transportation means and services to cover all mobility needs of the inhabitants and regular visitors, so that owning a car is not necessary. Second, the concept includes measures to support travelers in changing their mobility habits towards the use of sustainable transportation modes. These measures address travelers who are

© Springer Nature Switzerland AG 2019
H. Krömker (Ed.): HCII 2019, LNCS 11596, pp. 295–306, 2019.
https://doi.org/10.1007/978-3-030-22666-4_22

still using their own car. These measures are derived from psychological theories, such as the 'Theory of Planned Behavior' [1]. Third, the mobility concept describes physical and digital services that eliminate the need for a ride or at least reduce the distance, such as coworking spaces. All parts of the mobility concept are supposed to be developed based on the needs of inhabitants, local businesses and visitors.

The first part of the concept, regarding sustainable transportation means and smart mobility services, requires considering traveler characteristics and needs. In order to identify where action is needed, we had to answer the question "Which travelers' needs are not covered by the currently available sustainable transportation means and services in the city of Kaiserslautern?".

Therefore, we wanted to compare the travelers' needs with the characteristics of the available transportation means. We faced four challenges in eliciting traveler needs, which are addressed in Sect. 2. We decided to use provisional personas to resolve the challenges.

Personas are fictitious persons who represent a real group of people. They summarize all their relevant characteristics of persons. "Provisional personas are structured similarly to real personas but rely on available data and designer best guesses about behaviors, motivations, and goals." [2]. Since we could not find a persona template serving our purpose, we built an own persona template. Our contribution lies in the aspects being considered in the template and not in the design of it.

In Sect. 3, an introduction to personas in the area of mobility and potential characteristics for persona templates are presented. Section 4 presents the template and describe its elements. Thereafter, we describe how we used the template in Sect. 5. In Sect. 6, we discuss the template and present an evaluation plan. Finally, the conclusion summarizes this publication.

2 Motivation

In the best case, the elicitation of travelers' needs is done with a great number of diverse travelers of this urban district. Since the district is currently under construction, there are no inhabitants and employees yet who could be asked about their needs. Even before the construction work started, there had been no inhabitants and employees for several years. Having no travelers in the district yet is just one challenge in the elicitation of travel needs. Even if there were inhabitants and employees, they would have a hard time expressing their future travel needs.

The travel needs in the year 2029 (the year the mobility concept is addressing) differ from nowadays, because the life situation will change until then. Employees will probably not need to go to their office every day, if they even have their own office at all. Their working hours will be more flexible, so they will be more flexible regarding when to arrive at and leave from the office. We learned from other projects that many people find it hard to imagine future life. When asking people to imagine a normal day in the future, they often do not consider that the way of living, the available technologies and legal regulations will have changed, even when they are told to consider this.

Another challenge is that there will be diverse groups of travelers in the district with different needs. Appropriate segmentation of the travelers into groups is necessary to make sure that the needs of all groups are considered. There are several ways to perform segmentation [3]: Travelers could be divided according to their primary transportation mode, their activities such as commuting, or their destination. Another option is categorization according to socio-demographic characteristics, such as age and gender. People's way of life and attitudes could also be used as classification criteria. All these classification criteria were not sufficient for our purpose since there is too much variation within the suggested groups. For instance, separation by age just gives an idea about the needs, but individuals of the same age might have very different travel needs.

Furthermore, the suggested segmentation into traveler groups does not consider the use of digital services. Smart devices and software applications, among others, are necessary for planning, booking, and ticketing a trip. The requirements elicitation process in the area of software engineering needs to be integrated into the elicitation process in the area of mobility – or the other way around (Table 1).

Table 1. Overview of challenges encountered in the elicitation of travelers' needs in the project Enstadt: Pfaff

Challenge	Explanation
Uninhabited district	No habitants live and have lived there. Thus, there is nobody who could be asked about his travel needs
Consideration of future travel needs	The travel needs in 2029 will be different from those of today since the working and living situation will change
Broad range of traveler groups	The needs of several traveler groups need to be considered and documented separately
Combination of travel needs and requirements regarding digital services	The elicitation processes for requirements regarding transportation means and regarding digital services need to be combined

We decided to use provisional personas since they resolve the above-mentioned challenges. Provisional personas help to segment travelers, describe everyday life in future, consider aspects of living in the district such as close by travel destinations and commitment to the district community, and consider aspects regarding digital services. With the help of provisional personas, we were able to start working on the question "Which travelers' needs are not covered by the currently available sustainable transportation means and services?" We searched for a persona template that provides a comprehensive picture of travel needs. As we did not find a persona template appropriate for our purpose, we developed our own template.

3 Related Work

Personas were first described in 1999 in 'The Inmates Are Running the Asylum' by Alan Cooper [4]. Based on his basic idea, a persona template was developed, which initially contained general characteristics, such as:

- Name
- Age
- Family status
- Place of residence.

In the field of mobility, persona templates are used, for instance, in projects for the development of an open mobility platform [5], for improving the accessibility of public transport for disabled persons [6], and for the development of apps for public transport [7].

One approach for developing persona templates in the field of mobility is introduced by Baumann [8]. He suggests starting with the collection of as many user's characteristics and events relevant to the domain. Then, the characteristics are sorted into clusters, brought into a sequence starting with basic characteristics, and validated by domain and usability experts.

In the following, a collection of characteristics being part of persona templates in [5–8] is presented. The persona templates in these references have in common that the projects they are part of address certain transportation modes or apps. For each characteristic, a single reference is denoted but this characteristic might be mentioned in other references, too.

There are characteristics, which could also be part of persona templates used in domains other than mobility. However, the content of these characteristics (i.e. what is filled in in the template) refers to a specific domain.

- Preferences (e.g. barrier-free access to vehicles [5])
- Expectations (e.g. arrival on time [5])
- Goals (e.g. find cheapest transport option [7])
- Needs (e.g. something to travel safely and conveniently [6])
- Frustrations (e.g. dirty public transport [7])
- Challenges (e.g. mobility service not tailored to visually impaired persons [6])
- Impairments/restrictions (e.g. partially sighted [6], taking small child along [5])

These characteristics overlap and their interpretation depends on the template designer, for instance, in [7] 'arrival on time' was mentioned as an attribute of goals and not of expectations.

There are also characteristics that are specific to the domain of mobility:

- Used transportation modes (e.g. city bus, rental bike [8])
- Motivation for using a transportation mode (e.g. for using public transport: safe late night travels [7])
- Motivation to travel independently (indicated on five-point scale [6])
- Travel frequency (most weekdays [7], depicted on five-point scale [6])
- Knowledge of the area (e.g. poor knowledge [5])

- Knowledge of the transportation system (e.g. good knowledge [5])
- Tickets (e.g. monthly ticket, electronic ticket [5])

Tickets can be distinguished according to the ticket level, ticket type, ticket validity, ticket issued by, seating mode, class, and luggage [8]. In one persona template, the characteristic 'IT savyness' is included [6], which relates to tickets, if they are bought online.

Personas mostly include a short continuous text. The personas developed by Baumann (2010) only consists of the continuous text and a picture. Thus, the text can contain many characteristics. The text can be used to emphasize characteristics mentioned in other parts of the persona or add further information to it, such as:

- Daily life (e.g. going to the office [5])
- Activities during the trip (e.g. listening to music [7])
- Destinations (e.g. university, office [7])
- Purpose of travel (e.g. picking up child from school [5])
- Personality (e.g. spontaneous [5])
- Hobbies (e.g. reading [5])
- Attitude towards transportation mode (e.g. likes busses [5])

The literature about traveler segmentation partly confirms and complements the mentioned characteristics. In the introduction, we mentioned several characteristics for traveler segmentation. Summarizing, these options were socio-demographic characteristics, destination, activities, used transportation mode, way of life, and attitudes [3].

4 Description of the Template

We developed the persona template based on the characteristics mentioned in related work and on our own experience in the fields of mobility, psychology and software engineering. We selected parts of the characteristics described in related work based on our experience from projects about mobility and tried to reach a trade-off between a large number of characteristics and a template that can be completed and read within a reasonable amount of time.

Our persona template is illustrated in Fig. 1. An example of a persona is illustrated in Fig. 2. Within the template, similar data is grouped together and part of the data is visualized on scales. This facilitates the comparison of different personas. We de-scribe the elements of the template in clusters and explain the relation to travel needs and requirements.

4.1 Personality

We included four of the Big Five personality traits [9]: extraversion, openness, agreeableness, conscientiousness. Personality traits are the basis for concrete mobility requirements.

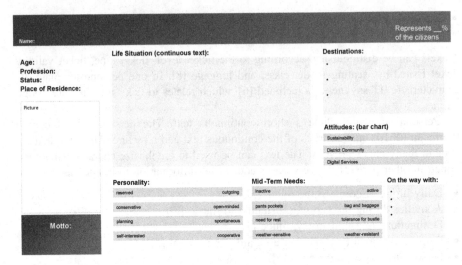

Fig. 1. Persona template

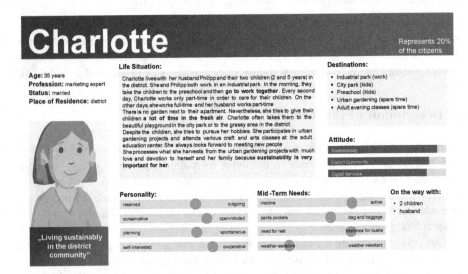

Fig. 2. Example of a persona based on the described persona template

- An *extravert* person is outgoing and probably enjoys conversations with others. Therefore, an extravert person is likely to enjoy ridesharing.
- An *open* person is more likely to try a new transportation mode compared to a conservative person (i.e. scoring low on openness).
- A person scoring high on *conscientiousness* probably likes to plan her trip beforehand and not spontaneously.
- A person scoring high in *agreeableness* is probably more cooperative and less self-interested. Such a person might accept a longer detour when giving others a lift to fulfill their needs, too.

To facilitate understanding of the personality traits, we did not label the scales with, for instance, 'low in agreeableness' and 'high in agreeableness', respectively. Instead, we selected an adjective that describes the trait. Traits can be described by a combination of adjectives. Selecting a single adjective is a limitation, which we prefer over abstract or ambiguous terms such as agreeableness. The fifth trait 'narcism' is not part of the persona because we do not see any clear relation to mobility.

4.2 Mid-Term Needs

A person's personality hardly changes over their lifetime, so we also added characteristics of a person according to their on a stage of life. One characteristic is the activity level. An elderly person is likely to score low on this scale. Such a person probably has the need to get from A to B without much physical activity. Thus, using a bike is not an option. In addition, we extended the persona template with a scale for transported goods that are carried during common trips. Some people travel just with what fits in their pockets, while others generally travel with a backpack. The bipolar scale 'need for rest & tolerance to bustle' expresses which needs the person has towards stimulation, crowdedness, and noisiness of transportation means. This scale is related to the personality trait 'extraversion'. Nevertheless, we consider it also as a mid-term need because we assume that the value on this scale varies during one's life. Children and older persons are more likely to feel overstimulated in a bus as their cognitive capacities might not be sufficient to deal with overstimulation. Hence, the button on this scale would be closer to 'need for rest'. A person who is more tolerant to bustle might not enjoy bustle but will be able to deal with it well. The last property we added was weather tolerance. This indicates whether a person is generally sensitive to certain weather conditions. For example, a sensitive person does not like to be outside in the rain, in a storm, or in the heat. As a consequence, for such persons the transportation modes need to offer protection from rain, wind, or heat. A weather-resistant person probably has no or only few requirements on a mobility service regarding the weather and would not mind taking the bicycle on a rainy day.

4.3 Attitude

The persona template includes attitudes towards sustainability, the district community, and digital services. These attitudes are relevant in the context of our Smart City project. A person scoring low on the sustainability attitude scale might require some incentive in an application for smart mobility to be motivated to use sustainable means of transport. A high score in district community reflects that the persona con-siders a good community in the district as important. This is an indicator that such a person would be willing to give someone in the district a lift. A low score in digital services indicates that such a person needs to be convinced of the benefits of digital services or that analog services need to be offered, too. In our project, attitudes are especially important for the second part of the mobility concept, which is about sup-porting travelers in changing their mobility behavior. According to the Theory of Planned Behavior [1], attitudes partly determine behavior. This also seems to be the case for mobility-related behavior; Haunstein and Hunecke [10] found an influence of attitudes towards cars and public transport on the usage of eco-friendly transportation modes.

4.4   Life Situation

A description of the life situation (in continuous text) is also part of the persona template. The continuous text includes, among other things, information about the socio-economic status, which expresses requirements regarding the fare. Also included are daily activities such as working, hobbies, and other leisure activities, which are the reason why people need mobility services. In addition to this description of everyday life, the text can describe other important values of the persona and can be used to extend the mid-term needs, the personality, or the attitudes. Moreover, the acceptance of innovative technology can be expressed by the text.

4.5   Destinations

There are some more elements in the persona template. 'Destinations' not only includes regular destinations but also characteristics such as the purpose of the travel (e.g. work). Information about the distance between the residence and the destination could be added, as well. However, the actual driven kilometers depend on the route a certain mode of transportation takes. For instance, a bus mostly does not take the direct way to a destination but a pedestrian mostly can do so. The distance did not have an added value in our project, wherefore we did not include it in Fig. 2.

4.6   On the way with

The 'on the way with' provides information on who the persona travels with on a regular basis. Such travel companions – who could be children, friends, but also a dog – lead to requirements on a transportation mode, such as available seats and the possibility to have a conversation.

4.7   Motto and Picture

The picture, which could be a photo or a drawing, should fit to the characteristics of the persona. The picture and the motto, which is a short summary of the key characteristics of the persona, ease to distinguish personas at a glance.

4.8   Represents __% of the Citizens

The estimation 'represents ___% of the citizens' reflects how many citizens in the district are similar to the persona. There are many characteristics that could be used to decide whether the persona is similar to other citizens. These characteristics are segmentation criteria. The user of the template can decide which characteristic should be used as a segmentation criteria. We used a combination of the employment status, family status, and residence as our main segmentation characteristics.

5   Usage of Template

Ideally, the template should be filled with consolidated data from real traveler groups. The template suggests which data to elicit from travelers; for instance, via interviews. Then, the template can be completed on a print-out or in a digital version. Our template is currently available as a print-out and a PowerPoint template in both German and English. We recommend taking a look at an already existing persona, when completing our template in order to get an idea of how to use the scales.

We already mentioned that we used provisional personas, which were created by researchers with a background in software engineering, mobility and psychology. We used the personas in a workshop for identifying mismatches between the travelers' needs and the currently available transportation modes. The mismatches were validated afterwards. During the workshop, we derived requirements from the needs that are directly or indirectly part of the persona. For instance, Charlotte has the need to take her two children with her; the derived requirement is that the mobility service has to transport two children safely. Requirements on a mobility service (cf. Fig. 3) also emerge in combination with the situation on a certain day (e.g. feeling tired, rainy and windy), local particularities (e.g. incline), the purpose of the trip (e.g. shopping).

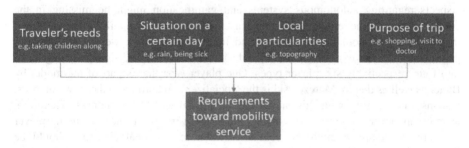

Fig. 3. Description of the requirements derivation process. Traveler's needs are one element in this process.

We separated these other aspects from the persona template, so that we can go through different scenarios with the same persona. This is also the reason, why we did not include the element 'needs' or 'goals' in the template, like other persona tem-plates do. In our opinion, needs such as arriving on time depend on the scenario. For a meeting, arriving precisely on time is important, but when traveling for touristic reasons, arriving a few minutes later is mostly not that critical. In the workshop, the participants did not fill in the persona template but worked with previously created personas. The participants were able to use the personas without much explanation. They mentioned that the personas supported them in considering a situation from a different point of view. In this first workshop, we found that tolerance towards different weather condition should be added to the template, which we did afterwards. In the future, we will replace the fictive content with data from potential travelers in the district and then conduct another workshop.

6 Discussion

The persona template for mobility was created in the context of a project in Germany, so the template was originally written in German. Therefore, the template might not apply well in projects taking place in a very different cultural setting. Moreover, the terms in the template might be less comprehensible in the English version. The choice of characteristics for the template is probably also influenced by our experience in mobility projects and the scope of the current project. For instance, we did not include the characteristic 'transportation mode', since it did not fit to our purpose for using personas. Therefore, use of the template in other mobility projects might require some elements to be modified, especially when considering travelers with special needs. For instance, the attitudes in our template might be irrelevant for other projects outside the Smart City context, while other attitudes might have to be added to the template. We consider adjustment of the template a regular step in the persona creation process. Nevertheless, the template should already provide a good basis for the creation of personas, so the number of adjustments is expected to be small.

Therefore, we plan to evaluate the completeness of our template. We consider the template complete if all the information about a traveler that is relevant for creating or enhancing smart mobility services can be filled in and extracted from the template. Aspects regarding autonomous systems and gamification might be missing in the current version of the template. It seems reasonable that travelers will differ in terms of trust towards autonomous cars, buses, and trains. Furthermore, we did not consider gamification. According to Bartel [11] and Marczewski [12], users can be categorized into four, respectively six, player types. One player type that is part of the model by Bartel as well as that by Marczewski is the 'socializer'. According to these two authors, persons who identify with this player type are motivated by relatedness. Therefore, team competition is one way to motivate this kind of person. Hence, including player types in the template might be beneficial when motivational elements should be included in a mobility application. Adding more elements to the template might increase its completeness, but more information might also have a negative impact on usability. A trade-off needs to be achieved between completeness and the usability of the template.

We plan to evaluate the template's completeness by filling in real data from travelers, elicited, for instance, via interviews. We would like to ask members of different projects to fill in the template. If the project member filling in the template cannot document all the information considered relevant for a persona by the project consortium with the help of the template, the template will be considered incomplete. Furthermore, the template will be considered incomplete if users of the created persona miss information. All information missing in the current version of the template should be discussed regarding its relevance for other projects. If the information is considered relevant for a high number of other projects by a group of experts, an element for filling in this information should be added to the template.

Besides that, we want to evaluate whether the template is comprehensible and can be filled in within an acceptable amount of time. Comprehension can be measured in three ways. One indicator for comprehension is whether the persona created with the template is reasonable; that is, whether the single elements of the template fit together. For instance, if the persona works as a social worker, has a high score on 'district community' and 'outgoing' but the text says 'dislikes social events and afraid of meeting new people', we consider the persona not reasonable. The persona might not be reasonable due to a lack of experience in creating personas or because the template is not comprehensible. Another indicator for comprehension is the number of questions asked while filling in the template. The third indicator is users' rating of comprehension.

Moreover, it should be evaluated whether filling in the persona template can be completed within an acceptable amount of time. The absolute amount of time is not important to us, since some template users might enjoy spending time on creating the persona, which makes it hard to determine whether the time spent is too long. In-stead, we consider the time as acceptable if the template users themselves consider it that way. If the participants perceive the time as too long, it should be investigated how the template could be improved. Maybe some elements need to be reduced or the inter-action with the template needs to be improved.

So far, we have not evaluated the template except for using it in two workshops with fictive data. Completing the template with fictive data is not an ideal option but served our purpose of getting into the project and identifying some mismatches be-tween needs and available transportation modes at an early stage in the project. We plan to replace the fictive data of the provisional personas with real data of travelers as soon as such data is available.

In the first step of the evaluation, we plan to address comprehension. We want to ask people with experience in the creation of personas to create one based on the German version of the template, since this version is more advanced. The participants will be free to choose a data set from one of their projects as a basis for the persona. The participants will not receive any introduction to the template. Instead, a member of the evaluation team will be present to answer questions emerging while filling in template. These questions will be noted and analyzed to check whether they indicate poor comprehension of the template. It might be possible to improve comprehension by replacing terms in the template with terms that are easier to understand and less ambiguous. Another option could be to develop some short instructions for the tem-plate.

7 Conclusion and Future Work

We developed a persona template for the creation of personas in the field of mobility. We explained our motivation for creating a new persona template. Our contribution lies in the aspects being considered in the template and not in the design of it. The template consists of several elements, which are partly derived from literature about user seg-mentation and templates used in practice. These elements can be grouped into these categories; basic data (such as name and age), life situation, personality, mid-term needs, destinations, travel companions, attitudes and a motto. Personas created with the template have been successfully used in two projects to date. A proper evaluation will be conducted in the future according to the presented evaluation plan. In the future, we also plan to investigate whether it is an option to let travelers fill in the template instead of asking them to answer interview questions. The expected advantages would be that less time would be needed to elicit travelers' characteristics and that this procedure would be more enjoyable compared to interviews. The template is publicly available (https://doi.org/10.5281/zenodo.2554034). We would welcome use of the template by others and would appreciate any feedback on completeness and comprehension.

Acknowledgements. Parts of this work have been funded by the 'EnStadt: Pfaff' project (grant no. 03SBE112D and 03SBE112G) of the German Federal Ministry for Economic Affairs and Energy (BMWi) and the German Federal Ministry of Education and Research (BMBF).

References

1. Ajzen, I.: The theory of planned behavior. Organ. Behav. Hum. Decis. Process. **50**, 179–211 (1991)
2. Cooper, A., Reimann, R., Cronin, D.: About face 3. The essentials of interaction design. Alan Cooper and Robert Reimann. Wiley; Chichester: Wiley [distributor], Hoboken (2007)
3. Hunecke, M.: Mobilitätsverhalten verstehen und verändern. SMV. Springer, Wiesbaden (2015). https://doi.org/10.1007/978-3-658-08825-5
4. Cooper, A.: The Inmates are Running the Asylum. Sams Publ, Indianapolis (1999)
5. VDV, D.V.: Definition und Dokumentation der Nutzeranforderungen an eine offene Mobilitätsplattform. VDV-Mitteilung 7046 - 03/2018 (2018)
6. Sheth, R.: riddhi. http://riddhi.me/projects/kaikki
7. Quattrucci, T.: Public Transport Victoria App Redesign: A UX Case Study. https://medium.com/@tony.quattrucci/public-transport-victoria-a-ux-case-study-de6ae49e8235
8. Baumann, K.: Personas as a user-centred design method for mobility-related services. Inf. Design J. **18**, 157–167 (2010)
9. John, O.P., Srivastava, S.: The big five trait taxonomy: history, measurement, and theoretical perspectives. In: Handbook of personality: Theory and Research, 2nd edn., pp. 102–138. Guilford Press, New York (1999)
10. Haustein, S., Hunecke, M.: reduced use of environmentally friendly modes of transportation caused by perceived mobility necessities: an extension of the theory of planned behavior. J. Appl. Soc. Pyschol. **37**, 1856–1883 (2007)
11. Bartle, R.A.: Designing virtual worlds. New Riders Pub, Indianapolis (2004)
12. Marczewski, A.: Even Ninja Monkeys Like to Play. Gamification, Game Thinking and Motivational Design. Gamified UK, [S.l.] (2015)

Understanding the Usefulness and Acceptance of Adaptivity in Smart Public Transport

Christine Keller[1(✉)], Susann Struwe[2], Waldemar Titov[1], and Thomas Schlegel[1]

[1] Institute of Ubiquitous Mobility Systems, Karlsruhe University of Applied Sciences, Moltkestr. 30, 76133 Karlsruhe, Germany
Christine.Keller@hs-karlsruhe.de, iums@hs-karlsruhe.de
[2] TTI GmbH - IBIZ, Willy-Andreas-Allee 19, 76131 Karlsruhe, Germany
info@ibiz-innovation.de
http://iums.eu, http://ibiz-innovation.de

Abstract. Adaptive passenger information for enhanced mobility experience may be the next step towards a smart public transport. In our research project, we explored adaptive passenger information and investigated options to increase the intelligibility of adaptive features. We set up an online questionnaire to study the acceptance of adaptivity in public transport information systems. In this paper, we describe our approach to adaptivity in public transport, the design of the questionnaire and we discuss results of our study.

Keywords: Smart public transport · Passenger information ·
Adaptive systems

1 Introduction

The public transport domain has been significantly transformed by digitization. Smartphones have created a personal access to public transport information and digital ticketing has made public transport accessible more directly than before. But also the availability of realtime information and the introduction of public digital information systems have changed the usage of public transport. As a next step, the application of Internet of Things (IoT) technologies for public transport is discussed, which could, among other things, leverage context-aware applications based on rich sensor data [9].

In addition to sensor data, digitization also generates real-time information on vehicles, schedules and on traffic situations that affect public transport. Considering this large amount of available data and increasingly intelligent ways to handle this information, smart public transport systems are within our reach [8,15]. The adaptation of passenger information systems to current situations plays a critical role in making public transport systems smart, more efficient and more attractive for their passengers. Passenger information systems turn

© Springer Nature Switzerland AG 2019
H. Krömker (Ed.): HCII 2019, LNCS 11596, pp. 307–326, 2019.
https://doi.org/10.1007/978-3-030-22666-4_23

into personal mobility assistants, that not only know the overall public transport situation and take delays, detours or even the amount of passengers in single vehicles into account, but on the other hand, they react according to the situation of the user, their current position, their goals and preferences, enabling a personalized mobility experience [22].

Additionally, interactive devices permeate our surroundings and, increasingly, public space and public transport [12]. Not only are most passengers carrying their own smartphone or other mobile computing devices, but public displays and public information systems are also familiar in cities and in public transport by now [3,8,19]. Intelligent or advanced traveller information systems leveraging these technologies are becoming an important component of the services for public transport agencies [4]. Besides improving core factors of public transport, such as punctuality or efficiency, Camacho et al. also argue for the development of more passenger-centric services that additionally enhance the passenger's experience [5]. From our perspective, an adaptive and smart passenger information system tackles both - the core factor of high-quality passenger information as well as enhancing the passenger's experience by adapting to their situation and actively supporting passengers according to their needs.

However, adaptive systems are not usable just because they are adaptive [21]. The acceptance of adaptivity depends on many factors and may vary between users. The design and development of an adaptive passenger information system should be approached carefully, in order to maintain usability and usefulness.

In our research project, we explore the application of adaptivity in passenger information systems of public transport. Our goal is to improve passenger information on every step of their trip and in any situation. As a part of this project, we investigate the integration of semi-transparent, multi-touch enabled displays in passenger information systems. These displays can be installed in public transport vehicles instead of windows or at stops and stations and serve as an interactive point of information. We also consider mobile applications and the possibility of multi-device interaction. In this setting, we explore suitable modeling of adaptation and the realisation of an adaptive and interactive passenger information system using said devices. In order to keep this adaptive system usable and useful, we also developed options to increase the intelligibility of adaptive features. In a first evaluation phase, we designed an online questionnaire to study how people would rate different adaptation options in an passenger information system.

The rest of this paper is structured as follows: in Sect. 2, we summarize our research of similar systems in related work. We then describe our development of adaptive features as well as intelligibility features for an adaptive passsenger information system in Sect. 3. We present our take on prospective users, multi-device interaction as well as different categories of adaptation we considered. In the following Sect. 4, we describe the design of our study, bringing the dimensions of device combinations, adaptivity categories and the variation of public transport situations, persona and multiple features together. In Sect. 5, we describe the results of our study, followed by a discussion and outlook on future analyses in Sect. 6.

2   Related Work

We conducted our investigation of related work focusing on adaptive, context-aware and interactive passenger information systems for public transport that involve public displays, smartphones or other media in public transport vehicles. The usability and usefulness of adaptive systems is also of particular interest to us, specifically in the public transport domain or in similar settings.

A very early context-aware system, which was actually deployed, is the GUIDE system by Cheverst et al. [6,7]. It did not utilize public transport but as a context-aware tourist guide, it provided information customized to the user's context. Being developed in the pre-smartphone era, it was implemented on a pen-based tablet PC and the communications network was specifically engineered for this system. Information and functionality were tailored based on the context categories of personal context, related to the user, and environmental context. The GUIDE system adapted the presentation of information by sorting the displayed lists of locations and points of interest. Cheverst et al. also implemented indications on adaptivity. When the user was presented a list of interesting locations, a note was displayed on the sorting rules, that were based on context. In their evaluation, the authors report a high acceptance rate for the overall system. The context-aware recommendations were found useful by the participants of the study. The authors also describe that the restriction of information by context was considered frustrating, a finding that emphasizes the careful design of adaptive systems. The work of Cheverst et al. is interesting, since it not only considers location as a context factor to adapt to, but considers the situation of the user in more detail, similar to our approach.

Tumas and Ricci describe a personalized and mobile city transport advisory system with the rise of smartphones in 2009 [22]. They focus on the implementation of a routing algorithm that computes personalized alternatives for a travel request. The personalization of their approach is based on travel preferences the users enter while entering the travel request in their mobile client and the adapted presentation of the responses is a ranking based on the user's preferences. While they describe that in a usability testing with students no usability issues were found, they do not report on the acceptance of the personalized ranking of results or other details.

Handte et al. present an application called Urban Bus navigator (UBN) [13]. This novel service provides micro-navigation and crowd-aware route recommendation for bus riders. The UBN system is built upon a distributed IoT infrastructure which enables the passenger's smartphone to interact with buses in real-time and allows buses to register the presence of on board passengers. The micro-navigation service provides a context-aware guidance of passengers along a bus journey by recognizing boarded bus vehicles and tracking the passenger's journey process. The crowd-aware route recommendation collects and predicts crowd levels on bus journeys to suggest better and less crowded routes to bus riders. The UBN system was tested collecting qualitative feedback from real-world bus users. The results indicate reduced cognitive effort for managing bus journeys, increasing motivation of using bus transport and better accessibility of travel information.

Chow et al. developed an adaptive mobile application for planning trips in Hong Kong public transport [8]. The system considers the location of the user via their mobile phone's GPS sensor as well as their walking speed, measured by a wearable device. The application supports planning as well as re-planning of trips due to real-time information, for example in case of missed buses, which is one of our use cases, too. The authors implemented are commendation algorithm that computes trips considering the walking speed of the user, as well as the real-time transport situation. However, walking speed and location are the only context factors taken into account, whereas we considered some more factors. Also, the authors did not yet publish an evaluation on the acceptance or usability of their system and its adaptation.

Abu-Issa et al. describe the development and test of a recommendation system in an Android app [1]. This system considers several context factors and then proactively suggests matching points of interest to the user. The user can then navigate to these points of interest. The authors report a high acceptance rate in their survey of test users. Their approach of using a proactive recommender system that suggests items to the user without specific user request is comparable to the approach of active adaptation that we chose to evaluate in our survey. However, the authors do not report specifically on the acceptance of this feature and they did not compare this feature to any passive alternative.

In a recent study, Oliveira et al. investigated the experiences and requirements of passengers in rail travel [17]. Their findings resulted in a experience visualisation as a customer journey experience map, the identification of critical points in rail travel, such as ticket collection or the finding of a seat, and some indications where technological innovations might mitigate negative experiences of passengers. Their findings, among other things, point towards the passenger's need for guidance in unknown settings - stations or coaches alike. Subsequently, the authors developed four personas for rail transport in the UK and investigated the user's openness towards technological advances in electronic ticketing and seat reservations [18]. Their paper proposes a system that, for example, directs passengers towards free seats and informs about occupancy levels of trains. They found that passengers highly value supplementary information and they emphasize the importance of correct information. Supplementary information and guidance on trips are goals of our system as well, especially considering the question if adaptation can enhance the aspect of guidance for public transport passengers.

An analysis of context-aware systems in general shows that successful adaptive user interfaces are still hard to find in practice [11,21]. However, incomprehensible adaptions can have an adverse effect on the usability of a system. To reduce this negative effect of adaptive behaviour, Paymans et al. have attempted to help users to build adequate mental models of such systems [11]. They developed a user support concept and applied it to the adaptive user interface of a context-aware mobile device. The authors evaluated their approach with users and reported that the user support improved the ease of use, but unexpectedly it reduced learnability. In addition, the support concept only provided real-time information about active modality and contextual factors, but not for system adaptations. Following the example of Paymans et al., we also developed features that are designed to help users understand the adaptive user interface of our system.

Lim et al. [16] present another approach that provides explanations to increase the intelligibility of an adaptive system. They examined the effectiveness of different types of textual explanations (why, why not, how to, and what if) in a controlled study with over 200 participants. The authors developed a web-based infrastructure that provides a functional input-output interface of an intelligent system prototype that provides different types of explanations. They found that providing reasoning trace explanations for context-aware applications to novice users can improve user's understanding and trust in the system, but their findings were not further tested in real-world-settings.

3 Adaptation for Smart Public Transport

Our goal is to better understand how adaptive mobility systems can support passengers in public transport and to design an adaptive passenger information system based on that understanding. We want to know how passengers accept different kinds of adaptations. In one step further, we take a look at intelligibility features for adaptations and assess their effectiveness. As a basis for further research and the development of several prototypes of an adaptive passenger information system, we analyzed and identified contexts and types of adaptation that are relevant in public transport scenarios. We then developed adaptation scenarios for different public transport situations and adaptation categories. We designed mockups for those scenarios and developed an online questionnaire to evaluate responses towards these adaptations in a first evaluation phase.

3.1 Personas for Adaptation

In order to design adaptive passenger information systems for public transport, we used the persona method to analyze and illustrate the prospective users of such a system. This method has been applied to the public transport domain before, for example by Hoerold et al., and only recently by Oliveira et al. [14, 18]. We built on results by Hoerold et al. as they describe personas for passengers in german public transport specifically. During the analysis phase of our project, we refinded and extended those personas. We classified passenger types according to their type and frequency of public transport usage, their knowledge of the region and of the public transport system and their smartphone usage.

For our first evaluation phase of adaptation scenarios, we focused on four personas. We chose them based on different requirements the personas would have towards an adaptive passenger information system. We selected the commuter persona, the pupil or student, the casual user and the power user. Commuter and pupil/student have in common that they have distinct regular trips and times. They have good knowledge about the trips they are taking regularly. The power user and casual user both have no regular work or school trips but use public transport for alls kinds of other reasons. They are distinguished by the frequency they use public transport and by their familiarity with the public transport system. While a power user is using public transport very often and is very comfortable in doing so, a casual user is using it only once in a while and may need more assistance, because they are not as familiar with it. In our study, we investigated if the assessment of adaptations would be different between people that match different personas.

3.2 Devices and Interactions for Adaptation

Our adaptive passenger information system is designed with different interactive devices in mind. We are considering apps for smartphones, but also public digital information systems on platforms and in vehicles. In our research project, we are studying the application of semi-transparent, multi-touch enabled displays in passenger information systems. When placed at stops and platforms, we call them public displays, while it is also possible to use such a display in a public transport vehicle as a window, which we call a smart window.

In addition to designing adaptations for smartphones and for public displays or smart windows, we are also considering possible interactions between public information systems and personal devices, in order to achieve an understanding of smart public transport systems as a whole. We designed adaptation scenarios for each of the devices independently but also developed some scenarios that consider the interaction between a smartphone and a public display or a smart window.

3.3 Public Transport Situations for Adaptation

As a basis for the development of our adaptation scenarios, we identified situations in which passengers might need information about their journey or about journey-related topics. A public transport journey can roughly be divided in three parts: pre-trip, on board and post-trip, following, for example, Oliveira et al. [18]. We would rename the "on board" part to "on trip", since we are considering trips with multiple legs, where a passenger can be at an interchange during their trip and is not on board a vehicle. Depending on the situation in public transport, the information need of passengers changes. We therefore identified several information needs for passengers in public transport, beginning with the information on a journey and reminders, for example for planning or starting a trip. On trip, passengers need information on interchanges, delays or disruptions. We developed adaptation scenarios for each of those situations.

3.4 Adaptation Categories

In our first evaluation phase we explored, if different kinds of adaptations were rated differently. For the development of our adaptation categories, we took six categories of adaptation into consideration. These categories are based on the categories of context-awareness described by Alegre et al., who proposed an extension to the categorization of Dey and Abowd [2,10]. The categories describe the subject of the adaptation in a context-aware system, which can be:

- the *presentation of information*, including the modification of the presentation of information
- the *active execution of a service*, where the system autonomously executes a service
- the *passive execution of a service*, where the system proposes the execution of a service to the user
- the *active configuration of a service*, where the system autonomously configures a service
- the *passive configuration of a service*, where the system proposes the configuration to the user
- tagging context to information for a better understanding of this information

We excluded the last category, because it does not directly result in adaptations that the users become aware of. In our study, we specifically investigated the difference between active and passive adaptations, in both categories, the *execution of a service* and the *configuration of a service*. A pair of scenarios with active and passive adaptations is described in Subsect. 3.6.

3.5   IFeatures - Understanding Adaptation

Incomprehensible adjustments can negatively impact the usability of a system. Therefore, we examined, if additional explanations make the system adjustments more intelligible. We also analyzed, which kind of explanations are more understandable to the users and which they would prefer. We developed three versions of intelligibility features (which we call IFeatures) based on the Why and Why Not Explanations of Lim et al. [16] and the support concept of Paymans et al. [11]. The intelligibility features provide explanations about a system's adaptation decision. We distinguished the following three versions of IFeatures and applied them to every adaptation scenario (Table 1):

Table 1. Versions of intelligibility features

Version	Description
Version 1	Display icons
Version 2	Display of icons and textual explanations
Version 3	Textual explanations

3.6   An Adaptive Passenger Information System

Our basic design of the adaptive passenger information system uses three different types of context for adaptations. We consider personal context, spatial context, interaction context and socio-technical context, following a categorization of Schlegel and Keller [20]. We assign users to personas that match their personal context. Personas are a first step to filter scenarios and adaptations for a user. Based on a user's history of trips, regular trips and times can be identified. They serve as a context factor for system adaptation. In the personal context of a passenger, their calendar and appointments are also considered.

The interaction context is determined by the available interactive devices that can be part of the adaptation. Adaptation based on interaction context fundamentally depends on availability of the devices. However, in further phases of our project we plan to broaden our scope and additionally consider usability factors in adaptation decisions based on interaction context.

As spatial context we consider the location of a user. There are different forms of location that are considered here. The absolute location of a user in terms of GPS positions is a basic context factor and on top of that, the location of a passenger in relation to public transport facilities are relevant to the system. The location at a certain stop point as well as in a certain public transport vehicle is used in some adaptations. Furthermore, the direct position of a user in front of a public display or a smart window is also relevant for some of the adaptations.

The socio-technical context considered in our scenarios, comprises of different public transport situations, as described in Subsect. 3.3. Based on these situations, the system can identify the passsenger's information need.

We developed 21 adaptation scenarios in total. We tried to comprise them covering different aspects of devices, situations and adaptation categories each, as shown in Fig. 1. Ten of these scenarios are part of a pair of active and passive adaptations. We will introduce three exemplary adaptation scenarios in the following.

Scenario	Situation	Device	Adaptation Category
1	Reminder	Smartphone	Active execution of a service
2	Change	Smartphone	Active execution of a service
3	Information on a Journey	Smartphone, Public Display	Presentation of information
4	Delay	Smartphone, SmartWindow	Active execution of a service
5	Delay	Smartphone, SmartWindow	Passive execution of a service
6	Disruption	Smartphone	Active execution of a service
7	Disruption	Smartphone	Passive execution of a service
8	Information on a Journey	Public Display	Active configuration of a service
9	Delay	SmartWindow	Active execution of a service
10	Disruption	Public Display	Passive execution of a service
11	Information on a Journey	Smartphone	Active configuration of a service
12	Information on a Journey	Smartphone	Passive configuration of a service
13	Change	Smartphone	Active execution of a service
14	Change	Smartphone, SmartWindow	Active execution of a service
15	Delay	Smartphone	Active execution of a service
16	Delay	Smartphone, SmartWindow	Active execution of a service
17	Delay	Smartphone, SmartWindow	Passive execution of a service
18	Disruption	Smartphone, SmartWindow	Active execution of a service
19	Disruption	Smartphone, SmartWindow	Passive execution of a service
20	Disruption	SmartWindow	Passive execution of a service
21	Disruption	Public Display	Passive execution of a service

Fig. 1. The adaptation scenarios for our first evaluation phase.

Fig. 2. Mockup of scenario 3: information on a journey

Scenario 3 - Information on a Journey: This scenario is based on the combination of a smartphone app that serves as a travel companion and a public display at a stop point. The intended user group is the group of commuters and pupils or students that are traveling on a recurring trip. The scenario is taking place in the morning, when the passenger is leaving their home and is on their way to school or to work. As a prerequisite, the app has been started and a trip was chosen or set. As the passenger approaches the stop and the public display, they can see that the information on the vehicle they will be taking is highlighted, as is shown in Fig. 2. In the textual description of the scenario, we describe that additional information is then displayed. This is information on the occupancy of the coaches of this train as well as the information on how many coaches there are. The detailed information on this vehicle is presented and highlighted based on the trip of the user, known by their smartphone app.

Scenario 16 and 17 - a Delay: The scenarios 16 and 17 address all personas and include a smartphone and a smart window. They are a pair with active and passive adaptation of the category *execution of a service* and are therefore based on the same basic situation. In this situation, the participant wants to visit someone using public transport. In their travel companion app they started the trip they want to take and have already boarded their tram where they are seated next to a smart window. The tram is experiencing a delay and from this point onward, the scenario description differs.

In scenario 16, displayed in Fig. 3, a notification opens on the smart window that informs the user of the delay. Since the intended next connection might be missed, three alternatives are presented to the user. They can choose one and transfer the trip to their smartphone. In this scenario, a service is actively executed by the system. The passenger's destination and other information on their trip is used for automatic planning of alternative routes.

Fig. 3. Mockup of scenario 16: a delay with smartphone and smart window

Fig. 4. Mockup of scenario 17: a delay with smartphone and smart window

Scenario 17 on the other hand, is passively executing a service, which is shown in Fig. 4. The user is presented with the information on the delay and can then choose to get alternative routes. They can transfer their destination from their smartphone to the smart window and a list of alternatives is displayed. A chosen alternative trip can also be transferred to their smartphone for further use in their travel app. Instead of a system that proactively uses the passengers data to plan ahead and directly present results, the passenger in scenario 17 is asked and then can take action to re-plan their trip. Consequently, scenario 17 requires more user interaction than scenario 16.

Fig. 5. Mockups for scenario 17 using IFeatures 1, 2 and 3

We developed IFeatures for each scenario and included them in the mockups for comparison. Figure 5 shows the three different IFeatures for Scenario 17. In order to compare the IFeatures, only the relevant part of the original image is shown.

4   Questionnaire Design

For our study we designed an online questionnaire with adaptive and randomized questions. The first questions for all participants are about age, occupation and about public transport and smartphone usage. We ask about the confidence of the participants regarding public transport, but also regarding their smartphone usage. Based on the given answers, we are able to present the participants public transport situations relevant to their public transport experiences. Participants that use public transport regularly to get to work, school or university are categorized as commuters and pupils/students, where other participants are grouped in the generic group of users. If participants did indicate that they do not possess or use a smartphone, they are only presented with scenarios where no smartphone is involved.

For each scenario that is presented to a participant, the textual description and the visual mockup is displayed. We ask a question on the usefulness and comprehensibility of this adaptation and then present versions of the mockup using the IFeatures, each alongside the base version without IFeature. Each IFeature is presented with questions on their intelligibility and usefulness. The order in which the IFeatures were shown is chosen at random. After all three IFeature versions, the participant is shown an overview over the base version and each IFeature version and asked to rate them in comparison.

Normally, all participants are presented three different scenarios which are chosen at random. When one of the randomly selected scenarios is one half of an active/passive pair, we overwrite one of the other random scenarios and present both parts of such a pair of scenarios. After two scenarios of an active/passive pair are presented, we ask a question comparing these two adaptations. The textual description and mockup of both adaptation scenarios is then shown for comparison.

Participants in the pupil/student or commuter group are first presented with two scenarios for pupils/students and commuters. Afterwards, one scenario of the general user category is shown. If this scenario is part of an active/passive pair, the participant will be shown the second part of this pair as a fourth scenario.

5   Results

We distributed the link to the online questionnaire to different users and mobility groups. We used different media and channels to reach out to a highly diverse audience. The only feature the study participants have in common is the potential usage of public transport in their local region.

After three weeks, 133 questionnaires were completed and 213 questionnaires were not completed, meaning participants aborted at different stages of progress. The age of participants was widely distributed, starting at 15 to 19 years of age and going up to 65 and older. The majority of participants, 66% are between 20 and 39 years old.

In order to assess the familiarity with the local present public transport system, we asked about frequency as well as purposes of public transport usage. The frequency of public transport usage was quite evenly distributed. Leisure and daily routes to work or school/university were most frequently reported as purposes of public transport usage. Asked about their familiarity with public transport, over 90% reported at least moderate familiarity with their local public transport system.

Regarding media and smartphone usage, also over 90% of the respondents stated that they possess and use a smartphone. Asked about their confidence in smartphone usage, only about 5% were not very confident or not confident. We also asked how confident they feel about buying tickets for public transport using an app, where 12% indicated they feel unconfident or very unconfident and 32% feel very confident. Overall, we can safely say that the participants of the study were rather familiar with smartphones and apps.

Since the adaptation scenarios shown to the participants depended on some of the answers they provided before, the scenarios were shown at different rates. In total, 504 adaptation scenarios were shown to participants. One scenario that was only shown to students and pupils was only rated five times, while 8 scenarios were shown between 30 and 40 times. On average, each scenario was shown 24 times. 41% of our respondents were commuters and 18% were students or pupils, which leaves 41% in the user category.

5.1   Adaptive Smart Public Transport

We asked participants to rate usefulness and comprehensibility of adaptation scenarios. Overall, the ratings of the adaptation scenarios were better than we expected. Figure 6 gives an overview over all ratings of our scenarios. Usefulness and comprehensibility werde both rated on a scala from 1 as *very useful* or *very comprehensible* to 5 as *not useful at all* or *not comprehensible at all*.

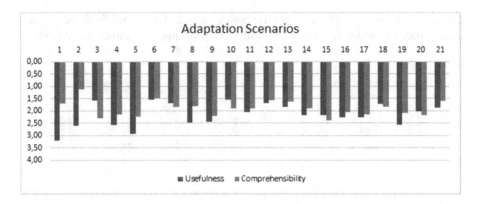

Fig. 6. Overview over ratings of usefulness and comprehensibility of all scenarios

Usefulness of Adaptation. On average, usefulness was rated as 2.16. Scenario 6 was considered the most useful with a score of 1.53. Scenario 6 only considers a smartphone app and the situation is a disruption of a service. It is a scenario for commuters, pupils or students. The user is notified that their usual tram is out of service. The smartphone application chooses an alternative and gives directions on which tram to use and when.

Scenario 1 was rated the least useful with a score of 3.20. Scenario 1 is a scenario based on regular trips of commuters, students or pupils. In this scenario, the user is reminded via their smartphone that they have to leave home in five minutes if they want to use their usual trip.

Comprehensibility of Adaptation. Comprehensibility was rated 1.91 on average. Scenario 2 was rated as the most comprehensible scenario, scoring 1.13. This scenario is a reminder via smartphone to get off the train at the next stop. Scenario 15 was rated as the least comprehensible scenario with a rating of 2.39. In this scenario, a notification on their smartphone alerts the user to alight the vehicle at the next stop instead of the one after that and to take a certain tram at the next station. We assume that this is rated as not very much comprehensible because the notification does not indicate why the system suggests this change of plans. Overall, comprehensibility tends to be rated better by persons with better knowledge of public transport, a result that is not surprising to us. However, it leaves the question how to better reach and support people with less familiarity to public transport.

We also looked at averages of ratings by devices and situations, as shown in Fig. 7. The group of scenarios with smart window or public display both contained those scenarios that used smartphones together with a smart window or a public display as well as the scenarios that featured only the displays. Overall, the usefulness of scenarios with public displays was rated best and scenarios with smartphone only were rated second. However, in terms of comprehensibility, the smartphone only scenarios scored best, which can probably explained by higher familiarity of most people with smartphones on contrast to public displays and smart windows, which are more unfamiliar to passengers in public transport.

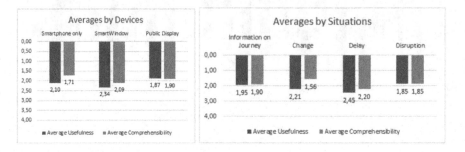

Fig. 7. Average ratings by devices and situations used in scenarios.

With regard to the situations, scenarios that take place during interchanges were rated most comprehensible, while scenarios of situations with disruptions were rated best regarding usefulness. Disruptions are situations when passengers need much information, because their trip is not only late, but they possibly can not carry on as planned. We think that adaptive systems can support passengers best when they need dynamic and up-to-date information the most and this result supports our theory.

In both categories, usefulness and comprehensibility, ratings of participants with moderate to low confidence in smartphone usage are worse than ratings of participants with more confidence. Since the numbers of respondents in the first category is quite low, we have to look into this connection more deeply, gathering more data.

5.2 Active vs. Passive Adaptivity

We had five pairs of scenarios based on the same situation and using the same devices that differed in the realization of the adaptation. One part of a pair implemented an active execution or configuration of a service, whereas the second part implemented the same adaptation in a passive way. An example of such a pair is explained in Subsect. 3.6. After seeing and rating both scenarios independently, participants additionally were asked to compare those scenarios directly.

Figure 8 shows the average rating of all scenarios using active adaptation and all scenarios using passive adaptation when rated independently. These groups do include the paired scenarios but also all scenarios using active or passive adaptation that were not part of a pair. Active adaptation scenarios were rated slightly better regarding comprehensibility than passive adaptation scenarios, which came as a surprise for us. We plan to look into this in more detail in future work. Passive adpatation scenarios were rated as more useful on average, however.

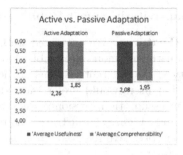

Fig. 8. Average usefulness and average comprehensibility of scenarios using active or passive adaptation.

Fig. 9. Numbers for preference of scenarios when comparing pairs of active and passive adaptation, pair 4, 5, pair 6,7 and pair 11, 12.

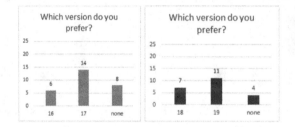

Fig. 10. Numbers for preference of scenarios when comparing pairs of active and passive adaptation, pair 16, 17 and 18, 19.

Regarding the comparison of pairs, shown in Figs. 9 and 10, the pair of scenarios 4 and 5, which are in a situation of delay considering smartphone and smart window, is the only pair where the active adaptation is mostly chosen over the version with passive adaptation. For all other pairs, participants mostly chose the passive version over the active version. We are planning to look into these results in more detail in the future, with more detailed evaluations of active and passive adaptations in passenger information systems.

5.3 IFeatures for Adaptivity in Smart Public Transport

Comprehensibility of IFeatures. We examined wether additional explanations make the system adjustments more intelligible. For this reason, we evaluated the comprehensibility of the different versions of the intelligibility features (IFeature). We assumed that users would prefer the reduced representation of version 1. In contrast to our assumption, version 2 was rated best for comprehensibility among all participants. Figure 11 illustrates the results for the scenarios 7, 11, 16 and 17. These scenarios have received the most replies in the dynamic compilation of the questionnaire.

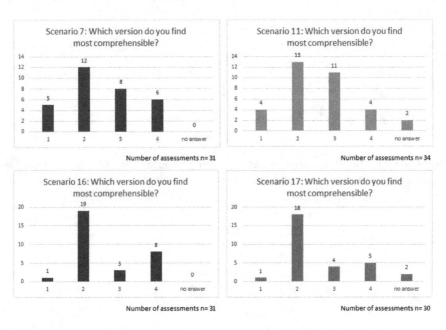

Fig. 11. Results for the scenarios 7, 11, 16 and 17 in terms of comprehensibility

We compared the three IFeature versions and an additional version 4 without IFeature. In Scenario 17, we had 30 participants who answered the question of comprehensibility. For the question which version is most comprehensible, version 2 received the highest approval with 60%. Version 4, the presentation without IFeature was rated with only 16.67% followed by version 3 with 13.33%. Version 1 received the lowest approval with 3.3% while 6.67% did not answer the question. A similar tendency could be found in all 21 scenarios. The combination of icons and text (version 2) were preferred by most participants.

Comprehensibility of IFeatures Related to Devices. We also differentiated the scenarios according to the three interactive devices: public display, smart window and smartphone. We summarized the results to compare wether there are any differences between the devices in terms of comprehensibility. Figure 12 illustrates the results.

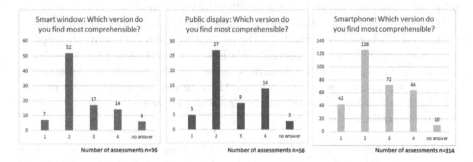

Fig. 12. Summarized results for the three devices: public display, smart window and smartphone in terms of comprehensibility

The summary of scenarios with smart windows resulted in 96 replies. For the question which version is most comprehensible, version 2 received the highest approval with 54.17%. Version 3 was rated with only 17.71% followed by version 4 with 14.58%. Version 1 received the lowest approval with 7.29% while 6.25% did not answer the question. This comparison also shows that version 2 was the most comprehensible for the participants. We conclude that the preference of a type of intelligibility feature does not depend on any device and will try to reproduce and support this conclusion in our future work.

6   Discussion and Outlook

In this paper, we presented a first analysis of the results of our study and have already drawn some conclusions on future work. Due to time and space constraints, we were not able to report fully on all results we were able to get from our data and hope to discuss those results in some future publications in greater detail. However, we could examine the usefulness and comprehensibility of our scenarios and found indications that adaptive information in situations with disruptions might be most interesting to investigate in the future. We also will look into the preference of passive adaptations over active adaptations when we perform studies with our prototypes.

The comparison of the different versions of the intelligibility features for comprehensibility showed that the combination of icons and text (version 2) received the most approval from the participants. In the next steps, we have to examine these results in real-world applications as well.

Unfortunately, some of our research questions could not yet be answered based on the results of our questionnaire, due to a lack of data. We therefore plan to conduct this study in a longer time period and with intensified efforts on the acquisition of test persons. We are planning in particular to reach more persons that are unconfident using their smartphones and apps and to examine their understanding and assessment of adaptation scenarios. On top of that, we plan to conduct separate studies with a greater focus on the comparison of active and passive adaptations and on the intelligibility features.

However, we have seen that the usefulness of adapatation in passenger information systems is rated positively by the participants and also comprehensibility is assessed better than we expected. This is a very good basis for future developments. We now strive to implement some of the adaptations we designed for this study as prototypes and to evaluate those in our laboratory. This evaluation of our adaptation scenarios is the first basis for our selection of scenarios for implementation.

Acknowledgements. This work was conducted within the scope of the research project "SmartMMI - model- and context-based mobility information on smart public displays and mobile devices in public transport" and was funded by the German Federal Ministry of Transport and Digital Infrastructure as part of the mFund initiative (Funding ID: 19F2042A). We would like to thank Nadine Vollers for her excellent contribution to the project.

References

1. Abu-Issa, A., et al.: A smart city mobile application for multitype, proactive, and context-aware recommender system. In: 2017 International Conference on Engineering and Technology (ICET), p. 1–5, August 2017. https://doi.org/10.1109/ICEngTechnol.2017.8308181
2. Alegre, U., Augusto, J.C., Clark, T.: Engineering context-aware systems and applications: a survey. J. Syst. Softw. **117**, 55–83 (2016). https://doi.org/10.1016/j.jss.2016.02.010.   http://www.sciencedirect.com/science/article/pii/S0164121216000467
3. Alt, F., et al.: Designing shared public display networks – implications from today's paper-based notice areas. In: Lyons, K., Hightower, J., Huang, E.M. (eds.) Pervasive 2011. LNCS, vol. 6696, pp. 258–275. Springer, Heidelberg (2011). https://doi.org/10.1007/978-3-642-21726-5_17
4. Camacho, T.D., Foth, M., Rakotonirainy, A.: Pervasive technology and public transport: opportunities beyond telematics. IEEE Pervasive Comput. **12**(1), 18–25 (2013). https://doi.org/10.1109/MPRV.2012.61
5. Camacho, T., Foth, M., Rakotonirainy, A., Rittenbruch, M., Bunker, J.: The role of passenger-centric innovation in the future of public transport. Public Transp. **8**(3), 453–475 (2016). https://doi.org/10.1007/s12469-016-0148-5
6. Cheverst, K., Davies, N., Mitchell, K., Friday, A.: Experiences of developing and deploying a context-aware tourist guide: the guide project. In: Proceedings of the 6th Annual International Conference on Mobile Computing and Networking, MobiCom 2000, pp. 20–31. ACM, New York (2000). https://doi.org/10.1145/345910.345916
7. Cheverst, K., Davies, N., Mitchell, K., Friday, A., Efstratiou, C.: Developing a context-aware electronic tourist guide: some issues and experiences. In: Proceedings of the SIGCHI Conference on Human Factors in Computing Systems, CHI 2000, pp. 17–24. ACM, New York (2000). https://doi.org/10.1145/332040.332047

8. Chow, V.T.F., et al.: Utilizing real-time travel information, mobile applications and wearable devices for smart public transportation. In: 2016 7th International Conference on Cloud Computing and Big Data (CCBD), pp. 138–144, November 2016. https://doi.org/10.1109/CCBD.2016.036

9. Davidsson, P., Hajinasab, B., Holmgren, J., Jevinger, Å., Persson, J.A.: The fourth wave of digitalization and public transport: opportunities and challenges. Sustainability **8**(12), 1248 (2016). https://doi.org/10.3390/su8121248

10. Dey, A.K., Abowd, G.D.: Towards a better understanding of context and context-awareness. In: Computer Human Interaction 2000 Workshop on the What, Who, Where, When, Why and How of Context-Awareness (2000)

11. Paymans, T.F., Lindenberg, J., Neerincx, M.: Usability trade-offs for adaptive user interfaces: ease of use and learnability, pp. 301–303, January 2004. https://doi.org/10.1145/964442.964512

12. Foth, M., Schroeter, R.: Enhancing the experience of public transport users with urban screens and mobile applications. In: Proceedings of the 14th International Academic MindTrek Conference: Envisioning Future Media Environments, MindTrek 2010, pp. 33–40. ACM, New York (2010). https://doi.org/10.1145/1930488.1930496

13. Handte, M., Foell, S., Wagner, S., Kortuem, G., Marrón, P.J.: An internet-of-things enabled connected navigation system for urban bus riders. IEEE Internet Things J. **3**(5), 735–744 (2016). https://doi.org/10.1109/JIOT.2016.2554146

14. Hörold, S., Kühn, R., Mayas, C., Schlegel, T.: Interaktionspräferenzen für personas im öffentlichen personenverkehr. In: Eibl, M. (ed.) Mensch & Computer 2011: überMEDIEN|üBERmorgen, pp. 367–370. Oldenbourg-Verlag, Chemnitz (2011)

15. Keller, C., Brunk, S., Schlegel, T.: Introducing the public transport domain to the web of data. In: Benatallah, B., Bestavros, A., Manolopoulos, Y., Vakali, A., Zhang, Y. (eds.) WISE 2014. LNCS, vol. 8787, pp. 521–530. Springer, Cham (2014). https://doi.org/10.1007/978-3-319-11746-1_38

16. Lim, B.Y., Dey, A.K., Avrahami, D.: Why and why not explanations improve the intelligibility of context-aware intelligent systems. In: Proceedings of the SIGCHI Conference on Human Factors in Computing Systems, CHI 2009, pp. 2119–2128. ACM, New York (2009). https://doi.org/10.1145/1518701.1519023

17. Oliveira, L., Bradley, C., Birrell, S., Davies, A., Tinworth, N., Cain, R.: Understanding passengers' experiences of train journeys to inform the design of technological innovations. In: Re: Research - the 2017 International Association of Societies of Design Research (IASDR) Conference, Cincinnati, Ohio, USA, pp. 838–853 (2017)

18. Oliveira, L., Bradley, C., Birrell, S., Tinworth, N., Davies, A., Cain, R.: Using passenger personas to design technological innovation for the rail industry. In: Kováčiková, T., Buzna, Ľ., Pourhashem, G., Lugano, G., Cornet, Y., Lugano, N. (eds.) INTSYS 2017. LNICST, vol. 222, pp. 67–75. Springer, Cham (2018). https://doi.org/10.1007/978-3-319-93710-6_8

19. Sahibzada, H., Hornecker, E., Echtler, F., Fischer, P.T.: Designing interactive advertisements for public displays. In: Proceedings of the 2017 CHI Conference on Human Factors in Computing Systems, CHI 2017, pp. 1518–1529. ACM, New York (2017). https://doi.org/10.1145/3025453.3025531

20. Schlegel, T., Keller, C.: Model-based ubiquitous interaction concepts and contexts in public systems. In: Jacko, J.A. (ed.) HCI 2011. LNCS, vol. 6761, pp. 288–298. Springer, Heidelberg (2011). https://doi.org/10.1007/978-3-642-21602-2_32

21. Schneider-Hufschmidt, M., Malinowski, U., Kuhme, T. (eds.): Adaptive User Interfaces: Principles and Practice. Elsevier Science Inc., New York (1993)
22. Tumas, G., Ricci, F.: Personalized mobile city transport advisory system. In: Höpken, W., Gretzel, U., Law, R. (eds.) Information and Communication Technologies in Tourism 2009, pp. 173–183. Springer, Vienna (2009). https://doi.org/10.1007/978-3-211-93971-0_15

Challenges for Local Authorities in Planning and Implementing Sustainable and User-Oriented Mobility Measures and Services

Sebastian Spundflasch$^{(\boxtimes)}$ and Heidi Krömker

Technische Universität Ilmenau, PF 10 05 65, 98684 Ilmenau, Germany
sebastian.spundflasch@tu-ilmenau.de

Abstract. Mobility has changed its face a lot in recent years. Today, mobility planning represents a major challenge for cities, especially when it comes to developing innovative and sustainable services and measures. The topic of mobility is enjoying great popularity in research. Numerous projects and initiatives are attempting to support cities in taking a step towards sustainable urban mobility. But it is important to understand the problems and challenges that mobility planners are faced with when shaping sustainable mobility. Only in this way future support activities can be tailored even more to the requirements and specific needs of mobility planners. The paper presents 15 challenges that cities and mobility planners face when developing mobility solutions. The challenges were identified in working together with different cities in the frame of the CIVITAS SUITS (CIVITAS SUITS is a R&D project funded by the European Union in the frame of the Horizon2020 program. The aim of the project is to increase the capacity of small-to-medium cities to plan and implement sustainable mobility. Further information on: www.suits-project.eu) project. Subsequently the challenges were validated with experts from the mobility sector.

Keywords: Mobility planning · Local authority · Sustainable mobility

1 Introduction

The mobility sector has changed its face considerably in the last 20 years. While in the past mobility consisted mainly of individual traffic and public transport, nowadays there are numerous alternative mobility offerings on the market. Thanks to the possibilities of innovative information and communication technologies and the increased awareness of the population for the topic of sustainability, these offers are enjoying increasing popularity. In addition, people's mobility behavior and the corresponding needs have changed. A comparison of the results of numerous studies on the subject of mobility in change [1] shows that young people in particular are less strongly fixed to one means of transport but rather use different means of transport adapted to their situation. Public transport can gain passengers through this development and cycling is also enjoying increasing popularity. Technological progress especially in the field of mobile devices, makes it possible to combine the different transport modes and thus to adapt mobility to one's own mobility needs.

© Springer Nature Switzerland AG 2019
H. Krömker (Ed.): HCII 2019, LNCS 11596, pp. 327–339, 2019.
https://doi.org/10.1007/978-3-030-22666-4_24

The mobility of the future will be much easier, more flexible and more individual for users. The vehicle of the future will be used on demand with shared ownership [2].

The changing mobility needs of transport users, numerous new mobility offers, technologies and data as well as the increased importance of the topic of sustainability are making mobility planning increasingly complex and facing cities and their mobility planners with major challenges. The aim of numerous efforts at national and European level, e.g. in the framework of the CIVITAS[1] initiative, is, among other things, to support local authorities and in particular mobility planners in these challenges and to push forward the shift towards sustainable mobility. The various activities in recent years have created a large pool of knowledge, with numerous guidelines, good practice examples, project reports, etc., which should support the mobility planners in shaping innovative sustainable mobility. However, when working with cities in the frame of the CIVITAS SUITS project it became clear that this knowledge is used too little. Smaller cities in particular do not have the necessary capacities to grasp and subsequently implement this knowledge. Employees do not have time to deal with extensive guidelines that are difficult to put into practice. Available good practice examples leave many questions unanswered, making it difficult to implement the proposed solutions in the own local context. Most of the information gathering of the mobility planners takes place through the direct exchange of experiences between cities, as well as at congresses and through the media. The lack of use of the existing support materials is also due to the fact that they are not tailored enough to practical needs and usually suit an academic audience instead addressing the questions raised by mobility planners.

From a user-centered point of view, it is therefore important to understand the challenges and problems that mobility planners face in their daily business and to take them into account when tailoring support offerings.

2   Challenges in Mobility Planning

In the context of the CIVITAS SUITS project, which aims at increasing the capacity of local authorities in mobility planning, the challenges local authorities faced with when planning and implementing sustainable mobility were examined in cooperation with 8 European cities. Gaps and barriers as well as drivers and enablers to mobility planning were analyzed in different workshops together with city representatives. The challenges presented in this paper show an important result of working with the project cities and are intended to provide an understanding of the challenges local authorities face when dealing with mobility issues. These challenges may serve as an important starting point when it comes to implement change processes and to support cities with their activities in a targeted manner. The following Table 1 provides an overview of the identified challenges, that will be described in more detail in the following.

---

[1] CIVITAS is a network of cities aiming on a cleaner and better transport in Europe. Also under this guise are a number of research projects on sustainable mobility carried out which are funded by the European Union.

Table 1. Overview of the identified mobility planning challenges

No	Challenge
1	Sustainability thinking
2	Institutional cooperation
3	Systematic staff deployment and – development
4	Effective project management and monitoring
5	Knowledge management/knowledge transfer
6	Understanding and applying innovative financing methods
7	Innovative procurement
8	Understanding political interests and affecting political decisions
9	Understanding legal and regulatory framework
10	Citizen participation
11	Estimating the feasibility and acceptance of measures
12	Interaction and cooperation with business partners
13	Identification and utilization of synergy
14	Use of innovative technologies and data collection methods
15	Application of research knowledge and adaption of good practice examples

2.1 Internal Challenges for Local Authorities

Some of the identified challenges can be attributed to the LA's internal organisation. These are challenges that are not linked directly to the development of concrete mobility schemes. Rather, they form the organisational prerequisites and the necessary attitude local authorities need when aiming on the development of innovative and sustainable mobility. Figure 1 shows by way of example that the development of mobility measures requires the interaction of numerous departments within the local authority. In the following, the challenges that the departments must face together in order to successfully plan and implement sustainable mobility are described.

Sustainability Thinking. To shape sustainable mobility, it is important to anchor a sustainable thinking among the staff working on mobility issues, but ideally across the entire authority. This is one of the biggest challenges as sustainability thinking is nothing that can be dictated by leadership. Employees must be sensitised for the topic and its importance. This will only work if the issue is anchored on the agenda as an elementary component in the development of mobility strategies and measures. If the staff should not only adapt good practice examples to mobility, but also act innovatively, it is important to understand the basic principles of sustainability. In the mobility sector, the term sustainability is closely related to the shift towards clean and environmental friendly transport. But there's a lot more to it than that. The outcome of the United Nations General Assembly 2005 [3] names three pillars for sustainable development: Social, Environment and Economic. The SUMP Guideline [4] specifies these principles for the planning of sustainable mobility and mentions e.g. focus on the people, accessibility and quality of life, health- and environment quality, participative planning involving all relevant actors, development of cost-effective solutions,

Fig. 1. Mobility planning requires the cooperation of various departments within the LA

evaluation of effects and establishment of a learning and improvement process and so on. The basic idea and the different principles must be internalized by the staff in order to apply them in the development of sustainable mobility solutions.

Institutional Cooperation. The development of mobility services and - measures is often very complex and touches different topics and areas of responsibility within the local authority. Thusan intensive cooperation between different administrative areas like shown in Fig. 1 is required. A key success factor is the willingness of the different departments to cooperate and to jointly implement mobility strategies and specific measures.

This requires the:

- development of a common vision
- bundling of competencies
- recognition and exploitation of synergy effects
- allocation of capacities
- definition of roles and responsibilities

The local authority of Stuttgart as well as West Midland Transport Authority for example have addressed this challenge through the creation of a Strategic Steering Committee that promotes cooperation between the departments and includes them in the planning of projects right from the start.

Since mobility services may not necessarily end at the city borders inter-municipal cooperation is playing an increasingly important role. The joint development and management of mobility services can greatly increase efficiency and lead to more sustainable concepts [5].

Systematic Staff Deployment and – Development. As already mentioned, the field of mobility planning has become increasingly broad, complex and difficult to penetrate.

Although an incredibly large pool of knowledge and experience has been published and is available on the topic, it is a big challenge, especially for small cities, to further develop the technical know-how and the expertise needed to plan and implement innovative services and mobility with a focus on sustainability.

Large cities have an obvious strength in that they more easily can pull together horizontal teams with the necessary skills to implement demanding mobility schemes. Smaller cities, where the mobility departments are by nature smaller, need to think carefully which competencies they develop within the department and which expertise shall be bought in. To date, projects that require expertise in innovative subject areas are often outsourced. Outsourcing per se is not necessarily wrong. Departments cannot know everything by themselves and it is better to involve experts. Almost all cities use to rely on some external agencies for certain tasks related to mobility questions. However, mobility planners need materials that supports their basic understanding of the different subject areas. Cities must look at their long-term vision and consider the direction in which they want to develop their staff, especially in view of the demands that innovative and sustainable development of mobility services entail.

Effective Project Management and Monitoring. Although the importance of project management is clear, insufficient project management is still a big barrier and often leads to serious delays or even the failure of projects. This applies equally to the subject of mobility planning. During the planning phase, the biggest problem seems to be an over-ambitious planning in combination with a lack of experience on innovative topics which leads to unrealistic plans. Another major challenge is the systematic monitoring and the early detection of problems and deviations, often caused by the fact that many departments work on specific projects, some of which have different views on the matter. Especially in smaller cities there seems to be still great potentials for improvement in the project management discipline.

In the frame of the MAX project "Successful Travel Awareness Campaigns and Mobility Management Strategies", the Guideline "MAX-Sumo - Guidance on how to plan, monitor and evaluate mobility projects" [6] was developed to support the project management for mobility projects. Another interesting Guideline developed in the frame of the GUIDEMAPS European Union project is "Successful transport decision making – A project management and stakeholder engagement handbook" [7] which provides useful models and approaches to improve project management in the mobility sector.

Knowledge Management/Knowledge Transfer. Knowledge Management and Knowledge transfer are very important but also very challenging tasks. To make them work properly, it is not enough just to purchase some software that facilitates this matter. It has to be supported by all employees, permanently maintained and optimized, which is a lot of effort. The implementation of an effective knowledge management within an organization is associated with numerous challenges. But an effective knowledge management greatly improves the capacity of an organisation and enhances its ability to innovate [8]. Especially since the planning and implementation of mobility measures depends to a large extent on experiences this topic plays an important role on the way to becoming a learning organisation.

Understanding and Applying Innovative Financing Methods. The aspect of financing plays one of the most important roles in the planning and implementation of mobility projects as it ultimately decides whether a measure is implemented or not. Of course, there are significant differences between cities in structurally strong regions where money does not play a major role and weak regions where money is a key issue.

However, mostly there is only little awareness of innovative or alternative financing options that go beyond the pure acquisition of municipal or federal funds. Insight from working with the project cities made clear, that financing is one of the most important issues in mobility planning, but at the same time it is also the area with most conservative action.

So the main challenge especially for cities with limited financial resources is to increase the capacity to identify, adapt and apply alternative/innovative financing methods for projects for which there is no funding available or urban funds are insufficient. Abig reason for the poor use of innovative financing methods like Congestion Charge, Crowdsourcing or municipal green bonds is, that there is often not enough space to try things out. So far there is only little experience on this topic in the mobility context. Accordingly, the risk of making mistakes or failing is comparably high. In the frame of the SUITS project, a guideline on innovative financing is currently being developed which will inform about new funding models and opportunities for new business entries.

Innovative Procurement
Procurement summarizes the activities of acquiring goods, works or services from external sources. Procurement is often seen as a necessary evil, but offers great potential in terms of sustainable mobility. The new EU's Procurement reform (LINK) enables goods and services to be procured in a sustainable manner. While the lowest price criterion has been the most important award criterion to date, criteria such as life cycle costs, pollution reduction, energy consumption or external transport costs are playing an increasingly important role after the reform. This makes it possible for cities to encourages bidders to pay more attention to the issue of sustainability. Cities may encourage suppliers to become creative and think of more sustainable solutions. The reform opens up completely new possibilities for bidders. Not necessarily the lowest price but the most creative solution that can receive the award.

The new possibilities in procurement requires close cooperation between the mobility planning departments and procurement departments in order to fully exploit the potentials.

The SUITS project is currently developing a guideline and decision support tools to help cities understand and take advantage of the new opportunities.

2.2   Challenges Arising in the Development of Mobility Measures and Services

In addition to the challenges described in the last chapter, which mainly concern the internal organisation of the Local Authority, there are a number of challenges that play an important role when it comes to the development of concrete services and measures. In this process, the Local Authority is operating in the tensions field between citizens, politics and business partners. These stakeholders have certain interests, some of which are contrary to each other and must be taken into account by the local authority. Only in this way a successful implementation of measures and services is possible (Fig. 2).

Fig. 2. Actors and conditions influencing the planning and implementation of mobility projects by the local authority

Furthermore, the available knowledge on the various mobility topics plays an essential role and the capacity of the administration to explore and operationalise this knowledge.

The sources from which this knowledge is taken are very different. The findings from various EU research projects, available on web resources like CIVITAS or ELTIS[2] in a big pool of guidelines, project reports and good practice examples seems to be used relatively little. Inspiration and exchange of experiences occur primarily through personal exchanges with other mobility planners, through content presented at meetings and congresses, and through media reports. A particular challenge for the mobility planners is to implement this knowledge in the specific local context. For this, a precise understanding of the framework conditions of the own city is important.

Following the challenges occurring in the mentioned tension field will be described in more detail.

Understanding Political Interests and Affecting Political Decisions

Receiving strong political backing is the prerequisite for the successful implementation of a mobility measure but at the same time one of the biggest challenges for mobility planners in local authorities. No matter how well planned a project is, without political support it will not be implemented. Complex projects and the absence of appropriate

---

[2] The webpages www.civitas.eu and www.eltis.org contain knowledge and numerous materials that have been developed in the framework of various EU projects on sustainable mobility.

experiences are considered as risky by decision-makers. Accordingly, it is difficult to receive support for such projects. Moreover, political moods are often unstable and depend on numerous factors. Mobility planners need to recognise this and propose the right measures at the right time.

Decision makers often have to meet decisions, on the basis of extensive information material or expert reports which are often not easy to access and to grasp in their entire scope. A potential is therefore the preparation of the information materials for decision makers in a user-friendly way. This can significantly influence decisions.

Understanding Legal and Regulatory Framework
The development and implementation of mobility projects often affect different policy areas. Accordingly, a lot of legal and regulatory frameworks must be considered. Depending on the measure this can be very challenging, for example when it comes to data protection issues.

Special challenges arise when changes in the legal framework become necessary in order to implement certain measures. This can lead to major delays in planning and implementation, and even to measures not being implemented as planned.

Citizen Participation
The active and early involvement of citizens in the development process of certain mobility schemes can have positive effects on later acceptance.

The present challenge includes both, the active involvement of users in order to understand their requirements and possible concerns, but also the proactive information of citizens on contents and backgrounds of projects. This is particularly important when certain measures aim to influence or change the mobility behaviour of citizens. Especially when planning innovative mobility services, it is important to tailor them to the local conditions and to the user needs.

Although the importance of involving citizens is widely known, the effort is often spared. Reasons are for example:

- The lack of experience and knowledge on participation methods
- The difficulty of estimating the required effort and the potential benefit in advance
- The difficulty of considering citizens' problems and requirements, which are often contradictory, in projects that are complex anyway

Moreover, citizens taking part in participation processes do not sufficiently reflect the opinion of the general public. Mostly citizens only look at the individual measures and at those that affect them personally. Usually they don't see projects in the big picture.

Local Authorities must consider different user groups, they have to think about how to reach hard-to-reach user groups, older people, people without mobile phones.

The main challenge is to raise awareness and to show benefits, reasons and background information to the citizens. Because "only when there is sufficient public support for change, the action will take place" [9].

Estimating the Feasibility and Acceptance of Measures
Another big challenge is to estimate the feasibility and future acceptance of measures.

Especially with innovative mobility schemes or services where there is only little experience available, the try-out of downscaled versions before hand is an appropriate mean to gain a better understanding for upcoming problems and thus be able to make predictions for workability and acceptance. This may also facilitate obtaining political support. The involvement of stakeholders during feasibility study is essential for success [10]. But especially in the development of mobility services in which external business partners play a major role, this is a particular challenge.

Interaction and Cooperation with Business Partners

The interaction and cooperation with business partners has become an increasingly important aspect in recent years, especially with regard to new mobility schemes like sharing services, offered by external providers.

From the city site, conditions must be created that make it attractive for providers to offer their services in the city. On the other hand, cities must communicate to suppliers exactly what they need and what effect they are aiming on. Local authorities and business partners need to understand each other's interests, define common goals and if necessary, find compromises. The challenge here is to jointly tailor services according to the user needs and that take into account the cities local conditions.

Identification and Utilization of Synergy Effects

Synergy effects between different mobility measures, schemes or services are a very important issue in the context of sustainable mobility planning especially as the aim is to develop holistic solutions. The SUMP Guideline [5] refers several times to the identification and consideration of synergy effects. In practice, however, this represents a major challenge for the cities. Synergies can be positive, when for example different mobility measures contribute to the same objective and enhance each other, but negative synergies can also occur, if measures torpedo each other or work in opposite directions. The challenge is to identify these synergies, to exploit multiplier effects and to eliminate mutually distracting effects.

Use of Innovative Technologies and Data Collection Methods

Information and communications technology (ICT) is seen as one of the biggest enablers towards mobility planning. However, data collection by means of innovative technologies and the use of this data for mobility planning is still a big challenge for cities and mobility planners. Many mobility departments, especially in smaller cities, lack detailed knowledge of which technologies exist on the market and which data can be collected, how this data can be visualised and how it can be used for mobility planning.

The work with the project cities demonstrated that there is a great interest in the use of innovative technologies and data collection methods. But the full potential seems to be far from being fully exploited. Reasons for this are a lack of technical knowledge on the part of the mobility planners, a non-transparent market, high prices and the question of cost-benefit ratio that is, in many cases, difficult to justify to decision makers.

New technologies for data collection are constantly coming onto the market and it's hard to keep track of what data is really needed and which technologies are suitable for own purposes. However, it is not always necessary to collect own data; often the data is available at other bodies or institutions or it can be obtained through cooperation with local traffic management centres or phone operators.

However, the main challenges here for the cities are to understand the technical possibilities for the collection of mobility data and the application of this data for the systematic planning and evaluation of mobility measures.

Application of Research Knowledge and Adaption of Good Practice Examples
Numerous projects on the subject of sustainable mobility have generated numerous findings and research knowledge, which is available in large volumes of guidelines on the various subject areas. A major challenge for all cities is to put this knowledge into practice. Guidelines are to extensive to read and sometimes hard to adapt. The work with the cities in the project has shown that their interest is mainly in good practice examples from other cities. But available good practice examples from different research projects do not necessarily address the problems that arise when trying to implement measures in the own city. What proves to be a good practice in one city does not necessarily mean that it will succeed under different conditions. Good practice examples have great support potential, but in the future they must be better tailored to the knowledge interests of mobility planners.

3 Evaluation of the Challenges

In order to validate the identified challenges, 46 mobility experts, including mobility planners, mobility consultants and academics, were asked in an online survey to assess the relevance of the identified challenges in the planning and implementation of sustainable mobility measures. Furthermore, in case they have experience working with concrete cities, they were asked to assess how the cities already cope with the different challenges in the frame of their actions. The results are briefly presented in the following.

3.1 Importance of Challenges

The experts were asked: *From your experience, please rate the general importance of the following challenges for cities when planning and implementing sustainable mobility measures!*

The importance of each challenge should be assessed by using a five-point Likert scale. It was also possible to add any missing challenges. However, no additions were made on this by the experts. Figure 3 provides an overview of the results.

The answers show that the ratings of the individual challenges are quite close to each other in view of the mean value. In addition, more or less all challenges are considered to be important for the planning and implementation of mobility measures and services. Only the challenge of *Innovative procurement* is considered to have a slightly lower significance in comparison to the other challenges. *Institutional cooperation, Citizen participation* and *Effective and efficient project management* are seen as particularly important, which also coincides with the experiences of the discussions with the mobility planners.

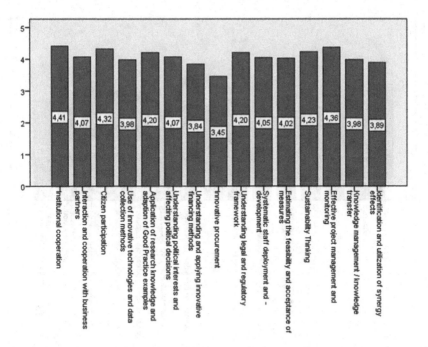

Fig. 3. Mean values of the importance rating of the identified challenges. Using a 5 point Likert scale (5 = very important 1 = not important at all)

3.2 Estimation on How Cities Cope with the Challenges

In addition, the experts were asked whether they work for a city or whether they could gain concrete experience from working with a particular city. In this case, the experts were asked to assess: *How well you think, the city you indicated coped with the individual challenges in their mobility planning activities?* A 5-point Likert scale was also used here. Figure 4 show the results.

The results show that in terms of the attested performance of the cities, there is still room for improvement in all the challenges. The performance of the challenges *Institutional cooperation, Understanding political interests and affecting political decisions* as well as the *Understanding of the legal and regulatory framework* are rated as comparatively good. It is interesting to note that the challenge *Innovative procurement*, for which the importance was rated relatively low in the first question, was rated to be low in terms of cities' performance in this question. The challenge use of *Innovative technologies and data collection methods* scores relatively poorly, too. This is consistent with the experience of working with the project cities. They confirmed, that there is a great interest in the use of innovative technologies and data for mobility planning, but very often the appropriate know-how and often also the necessary funds are lacking in this area.

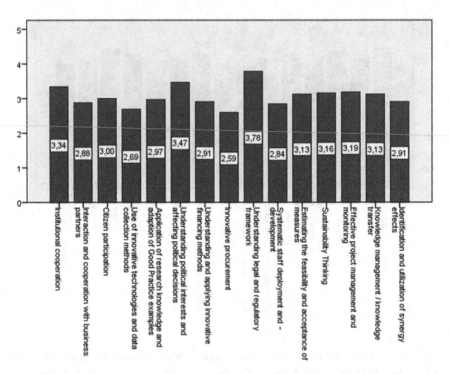

Fig. 4. Mean values of the performance rating using a 5 point Likert scale (5 = strong performance 1 = poor performance)

4   Conclusion

The challenges that cities face in the context of mobility planning are diverse. The field of mobility is constantly growing and becoming increasingly complex. Especially smaller cities with small mobility departments and very few personal are not able to build up in-depth knowledge in all relevant subject areas.

The topic of sustainable mobility enjoys a high priority. Numerous research projects funded by the European Union and national government have produced a large pool of knowledge that can hardly be grasped. The authorities lack the capacity to penetrate and implement this knowledge. The available knowledge resources are to comprehensive and too little focused on the knowledge interest of mobility planners. Some of this knowledge is simply too theoretical, barriers and problems faced by planners within the administration are not sufficiently taken into account. Existing support initiatives usually aim to provide knowledge on specific areas such as clean vehicles, mobility as a service or walking and cycling. But the challenges identified in the SUITS project mirror the questions coming up in every project: How can I get political support for my projects? How can I get administrative departments to work more closely together? How can I test solutions in advance? How can projects be financed even if no municipal funds are available? Which technologies and data do I really need to optimise my mobility planning? These questions and challenges are not or only insufficiently addressed in guidelines or in good practice examples which have been implemented in the frame of research projects.

The challenges presented in this Paper aim to provide a better understanding of the concerns of the mobility planners, showing their knowledge interest and the tension field moving in when planning and implementing mobility solutions. They should serve as a base, enabling future projects to support local authorities in a more effective way.

References

1. Schönduwe, R., Bock, B., Deibel, I.: Alles wie immer, nur irgendwie anders? Trends und Thesen zu veränderten Mobilitätsmustern junger Menschen. Innovationszentrum für Mobilität und gesellschaftlichen Wandel (InnoZ), Berlin (2012). https://www.innoz.de/sites/default/files/10_innoz-baustein.pdf. Accessed 22 Feb 2019
2. PWC Report: Five trends transforming the automotive industry (2005). https://www.pwc.com/gx/en/industries/automotive/assets/pwc-five-trends-transforming-the-automotive-industry.pdf. Accessed 22 Feb 2019
3. United Nations General Assembly: 2005 world summit outcome, resolution A/60/15. http://www.un.org/en/development/desa/population/migration/generalassembly/docs/globalcompact/A_RES_60_1.pdf. Accessed 21 Nov 2016
4. Guidelines: Developing and implementing a sustainable urban mobility plan. Rupprecht Consult (2014). http://www.eltis.org/sites/default/files/guidelines-developing-and-implementing-a-sump_final_web_jan2014b.pdf. Accessed 22 Feb 2019
5. VNG International: Inter-municipal cooperation introduction guide - approach to a successful IMC (2010). http://www.vng-international.nl/wp-content/uploads/2015/06/IMC_EN.pdf. Accessed 22 Feb 2019
6. MaxSumo-Guidance on how to plan, monitor and evaluate mobility projects (2009). http://www.epomm.eu/index.php?id=2602. Accessed 22 Feb 2019
7. Guidemaps - Successful transport decision-making. A project management and stakeholder engagement handbook (2012). https://civitas.eu/content/guidemaps-successful-transport-decision-making-project-management-and-stakeholder-engagement. Accessed 12 Feb 2019
8. Bornemann, M., et al.: An illustrated guide to knowledge management (2003)
9. Banister, D.: The sustainable mobility paradigm. Transp. Policy **15**, 73–80 (2008)
10. Wesley, F., Seaton, S.: Determining stakeholders for feasibility analysis. Ann. Tour. Res. **36** (1), 41–63 (2009)

Monitoring Road Surface Conditions for Bicycles – Using Mobile Device Sensor Data from Crowd Sourcing

Waldemar Titov[✉] and Thomas Schlegel

Institute of Ubiquitous Mobility Systems, Karlsruhe University
of Applied Sciences, Moltkestr. 30, 76131 Karlsruhe, Germany
{waldemar.titov, iums}@hs-karlsruhe.de
http://iums.eu

Abstract. This paper introduces an approach for monitoring cycleway conditions by collecting crowdsourced data from mobile devices. To collect the data, an application was developed and optimized to be used by many cyclists. The application uses acceleration and gyroscopic sensors to collect and upload road roughness data into a classification platform. A classification model classifies the monitored routes into three quality classes and synchronizes the results with the application. The methodology shows how to collect and classify road surface conditions of cycleways. By using the K-Nearest Neighbor machine learning algorithm as a classifier, we were able to achieve a forecast accuracy above 90% on average. We report on our experiences with classification accuracy of four different classifiers as well as the experimental evaluations of the system. The results support the potential development of a community portal that provides detected cycleway conditions from the up-to-date mobile crowdsensing application.

Keywords: Monitoring cycleway conditions · Mobile sensors · Crowdsensing

1 Introduction

Mobility and transportation are essential to the daily lives of people, spending on average more than 45 min on roads and in traffic [1]. Therefore, monitoring road conditions has received a significant amount of attention. However, since roads in urban regions are still mainly used by motorized means of traffic [2], highly automated sensor vehicles conduct pavement condition assessments. To boost non-motorized and thus emission-free means of transport, three major improvement tasks were recognized: first the motivation for non-motorized transport [3], second the increase in safety [4] and third the increase in traveling comfort [5].

While politicians and municipalities have found success in increasing motivation and safety for non-motorized means of traffic, little work has been done to evaluate the traveling comfort. In the current study, we examine the possibility of using a smartphone crowdsensing approach to monitor the conditions of cycle path surfaces capture the aspect of travel comfort of cyclists. Compared to the traditional way of collecting pavement data, assessing and monitoring road conditions by deploying the crowdsensing

© Springer Nature Switzerland AG 2019
H. Krömker (Ed.): HCII 2019, LNCS 11596, pp. 340–356, 2019.
https://doi.org/10.1007/978-3-030-22666-4_25

approach results in five major benefits. First, the quantity of participants collecting data. Second, the frequency of road conditions being monitored. Third, the almost unlimited mobility of cyclists' use of cycle path networks, consisting of many narrow routes. Fourth, the reduction of costs for municipalities. Fifth, the crowdsensing approach promotes civic engagement, unity and identity of their participants.

2 Related Work

During the preparation of this project, few examples could be located describing the determination and classification of road surface conditions for non-motorized traffic routes utilizing mobile devices. Much more work has been done either describing the classification for routes mainly used by motorized means of traffic or utilizing specially build sensor boxes to evaluate the quality of road surfaces. Nevertheless, some elaborations on related topics have been compiled and presented below. Some previous studies have focused weather conditions and seasonal variation to evaluate cycling comfort [6]. These rather temporal influencing factors should not be the focus of this study. Instead, spatially differentiated influencing factors should be considered. A study conducted in 2008 estimates the influence of local factors like: the routing and the shape of cycleways on cycling at about 70% [7].

According to the General German Bicycle Club e.V., (ADFC) studies have shown that most people would use the bicycle more often if they were to experience cycling more positively such as road space and routes more suited to cycling [8]. The recommendations for cycling facilities in Germany (ERA) include enabling fast and direct routes, the consideration of individual user groups like racing, - and electro bikes [9].

A study carried out at the University of Maryland confirmed the assumption that commuters can be encouraged to use the bike by taking off-road journeys [10]. This study conducted an online survey of students who live within a five-mile radius of the campus. One of the biggest motivations for cycling owners was a separate bike lane, other considerations were better lighting and a good map with local bike paths. When asked what prevented students from biking, the majority responded with concerns of safety on the road and poor cycleway conditions.

Through studying the related work for determining and monitoring road surface conditions for cycling paths, two different approaches stand out. Monitoring cycling paths conditions using dedicated sensor boxes, and monitoring cycling paths using sensors built in to modern smartphones.

2.1 Classification of Road Surface Conditions Using Sensor Boxes

An approach to detect potholes using GPS and accelerometer data from dedicated hardware devices mounted in taxi cabs has already been successfully explored. A system called the Pothole Patrol gathers data from vibration and GPS sensors which then processes the data to assess the road surface conditions. The sensor devices have been deployed on seven taxis running around the Boston Metro area. Using a simple machine-learning approach, the authors could show the ability of the system to identify potholes and other severe road surface anomalies from accelerometer data. Data from thousands of kilometers of taxi drives demonstrated that over 90% contained road anomalies needing repair [11].

Another study investigates road conditions by using customized embedded devices with microphone and accelerometer sensors. Unfortunately, the study does not explain how the collected microphone data is used for monitoring road surface conditions [12].

A further study is focused on detecting bad road surface quality, such as cracks, potholes, and obstacles to prevent accidents, particularly of cyclists. Low-cost ultrasonic sensors were used to periodically measure the distance from the system to the road to estimate the road surface condition. Through experimental evaluations, the study showed that the monitoring system can detect obstacles and holes in the front area of a bicycle [13].

2.2   Classification of Road Surface Conditions Using Smartphones

Monitoring and classifying cycleway road conditions using bike-mounted smartphones has been a rising topic in the last few years. The reason for the rise is the availability and user acceptance of mobile devices, and general market penetration. Modern smartphones are equipped with many highly sensitive sensors and have a high computing capability. These factors make it possible to use these mobile devices for many mobile applications such as monitoring cycleway road conditions.

The next study discussed an embedded surface road classifier for smartphones used to track and classify routes on bikes. The authors address the problem of the quantity of accelerometer data that would have to be uploaded along with GPS tracks for server classification. Their approach is to classify cycling paths online with an embedded classifier, that has been trained off-line, which makes the data upload unnecessary. This way the accelerometer data of a bicycle-mounted smartphone is collected, labeled, and a classification model is learned, which then again is deployed on the smartphone. The results indicate the requirement of moderately fast smartphones to conduct the classification on mobile devices. However, the system was able to detect larger uneven-ness', bumps and classify only short segments of tracks as satisfying. Also, the power consumption of the learning algorithm running on the smartphone is still questionable, mostly ruling out every crowdsensing approach [14].

Another study examines the suitability of smartphone sensor data for road condition determination. Two android devices with different sensors and computing capacities are evaluated in terms of data quality and data density. The approach of the study is to capture the acceleration data of a smartphone attached to a vehicle and upload the data on to a classification server for real time road classification. The results indicate a relatively high unreliability of mobile data connection. The authors suggest using a time shifted upload method instead relying on the availability of suitable mobile data connection. Furthermore, the measurement results are distorted by shock absorbers and vibration generated by the vehicle. Nevertheless, the basic use of smartphones to determine the quality of roads is confirmed [15].

A further study investigates the monitoring of motorized road and traffic conditions in cities using mobile devices. Data from location sensors, accelerometer and microphone is collected to evaluate whether and how road conditions have a negative impact on traffic management. The focus of the work is first the detection of potholes and bumps on traffic routes and second the estimation of the resulting effects such as dodging, strong braking and honking of the vehicles [4].

The context of the rapid growth of networked mobile sensors that record data in everyday life and share it via the internet is described in the next study. The term "mobile crowdsensing" is introduced and the benefits and challenges of such a development explained. Finally, the challenges and the lack of a single data infrastructure are identified [16].

2.3   Summary

Comparing both approaches to monitor cycleway road conditions using smartphones instead of sensor boxes promises to be a widely accepted approach. While sensor boxes have the ability to collect data that mobile devices cannot, such as particulate matter, temperature and humidity, the advantages of mobile devices make it an attractive platform to use for monitoring cycleway conditions. The benefits of mobile devices are listed below:

- The ability to collect data on light incidence, curviness, decelerations, changes in altitude and the roughness of surface conditions by applying different sensors
- High user acceptance and thus high market penetration of smartphones
- No additional costs and power supplement for sensor infrastructure
- High potential for participating in crowdsensing projects to enforce civic engagement

To ensure data quality using a smartphone it is necessary for the mobile devices to be completely rigid of the bike handlebars. Out of these considerations our approach is presented in the next chapter.

3   Monitoring Cycleway Conditions Using Crowdsourced Data from Mobile Devices

To implement our full-scale crowdsensing approach for monitoring road conditions of cycleways, the system must be transparent and intelligible for the user. Beyond intelligibility the user interaction with the system for the execution of a service must have a low-threshold for being executed regularly [16]. Furthermore, the execution of a service must be reliable and not perceived as an impedance. A major disruptive factor for users of mobile devices is the power consumption of an executed service, e.g. running a machine learning classification algorithm on the smartphone [14]. In most cases, energy draining applications are often viewed as more disruptive then mobile data draining applications [23, 24]. In the best case, the execution of a crowdsensing service is not even noticeable or embedded into another beneficial service.

All above listed considerations flowed into the conception of the project presented here. We are aiming to create a community portal to evaluate the comfort for bicyclists based on crowdsourced data. Generally, there are two approaches to monitor cycleway road conditions: event-based, which only detects locations with surface irregularities such as potholes; and continuous road classification. In the current work, the second approach was chosen to provide a wide overview of cycleway road conditions because this method collects continuous data through the entirety of the recording process. More specifically, we introduce a crowdsensing approach for monitoring road surface conditions of cycleways. Thereby we use the smartphone as a sensor platform and derive data from different sensors build in the most modern mobile devices.

A two-stage condition recording and evaluation procedure was developed for monitoring purposes. Furthermore, an Android data collection application called GyroTracker was designed, implemented and evaluated. In the first stage, data is collected via mounting the smartphone on the bike's handlebar using a bicycle navigation mount. Once the app is started, the location service of the device will run in the background. Upon completion, user's own location will be displayed on a google map. Due to the high spatial relevance of the measurement data, the data recording can only begin after the device's location is available. Once located, users can start recording of both location and sensor data. During the bike ride, GyroTracker reads the four sensors: location, gyroscope, acceleration, and linear acceleration and stores the data into the smartphones database (Fig. 1). More details and purposes of each used sensor type is provided in Sect. 4.

After the completion of a recording, the measurement-id is incremented in the background, ensuring the identification of each recording. After finalizing the recording, the data is uploaded to the classification server (Fig. 2). On the server, the data is cleaned and further processed. To ensure the assignability of sensor data with the location data, the location data must be interpolated.

Fig. 1. Collecting data via GyroTracker.   **Fig. 2.** Uploading data after collection.

In the second stage, the uploaded data is classified by a machine-learning algorithm into three quality classes (Fig. 3). More information about the classification algorithms is provided in Sect. 5. After classification, the classified tracks are synchronized with the application GyroTracker to visualize the classification results (Fig. 4).

Based on the described approach, measurements are repeatedly evaluated and changes in infrastructure become immediately visible. The conditions of bike paths are displayed on the community portal for cyclists, aiming to promote the non-motorized means of individual traffic.

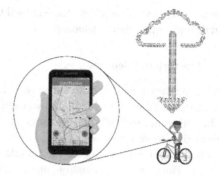

Fig. 3. Classification of the data on the server.

Fig. 4. Synchronization of the classified data.

4 Implementation of the GyroTracker Application

Due to many sensors, high computing power, high global market share of over 75%, relatively low cost and the easy accessibility of the programming API, Android smartphones are well suited for being used as sensor platforms. It is possible to realize first measurement results after a short time of implementation. For the determination of cycleway road quality, the crowdsensing data collection application GyroTracker was developed. To address the widest possible user group, the support level has been set to API level 17, allowing 97–98.4% of android smartphone users to use this application on their smartphone [25, 26].

4.1 Functionality

To achieve a greater chance of usage and to correspond with the guideline imposed in the concept, the data collection was embedded into a navigation layout as the beneficial service for users. Therefore, the central element of the application, is a map view of a user's location. Google Maps was chosen as the map provider for GyroTracker, for its up-to-date maps, accuracy and the relatively simple implementation in an Android environment.

In the next step the map layout is created. The map is automatically centered on the last known location and enlarged to zoom level 15. As soon as the system has prevailing coordinates, the view is updated to the new location. This way, only the required map tiles are downloaded which reduces the mobile data consumption. Furthermore, a separate location button has been created, allowing the user to manually move the view to the current location at any time.

While recording a track, the current position of the smartphone is focused, and automatically updated to a new map section at any new location. Zooming in or out of the map is still possible during a measurement. Furthermore, all coordinate pairs are stored in an array list during the measurement. These are used to mark the path covered on the map with a polyline. The polyline is still available after completion of the measurement ride and is automatically removed from the map at the beginning of a new measurement. This allows the recorded track to be traced after the measurement is stored in the smartphone database.

4.2 Operational Sensors

During the measurement, different sensors are used. The sensor technology built into the smartphone can be divided into two categories: hardware- and software-based [17]. In the following, the individual sensors and their functions during a measurement are presented in more detail.

The location sensors locate the mobile device during the data acquisition phase (Fig. 5). Thus, in the evaluation phase, the recorded sensor data can be uniquely assigned to their coordinates. To determine the location of a smartphone, a combination of satellite-based global navigation system (GNSS) data, Wi-Fi and mobile data is used [18]. However, the methods of the recipients used to determine the location are different. While the accuracy of a Wi-Fi location is 2–3 m, the accuracy of the GNSS location is officially stated as 10 m, but usually more accurate in open terrain [19].

In accordance with Fig. 6, the hardware-based gyroscope measures the rate of rotation of the smartphone in three dimensions to capture the relative change in position [20]. Thus, potholes and other bumps in the road surface can be detected.

$g \approx 9.81$ m/s^2

Fig. 5. Location sensors.   **Fig. 6.** Gyroscope.   **Fig. 7.** Acceleration.   **Fig. 8.** Linear acceleration.

The hardware-based accelerometer measures the acceleration forces in three axes including gravitational force (Fig. 7). The individual measured values are combined into a 3-dimensional vector, which indicates the direction and strength of the currently acting force [21]. The main task of this sensor is to detect uneven road surfaces such as potholes, transverse and longitudinal bumps, curbs and dropped curbs. Additionally, the measured values are used during the evaluation phase to validate the linear acceleration data.

The software-based linear accelerometer measures the instantaneous linear acceleration acting on the mobile device (Fig. 8). The peculiarity of this sensor is that the gravitational force $g \approx 9.81$ m/s^2 prevailing on the earth is mathematically eliminated. The measured values are recorded by the previously presented accelerometer and internally reduced to linear acceleration by using the Kalman high-pass filter [22]. The acceleration sensors behave as follows:

$$acceleration = linear\ acceleration + gravity \tag{1}$$

During a measurement, both accelerometers are recorded to be able to validate the results against each other in the evaluation phase. However, the evaluation phase revealed different sample rates. Since the sensors used are streaming sensors, android's sensor manager framework does not offer an opportunity to set a specific sampling rate for all three sensors, so each sensor has its own sampling rate [17]. While the gyroscope has a sampling frequency of approximately 150 Hz (six times a second), the accelerometer is 50 Hz (50 times per second) and the linear accelerometer is as low as 20 Hz (100 times per second). This challenge could be solved successfully by correlating the values using their timestamps. For this purpose, in addition to the sensor values, the associated timestamps were prompted and stored in the database. The additional saving of a timestamp ensures the subsequent assignment of the measured values with their associated geolocation.

Due to limited space on a smartphone screen, only the values of two out of the three used sensors are displayed on screen. The gyroscope and the linear accelerometer were selected as the most meaningful due to the already eliminated gravitational force. To display the most recent sensor values, text fields were created at the bottom of the screen. These text fields are overwritten each time the sensor values changed, so the most recent value of the sensor is always visible.

During the recording of a route, the sensor values with their accuracy specifications and the timestamp are written to the SQLite database. The individual measurements can be distinguished from one another on the database by the variable measurement-id.

4.3   User Interface

To increase the convenience of crowd users, the layout of GyroTracker was intentionally left uncomplicated. After opening the application, the user is situated in the main view, from which a quick start of the measurement is possible. As previously mentioned, the central element is the Google Map, which simultaneously displays a user's current location. Above the map is the application name. Below the map, the current sensor values are displayed. While the values of the 3-dimensional gyroscope are shown on the left side, the values of the linear accelerometer are on the right side. Furthermore, the 'My Location' button, the button for starting the recording as well as the upload button are located on the map view.

For a better overview of which processes are currently running in the background of the application, the respective buttons are highlighted. Thus, the button to start and stop the measurements during a measurement is highlighted in red (Fig. 9a). The upload button is also highlighted in red while the data upload takes place (Fig. 9b). The same red color, used for highlighting background tasks, is used to mark the path during a measurement ride (Fig. 9c). Thus, the user encounters only two colors in the application so a learning effect of the processes can occur.

(a)   (b)   (c)

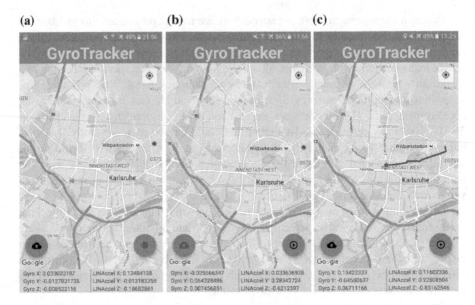

Fig. 9. a. Active measurement in the background. b. Active upload in the background. c. Polyline displays recorded track. (Color figure online)

5   Classification of the Collected Acceleration Data

Machine learning describes an artificial generation of knowledge from experience. Thus, an artificial system, after completion of the learning phase, can generalize the learned examples and apply them independently. In the present case, patterns and laws are recognized in the recorded training data, and these are later applied to the real measurement data. However, before being classified by a machine learning algorithm the algorithm is trained by a training data set with a distribution of 70/30.

5.1   Trainings Dataset

To create a trainings dataset to teach our classification algorithm, a definition of the perceived cycleways quality was needed. Therefore, a two-step approach was chosen. First, tracks of cycleways with different pavements and different roughness' were recorded. Second, the cyclists and in-line skaters were questioned on site to classify the pavements into three quality classes, as they are most effected by the rough roads. In this experimental study we gained information about the actual perceptions of the bike road users. The survey took place in Karlsruhe, the city with the second largest German biking community [27]. The results of the survey indicate a flat concrete superstructure road and a smooth paved road as ideal for cyclists and in-line skaters (Fig. 10a). A conditionally suitable road is defined as a road covered with coarse asphalt with partially used tar patches, a coarse asphalt surface and rarely to regularly occurring potholes (Fig. 10b). An unsuitable cycleway is defined as an unpaved gravel road or pebble path as well as an uneven track with many transverse and longitudinal bumps (Fig. 10c). The results gained out of this experiment were used to label the recorded data in the training data set prior to the model training process.

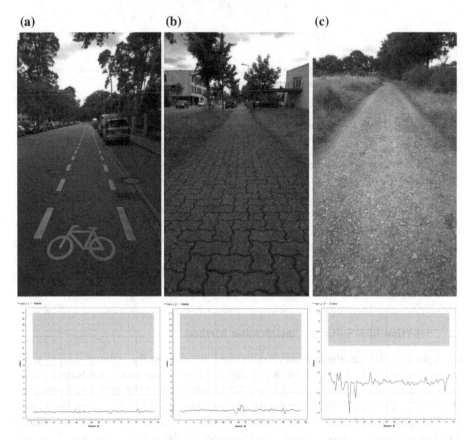

Fig. 10. a. Image and corresponding graph of rotation speeds occurred cycling a cycleway with quality class 1. b. Image and corresponding graph of rotation speeds occurred cycling a cycleway with quality class 2. c. Image and corresponding graph of rotation speeds occurred cycling a cycleway with quality class 3.

In Fig. 11a, an illustration visualizes the evaluation of the x-variable from the gyroscope sensor for one representative track from each class. Figure 11b and 11 contrast the y- and z-variables of the linear accelerometer with one representative track from each class, while Table 1 shows the numerical evaluation of the measured data for one representative track from each class.

(a)   **(b)**   **(c)**

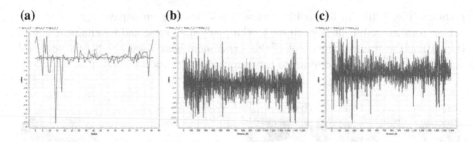

Fig. 11. a. Amplitude of x-variables gyroscope. b. Amplitude of y-variables linear acceleration. c. Amplitude of z-variables linear acceleration.

Table 1. Measured acceleration data of the 3 different classes.

Sensors	Class 1	Class 2	Class 3
Gyroscope x- variable	Min: −0.32 Max: 0.48 Avg: −0.01	Min: −0.59 Max: 0.99 Avg: 0.02	Min: −7.58 Max: 2.50 Avg: −0.03
Linear acceleration y- variable	Min: −5.02 Max: 6.98 Avg: 0.11	Min: −6.59 Max: 12.14 Avg: 0.04	Min: −15.45 Max: 19.74 Avg: 0.16
Linear acceleration z- variable	Min: −10.60 Max: 10.03 Avg: −0.11	Min: −20.09 Max: 23.50 Avg: −0.88	Min: −46.32 Max: 35.56 Avg: 1.33

5.2   Selection of a Suitable Classification Method

In preparation for the algorithmic classification, the data in the training dataset were labeled according to their classes and made available to the model for training. The training of the model using the trainings dataset was carried out offline using Rapid-Miner, an open source system for data mining. The uploaded files from the Gyro-Tracker application were directly imported into RapidMiner and further processed. RapidMiner supports classification algorithms such as Naive Bayes, Decision Tree and Neural Net, allowing for efficient evaluation of collected sensor data. To ensure model correctness, the resulting model was validated ten times, which means that the input data are divided into ten parts in the first step. In the next step, a model is trained on nine of the ten sub-data and applied to the tenth remaining part. This process iterates ten times through the entire data set and averages the model accuracy determined for each iteration. These averaged values can be found in Table 2 for each classification algorithm.

As Table 2 demonstrates, each classification algorithm has its advantages and disadvantages. While tracks labeled class 1 has been satisfactorily recognized only by Naive Bayes and K-Nearest Neighbors (K-NN), data labeled class 2 has been well recognized by all but Naive Bayes. Meanwhile, tracks labeled as class 3 roads have only been successfully identified by K-NN. The reason for such a high instability of the accuracies is the high instability of the sensor values. The extremely sensitive

Table 2. Forecast accuracy of the cycleway classification of four selected methods.

Method	Accuracy	Class 3	Class 2	Class1
Naive Bayes	39.42 ± 0.53	69.56	5.57	90.48
Neural Net	52.28 ± 1.35	20.87	87.24	0.10
Decision Tree	66.21 ± 1.27	51.87	94.73	0.41
K-NN	91.12 ± 0.78	90.38	92.68	87.42

gyroscope and the accelerometer provide values in all areas. Often, the road class changes several times within one second, which only corresponds to reality in a few cases. In view of this, the validation result of the method K-Nearest Neighbor with 91.12% and a standard deviation of ±0.78% were found an acceptable value and therefore implemented as the standard classification method on our classification server.

6 Experimental Evaluation and Results

To evaluate the developed cycleway road surface classification, an experimental evaluation on and around campus of the Karlsruhe University of Applied Science was performed. Students and volunteers were asked to participate in the data collection process. Study participants could participate in two different ways. First, while commuting from home to university by mounting their smartphones on the bike's handlebar. And second, by taking the Segway Personal Transporter (PT) for a ride. Figures 12 and 13 show participants riding a bicycle on separated cycleways as well as bike lane collecting data with the GyroTracker application.

Fig. 12. Participant cycling a cycleway collecting data for cycleway road classification.

Fig. 13. Participant cycling a bike line collecting data for cycleway road classification.

Figures 14 and 15 a, b and c show participants adjusting and riding a Segway PT on cycleways and shared paths collecting data with the GyroTracker application.

(a)

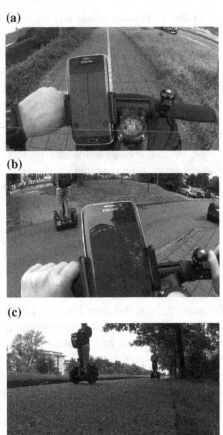

(b)

(c)

Fig. 14. Probands preparing data collection application for a measurement ride.

Fig. 15. a. proband collecting data on a two-way cycleway. b and c. two probands collecting data simultaneously.

During the evaluation, all participants were equipped with Samsung Galaxy S6 Edge with the android version 6.0.1 and 4 × 2.1 GHz processor devices. Deviating from [15], our evaluation focus is on the method used for classification of collected data, therefore the collection of measurement data by further devices was omitted.

In the short period of the experimental evaluation, a total of 21 measurements of different cycleway road qualities was carried out. The result was a dataset with over 510,000 individual measuring points. Afterwards, the twelve pavements most frequently used on cycle paths were selected and classified with our pre-trained model.

The results show that our trained model with the K-NN as the classification algorithm has stable results in all three classification classes. The accuracy of the selected classification method proves reasonable results at 90.33%. According to Table 3 only 3,308 of the 34,574 data points were detected incorrectly.

Table 3. Result of the modeling of the classification method K-Nearest Neighbor.

Accuracy: 90.33 ± 0.78% (micro: 90.33%)				
	True bad	True okay	True good	Class precision
Prediction bad	8,936	712	161	91.10%
Prediction okay	692	16,738	716	92.24%
Prediction good	353	674	5,592	84.48%
Class recall	89.53%	92.35%	86.44%	

To increase the accuracy of the models and to find out what pattern were recognized in the dataset, the input data was varied several times during model creation. When the data collection is conducted with a bicycle, the forces acting in the direction of travel are given the highest value, while the forces acting crosswise to the direction of travel can be seen more in connection with steering and pedaling (Fig. 16a). To eliminate these influencing factors in the validation phase, the y-and z-variables of the gyroscope and the x-variable of the accelerometer were excluded from the modeling. However, with this manipulation no improvement could be achieved, on the contrary, the accuracy of the models deteriorated rapidly. The model deterioration indicates that the 3-dimensional data appears in a certain pattern, which was recognized by the model and applied while classifying the tracks. A different pattern of the data was found when a Segway PT was used collecting the data. The forces acting transverse to the direction of travel would play a more significant role because the wheels are parallel and next to each other instead of behind each other like in the case of a bike (Fig. 16b).

(a)   **(b)**

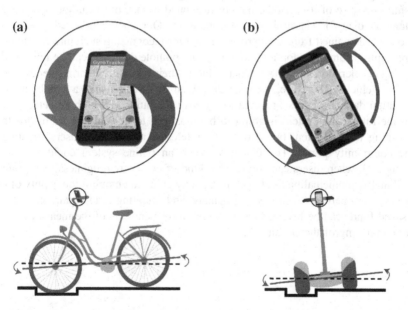

Fig. 16. a. Rotation forces on longitudinal axed bike captured by gyroscope. b. Rotation forces on cross-wise axed Segway captured by gyroscope.

The International Roughness Index (IRI) is a standard global index of road roughness [28, 29]. The results of our study show that the IRI and the vertical component of the linear acceleration values have a high correlation. Thereby the vertical component of linear acceleration is calculated as the sum of modulus of y-values and modulus of z-values of the linear acceleration data (Formula 2).

$$\sum |y - values| + |z - values| \tag{2}$$

7 Conclusion and Future Work

This paper introduces an approach for cycleway road classification using crowdsourced data derived from on bicycles mounted smartphones. The data collection and contribution to the system is done by GyroTracker, an application specially developed for the purposes of volunteers crowdsensing the road surface. The classification of the collected data is done by a machine learning algorithm in the cloud, where the data is uploaded after collection. After the data is classified the track sequences are synchronized from the cloud to the user's smartphone.

The evaluation of the system showed that variations of speed during the measurement process have a large influence on the recorded sensor data, which has been observed in previous research [27]. This factor can be minimized by additionally considering users' current speed in the model. Further improvement of the classification results can be achieved by filtering out data noise caused by users when pedaling and steering. In particular, pedaling can be recognized in the data by machine learning algorithm because of its periodic nature. In related works, other authors rely only on acceleration data for classifying road conditions. Our approach is using additional rotation speeds gained from the gyroscope data for recognizing such unwanted events, like pedaling. From our adoption, the bump and pothole detection can be improved by analyzing acceleration and rotation data, due to their certain regularity, depending on the used vehicle. Therefore, we introduced an approach using a combination of acceleration data and gyroscope data to improve the bump and pothole detection.

In the future, the system is going to be expanded to not only be useful for those contributing to the system but also for all interested and affected users, creating an online community portal. The benefit of such an online system could be a route planning tool, where users can adjust what kind of cycleway roughness they prefer to cycle. Finally, municipalities and cycling-friendly cities can monitor the quality of their cycleway route network for detecting damage and initiating road repair, free of cost. The saved funds can be invested in the actual implementation of the measures instead of the monitoring of the conditions.

References

1. AAA-Newsroom. https://newsroom.aaa.com/2016/09/americans-spend-average-17600-minutes-driving-year/. Accessed 21 Aug 2018
2. BTS. https://www.bts.dot.gov/content/commute-mode-share-2015, last accessed 21/08/2018
3. ERA – Forschungsgesellschaft für Straßen- und Verkehrswesen (Hrsg.): Empfehlungen für Radverkehrsanlagen (ERA), Ausgabe (2010)
4. Mohan, P., Padmanabhan, V.N., Ramjee, R.: Nericell: rich monitoring of road and traffic conditions using mobile smartphones. In: Proceedings of the 6th ACM Conference on Embedded Network Sensor Systems, pp. 323–336. ACM, November 2008
5. RASt – Forschungsgesellschaft für das Straßen- und Verkehrswesen (Hrsg.): Richtlinien für die Anlagen von Stadtstraßen (RASt) Edition (2007)
6. Bosselmann, P., et al.: Sun, Wind, and Comfort a Study of Open Spaces and Sidewalks in Four Downtown Areas (1984)
7. Thomas, T., Jaarsma, C.F., Tutert, S.I.A.: Temporal variations of bicycle demand in the Netherlands: The influence of weather on cycling (2009)
8. The Transport Policy Program of the ADFC: Allgemeiner Deutscher Fahrrad-Club e.V. Berlin (2014)
9. F. f. S.-. u. V. (FGSV): Empfehlungen für Radverkehrsanlagen ERA, Köln: FGSV Verlag (2010).
10. Akar, G., Clifton, K.: Influence of individual perceptions and bicycle infrastructure on decision to bike. Transp. Res. Rec.: J. Transp. Res. Board **2140**, 165–172 (2009)
11. Eriksson, J., Girod, L., Hull, B., Newton, R., Madden, S., Balakrishnan, H.: The pothole patrol: using a mobile sensor network for road surface monitoring. In: Proceedings of the 6th International Conference on Mobile Systems, Applications, and Services, pp. 29–39. ACM, June 2008
12. Mednis, A., Elsts, A., Selavo, L.: Embedded solution for road condition monitoring using vehicular sensor networks. In: The 6th IEEE International Conference on Application of Information and Communication Technologies (AICT 2012), pp. 1–5, October 2012
13. Taniguchi, Y., Nishii, K., Hisamatsu, H.: Evaluation of a bicycle-mounted ultrasonic distance sensor for monitoring road surface condition. In: 2015 7th International Conference on Computational Intelligence, Communication Systems and Networks (CICSyN), pp. 31–34. IEEE, June 2015
14. Hoffmann, M., Mock, M., May, M.: Road-quality classification and bump detection with bicycle-mounted smartphones. In: Proceedings of the 3rd International Conference on Ubiquitous Data Mining, vol. 1088, pp. 39–43, August 2013. CEUR-WS.org
15. Lauer, A.J., Jochem, A., Zipf, A.: Straßenzustandsermittlung durch Klassifikation mobiler Sensordaten von Smartphones (Doctoral dissertation, Master's thesis, Abteilung Geoinformatik Geographisches Institut, Universität Heidelberg)
16. Ganti, R.K., Ye, F., Lei, H.: Mobile crowdsensing: current state and future challenges. IEEE Commun. Mag. **49**(11), 32–39 (2011)
17. Sensors Overview on Android Developers (2018). https://developer.android.com/guide/topics/sensors/sensors_overview.html. Accessed 12 Dec 2018
18. Location Manager on Android Developers (2018). https://developer.android.com/reference/android/location/LocationManager.html. Accessed 12 June 2018
19. Wiki Open Street Map, accuracy of GNSS- data (2018) http://wiki.openstreetmap.org/wiki/DE:Genauigkeit_von_GPS-Daten. Accessed 14 June 2018

20. Intel Developer Zone: Anleitung für Software-Entwickler zu Sensoren auf Android*-Tablets mit Intel® Atom™ Prozessor (2018). https://software.intel.com/de-de/android/articles/software-developers-guide-for-sensors-on-intel-atom-based-android-tablets. Accessed 03 June 2018

21. Heller IG mbH, Zustandserfassung und -bewertung (2018). https://www.heller-ig.de/dienstleistungen/zustandserfassung-und-bewertung/. Accessed 29 May 2018

22. Sensor Event on Android Developers (2018). https://developer.android.com/reference/android/hardware/SensorEvent.html. Accessed 12 June 2018

23. Pihkola, H., Hongisto, M., Apilo, O., Lasanen, M.: Evaluating the energy consumption of mobile data transfer—from technology development to consumer behaviour and life cycle thinking. Sustainability **10**(7), 2494 (2018)

24. Tawalbeh, M., Eardley, A.: Studying the energy consumption in mobile devices. Procedia Comput. Sci. **94**, 183–189 (2016)

25. GlobalStats. http://gs.statcounter.com/os-market-share/mobile/worldwide. Accessed 28 Jan 2019

26. Android API Version Distribution (2019). https://developer.android.com/about/dashboards/index.html. Accessed 28 Jan 2019

27. SWR. https://www.swr.de/marktcheck/ranking-der-fahrradfreundlichsten-staedte-karlsruhe-auf-platz-1-in-bw-mainz-in-rp/-/id=100834/did=19572068/nid=100834/1wscb2e/index.html. Accessed 27 Jan 2019

28. Du, Y., Liu, C., Wu, D., Jiang, S.: Measurement of international roughness index by using Z-Axis accelerometers and GPS. Math. Probl. Eng. **2014**, 10 pages (2014). Article ID 928980. https://doi.org/10.1155/2014/928980

29. Sayers, M.W.: On the calculation of international roughness index from longitudinal road profile. Transp. Res. Rec. **1501**, 1–12 (1995)

Spatial Features and Elements Affecting Indoor Wayfinding—A Case Study in a Transit Hub

Dadi An, Junnan Ye, and Wei Ding[✉]

School of Art Design and Media,
East China University of Science and Technology,
M Box 286, NO. 130 Meilong Road, Xuhui District, Shanghai 200237, China
61915633@qq.com, yejunnan971108@qq.com, dw6789@163.com

Abstract. Indoor spatial features and elements have been identified as essential points that affecting wayfinding performance of a complex building. This research chose Shanghai south railway station to have a comprehensive understanding on the influence of spatial features and elements on wayfinding. Combination methods of tracking observation and space syntax analysis were introduced.

Decision-making points refer to the positions of access, turning, or going up and down the stairs. The complicity or symmetry of architecture plans are the main causes of dis-orientation. Signage plays a decisive role during the wayfinding process in an unfamiliar environment, especially in the underground corridor of a transit hub. With the combination of subjective observation, isovist analysis can be used as a powerful tool for describing the spatial features, and the plan of the architecture plays an important role on the passengers' mobility.

To conclude, indoor wayfinding performances of a complex building are influenced by both spatial features and elements. Spatial features can determine the indoor mobility pattern; functional elements contribute to the main impression; and the visual elements (especially signage) significantly impact the capture of spatial knowledge during wayfinding.

Keywords: Wayfinding · Spatial configuration · Signage ·
Decision-making point · Space syntax

1 Introduction

1.1 Wayfinding Factors

As the scale of architectures are getting bigger and complex, wayfinding within a public building such as airport, railway station, shopping mall, or hospital often proves to be a challenge for most pedestrians. The indoor spatial features and elements have been identified as decision-making points that affecting wayfinding performance of a complex building. Decision-making points are the stop points where people pause and confirm their direction, percept the space, choose the destination path in the process of wayfinding. Whether the wayfinding process can be completed or not depends on every

© Springer Nature Switzerland AG 2019
H. Krömker (Ed.): HCII 2019, LNCS 11596, pp. 357–367, 2019.
https://doi.org/10.1007/978-3-030-22666-4_26

single decision-making point during the journey. A decision-making point is not only at the intersection or turning point in the path, but also at any position in the wayfinding process that may confuses pedestrians. In general, the more decision-making points in a building space, the more difficult it is to find a way.

Eaton (1991) suggested that the determination of decision-making point is closely related to environmental information, including the effectiveness of the information, relationship between environmental and pedestrian and the personal wayfinding ability of pedestrian. The investigation and analysis of environmental information is particularly necessary to understand the impact on wayfinding behavior. The significance of environment information survey is to help people find the decision-making points and the significant features of spatial cognition. This study divides environmental information factors into spatial features and spatial elements.

The spatial feature includes whether the long or short of an atrium, wide or narrow of a corridor, bright or dark of a room, which is the primary factor to form the spatial image. The characteristic factors of a space such as the volume and size, the shape and proportion, the enclosure and openness, and axial symmetry are the spatial configuration that people initially percept. Moreover, in the process of transfer, pedestrians and passengers usually form memories of the facilities with practical functions, mainly including signages, lighting, arts, advertisements, seats, toilets, ticket machines, convenience stores, newsstands, ATM information service stations and gates.

1.2 Related Works

Pioneering study (Weisman 1996) distinguished the four general classed of environmental variables impacting wayfinding process: visual access, the degree of architectural differentiation, the use of signage or room numbers, and plan configuration. Follow-up studies enrich the theory of wayfinding and closed the gap between architectural design and spatial cognition by adopting both subjective and objective research. Passini (1984) suggested that physical spatial features do not form an impression in the brain, but the function itself is memorable. Three primary ways of human wayfinding were summarized: landmark-navigation, route-navigation, and map-navigation. Human wayfinding require successful wayfinding information to obtain perception of the environment through different levels. Research on the influence of wayfinding behavior (Montello et al. 2006) conclude four main aspects including spatial differences, visual accessibility, complexity of plan and guiding signs.

With respect to methodology, some studies tried to identify some fundamental aspects of a building that impact wayfinding by empirical experiments of wayfinding behavior (Barton et al. 2012; Hölscher and Brösamle 2007), or investigation of spatial perception by adopting tracking observation, spatial layout analysis, and questionnaires to understand the importance of spatial factors affecting wayfinding and orientation (Dogu and Erkip 2000). Meanwhile, a geometric model was also proposed in order to calculate the decision-making point and to create a network of nodes as wayfinding assistances (Makri and Verbree 2014). Space syntax is also a common research methodology in the field of spatial cognition and legibility. Some researches evaluate the wayfinding ability by analysis of spatial configuration and visual form (Abdelbaseer 2012). Long and Baran (2012) obtained the measurement of intelligibility at a neighborhood scale. Frank and Van (2014) explored the potential use of space syntax to evaluate wayfinding performance in underground space. Most empirical studies were carried in both urban environments or indoor spaces while few relevant researches in a transit hub in China were found.

2 Research Design

2.1 Investigation Site

This research chose Shanghai south railway station as a case study (Fig. 1), which is a typical multifunctional transit hub including railways, three metro lines, two bus centers, and a coach station. All these multimodal transports are connected by a winding underground corridor with over 20 exits. As one of the most important transfer hubs in Shanghai, the average daily flow of passengers reaches 300,000 in Shanghai South Railway Station. However, the problem of spatial legibility in the process of wayfinding becomes increasingly prominent. Volunteers at the station had to answer more than 100 questions a day about directions. Through the investigation and analysis of the current situation of the site, it is found that the current problems of Shanghai south railway station are as follows:

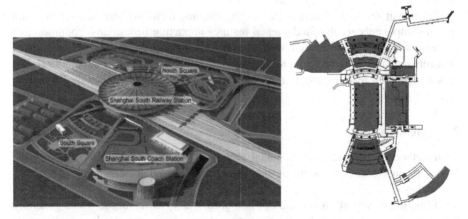

Fig. 1. Study areas of Shanghai South Railway Station.

- Circular symmetrical buildings reduce pedestrians' orientation;
- Station exits are located on the ground floor, making it difficult for pedestrians to find their way;
- Numerous passageways and exits lead to confusion of pedestrian positioning;
- There is a large amount of traffic at the main node, and the readability of signage is not enough;

- The coach station and the bus station are located on the south and north sides of the main building respectively, which are connected by a long, irregular shaped underground passageway.

Table 1. Spatial features and elements.

Categories		Contents	Descriptions
Spatial features		Layout, scale, direction, proportion, etc.	Fundamental factors impacting spatial cognition, location, and movements
Spatial elements	Functional elements	Toilet, shops, kiosks, escalators, ticket vending machines, etc.	Functional elements related with passengers' transfer
	Visual elements	Lighting, color, planting, arts, signage, etc.	Visual elements during passengers' mobility

According to the above problems, it is necessary for this study to observe and analyze the spatial characteristics and elements of Shanghai South Railway Station, as well as the wayfinding behavior of pedestrians, so as to find out the rules of transit hub design, to improve the legibility of transfer space and the enthusiasm of people to participate in the use of public transportation.

For transit hubs, the interior plan configurations, including size scale, form, and spatial components are the fundamental features impacting pedestrians' cognition and movements. Besides, compared with physical configurations, functional and visual elements remain the more profound impressions in people's mind (Passini 1984). We took both spatial features and elements as environmental factors affecting wayfinding behavior (Table 1).

2.2 Methods

Both methods of observation and spatial analysis were introduced in this study to have a comprehensive understanding on the influence of spatial features and elements on wayfinding.

Tracking observation is a kind of non-participatory observation, which generally hopes to record the real behavior of pedestrians in a completely natural state. Researchers track and observe all wayfinding behaviors of pedestrians from the starting point to the destination, such as turning, stopping, and looking for signs. In order to record the detailed details of the behavior of the test subjects, it is usually adopted to track and record the video after the test subjects. If necessary, interview and questionnaire survey can be conducted to obtain more detailed information after follow-up observation, such as the visual perception and the distance in the transfer process, or the difficulty of wayfinding.

Space Syntax theory, first forwarded by Hillier and is team (Hillier and Hanson 1984), has been developing more than thirty years. It has been widely applied in the field of urban study and complex buildings to assess the spatial structure of street network, parks, neighborhoods communities, offices, universities, shopping malls, hospitals, museums, railway stations, and other public facilities. According to space syntax, analysis of spatial structure includes three basic concepts: the axial line, the convex space, and the isovist field. The axial line analysis adopts the longest sight line within a space to represent people's movement in straight lines along streets, rooms and corridors. The convex map means all points within a space that can be joined to all others without passing outside the boundary of the space (Hillier 1985). In later development, convex space analysis turned to point depth and the all-lined analyses. Isovist analysis offers a way of geometrically describing the spaces and forms of a space which can be seen from a particular position. It is part of visibility analysis and combines a consideration of environmental factors and human experiences, such as the relationship between space form and visibility, as well as its impact on peoples' movement. Therefore, isovist analysis is an effective method to quantify the relationship between vision and behavior and commonly used for orientation and wayfinding.

This study applies both objective (tracking observation) and subjective (space syntax) methods to survey the indoor environmental feature of a complex building and the impact on human wayfinding behavior. The main purposes of this study are: (1) to understand the main factors of spatial features and elements that influence indoor wayfinding behavior; (2) to determine impact of signage on pedestrian within an unfamiliar environment; (3) to exam the efficiency of space syntax to support the investigation on indoor wayfinding performance.

3   Results and Discussion

3.1   Tracking Observation

As can be seen from the plan, Shanghai south railway station is a circular building with two squares from north to south. The symmetrical distribution of the space may cause confusion in the direction. The underground floor is connected to the traffic mode through the corridor. The space of the corridor is not high enough, and artificial lighting is adopted to achieve general visibility. There are plenty of shops and restaurants along the corridor. As the main transfer corridor, the underground passageway and the ground floor are connected by multiple elevators, escalators and stairs. According to the field observation, the station has a huge number of passengers during the day, and the escalator is the most popular vertical mode of transportation for pedestrians and passengers. Obviously, the ends of the main passageway connecting the railway station entrances and exits with metro line 1 and 3 are the key areas of wayfinding and spatial cognition.

In the station, especially the underground passage, there are complete catering shopping and other service facilities. The shop's signboard lighting and color are eye catching, which may have certain influence on the direction identification and wayfinding. There are three main types of functional facilities; basic functional category, including shops, restaurants, banks, ATMs and kiosks; ride functional category, including information office ticket office brake automatic door; and transportation functional category, including elevators, stairs and escalators. In addition, like most airports and stations, Shanghai south railway station adopts guidance signs recommended by national standards. These signs reasonably set in the intersection of the entrance and exit of the elevator and other important decision-making points, played a necessary role in wayfinding and providing information.

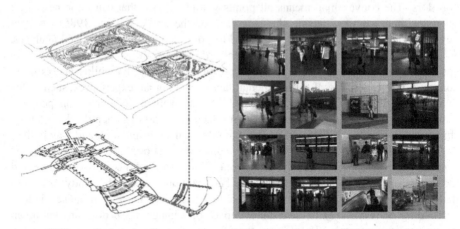

Fig. 2. Record of a passenger's mobility, including his moving route (left) and main behaviors (right).

This study invited twenty-six volunteers (average age 21; equal numbers of male and female) to achieve a wayfinding task from one of the railway station exit to coach station within Shanghai south railway station. Some basic demographic information about our subjects was collected. All participants has been well educated with a bachelor's degree or higher, and are rarely visiting or not familiar with Shanghai south railway station. Each of the behaviors, such as moving, pause, turning, looking around, or asking for directions, as well as the moving routes, were recorded by a non-participant observation. Some detailed information was also collected including:

- Origin-destination;
- Mobility routes;
- Time;
- Main behaviors;
- Achievements;
- Assessments.

Figure 2 shows one of the participants' completion of wayfinding task. Participants were asked to walk from the exit of the train station to the ticket office of the coach station in the south square. According to the path and behavior record, the subject completed the transfer task, but made several mistakes. After getting out of the railway station, he hesitated at the entrance of subway line 1 for a moment and observed the sign. Then he went through the transfer channel of B1 floor and entered the underground passage of south square. By asking, he turns into the passageway that connects the coach station, walks to the end but turn back, then go up to the opposite side of the street by the elevator of the main passageway, cross the street, enter the waiting hall from the ground entrance of the coach station, and finally arrive at the ticket office.

Table 2. Record of behavior (4 of 26).

Participants	Time	Main behaviors	Achievement	Assessment
A	9:27	Look at the signage	Yes	Smoothly
B	12:25	Look at the signage, pause	Yes	Normal
C	15:20	Look at the signage, pause, ask for direction, return	Yes	Tedious
D	11:09	Look at the signage, pause	Yes	Normal

This subject had a longer path and took slightly longer time than others (Table 2). The information of wayfinding reference by most subjects is basically the same. In addition to looking through the end of the corridor to identify the location at the beginning, the other process is almost supported by guidance sign information. Figure 4 records all the signs observed during wayfinding. It is interesting that at the turning point of stairs and escalators, all participants carefully observed directional signs, while in the process of long-distance linear walking, it is also necessary to further determine with the help of signs. The utilization rate of suspension type of guidance signs is very high, but not those on the ground. More detailed information signs are frequently observed during the pause.

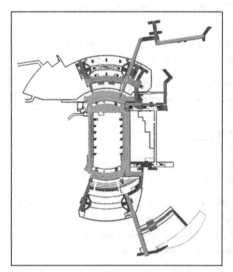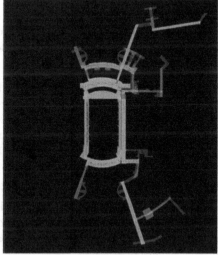

Fig. 3. Isovist analysis of the underground corridor in Shanghai South Railway Station. (Color figure online)

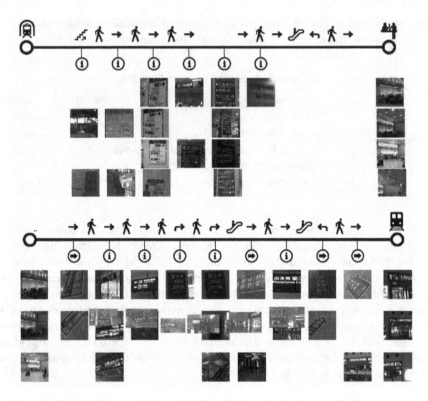

Fig. 4. Signage is the main information source during wayfinding.

During observation, the behavior of looking at signage almost took place in every step of every participant within the task. However, there remains a series of problems:

- Insufficient continuity of information content of guiding signs;
- Lack of surrounding public transport information and entrance and exit information content;
- The small font of the sign makes the visual distance shorter;
- The location of some signs are unreasonable;
- Some informational signs are less readable;
- Signs are severely disrupted by advertisement.

Signage plays a decisive role during the wayfinding process in an unfamiliar environment, especially in the underground corridor of a transit hub where is difficult to orientate in the case of catching trains. The continuity of signage information requires special attention when signposting as a spatial knowledge supplier.

3.2   Space Syntax

The isovist analysis is useful for analyzing the degree of visibility of a panoptical view of a space from a particular position, and weather an obstacle can increase or decrease peoples' view. In the software of Depthmap, we can calculate the degree of integration of each isovist root within a room. Spaces are divided into small grid cells. The more precise of the division, the accurate the results will be. With the purpose of understanding the spatial features of Shanghai south railway station, we also adopted the isovist analysis of the Space Syntax theory and applied in quantitative analysis.

The underground space of Shanghai south railway station can be translated into accurate mathematical diagrams and the visual access can be colored in different degrees of hues (Fig. 3). We can describe the spatial environment in terms of comparing the conditions of different points within the same space.

The cold color indicates a lower level of visibility. Clearly, color of blue, green and cyan obviously dominate in the result, implying the wayfinding performance of the underground passageway perform poorly. The green corridor means the spaces are not well visible and can orientate themselves poorly but better than blue ones. Especially at the end of south square and north square underground corridor, there are several segregated spaces connecting to street-level stairs and escalators. The segregation of these irregular turning points is the result of the poor configuration of the underground corridor of Shanghai south railway station. In addition, four semicircular branch way also appear blue, indicating the lowest visibility. The poor visibility of these areas makes the passageway difficult to access.

The color of red, orange and yellow indicate a higher level of visibility. In Shanghai south railway station, the areas of visual accessibility are only a few and concentrated, and are mostly located near the exit of the railway station. Coincidentally, these locations also serve as connections between train entrances and taxi stands, subway stations, and north or south squares.

4   Implication and Conclusion

4.1   Decision-Making Points

During the wayfinding task, all subjects tended to choose shorter path and time to walk towards brighter space, and all of them tended to complete the pathfinding task mainly depending on their personal ability. It is essential to emphasize the decision point of the indoor environment. For transit hubs, decision points refer to the positions of access, turning, or going up and down the stairs, such as;

- Intersections;
- Linkage spaces (e.g., main entrance, metro exit);
- Longitudinal connections (e.g., elevator hall, escalators, stairs);
- Platforms (e.g., train platform, taxi stands);
- Entrances of each transport mode;
- Bends of linear spaces.

Decision points are the important evidences of signage settlement, information arrangement, and should be highlighted in map design as well.

4.2   Spatial Features

All participants described the spatial characteristics of Shanghai south railway station as long and complex. In this huge space with multiple functions, the pedestrians who visit for the first time often feel confused. Because most passengers do not have basic concept to the whole space, long and symmetrical corridor space with similar floor, interior design and shops largely weakened spatial individual characters. Field observations have shown that many pedestrians are unable to locate themselves quickly even after looking at the floor plan.

The complicity or symmetry of architecture plans are the main causes of disorientation. For Shanghai south railway station, the circular spatial structure with same north-south squares makes passengers confused. On this occasion, the differences of interior design, such as color, material, lighting, or landmarks would contribute as reference substances for direction identifying or memory points during wayfinding.

4.3   Signage

In transit hubs, the guidance sign system is the main and most important way for pedestrians to determine their orientation and find their destination. The accuracy and continuity of signage and the rationality of sign position are particularly important. Incorrect sign information can cause pedestrians' confusion.

Directional sign, providing orientated information at decision-making points, is highly applied by pedestrians. Directional sign should be set along the pedestrian path at the critical decision-making point turning point. Directional sign should be set according to the principle of hierarchical. The design of signage system based on the user's demand satisfies the pedestrian's wayfinding behavior and plays a positive role in sustainable mobility. Effective signage provides useful information at a suitable position for pedestrians to quickly obtain useful path information.

In a word, wayfinding behavior is affected by various factors, such as age, gender, education background, personality and habits of pedestrians, as well as the spatial and environmental characteristics and elements such as signage system (Montello and Sas 2006). In transit hubs, pedestrians and passengers are comprehensively affected by the spatial features, visual elements and functional facilities of the spatial environment, as well as the guidance sign system, so as to form a complete spatial cognition and meet people's needs for safe, convenient and comfortable transfer.

5   Conclusion

To conclude, indoor wayfinding performances of a complex building are influenced by both spatial features and elements. Spatial features can determine the indoor mobility pattern; functional elements contribute to the main impression; and the visual elements (especially signage) significantly impact the capture of spatial knowledge during wayfinding in the context of unfamiliar environments.

Finally, in respect of methodology, with the combination of subjective observation, isovist analysis can be used as a powerful tool for describing the spatial features, and the plan of the architecture plays an important role on the passengers' moving patterns.

Acknowledgements. We would like to acknowledge the support by Shanghai Pujiang Program (18PJC022), Shanghai Summit Discipline in Design (DB18302) and the Fundamental Research Funds for the Central Universities. We would also like to thank the anonymous referees who provided useful insights on the earlier version of this paper.

References

Passini, R.: Wayfinding in Architecture, p. 111. Vand Reinholdn Nostra, New York (1984)

Eaton, G.: Wayfinding in the library: book searches and route uncertainty. RQ **30**(4), 519–527 (1991)

Montello, D.R., Sas, C.: Human factors of wayfinding in navigation. In: International Encyclopedia of Ergonomics and Human Factors, 2nd edn., pp. 2003–2008. CRC Press, Taylor and Francis, Ltd., London (2006)

Long, Y., Baran, P.K.: Does intelligibility affect place legibility? Understanding the relationship between objective and subjective evaluations of the urban environment. Environ. Behav. **44**(5), 616–640 (2012)

Hölscher, C., Brösamle, M.: Capturing indoor wayfinding strategies and differences in spatial knowledge with space syntax. In: Proceedings of the 6th International Space Syntax Symposium, Istanbul (2007)

Abdelbaseer, A.M.: Evaluating way-finding ability within urban environment. In: Proceedings of the Eighth International Space Syntax Symposium, Santiago De Chile (2012)

Makri, A., Verbree, E.: Indoor signposting and wayfinding through an adaptation of the Dutch cyclist junction network system. In: Proceedings of the 11th International Symposium on Location-Based Services, Vienna, Austria, 26–28 November 2014

Dogu, U., Erkip, F.: Spatial factors affecting wayfinding and orientation a case study in a shopping mall. Environ. Behav. **32**, 731–755 (2000)

Weisman, J.: Evaluating architectural legibility: way-finding in the built environment. Environ. Behav. **13**, 189–204 (1996)

Frank, V.D.H., Van Nes, A.: Improving the design of urban underground space in metro stations using the space syntax methodology. Tunn. Undergr. Space Technol. **40**, 64–74 (2014)

Hillier, B.: The nature of the artificial: the contingent and the necessary in spatial form in architecture. Geoforum **16**(2), 163–178 (1985)

Hillier, B., Hanson, J.: The social logic of space. Elementary Build. transform. (5), 176–197 (1984). https://doi.org/10.1017/CBO9780511597237

Smart Biking as Gendered Innovations?
The Case of Mobike in China

Hilda Rømer Christensen[✉]

University of Copenhagen, Copenhagen, Denmark
hrc@soc.ku.dk

Abstract. The aim of this paper is to scrutinize smart biking in the context of gendered innovations and sustainable transport. Based on media representations, interviews, reports and surveys, the paper situates the establishment of the Chinese biking company Mobike in the landscapes of recent innovation strategies. What are the gendered implications of shared biking and Mobike design and technology? Does Mobike provide a challenge to the car-centric developments of urban mobility in China? What are the broader lessons to be learned in terms of sustainable urban transport? The Mobike company rolled out its first waves of bikes in the streets in Shanghai and Beijing in 2016, since when smart biking has spread like wildfire, particularly in China, but also in other countries around the globe. This paper contributes to situating and analyzing innovative practices related to gender and sustainability. Mobike was invented by a woman, Hu Weiwei, who set out to address the pressing needs of urban residents in respect of more convenient, sustainable and efficient modes of transport. It is argued that Mobike should be seen as an experimental case of disruptive gendered innovations that feeds into the challenges of sustainable urban transport and social equality.

Keywords: Smart biking · Gender · Innovation · Urban transport · Sustainability · China

1 Introduction

"Bikes can change not only people and cities, but also the world. It is not only a symbol of peace, but also a weapon with which to fight climate Change. (Hu Weiwei, founder and president of Mobike on the reception of the UN Champion of the Earth award, UN Conference on Environment, Nairobi, December 2017).

In December 2017 the Chinese biking company Mobike was awarded a prestigious global environmental prize, the UN champion of the Earth Award for Entrepreneurial Vision. The prize ceremony was broadcast all over the globe and made a polished impression of a successful company, which over less than two years had pioneered shared and smart cycling provision not only in China, but also around the world.

Mobike, together with a couple of similar Chinese start-ups, have had a visible impact in Chinese cities for a few years, where shared biking has become a smart form of transport. Shared biking has changed urban China from a vanishing Kingdom of old bikes into a potential new Kingdom of smart bikes [1].

© Springer Nature Switzerland AG 2019
H. Krömker (Ed.): HCII 2019, LNCS 11596, pp. 368–377, 2019.
https://doi.org/10.1007/978-3-030-22666-4_27

Mobike is an interesting case in terms of the links between gender and innovation, as well as in the contexts of sustainable transport. Mobike was invented by Hu Weiwei, a young female graduate with a background in motor journalism. This contrasts with the general gender profile in innovation, where men dominate as technical inventors, entrepreneurs and investors. Besides Mobike and shared biking seem to have disrupted the former gender distinctions in biking culture in China because of its equal appeal to men and women. Last but not least, the new bike schemes have contributed to curbing CO_2 emissions and to changing modes of urban transport [2].

This paper fills a gap in studies of smart mobilities in urban China. More generally too there are glaring gaps in both China and the West when it comes to studies of the gender aspects of smart mobilities. For example, analysis from the West shows that regular users of shared cars comprise a rather exclusive group of mainly middle-class men and a minority of professional women without household obligations. Women who have family responsibilities or a more complicated transit pattern find themselves being referred to the cheaper commuter-travel alternatives to a greater degree than men. They have not been able to source more convenient but also more expensive forms of travel [3, 4].

At present shared or smart biking also figures as an under-researched field, especially in China, where by 2018 only a few reports had become available, mostly written by various university teams and commissioned by the dominant bike-sharing companies themselves, such as Mobike and Ofo [6, 7]. In addition, a number of more independent studies have focused on the contribution of shared biking to sustainable new value co-creation and ethical processes, showing how users become involved in the handling of free-floating bike fleets [8–11]. Most of these studies are based on quantitative data and address limited issues related to how to do business better and implement responsible governance. None of these studies explicitly concern gendered differences or preferences. Qualitative analysis of the cultural and social aspects of biking are by and large absent in both Chinese transport research generally and the field of smart biking in particular. In my view this is a vital field to research, not least bearing in mind the tremendous progress and revolutionary leaps made in China in respect of urban development and transport modes and practices in recent decades. It is well known that urbanization, city growth and new infrastructure have had dramatic effects on the qualities of everyday life, as well as social relations. What has been noted less is that these developments have produced new social (in)equalities in transport and mobility in respect of gender and class [12].

In the following I will examine the introduction of smart and shared biking in contemporary urban China. First I will situate the creation of Mobike in the current conceptual landscapes of disrupted and gendered innovation. Next I will explore the gendered implications of Mobike design and technology and ask how the scheme has contributed to a new biking regime in China. What are the broader lessons to be learned in terms of sustainable urban transport?

2   Mobike as Disruptive-Gendered Innovation

From a conceptual point of view, Mobike is aligned with the notions of disruptive and gendered innovation. The idea of disruptive innovation was coined by the American scholar Clayton Christensen, who in 1997 described it as providing "cheaper, easier to use alternatives to existing products or services often produced by nontraditional players that target previously ignored systems and/or are used in novel contexts and combinations" [13]. In this context the notion of disruptive innovation suggests a broad location of innovation processes in contrast to the weakness of the dominant ideas of innovation, which are separated from their social contexts.

Christensen's idea of disruptive innovation challenged the prevailing definition of innovation as exclusively linked to technology and market outcomes. It also underlined that, if innovation was to go hand in hand with the low-carbon society and environmentally friendly practices, it could not be brought about by isolated high-tech solutions. There was an imperative in this logic for innovation to include broader mobilizations to involve people and politics. Besides it was urged that profound changes in both everyday life and socio-economic systems were needed [14, 15]. From this perspective, disruptive innovation was understood as the invention of alternatives to high-tech solutions, such as refrigerators or alternative air-conditioning systems which use simple materials such as water or sand and were developed outside high-tech centers.

Mobike to some extent meets the criteria of disruptive innovation. On the one hand one can argue that Mobike provided a simple solution, having been invented by an outsider to hi-tech circles. The design process, on the other hand, differed from the more genuine examples of disruptive innovation. Mobike not only connected cutting-edge smart-phone technologies to low-level transport and bike technologies, it was also a creation embedded in the new innovation and work culture developed in the current era of globalization and the platform economy [16].

In addition, the creation of Mobike is aligned with the notion of gendered innovations which have specified and qualified a range of issues related to gender, power and (in)equality that tend to be glossed over by the notion of disruptive innovation.

Gendered innovations is a flexible term that implies both theoretical understandings and methodological and empirical or material practices. For instance, the US/European-based project *Gendered innovations* has brought the issues of the user perspective and user-driven innovation to the fore in addressing the unhappy consequences of gender-blind research [17]. The lack of a gender perspective and the more profound gender dimensions in technology development and design are regarded as a reason for the failure of many start-ups and indicate why they are not "changing the world" as predicted [18]. Bray, an anthropologist specializing in China writing along similar lines, argues that successful innovations must depart from the daily practices in the homes from the kitchen-table perspective and include the masses [20]. The kitchen-table perspective is also a gendered perspective, indicating the importance of including women's everyday perspectives. In general, and seen from these various perspectives, the fact that too many useless things are produced is also due to the lack of deeper understandings of the processes in which technology becomes useful and practical [16].

The notion of gendered innovations is a corrective to the main bulk of current research and innovation practices, as it argues that gendered innovations require a proactive account. Traditional forms of gender bias or gender blindness in research and innovation are seen as socially harmful and expensive, as well as leading to missed market opportunities [17: 8]. According to Shiebinger et al. gender should not be handled as an add-on after the failings of the product or the process have been revealed. In this respect, gendered innovations are presented as a "sophisticated method" that not only includes a focus on gender balance in research and innovation teams but more importantly comprises a focus on gender as an analytical category in innovation processes. This means that gendered implications, and not only those for women, should be reflected from beginning to end, from the conception of ideas via design processes to methods, and should be applied to the inclusion of gendered end-users. *Gendered Innovations* today form a dynamic online resource that continues to provide strong case studies showing how gender-inclusive research can harness creative power and how this leads to useful and better innovation in areas such as design, planning and production [20].

In the following I will merge and displace the notions of disruptive and gendered innovation into the hyphenated term "disruptive-gendered innovation". Disruptive-gendered innovation includes an assemblage of both digital and non-digital technologies in alignment with a notion of gendered producers and users. Moreover, disruptive-gendered innovations should meet every-day needs and create significant changes in terms of sustainability and gender equality.

The establishment of Mobike is a suitable empirical example of how disruptive-gendered innovations have come about. The data consist in a combination of Hu Weiwei's socio-biographic data and practices together with written reports, promotional materials, research articles, media materials and consultations with Mobike staff and experts and my own fieldwork in Beijing and Shanghai.

Motivated by the mundane experience of complicated channels of access to rental bikes in both China and Sweden, Hu Weiwei started to explore the potential of transforming smart-car mobility technologies for use in the field of shared biking. In so doing she transferred the knowledge she had obtained as a graduate in communication and from her career in car journalism and car-sharing possibilities into a new field of urban transport. This in turn led her to set up an interdisciplinary team of engineers and IT specialists in a start-up project aimed at developing a new shared bike system. In pursuing this goal, Hu Weiwei literally penetrated the complex structures of designers and factories, who tended to regard her ideas as crazy and unrealistic, and some of her team-workers even left the project. And often, as she later recalled, she was met with resistance and stereotypes: "Can a young female journalist lead her company to achieve her target? She won't be successful!" [21]. In the actual process, she managed not only to take the lead in the creation of a new design and, with her team, learn to overcome technical challenges, but was also able to obtain venture capital, which requires networking and capital and normally presents another obstacle for women entrepreneurs and innovators [16: 60].

As for intellectual property rights, the practice of the Mobike founder also resembles that of other significant women entrepreneurs in adopting a relaxed attitude towards patenting, one that echoes the current trend towards open innovation. At the time of my interview with her in 2017, she claimed to be upholding the intellectual property rights to the pioneering Mobike business model, yet the Mobike strategy has not been to sue for breach of patent but rather to keep at the cutting edge when it comes to innovation. In her own words, "The most effective solution is to be a pioneer all the time" [21].

From this perspective, Mobike finally provides a clear and so far rare example of disruptive-gendered innovation enabled by the transition from industrial to digital modes of production. From this perspective, Mobike becomes an achievement that signals the move from male-dominated industrial technologies to digital technologies "based on brain rather than brawn, and on networks rather than on a technical-professional hierarchy". It is a window of opportunity that has been predicted time and again by feminist science and technology scholars [22, 23]. It is also a shift that signposts new assemblages of women and technologies or machines that challenge existing hierarchies. In the case of Mobike, it is this new alignment that provides a promising turn towards more environmentally and user-friendly modes of urban transport and other devices.

3   Mobike Design: De-gendering and Democratizing?

Seen in historical perspective, smart bike sharing has evident advantages over earlier modes of renting bikes in China, which were often provided with government support. Mobike technologies were pioneers and first movers in the upcoming and thriving field of smarting and greening mobility where "smart" indicates more specifically the incorporation of the latest technology, internet and telecommunication devices into mobility services, while greening includes the effects on non-motorized and sustainable forms of transport and the enhancement of the low-carbon society [2].

Bearing in mind the non-engineering and non-technical background of the founder and later president of Mobike, Hu Weiwei, the launch of Mobike as a utility low-carbon business model was path-breaking in smart technological innovation. This made Mobike the first mover in the combination of bike technology with app-based GPS devices, which, for example, differed from the main competitor OFO's bikes, which at that time required a specific manual code to unlock the bike [11]. Mobike's technology introduced the ability to request track and pay for trips using mobile devices, thus initiating a change to the biking landscape and users' transport habits.

As for the appearance of the smart bike, the original "classic" Mobike bicycle had a rather simple and robust shape, with all its vital parts put in boxes and with wheels that could be easily dismantled for simple repair. As for design developments, Hu Weiwei went to various factories to find the most suitable materials. In the process she realized that the available "of the shelf bikes" in existing factories did not meet the require-ments. Mobike therefore established its own team of industrial designers who assisted in designing the hardware of the signature Mobike. The first Mobike models had a robust yet certain elegance in design and visibility, which tapped into the search for smart solutions and new lifestyles in the emerging Chinese middle classes. Not least the colors of the Mobikes in silver and orange quickly became iconic and were until recently used for all the various Mobike models. Also the unisex design of the Mobike contrasted with the second and third generations of docking station shared bikes, which often came in feminine coded forms and colors, including small plastic baskets at the front. This was another way in with the Mobike models introduced a different approach and design: while the basic color was silver, the orange Mobike logo represents both elegance and warmth in Chinese aesthetics. What is more, Mobike offered a model of bike which seemed to have no clear gender distinction or appeal and which in principle provided accessibility for all. As such Mobike bike models can be said on the one hand to have de-gendered the existing

Chinese biking designs with the introduction of a "one model fits all" design. This is also a contrast with earlier commercial landscapes, in which bike models and designs were highly gendered, being promoted, for example, in stereotypical and hierarchical displays in both bike shops and marketing. In the fashion-setting shops in Beijing and Shanghai, fancy sports bikes in black and grey shades intuitively associated with men and masculinity were displayed in the front of the store, while so-called women's bikes, which were cheaper and of lower quality and decked out in pale pink and blue colors, were stored in the rear of the shops. As for the new and less gendered Mobike models, it is significant that Mobike unintentionally entered a contested field of how best to meet various needs and ideas related to gender. Mobike initially received complaints from women customers over the lack of affordances in shape, access and bag transport. Mobike responded to such complaints by changing the original form and equipping its bikes with a robust bike basket for handbags, a model that has been copied by several of the other shared bike brands on the market.[1]

Mobike also seems to challenge the social and hierarchical order of cyclists that was widespread in pre-Mobike times, when cyclists were located in a hierarchical pyramid, with sports and mountain bikers at the top, fashion bikers in the middle, and mundane bikers who used their bikes as a daily means of transport at the bottom [1]. Here again shared biking denotes a both de-gendering and an anti-hierarchical move into the biking landscape, where Mobikes and other models have become popular among a broad range of the emerging middle classes, notably among their lower levels.[2] Yet new innovations might be underway: Mobike is the most expensive shared bike provider, which conforms to their relatively higher quality and its price sensitivity to women users with lower salaries.

Mobike bikes were from the very beginning intended to boost the quality and durability of average Chinese bikes. The prototype bike was made of aluminum and was predicted to have a durability of over four years in all weather conditions. This claim emphasized the idea of circular economy and sustainability in the durability of the bikes' life-cycle. The improved and more elegant models have also made Mobike bikes more attractive in the landscape of shared bikes, and in general they are regarded as being better in quality compared to other brands, not least OFO's bikes, Mobike's the main competitor.

---

[1] Presentation by Mobike's technical director, World Research Institute, Beijing, Dec. 2nd 2017.

[2] Consultation with Mobike staff, December 2018.

In their brief existence, the Mobike models have been subject to ongoing improvements and refinements both technologically and in design. The latest model, *Mobike Next Generation* launched in 2017, included both aesthetic and technological refinements [24]. It was presented as "industry-leading" and based on experimental scientific data that produced a better bike: "Thanks to the development of a lower resistance tire, a light frame and the new automatic gearing the most recent model offers Mobike's smoothest and most comfortable ride to date." All in all Mobike feeds into the role of the bike as a lifestyle icon, a phenomena which was a fact throughout most of the twentieth century in China. As part of the next generation of innovation, Mobike hired the world famous Japanese industrial designer Naoto Fukasawa, who has introduced the design for an even smarter concept bike and has emphasized how, throughout history, bikes have functioned as much as an expression of style as a smart means of transport [24].

Two reports issued in spring 2017 provided the first positive evidence of the contribution of shared bikes to the successful revitalization of biking in China, reports that used big data analysis to reveal user profiles and the context of biking. Among other things these reports measured the cycling environment against six dimensions which ranked cities according to use, parking, sustainability, health, service and social cultivation. The reports also demonstrated that shared bikes have mobilized both men and women, as well as different generations. In the cities that were reported in the survey, women accounted for between 40% and 50% of users, with Kunming and Tianjin at the top and Xian at the bottom. The age distribution also proved quite striking, with most users being between 18 and 45 years old, yet with a large proportion made up of the group between 30 and 45 years. Beijing was declared for the most biking-friendly environment and had the highest rate of biking. Shanghai was ranked first in terms of commuter biking. In addition it was reported that men over the age of 60 took the longest rides, primarily as leisure rides in holiday resorts such as Hainan Island in southern China.[3] As for the class and social perspectives, Shenzhen provided a good example: here it turned out that Mobike bikes were mostly used at night by public cleaners, as they were the only mode of transport available to them [21].

---

[3] Bike sharing report issued by IFO and Mobike, the dominant bike sharing platforms. 1. *Report on Bicycling in Major Chinese Cities in 2017*, by Ofo Bicycle CO and the Chinese Academy of Transportation Sciences, Tongji University. Report from Beijing Mobike Technology Ltd. and Beijing Tsinghua Tongheng Urban Planning and Design Institute. The report was reproduced in *China Daily* 14.4.2017: 'Sharing puts bikes back on the street', by Ouyangshijia@chinadaily. com.cn. The Ufo report provided the most adequate sociological evidence, looking at six dimensions and ranking cities with shared biking in a points system, including gender, age, time use/distance etc.

1. Use of bicycle, 20 points
2. Level of parking facility, 20 points
3. Level of energy saving and reduction of CO_2 emissions, 20 points
4. Level of contribution to health/leveraging health, 10 points
5. Level of service, 15 points
6. Level of social civilization, 15 points.

4 Closing Remarks

Innovation has become the new enthusiasm and mania in both developed and developing countries all over the world in recent decades, introducing a pressing need to locate and situate innovative practices. Mobike here qualifies as an experiment in disruptive-gendered innovations which might be capable of redressing major challenges and potentials in contemporary China. Low-carbon innovations in China, such as the introduction of Mobike and shared biking in general, are an issue of global significance in this regard, not least because of its potential for curbing the growing carbon footprints of China's economy and population. In addition this is an example of the potential created by China's spectacular social and economic growth and its support for the creation and implementation of low-carbon innovations.

Mobike has been promoted as a home-grown innovation and as a manifest sign of China's ability to be a first mover in the global race to innovate. It is a challenge to the hegemonic ideas of China and other developing countries figuring just as followers of Western technologies and developments.

The establishment of Mobike was an intervention which initially challenged the male monopoly in technological innovation and which seems to have enhanced gender equality in the field of everyday biking. At the same time, shared biking in the shape of Mobike also reproduces existing inequalities between developed urban and rural areas, where bike sharing is not provided or is inadequate and where the ownership of smart phones and the required subscriptions and capacities are distributed unequally, in some cases lacking entirely [25]. Even though the new systems seem promising, the long-term effects are yet to be seen in terms of, for example, the introduction of improved infrastructure and safety measures, pricing politics and easy online access. In addition there is the benefit of more family-and women-friendly all round designs and the possibility of transporting family members, whether children, elderly relatives or the disabled.

Shared biking represents a potential revitalization of biking abilities and also meets the demand for simple, convenient and smart modes of transport among Chinese citizens today. Seen from the outside, this bold innovation might have the potential to take Chinese cities in new directions in combining smart technologies with more sustainable forms of transport. Like many other innovations, Mobike and shared biking have introduced ambiguities and elicited both enthusiasm and anger. Mobike and other smart biking companies have contributed to bringing bikes back to urban citizens and revived the feeling of biking as freedom and autonomy. Yet shared biking has also produced problems related to the oversupply of low-quality bikes and urban disorder.[4] Another concern is the (mis-)use of personal data for commercial purposes by the major investors in shared bikes. Notwithstanding all the problems and challenges,

---

[4] In 2018, Mobike addressed this issue when they released a lifecycle plan (https://mp.weixin.qq.com/s?__biz=MzA5OTYzODY1Mg%3D%3D&mid=2247486869&idx=3&sn=aac18c3b097916b71201f 4b77d7944c7&chksm=90fe0186a789889023310a2e12dd23d3821edeeac263655bc8da7a33d87c72d 08d55aa93768d&mpshare=1&scene=1&srcid=0705yqq2qUkBD4Jzw7vgtI58&pass_ticket=RVH79 %2Bh14807gFqAtRpbrtRYRgghWp4CnxUbAEZMdrF0r1ox2EF2%2Bf8dtni7trfI&rd) based on the "3Rs" – Reduce, Reuse, Recycle.

however, I argue that Mobike and shared biking in China have provided a potential new avenue alongside the male-dominated, car-centric developments in Chinese transport and mobility in recent years. Even though it is too early to come to a final conclusion or learn lessons, for a brief moment Mobike created a shortlived, but strong alliance of users, including women and low-income residents, with market and government interests which qualifies it as a disruptive-gendered innovation. However, the case also demonstrates the fragility of such disruptive alliances, showing that a stronger grip on both smart innovation and governance is required to keep the bikes and their various gender implications on track.

References

1. Christensen, H.R.: Is the Kingdom of the bicycles rising again? Gendering cyclists in postsocialist China. Transfers Interdisc. J. Mobility Stud. **7**, 1–20 (2017). https://doi.org/10.3167/trans.2017.070202
2. China bike sharing and urban development report: Tshinghua Urban Planning and Design Institute/TUPDI) 2017, Beijing (2017)
3. Singh, J.Y.J.: Is smart mobility also gender smart? In: NOS–SH Workshop: Smart Mobilities in the Nordic Region, Oslo, August 2017. Unpublished paper
4. Christensen, H.R.: Gendering mobilities and (in)equalities in post-socialist China. In: Scholten, C., Joelson T. (eds.) Integrating Gender into Transport Planning: From One to Many Tracks, Palgrave 2019, pp. 249–269 (2019). https://doi.org/10.1007/978-3-030-05042-9_11
5. Weber, J., Kröger, F.: Degendering the driver. Special Section. Transfers **8**(1), 15–23 (2018)
6. Report on Bicycling in Major Chinese Cities in 2017, by Ofo Bicycle CO and the Chinese Academy of Transportation Sciences Tongji University
7. Report from Beijing Mobike Technology Ltd. and Beijing Tsinghua Tongheng Urban Planning and Design Institute (2017)
8. Ibold, S., Nedopil, C.: The Evolution of Free-Floating Bike-Sharing in China, 3 August 2018 (2017). https://www.sustainabletransport.org/archives/6278
9. China Daily 14.4.2017: Sharing puts bikes back on the street. By Ouyangshijia@chinadaily.com.cn
10. Yin, J., Qian, L., Singhapakdi, A.: Sharing sustainability: how values and ethics matter in consumers' adoption of public bicycle-sharing scheme. J. Bus. Ethics **149**(2), 313–332 (2018). https://link.springer.com/article/10.1007/s10551-016-3043-8
11. Lan, L., Ma, Y., Zhu, D., Mangalagiu, D., Thornton, T.T.: Enabling value co-creation in the sharing economy: the case of Mobike. Sustainability **9**, 1504 (2017). https://doi.org/10.3390/su9091504
12. Christensen, H.R.: The lure of car-culture: gender, class and nation in 21st century car culture in China. In: Women, Gender, and Research 1.205, pp. 96–110 (2015) https://tidsskrift.dk/index.php/KKF/issue/view/3107/showTocrr
13. Christensen, C.: The Innovator's Dilemma. Harvard Business School Press, Cambridge (1997)
14. Tyfield, T., Jin, J., Rooker, T.: Game changing China: lessons from China about disruptive low carbon innovation. Research report. Commissioned by NESTA: The National Endowment for Science Technology and the Arts, UK, June 2010

15. Willis, R., Webb, M., Wilddon, J.: The disrupters: lessons for low carbon innovation form the new wave of environmental pioneers. London NESTA (2007)
16. Poutanen, S., Kovalainen, A.: Gender and Innovation in the New Economy: Women, Identity and Creative Work. Palgrave Macmillan (2017)
17. Schiebinger, L., Klinge, I.: Gendered Innovations: How Gender Analysis Contributes to Research. Publications Office of the European Union, Luxembourg (2013)
18. Schiebinger, L. (ed.): Gendered Innovations in Science and Engineering. Stanford University Press, Stanford (2008)
19. Oldenziel, R.: Making Technology Masculine: Men, Women, and Modern Machines in America, 1870–1945. Amsterdam University Press (2004)
20. Bray, F.: Only connect: comparative, national and global history as frameworks for the history of science and technology in Asia, East Asian science, technology and society. Int. J. **6**, 233–241 (2012)
21. Gendered Innovations. https://genderedinnovations.stanford.edu/
22. Weiwei, H.: Why I created Mobike? 2017-02-08 A轮学堂 A轮学堂 in Yixi media. Speech/Interview by Hu Weiye, 17 Feb 2017 (translated and transcribed, March 2017)
23. Wajcman, J.: Feminist technology. Cambridge J. Econ., 1–10 (2009)
24. Wajcman, J.: Feminist theories of technology. Cambridge J. Econ. **34**(1), 143–152 (2010). https://doi.org/10.1093/cje/ben0
25. Mobike Next Generation: 22 September 2017 (PR presentation)
26. /Chart/Smartphone-User-Penetration-China-by-Age-2014-2020-of-mobile-phone-users/188732. https://www.emarketer.com

Gender Issues in the Digitalized 'Smart' Mobility World – Conceptualization and Empirical Findings Applying a Mixed Methods Approach

Barbara Lenz[(✉)], Viktoriya Kolarova, and Kerstin Stark

DLR Institute of Transport Research, 12589 Berlin, Germany
Barbara.Lenz@dlr.de

Abstract. While digitalization provides a range of new mobility options, neither gender specific requirements nor role and task specific issues are part of the recent developments. Against this background, this contribution is looking into gender issues in the digitalized 'smart' mobility world aiming to contribute to the development of future 'smart' mobility options, such as autonomous driving and sharing. A conceptual model which integrates a theoretical framework on gender specific mobility patterns and technology acceptance elements is proposed and mirrored with empirical evidences from recent works of the authors. The results suggest that it is above all the fit of smart mobility options with current everyday-life situations that influences the acceptance and use of those options. The gender differences that become obvious then are due to gender differences in the everyday-life situation between women and men.

Keywords: Gender equality · Smart mobility · Autonomous driving · Travel behavior · Technology acceptance

1 Introduction

Digitization provides quite a range of new mobility options, in particular for urban residents. Gender related concerns, however, have not been part of recent developments, neither did role or task specific issues of both women and men. New mobility concepts, such as car-sharing or bike-sharing for instance, do not provide possibilities to transport children or loads like household errands. At the same time gender issues are barely taken into account when it comes to considerations about the usefulness of new mobility options. This does also apply to automated driving being an option that lies ahead in a mid-term future. In this context one particularly striking example of unawareness of gender issues is the question, if users will be keen to carry out 'productive tasks' in the automated vehicle, as this has been a repeated argument to support positive expectations on the acceptance of automated driving. From other areas of technology use we know, however, since long that gender difference concerning acceptance and use of [new] technologies exist, so that gender differences are likely to be present also for the use of new ICT based mobility options, often tilted by the term 'smart mobility'.

© Springer Nature Switzerland AG 2019
H. Krömker (Ed.): HCII 2019, LNCS 11596, pp. 378–392, 2019.
https://doi.org/10.1007/978-3-030-22666-4_28

These observations have motivated the authors to report on some first insights from their empirical work on gender differences in the use of new mobility services on the one hand, and their work on the acceptance of automated driving on the other. To introduce a gender perspective our article starts with the elaboration of a theoretical concept that basically builds on Robin Law's analytical framework to understand gender and mobility (Robin Law 1999). This framework conceptualizes the context conditions for gender specific mobility patterns and the impact of role and societal norms; the authors will widen this concept by technology acceptance elements referring to Thompson et al. (1991) who reflect (a) the acceptance and use of new technologies and new service options, and (b) the importance of the fit that these technologies and options do provide or not to existing activity patterns and routines. The chapters that follow (i) describe the methodological approach which was applied in the empirical research; (ii) present and explain the empirical results; (iii) provide a discussion of the results by reflecting the empirical findings against the conceptual framework that the authors developed. This outcome will allow for (iv) conclusions and recommendations for further research.

2 Theoretical Approach: Integrating Analytical Framework and Acceptance Model

We developed a theoretical approach to support our considerations about smart mobility and gender and to mirror our empirical evidences that offers an analytical structure to understand gender specific travel behavior and link it to the gender-specific acceptance and adoption of new mobility concepts. To this purpose we merged the concepts that Robin Law (1999) and Thompson et al. (1991) provide into a consistent 'integrated concept' on the gender specific acceptance and use of new digitized mobility options. In the following we will briefly describe those concepts and elaborate on the purpose and nature of the integration we applied.

2.1 Law's Analytical Framework to Understand Gender and Mobility

The concept of 'gender' is the basis for Law's analytical 'framework to understand mobility and gender' (Fig. 1); the basic objective of her concept is to understand the causes that lead to the "gender processes (power relations, identity formation and transformation, and the meanings of various aspects of mobility) that shape varied mobility patterns" (Hanson 2010: 13). Law refers to gender as a category that structures social relationships "through the gendering of the *division of labour and activities*; *access to resources*; and the *construction of subject identities*". In addition to the social relations she identifies the meaning of activities (the '*symbolic code*'), and the "*built environment*, which includes the organization of land uses in space, and the physical design of sites, places and routes" as key factors (Robin Law 1999: 575; italics in the original publication). Those key factors do not only shape mobility behavior, but also other behavioral areas, so for instance individual consumption. Relevant in the context of mobility are the assignment of household tasks and the allocation of household resources, experience of embodiment, individually 'useful' and socially accepted mobility practices, and the structure and organization of land use. As those factors are largely linked to sex as an biological characteristic, also in the so-called 'modern' societies of industrialized countries, gender differences are mostly investigated and described as differences in attitudes and behaviors of women and men. But this restriction is also due to statistical categories and their application in quantitative surveys (Perez 2019).

The observed result is gendered travel patterns, that appear in different forms depending on the spatial and cultural context: For many industrialized as well as emerging countries and cities studies have revealed differences in daily mobility between men and women with respect to mode choice, complexity of travel patterns, trip length and trip duration (Minster et al. 2019; for overview on earlier publications see Hanson 2010: 12). On average car use has a higher share among men than among women; women combine more activities along their trips and have more escorting trips; men make longer trips in terms of both individual trip distance and trip duration, and daily trip length. Additional differences concern the kind of activity, and the time and the place of its accomplishment, in particular for shopping and work; while women have more shopping[1] trips than men, men have more trips to work and business trips than women (Rosenbloom 2006; MiD2008 and 2017). Although those differences have levelled off in the last two decades, imbalances were only mitigated but not eliminated, depending considerably on the stage of life (see, for instance, Crane 2007; Nobis and Lenz 2004).

Law is one of the very few authors who give particular consideration to the role and meaning of technology in the context of gendered mobility. She identifies variations in access and use of technology for mobility due to the uneven distribution of financial resources and of skills between women and men, but also due to gendered practices in the use of machines and a particularly "deep-seated and wide-ranging connection to gender distinctions" (p. 579) for transport technology. Against this background technology might not only be seen – as Law suggests in her scheme (Fig. 1) – as an issue of gendered *access*, but as an *outcome of access* to resources and skills and the gendered meaning of mobility practices as well thus generating a significant impact on the meaning of technological change in the context of mobility for women and men respectively. However, research on this issue has remained limited. Most research that addresses 'women and technology use' has concentrated on domestic or office space (Hartmann 1987; Wajcman 1994). Gender related mobility research has predominantly focused on existing mobility behaviors and patterns and the analysis of underlying causes. This more 'retrospective' approach has led to a certain neglect of more prospective issues as there are acceptance and appropriation of mobility related technology.

---

[1] Unfortunately travel surveys using the category 'shopping' usually do not distinguish between shopping for personal issues which may include shopping or window shopping as a leisure activity and making errands for household caring purposes.

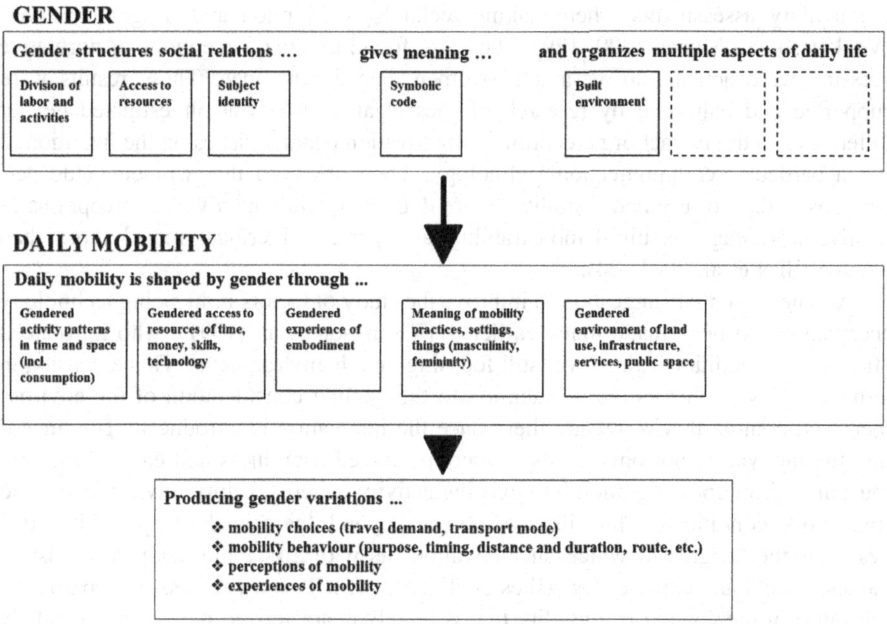

GENDER

Gender structures social relations ...			gives meaning ...	and organizes multiple aspects of daily life		
Division of labor and activities	Access to resources	Subject identity	Symbolic code	Built environment		

DAILY MOBILITY

Daily mobility is shaped by gender through ...				
Gendered activity patterns in time and space (incl. consumption)	Gendered access to resources of time, money, skills, technology	Gendered experience of embodiment	Meaning of mobility practices, settings, things (masculinity, femininity)	Gendered environment of land use, infrastructure, services, public space

Producing gender variations ...

❖ mobility choices (travel demand, transport mode)
❖ mobility behaviour (purpose, timing, distance and duration, route, etc.)
❖ perceptions of mobility
❖ experiences of mobility

Fig. 1. Law's analytical framework to understand gender and mobility (Law 1999: 576)

2.2 Acceptance and Fit of New Technologies

Access to and use of technologies have become even more important in the era of digitalization as it opens an additional 'gate' to virtual and physical places as well (cf. Hanson 2010: 7) as to new mobility options. As highlighted before most studies that address gender specific aspects and/or the gender gap in the use of new technologies refer to sectors other than transportation. An abundance of studies has been carried out on the acceptance of computers and the Internet, mostly in the context of the working world. As a theoretical basis to allow for quantification, Davis (1985, 1989) developed the technology acceptance model (TAM) whose general proposition for user acceptance of new technologies are "perceived ease of use" and "perceived usefulness" (Wixom and Todd 2005). The model has been improved over the years by means of several modifications and enhancements (for a comprehensive overview see Mazman et al. 2009, and Li, 2011). Particularly important improvements to the model were the integration of subjective norm, the examination of gender differences in the role of the original TAM constructs, and the related role of experience as proposed by Venkatesh and Morris (2000).

In their study about gender variations in the use of technological innovation, Venkatesh and Morris studied the factors influencing the use of a new technology: They found that men were more driven by instrumental factors (perceived usefulness), whereas women were more motivated by process (perceived ease of use) and social (subjective norm) factors. The authors concluded that "men only consider productivity-related factors, [while] women consider inputs from a number of sources including productivity assessments when making technology adoption and usage decisions" (Venkatesh and Morris 2000: 129). They also found that the perceptions of normative pressure were actually lower among women than among men. These results were supported and enhanced by research of Ilies et al. (2005) who investigated gender differences in the impact of perceptions of innovation characteristics on the intention to use a particular communication technology. They observed that women value perceptions of ease of use and visibility[2] more than men, while men value perceptions of relative advantage, result demonstrability and perceived critical mass[3] more than women (Ilies et al. 2005: 24).

Another essential suggestion to improve the study of factors influencing technology acceptance had been made much earlier by Thompson et al. (1991), who introduced "facilitating conditions", however still focusing on job environments. This extension in terms of relevant factors draws attention to the explicit consideration of the environment and context that is already there once the innovation is introduced. This means that the innovation not only needs to meet perceived usefulness and ease of use, but must also allow for integration into existing activity patterns, habits, environments and framework conditions. Thus its scope goes beyond 'perceived compatibility' that describes the "degree to which an innovation fits with a potential adopter's existing values, beliefs and experiences" (Ilies et al. 2005: 15). This aspect seems of particular relevance in the context of mobility that is largely characterized by routines for daily travel (Scheiner 2009; Wiles 2003). Routines are everyday habitual behavior characterized by regular actions; in many cases persons who apply routine behavior are not conscious of it (Canzler and Franke 2000). This explains why routines are usually not put into consideration, but relax everyday life from the obligation for permanent decisions, help to make everyday life less complex and lead to quite stable behavioral patterns (Giddens 1996). This applies in particular to travel behavior that has proven not being very much open to external influence (Franke 1999). Consequently we may assume that the arrival of new technological options and services in the mobility sector are not only an issue of technology acceptance in itself, but also of the possibility to integrate them into existing travel behavior.

2.3   Integrating 'Facilitating Conditions' into Law's Analytical Framework to Understand Gender and Mobility

To better understand the gender-specific implications of new technologies on mobility choices that are influenced by gender structures and gender specific needs and practices, we suggest a framework that integrates 'facilitating conditions', as proposed by Thompson et al. (1991), into Law's analytical framework. As Thompson et al. highlight the relevance of existing activity patterns and habits (i.e. 'routines'), the

---

[2] 'Visibility' refers to the degree to which the use of an innovation is apparent (Moore, Benbasat 1991).

[3] 'Perceived critical mass' refers to potential adopters' perception of whether an innovation has attracted a critical mass of users influence use intentions (Lou et al. 2002). Perceived critical mass is one of the most important determinants of the acceptance of groupware technologies and also influences intentions to use a communication technology (Ilies et al. 2005: 16).

integration should be made in the 'Daily' section, i.e. the section that refers in particular to activity patterns in time and space and mobility practices, but also to access to resources, experiences of embodiment and environment of land use, infrastructures, services and public space. As an outcome we get a modified framework to further conceptualize research that aims at understanding the interrelation of gender and digitalization-based mobility. With this approach, the acceptance of *smart* mobility can be understood as something that is gender specific and depends on how well it fits into existing mobility practices (Fig. 2).

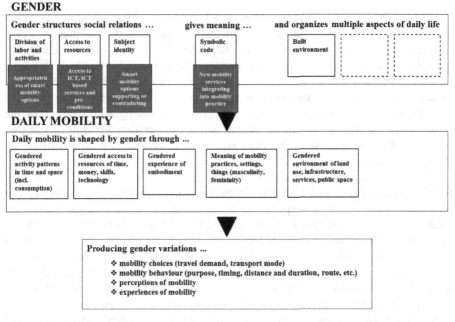

Fig. 2. Conceptual framework to explore the potential impact of 'Facilitating conditions' on gendered mobility, building on Law's framework to understand gender and mobility

3 Empirical Insights

In recent years we have investigated intensively acceptance and use of new mobility options, both as enhancement or complement to existing mobility options. In addition we explored perspectives of future mobility options as they might be created in particular by automated driving. With the aim to enrich existing findings we conducted another analysis of the existing empirical data applying the concept explained above, where we integrated the approach of Thompson et al. (1991) into the basic concept of Robin Law (1999) to understand gender and mobility. We then explored the main dimensions of the conceptual model, i.e. relevant facilitating conditions of user acceptance in the context of new mobility options of different kind. The outcome is a first step in getting deeper into gender issues of the digitized 'smart' mobility world. The particular questions which we address now are:

- Is there a gender gap regarding the acceptance of new mobility concepts?
- How or how much do smart mobility offers fit the requirements of different, larger parts of the population, including care-associated ones?
- How should offers be designed and tailored to attract a diverse clientele?
- How might smart mobility contribute to gender equality taking into account the diversity of needs?

The empirical evidences have been gathered (i) from quantitative studies of gender-specific mobility patterns, and (ii) from qualitative studies on the acceptance of new mobility concepts and automated driving. Regarding the insights from the qualitative studies on demand-responsive mobility services, gender-specific issues had not been a genuine topic. However, gender-related issues came up by themselves. Aspects such as vehicle design, capacity or the operating area of new mobility services were discussed with regard to issues like the exclusion of persons travelling with children or the possible use for running errands or getting to work rather than for club hopping on weekends. For the purpose of this contribution we re-analyzed the data focusing on these very issues.

The combination of quantitative and qualitative methods allowed quantifying mobility behavior and analyzing travel patterns on the one hand, and looking deeper into users' motives and perceptions in a more explorative way on the other. The quantitative analysis used the German National Travel Survey MiD 2017 (Mobilität in Deutschland 2008) (BMVI 2018) and focusses on exploring gender-specific differences in travel patterns and access to mobility tools. Additionally, we have analyzed data from a quantitative online survey on acceptance of automated vehicles and demand-responsive automated services conducted in Germany. The two qualitative studies were based on focus groups discussions (FGD) with potential users of new 'smart' mobility concepts including automated vehicles. Both studies have addressed the perceived advantages, concerns and requirements of users related to such concepts. While the first study addressed on-demand (automated) bus services, the second one focused on privately owned and demand-responsive automated vehicles. Table 1 gives an overview of the study design and sample characteristics of the considered empirical studies.

In addition we will refer to MiD2017, the representative National Travel Survey for Germany that empirically captures travel behavior of the population in Germany. The insights we get by MiD2017 will serve as a starting point and help framing the findings of our own studies.

The following chapters will present the findings that came out when we got into our empirical data against the background of gender issues as structured by our 'integrated concept'. We start by exploring those gender issues throughout the focus group discussions indicated in Table 1, then look at the quantitative data of our own online survey.

Table 1. Overview of the empirical studies and data sets used.

Year	Method	Topic	Sample characteristics
2019	Online survey (N = 380)	Willingness to purchase and use a privately owned automated vehicles (AV) or to use demand-responsive automated services; focus on commuting trips	Representative sample regarding age and gender for the population between 18 and 69 years old in Germany
2015	3 focus group discussions (N = 21; Berlin)	User requirements, evaluation and concerns related to automated vehicles (AV) and shared automated vehicles (SAV)	Average age 38y, one group with average age 63; average gender distribution: 56% men, 44% women
2018	3 focus group discussions (N = 13, Fuerstenwalde nearby Berlin)	Acceptance of automated vehicles (AV) and demand-responsive mobility services regarding user requirements in different user groups, e.g. women travelling alone or with children	

3.1   Appropriateness of Smart Mobility Options

Looking into current differences in the travel patterns of men and women in Germany using MiD 2017 again shows trends that had been uncovered already in previous studies on gender specific travel behavior in Germany. Earlier analyses had – first of all – ascertained differences in gender specific travel patterns and were able to relate them to gender-specific differences in the employment status and distribution of child care and household obligations (Nobis and Lenz 2004, see also Nobis and Lenz, 2019/*forthcoming*). The authors showed, for instance, that child care obligations were still mainly a duty of women/mothers, as that 74% of the trips with main purpose escorting children under 14 years old were made by women.

Against this background, we further explored how and how much this 'division of household and childcare labor' impacts the first dimension of the integrated framework which is the appropriateness of smart mobility options. Regarding the (re-)analysis of the empirical evidence our guiding question was how available smart mobility options comply with the needs of persons given their roles and obligations in households. In the case of persons being responsible for household and childcare particular aspects were related to being escorting/bringing a child, being underway with a child and needing transportation for household goods. The 'reference case' for our data was automated driving that might support or not individual or household needs.

A: Time use and time evaluation
Considering our qualitative data from 2015, the findings on future mobility services suggest that the possible delegation of escort rides to self-driving vehicles is especially attractive to persons with care obligations. Whether or not these vehicles are privately

owned or shared seems less relevant as long as the service provides a door to door connection and meets certain requirements, the most important one would be security, therefore registration is preferred. Besides security, parents stated openness to use an autonomous vehicle for getting their children only when the children are at certain age, are carried from one supervised place to another, and once parents have built trust in the reliability of the technology:

> P: Once I trust the car ... it [the car] might pick up the children without driver, so I can go home making dinner or something."

> P: I mean depending on age ... I was taking public transportation by myself at 12, so this would be similar thing - the kid can go in the car or the kid wants to go to the mall at the weekend or something and you go to meet them there. The car can take them there ... if I'm coming from work and the child is going from a supervised place to a supervised place separately.

Further considerable use discussed in the focus group discussions was related to using automated vehicle services for running errands, including picking-up shopping items from the grocery:

> P: I think that if you trust the car to drive without you in it, you can use it as a pick-up and delivery service - you can send it somewhere to pick something for you and bring it back to you, you can send it on errands and you can be productive doing something else; so for me it's ability to delegate a tool instead to pay a driver to go run these errands for you.

Thus, benefits regarding time use, less time fixes, and more flexibility in the daily life would be an important advantage of 'smart' mobility services for people with care and household obligations. This also means that there would be a potential for business models tailored for people with household and child care obligations, as long they meet the user group's particular requirements.

Looking further into perceived advantages of riding autonomously instead of driving manually, we found other potential differences concerning the evaluation of travel time and the quality of travel. Here again, gender-specific care obligations, such as taking children to school or to after-school activities represented important implication for individual's evaluation of different autonomous driving options. So, for instance, potential advantages mentioned by parents include spending more qualitative time with their children while travelling:

> P: I would feel like I would be more alert with my kids if my kids were in the car... It is just because when I'm driving I can't pay attention whether they are good back there.

B: Vehicle design, equipment and availability

Currently available smart mobility services regularly are oriented towards users that travel without children. Parents from the qualitative survey in Berlin (2019) stated a couple of reasons why car sharing was of no use for them. Car sharing systems e.g. usually lack child seats, and there is no way to check via app if an available vehicle is equipped accordingly or not. Moreover, the availability of new mobility services is often limited to central urban areas. Thus, a majority of the interviewees of those who were not living in the center, hardly use car sharing, simply due to that reason.

> P: [I]t becomes problematic exactly where the Car-Sharing areas stop where all these great super start-up projects go and where then really only the train goes out.

P: Also this Car-Sharing that has to be given back again that's what we had the other day, then you can rent it but you can only rent it inside the inner city district.

Moreover, the flexible combination of their regular means of transportation with digitized mobility options under special circumstances is a relevant matter. In particular, regular cyclists that also carry their children by bike or bike trailer see great potential for the case of tired children after a day trip or in sudden rainy weather. The vehicle, then, would require storage or bike transport option. In the same vein, the combination of mass transit in terms of integration in the regular fare system is seen as important requirement given the fact that additional fares on top of the already high expenses for tickets for themselves and their children is seen critical.

P: I have been noticing, [...] when you must pay extra for every bus trip each time for yourself and for the child – well, that motivates you to walk instead and save the money.

Conclusion for this dimension

Considering appropriateness of smart mobility options as a facilitating condition for user acceptance confirm that gender-specific obligations, such as household duties and childcare, may be a relevant factor of influence with respect to requirements on size, capacity, and operational areas of new mobility options, but also with respect to car use and travel time perception in services particularly offered once automated vehicles are available. To become attractive for the user group of persons travelling with children, future services should provide corresponding equipment to some extent and make this information accessible in real-time. Given the fact that families are increasingly moving to peripheral areas in order to obtain affordable rents, the limited business area of sharing services is becoming a major obstacle even to the simple fact to get into contact with new services. Finally, fares have to remain moderate for families as additional costs that become due for each family member might be a serious barrier to use new service options.

3.2   Access to ICT, ICT Based Services and Pre-conditions

Considering access to relevant resources as part of the facilitating conditions for the uptake of smart mobility services, several areas should be mentioned and explored: (i) Access to resources is related to physical access to services or mobility tools, (ii) to relevant experiences, and (iii) to knowledge and capabilities for using smart mobility options. The respective research question – taking into account the availability of data – focused on the following: Is there any difference in accessibility of ICT/smart mobility options in terms of physical access and capabilities from a gender-sensitive perspective? To this issue we regarded the possession of driving licenses.

The differences in driving license possession rates and car availability between men and women have continued to decrease since years (Nobis and Lenz 2019/*forthcoming*). Looking into the data from the German National Household Travel Survey MiD2017 Nobis and Lenz (2019/*forthcoming*) found that there is objectively no gender gap regarding possession of driving licenses and digital Simultaneously 78% of men and 71% of women have a car available at any time (including using the car as passenger and car-sharing services). This means that accessibility of resources does not largely differ between women and men. However, more impact might have availability of other than directly mobility-related resources, like household income or credit card ownership. Unfortunately there are no data available to connect these potential influence factors to the use of smart mobility options by women and men.

Another influencing factor that is directly linked to the availability of resources is experience with different kinds of technologies (vehicle, services, access). Our survey data analysis suggests that there are differences in the compatibility of smart mobility services and functions with existing experiences and beliefs between men and women. Relevant experiences as, for instance, individual pre-conditions affect the willingness to use 'smart' mobility options as they influence the perceived own ability to use the system and the evaluation of new options (Lüthje 2007). In other words, we suggest that even though objectively the access to relevant resources which enable using 'smart' mobility options hardly differ between sexes, relevant experience of men and women might be different and thus will affect the evaluation of new services. For instance, looking into usage of car-sharing services, as reported in MiD2017, shows that more men than women use carsharing at least once per moth (even though the percentage of people using car-sharing services in Germany overall is very low and concentrates on urban areas where car-sharing services are available). Moreover, looking into trips made by car in Germany, we find a trend that higher class vehicles are rather available to men than women, and mini and compact class vehicles rather to women than men.

Considering acceptance of autonomous vehicles shows that previous experiences with advanced assistance systems increased the willingness to purchase an automated vehicle (AV). In an online survey conducted by the authors in Germany, more men (36% of the men) than women (17% of the women) stated that they are willing to buy an AV and simultaneously more men (43% of the men; 32% of the women) had made the experience with advanced assistance systems in the past. Consequentially, the benefits of a fully assisted vehicle were much more conceivable to men than to women. In this case, the data showed clearly overlapping effects of experience and gender. The reasons behind these results can be multiple and might among others be related to gender-specific differences in the technology awareness (i.e., interest in technology), car use motives, but also vehicle ownership and vehicle type preferences (as reported above, for instance, higher class vehicles are rather used by men than women). Furthermore, technology awareness might not only influence seeking collecting experience with new technologies, but also the interpretation of experience made:

P: I have tried automated parking with my colleagues. I didn't knew it before that. I found it great, but I cannot imagine using it in my daily life. ... It worked, but I still somehow wouldn't trust it. When I drive, I would still like to park by myself.

Further analysis of the online survey data shows notable differences in the level of trust in the technology between men and women. Compared to women, higher share of men stated to trust that a computer can drive the car with no assistance (44% of the men; 27% of the women) and that the automated driving system provides more safety (35% of the men; 22% of the women). Here again, experience play an important role for increasing trust in the system. People who had made experience with advanced assistance systems tent to trust the technology more than people without any experience with such systems.

Conclusion for this Dimension

Considering the objective access to resources relevant for using 'smart' mobility options, such as driving license or digital devices, it seems that there is virtually no gender gap between the sexes. However, the perception and willingness to use vehicle automation functions differs between men and women resulting, among other factors, by differences in relevant experiences. Thus, enabling experience with new services and functions might contribute increasing trust and acceptance in 'smart' mobility services.

3.3 Smart Mobility Options Supporting or Contradicting Subject Identity

Subject identity in the context of smart mobility refers to the socially mediated and individually embodiment of expectations regarding the choice among mobility options. The acceptance and use of different mobility options, therefore, is also a matter of social expectation and association. How do persons perceive and evaluate different smart mobility options with regard to their accordance, i.e., compatibility, with their subject identity? With regard to one of the following examples ridesharing at night could be evaluated as inappropriate and not recommendable for a young woman, when the social expectation and public discourse suggest that it is risky. From our qualitative data we have a look into two aspects in this regard, *perceived security* and *driving pleasure*.

A: Perceived Security in Shared Vehicles

Although further research is needed, there are already some clear indications of a social expectation that rideshare might be a security issues for women in particular. A qualitative study on automated driving in public transport (Stark et al. 2019, in press) study found that a woman-specific limited sense of security was associated with ridesharing in smaller vehicles, especially at night, but interestingly not by women interviewed, but by individual men.

In the findings of the focus groups presented here we, too, do not find a special reluctance and perceived insecurity of women regarding the use of shared mobility services even if there is no driver and at night. Although some would feel more secure and comfortable with a driver around they would not give up using the service especially considering the alternative of walking home or waiting for the next regular bus.

P: Well yes, I would feel more secure with a driver on board.

I: But would you get on the bus say if it would be small vehicle with four seats and a stranger on it?
P: Yes.
I: And given there is no driver, how should the vehicle be equipped so that you would feel secure?
P: Well, there should be some kind of emergency button, [...] some easy way to get out in case you feel uncomfortable.

However, to support the perceived security not only for women but for vulnerable social groups the service could be tailored accordingly with vehicles designed and equipped e.g., with an emergency button or a hotline.

B: Driving Pleasure and Driving as Self-expression

Another aspect of subject identity which came out in the context of automated driving in the group discussion was the meaning of driving as a mean for self-expression as well as intrinsic motives for car use, such as enjoying driving. The loss of driving pleasure due to automated driving was mention often in the group discussions as an important disadvantage of the technology, mostly by men (e.g. *"The biggest disadvantage for me will be the loss of driving pleasure – I like driving")*. One of the women in the group discussion has referred to this aspect in a more distant manner: *"... people who enjoy driving might be bored".* Looking into the data from the online survey, however, there were no statistically significant differences between the sexes regarding stated pleasure of driving.

Further, the symbolic role of car ownership and car driving was addressed by single participants, again primarily men – *"I am thinking about like cars have a certain kind of 'cool factor', like some people like to drive [showing typical driver's expression to show up while driving]... And can you really do that in a self-driving car?"*. Another participant refer to driving as a way for self-expression: *"I think that many people identify themselves with their driving style, it is a way to express themselves. It is not by coincidence that some people drive fast cars."* Interestingly, when analyzing the data from the user survey regarding car-related attitudes, they were differences between men and women only on one aspect - the social expressive role of the car. On average, men agree that the brand of a vehicle is more important than its functionalities than women do.

Conclusion for This Dimension

Important aspects regarding how smart mobility options support or contradict subject identity include the perceived security in shared vehicles and the role of the car and car driving as a mean of self-expression. Because of the small sample of the group discussion and the contradicting results from the online survey, these results cannot be generalized. Also, they can be a result of role conformity behavior rather than underlying differences. However, they raise the important role of gender-related symbolic codes in the perception and evaluation of new mobility options as well as the role of social norms.

3.4   Symbolic Code/New Mobility Services Integrating into Mobility Practice

The guiding question used by the data analyses for this dimension was: How does a given smart mobility option represents the society's/the peer group's self-concept, life style, or values?

This dimension show high overlapping with the dimension of subjective identity. Symbolic codes can be associated with what the society defines as "masculinity" or "feminity" or the common sense of services appropriate for women and men. Looking into our empirical data, we could not directly found insights on this aspect. However, the points mentioned in the dimension related to subjective identity might provide some first direction of thought. For example automated/assisted vehicles was mostly not considered as 'real cars', because one cannot exercise his or her driving skills and also one is rather a passenger than having an active role while travelling.

4   Discussion and Conclusions

In this paper, we have considered to what extent new mobility options fit in with supposedly gender-specific everyday activities and how an ill-fitness impacts on the acceptance of new mobility options. For this purpose, a conceptual model was proposed which integrates theoretical approaches on gender-specific mobility and user acceptance. In this context, we explore the effects of social roles, (care) obligations as well as norms on daily mobility patterns. Moreover we discussed paths for a reasonable development of smart mobility options which consider the diverse requirements and needs of different social groups and the goal of providing mobility for all, instead of focusing on the minority of the early adopters' of smart mobility concepts. In doing so, we looked into the assumption of gender-related differences in the perception and evaluation of smart mobility options. We found that besides differences in activities and tasks, also not directly gender-related experiences and pre-conditions strongly influence how new mobility options are being perceived and evaluated.

References

BMVI (Bundesministerium für Verkehr und digitale Infrastruktur): Mobilität in Deutschland – MiD. Ergebnisbericht, Bonn (2018)

Canzler, W., Franke, S.: Autofahren zwischen Alltagsnutzung und Routinebruch. Bericht 1 der choice-Forschung. FS II 00-102, Veröffentlichung der Abteilung ,Organisation und Technikgenese' des Forschungsschwerpunktes Technik – Arbeit – Umwelt am Wissenschaftszentrum Berlin (WZB) (2000)

Crane, R.: Is there a quiet revolution in women's travel? revisiting the gender gap in commuting. J. Am. Plan. Assoc. **73**(3), 298–316 (2007)

Davis, F.D.: Perceived usefulness, perceived ease of use, and user acceptance of information technology. MIS Q. **13**, 319–340 (1989)

Franke, S.: Car Sharing zwischen Ökoprojekt und Mobilitätsdienstleistung. Zur Veränderung der Nutzungspraxis des Automobils. Unpublished Dissertation at Technical University Berlin (1999)

Giddens, A.: Leben in einer posttraditionalen Gesellschaft. In: Giddens, A., Lash, S. (eds.) Reflexive Modernisierung, pp. 113–194. Eine Kontroverse, Frankfurt (1996)

Hanson, S.: Gender and mobility: new approaches for informing sustainability. Gend. Place Cult. **17**(1), 5–23 (2010)

Hartmann, H.I. (ed.): Computer Chips and Paper Clips. Technology and women's employment, vol. II Case studies and policy perspectives. National Academy of Sciences – National Research Council, Washington D.C. (1987)

Ilie, V., van Slyke, C., Green, G., Lou, H.: Gender differences in perception and use of communication technologies: a diffusion of innovation approach. Inf. Res. Manag. J. **18**(3), 13–31 (2005)

infas, DLR: Mobilität in Deutschland 2008. Ergebnisbericht. Struktur – Aufkommen – Emissionen – Trends. Bonn, Berlin (2010)

Robin Law: Beyond 'women and transport': towards new geographies of gender and daily mobility. Prog. Hum. Geogr. **23**(4), 567–588 (1999)

Li, L.: A critical review of technology acceptance literature, Accessed 19 Apr 2011

Mazman, S.S., Usluel, Y.K., Çevik, V.: Social influence in the adoption process and usage of innovation: gender differences. Int. J. Behav. Cogn. Educ. Psychol. Sci. 1, 229–232 (2009)

Nobis, C., Lenz, B.: Gender differences in travel patterns. In: Research on Women's Issues in Transportation, Volume 2: Technical Papers, pp. 114–123, Chicago, Illinois (2004)

Nobis, C., Lenz, B.: Gender differences in using digital mobility services and being mobile. In: Abstract proposed to TRB "Women in Transportation" Conference, Irvine (2019, acceptance pending)

Perez, C. C.: Invisible Women: Exposing Data Bias in a World Designed for Men, London (2019)

Rosenbloom, S.: Understanding women's and men's travel patterns: the research challenge. In: Research on Women's Issues in Transportation: Volume 1 Conference overview and plenary papers, Conference Proceedings, Washington, DC, vol. 35, pp. 7–28 (2006)

Scheiner, J.: Sozialer Wandel. Raum und Mobilität. Empirische Untersuchungen zur Subjektivierung der Verkehrsnachfrage, Wiesbaden (2009)

Thompson, R.L., Higgins, C.A., Howell, J.M.: Personal computing: toward a conceptual model of utilization. MIS Q. 15, 125–143 (1991)

Venkatesh, V., Morris, M.G.: Why don't men ever stop to ask for directions? gender, social influence, and their role in technology acceptance and usage behavior. MIS Q. 24, 115–139 (2000)

Wajcman, J.: Technik und Geschlecht: Die feministische Technikdebatte. Frankfurt, New York (1994)

Wiles, J.: Daily geographies of caregivers: mobility, routine, scale. Soc. Sci. Med. 57(7), 1307–1325 (2003)

Wixom, B.H., Todd, P.A.: A theoretical integration of user satisfaction and technology acceptance. Inf. Syst. Res. 16(1), 85–102 (2005)

How to Integrate Gender Equality in the Future of "Smart" Mobility: A Matter for a Changing Planning Practice

Lena Levin[1,2]([✉]) [iD]

[1] VTI, the Swedish National Road and Transport Research Institute,
581 95 Linköping, Sweden
lena.levin@vti.se
[2] K2, the Swedish Knowledge Centre for Public Transport,
Bruksgatan 8, 222 36 Lund, Sweden

Abstract. Sustainable transport is one of the key challenges of the UN and EU to ensure to meet society's economic, social and environmental needs whilst minimising undesirable impacts. Sustainability planning may require changing the way we solve transportation problems. From the perspectives of the sustainability, we may assume that the emphasis should move in direction to changing the practice: but exactly what practice and who's practices are to be changed? One way is to investigate gendered mobilities. The main differences in mobility patterns between women and men at a general level, are found in modal choice and travel distance. Women's practices tend to be related to the most sustainable means of transport, while men's practices are related to more unsustainable transport. Relying on studies on transport planning including focus groups, interviews and workshops in Sweden, this paper ties the concepts of gender equality, to contemporary planning and sustainable "smart" mobility, and investigates in what way knowledge about gender equality is elaborated in regional planning practice. It appeared from the interviews that both gender equality and diversity were perceived as difficult in regional transport planning and that more knowledge and experience were needed. It was pointed out that there existed some knowledge but that there were no structures for how it could be incorporated into the planning process. Noticeable are conflicting practices, while policy on gender equality are attached to the planning there are still beliefs that transport planning can be gender neutral and free from social impacts. The smart mobility approach promises improvements of mobility and access opportunities for all.

Keywords: Gender equality · Knowledge production · Practice · Mobility · Transport planning

1 Introduction

Sustainable transport is one of the key challenges of the UN Sustainable Development Goals (UN 2018) and EU Sustainable Development Strategy (EU 2018) with the intentions to ensure that the transport systems meet society's economic, social and

© Springer Nature Switzerland AG 2019
H. Krömker (Ed.): HCII 2019, LNCS 11596, pp. 393–412, 2019.
https://doi.org/10.1007/978-3-030-22666-4_29

environmental needs whilst minimising undesirable impacts. A narrow definition of sustainable transport tends to favour technological solutions, while a broader definition tends to favour more integrated solutions, including travel choices, economic motives, institutional reforms, land use changes as well as technological innovation and users' perspective. Following the broader definition, we assume that sustainability may require changing the way people think about and solve every day transportation problems. Mobility and everyday movement differ between different groups. Women's practices tend to be related to the most sustainable means of transport (walking, biking, public transport), while men's practices are related to more un-sustainable transport (private car). In this paper a practice lens offers an opportunity to identify and problematise prevailing knowledge about gendered transport and the strategies/routines constituted through the everyday practice of transport planning.

The aim is to explore what gender equality may mean in contemporary transport planning. The following questions are guiding the analysis: How do the planners perceive gender equality generally and in relation to the planning practice? In what way do practice elaborate knowledge about gender equality and how do gender equality connect to future transport solutions? The data consists of interviews, focus groups and workshops with transport planners in Sweden. We consider the practice as the dynamics of everyday activities, how activities are generated and how they operate within different contexts and over time (Feldman and Orlikowski 2011; Shove et al. 2012; Shove et al. 2015; Watson 2012).

2   Theoretical Framework

The theoretical framework of this paper is inspired of practice theory, and gendered transport planning/gendered politics. The concept of practice is central to the understanding of social life and everyday routines. The logic of practice is how it is produced, reinforced and changed by situated actions with intended or unintended consequences. The world is seen as brought into being through everyday activities. 'The social is a field of embodied, materially interwoven practices centrally organised around shared practical understandings' (Schatzki 2001: 3; cf. Garfinkel 1967). From a theoretical view: structures, institutions, routines, etcetera cannot be conceived without comprehending the role of agency in producing them, and similarly agency cannot be understood as an isolated human action but rather be understood as always already configured by structural conditions (Feldman and Orlikowski 2011). Social actions and interactions are forming mutual constitution of e.g. an organisation or a specific job setting. However, it is important to note that mutual constitution does necessarily not imply equal relations.

This paper takes up the idea that social practice is connected to infrastructure across space and time (Shove et al. 2015; Watson 2012) and that chains of activities, discourses and interactions are important for the understanding of a contemporary practice. In taking this approach, we work with the proposition that the forms of mobility in a society are dependent on citizens' everyday practices and on the structure of the cities, landscapes, etc. Automobility, for example, can be understood as the outcome of interconnected patterns of social practices, such as working, going to school, shopping, and visiting friends and family. We also acknowledge that social practices are partly constituted by and embedded in material arrangements that cannot be completely controlled by individual transport users.

Litman and Burwell (2006) pointed out that conventional transport planning tends to assume that transport progress is linear, and that the progression consists of newer, faster modes displacing older, slower modes: e.g. walk → bicycle → bus → automobile → improved automobiles. At the same time, there is an emerging new sustainability paradigm that reflects a model which assumes that progress means that all modes, not just the newest, should be developed by improved conditions. Walk and bicycle are useful modes in urban city space while bus and train are useful modes in suburban and inter-city areas and improved automobiles are useful on certain distances in the city as well as rural areas. Improved transport does not necessary means faster travel or more mileage, but increased comfort, safety, cost savings and reduction of the total need for travel. '[I]n many cities, the most beneficial strategies may involve improving walking and cycling, more support for public transit, and restricting automobile travel in congested urban area' (Litman and Burwell 2006: 335). Furthermore, research and sustainable planning reflect that impacts, and objectives often interact, and solutions must reflect integrated analysis (Papa and Lauwers 2015; Yamini 2017).

It has been shown in research and practice that women and men still have different roles and responsibilities and that these differences affect travel choice and opportunities (Greed 1994, 2008, 2019). Noticeably women are still responsible for childcare, shopping and household work. In the same time, most women work outside the home. This means that women's journey-to-work looks more complex and multipurpose compared with the home-to-work and back again journey of the traditional (male) commuter (Greed 2008: 244). The main differences in mobility patterns between women and men at a general level, are found in modal choice and travel distance reported in travel surveys (cf. Law 1999; Levy 2013; Scheiner and Holz-Rau 2012) and obviously, women's practices tend to be related to the most sustainable means of transport, e.g. walking, bicycling, and public transport; while men's practices are related to more un-sustainable transport, e.g. longer journeys by private car.

Differences are in many cases related to driving licence and access to private cars (Blumenberg 2000; Rosenbloom 2006). General figures in transport planning do not often analyse gender differences, or various resources and cultures among citizens (cf. Greed 2008; Polk 2008). Levy (2013) means that while mainstreaming transport planning avoids recognising the social position of transport users, it is not ideologically neutral either. The notion of travel choice is central to the modelling process in transport planning. Levy grounds her critique on that transport policy and planning is predicated on several implicit assumptions. Such assumptions relate to gender relations and the structure of households, the division of labour in households, the control of resources and the power of decision-making – and these assumptions are basically informed by "western" and middle-class values. To these assumptions are added other transport-based biases – e.g. the focus on the journey to work and motorised transport, particularly on private cars which are unaffordable to most poor urban women and men. The affordability approach is still actual in the era of planning for automated private vehicles, regardless whether the automated system would be based on ownership or hired vehicles.

The genderless modelling is a problem but also the assumptions that women and men, boys and girls, can exercise individual choice. Levy points out that women tend to use cheaper modes of transport than men, particularly low-income women. Both poor women and men tend to walk more than the average of the population and especially poor women. Hanson (2010) and Sheller (2004, 2014) have emphasised the need for alternative to the ideas of genderless modelling of transport planning and a need for a more inclusive approach. Research have also argued for more explorative and inter-disciplinary studies instead of the instrumental quantitative studies promoted by governments and transport authorities (Aldred et al. 2015; Angeles 2017; Law 1999). Global as well as national and regional challenges today make it impossible to ignore diversity and the inequalities that arise from not having a plan for gender equality and diversity mainstreaming. According to CIVITAS (2014) evolution of household and parental models, developments in the labour market and new technologies with the new forms of work, as well as women's increased labour market participation and population ageing; are likely to extend the variety of mobility patterns and necessitate appropriate transport policies. Especially a need for combining attention to sustainability with attention to gender and age-specific mobility needs. Research shows that multiple categorisations play a role in the conditions under which a person lives. Geographical location, city or rural, inner city or suburb, residential area, educational level, socioeconomics (class), employment, race, colour, belonging to national minority, political view and appearance are examples of categories (cf. Crenshaw 1991). Several of these categories are protected against discrimination by the Treaty of Lisbon and collated in the 2030 Agenda.

In this paper the concept of mobility refers to people's movements outdoors, in terms of transportation to access desired places, activities, services, and people (i.e., destination dependent) or simply to move around alone or together with other people (i.e., destination independent). Both perspectives (i.e., destination dependent and independent) can be important for a person's health and wellbeing and impose independence and self-fulfilment (Jones et al. 2013; Mollenkopf et al. 2004; cf. Levin 2019). The subjective meaning of getting out and about includes the psychological and social benefits of mobility as a personal resource and the ability to take part in society. Metz (2000) and Kaufmann et al. (2004) have also identified another dimension of mobility: the potential to travel, which is sometimes discussed in terms of motility. In this paper the concepts refer to a transport and infrastructure planning context. The concepts are tied to the practice of planning for future mobility, and in this sense, also the motility approach might be relevant for understanding of who are included and who get access to a smart mobility solution.

The concepts of smart city and smart mobilities have been launched, as a long-term vision of a better urban area, aiming at reducing its environmental footprint and at creating better quality of life for citizens. However, the concepts of smart cities and smart mobilities constitutes a complex vision and it is a difficult topic to grasp. It involves both environmental, social and economic aspects and it needs both high-ranking technologies and intellectual and just behaviours. Smart mobility is permeated by ICT, to reduce congestion and support the optimization of traffic fluxes but at the same time, it seems as the current storylines very much reinforce and reproduce the "system of automobility" (Freudental-Pedersen et al. 2019). It has been stated that smart mobility is sometimes used as an "evocative slogan", with its core in vehicle technology and ITS, in consumer centred demand for mobility and the solution in new tecno-centric mobility services, which in many cases seems to lack fundamental connections with other central aspects of mobility planning and governance. Jeekel's (2017) suggestion is to broaden the approach towards sustainable mobility and towards smart city and city as a place, 'to become an active and comprehensive strategy helping to reach sustainable development' (p. 4306). It is also about citizens' opinions about liveable cities and quality of life, including e.g. accessible local public transport services (Benevolo et al. 2016). Bencardino and Greco (2014) interpret "smart cities" as "social cities" and asking for a system with new technologies that responds to a strategic project that start from the bottom, from the real needs of the citizens. The challenge is to urbanise technologies to make them useful to new urban needs. Talking about social cities instead of smart cities explores how media technologies can empower people to act and co-construct their mobility and the mobile environment in what may be called civic empowerment. Smart mobility is in the best case an approach that promises the improvement of mobility and opportunities for all.

3 Data and Methodology

This paper is about planning processes, focusing on how planning traditions, values and perceptions appear and may shape the future. The examples come from one region in the middle of Sweden with many industry-related transports where new service sectors have difficulty in claiming space in the transport system; and from one region in south-west of Sweden with both a heavy industry (e.g. cars, trucks) and growing information, service and tourism industries.

Data collections of four focus groups with 5–10 participants each, were carried out in 2014 (Thoresson and Levin 2014; cf. Levin and Faith-Ell 2014; Levin et al. 2016), document analysis and eight follow up interviews four years later (Levin 2019), and workshop data from an ongoing research project on social impact assessment connected to a new smart urban bus system (i.e. metrobus).

There are 21 geographical regions in Sweden, and they are responsible for producing regional plans (county plans) for the transport infrastructure sector twelve years ahead. These plans, which are updated every four years, form the basis for the national planning and economic allocation of resources in the transport area. This planning process was the back ground for a transdisciplinary research project during the years 2013–2015 focussing on how to integrate gender equality in regional transport planning. The focus group interviews were accomplished in the beginning of the project. They were moderated by a researcher, and another researcher took notes during the discussions. The sessions were recorded by a voice-recorder after the permission from the participants (cf. Codex 2019) and transcribed afterwards.

The two regions participating in the transdisciplinary research Region Dalarna (in the middle) and Västra Götaland (in the south west of Sweden) took part in a series of activities (e.g. focus groups, workshops, seminars). Before the recruitment of focus group participants, the researchers analysed planning documents from each region and noted topics relevant for the interview study. Questions were thematic and based on the topics from the document study, and participants were also invited to develop own questions and ideas during the focus group discussion. The issues that started the discussions were about how to consider gender equality and the fulfilment of the gender equality goals in the core of transport planning. The focus group discussions then centred on how they work today with gender equality and how they wanted to work with gender equality in the future.

A follow up study including eight telephone interviews were conducted during October 2018 with planners in Region Dalarna. Together with document analysis the interviews constituted an evaluation of the gender mainstreaming in the Region Dalarna four years after the transdisciplinary research. The work in Västra Götaland has developed in another direction and they are now participating in another project about developing tools and methods for social impact assessment in the planning of new smart mobility solutions in the city of Gothenburg (Levin and Gil Solá 2019).

In this paper we relate to the four focus group discussions (2014), follow up interviews together with document analysis (2018), and data from workshops and planning meetings in Gothenburg (2018–2019).

Analysis was based on the transcribed focus group talk and notes from interviews; and with qualitative discourse analysis approach. Here we use an active reading methodology. When we read a transcript, it is to find opinions and get an insight into various issues. An active reading means that we try to understand the content patterns, the ideational structure. What is it about and what perspectives does it bring about? (cf. Fairclough 1992). Especially interesting is the practice, i.e. the compilation of activities that the text/talk describes (Feldman and Orlikowski 2011; Shove et al. 2012). The excerpts presented here are translated from Swedish by the author.

4   The Nordic Context

The Nordic countries are usually at the top of the list in international comparisons on gender equality. Despite this top position we can (based on current research and the data analysis in our study) assume that Sweden struggles with operationalisation of the gender equality goals. Research show that in the both women and men are car drivers to a large extent, but men drive longer distances and women are more often travelling as passengers (Hjorthol 2000, 2001). The main differences occur in the older ages where fewer women than men have access to driving licence and private cars, but during the past few decades differences between older men and women have decreased (Hjorthol et al. 2010). Other analyses show that differences between men's and women's travel have remained constant over the past 20 years. Men spend more time in travel linked to business or education and less time in travel for household purchasing and service matters. Compared with women, men make approximately 20% fewer trips linked to unpaid domestic care and work. Men drive cars almost 80% farther than women do, according to evaluation of the gender equality objective in Swedish transport policy (Trafikanalys 2017: 28–29)[1]. In UK men are twice as likely as women to cycle to work and cycling tends to be dominated by younger adults. By contrast, in higher cycling countries and cities, gender differences are low, absent, or in the opposite direction. In some high-cycling countries, women tend to cycle more than men (Aldred et al. 2015).

Sweden's long tradition of gender equality work has resulted in six national gender equality goals and a set of national gender equality goals in transport politics (Rönnblom 2011, 2017; Vagland 2004, 2006). The current national gender equality goals in transport politics was adopted in 2009 (Government Bill 2008/09:93) and have since then been evaluated in several studies (cf. Faith-Ell and Levin 2012, 2013; Levin and Faith-Ell 2011, 2014, 2019; Joelsson and Scholten 2019; Smidfelt Rosqvist and Wennberg 2012; Smidfelt Rosqvist 2019; Wittbom 2009, 2018). Evaluations have identified difficulties in implementing goals in practice. For example, significant gender pay gap, lack of women in top management positions in businesses, and sharing the unpaid work on caring for children and relatives which disfavour women's life earnings from paid labour (Alnebratt and Rönnblom 2016; Göransson 2007). We believe that these inequalities are due to deep-rooted politics in all fields, political assemblies, and a need for more practice-oriented planning and governance.

The transport policy objectives in use today come from the Government bill 2008/09:93. In the bill, a functional goal regarding accessibility and a consideration goal regarding safety, environment and health were adopted. The functional goal also addresses gender equality. It is stated that the transport system should be equal, e.g. address the needs of men and women in an equal way, and that the work model, implementation and results of the transport politics should contribute to a gender equal society. The Governmental bill states that the strategy to reach gender equality should be gender mainstreaming, i.e. a strategy in which decisions in all political areas should have a gender perspective (EIGE 2012; EU: GAP and SDS; cf. Christensen et al. 2017).

5 Research Result

In this section we investigate the planning practice, from the perspective of the two regional planning contexts in Sweden: Region Dalarna och Västra Götaland. The region in the middle of Sweden (Dalarna) is characterised by industry-related transports from mining and forestry where new service sectors have difficulty in claiming space in the transport system; and the region in south-west of Sweden (Västra Götaland) is characterised by both a heavy industry (cars, trucks) and growing information, service and tourism industries.

---

[1] Transport Analysis is a government agency, providing decision-makers in the sphere of transport policy with evaluations and policy advice.

Based on previous research, we knew (when we started this project) that a basic and common understanding is needed when working with gender equality in planning contexts. The first step was to understand that gender equality has to do with their own activities. This was a reflexive phase. Thus, the research in the two regions started with an explorative focus group study considering gender equality as a concept and the participants' experience of the gender equality goals in the transport planning. The purpose of the focus groups was to investigate the current situation, and to give the research team insights of how the work has been conducted. The focus groups were also a stepstone to start the discussion and reflection within the organisation that aimed at more systematic work with the gender equality goals in regional transport planning. We introduced a transport related operationalisation of the national gender equality goals: for example, the notion of equal division of power and influence, which means that that women and men should have the opportunity to participate in transport planning and have the power on their own mobility; and the notion of economic equality i.e. that women and men must have the same opportunities and conditions as regards education and paid work which give economic independence throughout life. We suggested that the transport system should be seen as facilitating the achievement of gender objectives, then the design of the transport system is central to upholding various parts of daily life for women and men.

The participants in the focus groups were first invited to talk about their thoughts on gender equality in society as a whole, and then in connection with their own practice i.e. the regional transport and infrastructure planning with the perspective 10–12 years forward. In the conversation, several basic dimensions of gender equality were discussed as economic equality (income and labour market), unpaid home and care work, power and influence, and men's violence against women. The last theme was only indirectly touched by the participants addressing women's safety in connection with travel.

Several participants preferred to take the point of departure in the individual and referred to an idea that a central task for planning is to create the conditions for individual choice. Implicitly, gender equality is associated with equal opportunities to be able to make individual choices. However, the categories of woman and man are in these discourses often used without being problematised, for example by talks about the different needs, travel patterns, preferences or values of women and men as mirrored through individual travel patterns. It is about women travelling with public transport to a greater extent than men and performing travel patterns which are often linked to home and care work, while men travel longer distances and more by car; and also, that women value traffic safety higher than men do. There appeared elements of uncertainty about gender and travel patterns, especially about possible causes to the differences between women and men when it comes to everyday travel. Excerpt 1 (below) is centred on women concerning both journeys to work and journeys related to home and care work, and it raises the question to what extent the different travel patterns of women and men depend on individual choices, life situation, preferences and social structures. It is also an excerpt that highlights a deportment in relation to the issue of gender equality in the transport system that is representative of most of the focus group interviews and most of the participants.

Excerpt 1 (FG 4:13):

A: But is it not that women choose ... workplace closer to their home than men do to a greater extent, and then it becomes easier to choose another mode of transport [than the car], I think.

B: But it is also different depending on where you are in the life situation. It's ...

C: However, women do not travel as long as men do in any case and have shorter distances between housing and workplace.

Moderator: Yes, what is it then?

B: They don't think it's so fun to go that far, I don't know ...

C: On average, the women take more responsibility for the children. We talk about, with short trips, that you travel more in the chain [chain trip], maybe go and pick up the child on the way home and so.

There are participants who question the meaningfulness of the categories of women and men, in particular as significances for uniform groups. Several argue that the variations within the groups of women and men are so great that it is misleading to say anything about what women or men generally think or how they usually choose to travel. The categories represent an imaginary average, but it can also be a simplification if the variation is large. The participants who question gender as a category in this way refer to the importance of circumstances, the individual's life situation and that people are individuals more than the sex. But it is also appearing a tendency to gender blindness, as the users of transport systems are seen as genderless individuals.

A recurrent pattern in the focus group conversations was to relate gender equality to equity, usually in terms of "broadening the perspective" and thereby including gender equality as part of equality between all other groups, not just an issue about women and men. This means that women are treated as one of several marginalised groups in society and in a sense also equated with considerably smaller groups. The concept of equality refers to the equal value of all people, which in addition to gender actualises discrimination based on, for example, class, ethnicity, disability and age.

Excerpt 2 (FG 2:28):

A: I think as usual ... I always think it is difficult when the concepts sound almost identical, to keep apart. It would be better if it was black on one side and white on the other side, then it is easier to keep it apart, but yes ... But it is a wide concept I also know. And one would like to make the concept of gender equality in some sense of equality [for all] and think about ... It is not just this with women and men, but disabilities and ...

B: Children and the elderly ...

C: Children and the elderly, [people who have] difficulty with the language and so on. There is a lot to think about in the transport sector.

The conversations that sometimes relate to a broader equality concept give the impression of a desire not to overlook other discrimination and injustice, while at the same time risking reducing gender equality issues to a subcategory. Many times, the planners speak explicitly or implicitly about a transport system for all. One of them says:

Excerpt 3 (FG 4:12):
We do not speak so much perhaps [about] men and women, but to offer as many [people] as possible good travel options as possible. And it covers both sexes, it's not one or the other. It is to offer the residents and citizens sensible travel options. Then we cannot do that in all travel relationships, but that is another issue. And 'everyone' includes both children and the elderly and ... everyone, we do not work so much on men-women, in that way, but travellers and citizens, and this includes more.

That utterance was followed up by this utterance from another participant:
But it is ... I think an important issue when it comes to infrastructure, what you prioritize, it can be the geographical choice. Should one strengthen infrastructure near cities in rural areas, where do you place the money geographically? After all, it is a fairly big question I think among the politicians and the local government associations, to make the choices and the priorities. It is probably a much bigger issue than gender equality I believe.

Although this approach to gender equality recurs in the material, it is important to note that while some of the participants repeated this several times during a conversation not all participants expressed it actively. There are also participants who are openly critical of such an approach. Next citation expresses that women are not marginalised per se and argue for a more comprehensive approach.

Excerpt 4 (FG4: 19):
I do not really think about parallel gender equality with these groups [other discriminated groups] as... if you have the worse financial position or if you are a child [...] then to say that women are among them, it seems a little offhand to considering that, because it is half the population. [...] one cannot say that it is just about equality [among marginalised groups].

In three of four focus groups there were examples of how "specialists" characterise the conversation. It is particularly clear in one focus group, relatively prominent in two further and to a lesser extent in the fourth. These people are more familiar with gender issues, in many cases due to that they have worked with various forms of gender equality projects. They have the capacity to discuss gender equality in a different way than other participants, which in at least one case seems to characterize the conversation as a whole.

One participant expressed that gender equality can be regarded as a means of other goals in society, such as economic growth and citizens' independence. She also emphasized that gender equality should primarily be understood as a structural problem, not an individual. It's about economic equality; 'That public resources should be distributed equally to men and women'; and about 'creating structures that smooth out the distribution of unpaid home and care work'. She was one of the participants who were particularly familiar with gender issues and in this focus group the issue of individual versus structure became central.

Even though most of the participants in the focus groups emphasized that gender equality is an important issue that should have an impression in the planning process, descriptions indicated that equality in practice has a weak position in relation to other goals such as economy and road construction. It appears to be something with symbolic value rather than a real integrated analysis of consequences that are important for the prioritisation of measures.

Excerpt 5 (FG 4: 38):
But there is no one who raises this issue of gender equality [as a real objective] ... any discussion about that issue at all. It is in some way that it is done afterwards, so you write down; yes, 'this is good for gender equality', or something like that.

Excerpt 6 (FG 3: 9):
... I think it is very good if someone could work out better tools. We [...] do such aggregate impact assessments now and then we do this goal analysis [e.g. goals on gender equality] and one should also give an interpretation if the measure favours women or men or both and then we have a standard formulation which one always inserts 'when you build a new road it will probably benefit most men for they are driving cars but it is probably good for women if they go by bus', so we do.

A part of the focus group interviews concerned questions about their own room for maneuver, i.e. what the planners themselves think they can do. The view of this seems to vary between different participants and be dependent on what support is available in the form of tools, knowledge and methods, but also resources in the form of time and money. It happened that the participants expressed that the room for maneuver is limited, for example by referring to political goals or directives and to other authorities, or to the issue so cross-sectoral that it limits the possibility of change. One participant expressed that sometimes it is unclear who should take responsibility in infrastructure planning. Another participant said, about work at weekends (e.g. caring professions, retail trade and tourism workers):

Excerpt 7 (FG3: 11):
We talked about this with weekend workers [...] who are usually women then and women-dominated professions. There I feel that what we in the transport policy, what we can do ... we can offer and facilitate for them, and not complicate [...] but then that is the other aspect; what comes to mind in the great gender equality perspective that you... maybe... just this with raising wages for example or take a holistic approach around... the women's profession [...] it is as much as you can grasp here but i feel that we can do; we can facilitate their everyday lives.

This quotation relates to a couple of things mentioned above; partly the perceived conflict between satisfying unequal needs or contributing to a change in society, and partly that the concept of equality is perceived as so large and cross-sectoral that it becomes difficult to manage. Not least, it becomes difficult to translate into concrete implications for the planning of the transport system. In the last sentence, the participant explicitly makes his view on the scope for action. Next two excerpts (8 and 9) pointing into the future. First about how to enlarge the region to increase the labour market for citizens and what this is about:

> Excerpt 8 (FG 2):
>
> A: I think about this regional enlargement, I think this is an interesting concept. So, the regional expansion itself is... is about changing travel patterns all the time, we want people to travel longer, and it should go faster. The whole goal itself is to change a travel pattern because we believe, believe and know, in any case according to research, that if we can move, and a larger as well as critical mass [of people] can retrieve new labour markets, then it goes better as well, in growth, we earn more money, wages increase. So really, one can say that the regional enlargement right now benefits men, because most men travel these long journeys today.
>
> B: But then I think that regional enlargement can be achieved through digitalisation as well, not just through infrastructure. You can reach several parts of the region in a different way.

Speaker A probably mean that most men have already embraced regional enlargement and benefit from it since they already travel long journeys. The aim is to further increase the region for all citizens by more efficient and faster travel modes. Speaker B points out another possible solution with the digitalisation not relaying on traditional transport infrastructure. New technologies may increase mobility options but does not have to increase commuter distance every day. Digitalisation, media technologies, and so on can empower people to act and co-construct their mobility and the mobile environment in other ways than before and increase civic empowerment. People may work on distance and like speaker B suggests 'reach several parts of the region in different ways'. Smart mobility is in the best case producing well-functioning networks and opportunities for all. Still, when gender equality is considered, the fact is that many women cannot choose the times they would like to travel or places to visit, since they often are expected to be in a certain place during a certain time of the day (e.g. hospitals, schools, preschools, home care services).

There was a quite low awareness of what resources should be needed from the travellers to use new regional enlarging transport supplies, e.g. affordability and other aspects on access to faster (automated) high speed transport modes (e.g. long distance commuter trains, cable ways and super buses). Next utterance link to a discussion on social impact assessment which was developing in Gothenburg at the time for the focus group study.

Excerpt 9 (FG 1):

So when it comes to the future, I think so to say that one ought to get into it more and more in the societal planning, thinking so [...] when a municipality plans and builds new housing, like in Gothenburg, you really do such social impact assessment about what does it mean to build residential areas here? what is accessibility? That already, in some way eliminates the problems [...] what you then desire and see are such measures that promote gender equality and integration.

In all focus groups, there was a generally idea that there is a lack of knowledge, tools and methods for integrating and working with gender issues in infrastructure planning. None of the participants said that the support was satisfactory, although the view of how big this problem was varied between participants. The planners asked for several types of support and improvements to develop work on gender issues. It can be summed up: concretisation, objective or impartial facts, more knowledge of those who plan and decide, e.g. through education, and developed methods for analysing consequences and finding solutions. The first theme, concretisation, revolves around understanding the concrete implications for the transport system and responds to the difficulty described above that the concept of equality was perceived as a large area difficult to grasp. It simply requires a kind of translation from the general to the specific, according to the focus groups.

Excerpt 10 (FG1: 37):
But (..) when you start working with gender equality in [relation to] new issues, you need this specific knowledge and how it is linked to the [specific] area. For when we... when we have worked with gender equality and growth, we experienced that all people working in our organisations' target groups, yes, they have been on these gender equality courses, and talked about it at some general level. But then they go home and do not have an idea how they should connect this to their business. [...] So, this is... that you must be able to work in the area of gender equality in your area.

When we found this noticeably lack of knowledge and need for working methods, we decided to produce a simple handbook (Halling et al. 2016). However, despite this effort the discourse on need for knowledge, methods and tools continued in the follow-up study in Dalarna four years later. There were two planning documents which were evaluated from a gender impact assessment perspective. The document analysis showed that in the work with the regional system, data on the population have been integrated into themes such as road safety, education and work commuting, but the analysis of population data had weak contextual connection to gender equality and to diversity. In the documents, goal formulations were often repeated from national policy documents without having a regional disaggregation of data or their own analysis.

It appeared from the interviews that both gender equality and diversity were perceived as difficult in regional transport planning. It was pointed out that there existed some knowledge but that there were no structures for how it can be incorporated into the planning process. The biggest obstacle to further integrating these aspects was said to be that the planning does not really take the step of continuing to develop the methods and to improve the internal work after the research project was ended. Compared to other fields of work, the field of transport planning seems to lack skills:

Excerpts 11–13 (interviews):
We spend more time internally rather than external work. We must change our approach to internal work. (IP1K)
It's just politically, there are other trends. Old structures, that there is no tradition. There is a certain path way. (IP3K)

Skills supply. When you work in the municipalities and have a dialogue with them, you can point to large areas where it is big difference between women and men, and have a creative discussion about why it looks like this and what can be done. I have not seen such contexts in the transport field. (IP2M)

Some statements indicated that pioneering work was going on. Gender equality and diversity were described by some interviewees as "new" elements in regional transport infrastructure planning, compared to issues such as accessibility, economic growth and the environment. There, was a lack of developed routines for integrating gender equality and diversity into the work processes. Primarily a clearer focus is required – to integrate, for example, gender-disaggregated statistics at regional level and analyses of the region linked to the target formulations at national level (transport policy and national gender equality goals). An example of a method could be gender impact assessment (GIA), but also a broader social impact assessment (SIA) where different groups can be included based on the objectives stated in the Discrimination Act. These analyses should be strategic and goal oriented (cf. Levin and Faith-Ell 2019).

What does it mean to plan for increased sustainability in relation to gender, public transport, more cycling, regional enlargement, travel faster with more technically developed faster modes of transport (smart mobility solutions)? This is the topic for the ongoing process in the other region we have investigated. Västra Götaland is in the beginning of planning for a new rapid bus system called metrobus. Citation from the responsible planning officer after a workshop in October 2018:

Excerpt 14 (conversation):
Lack of knowledge about the needs and conditions of different groups: We need to describe where it is relevant to acquire more knowledge and conduct dialogue (e.g. around station design) and how it should be done. Here we think that the time geography approach can give us more knowledge about different groups' use of a place and how the metro bus system needs to be designed to contribute positively to the area from a social point of view. Who will benefit from the new metro bus system? New travel opportunities and thereby broadened work and study market for those who today lack realistic public transport connections and who cannot take the car. Important to think all day and the whole trip, (possibility of night workers to take the fast bus line, security at the stations when it is dark, etc.): Identify which are the target points for workplaces? GIS? How many people are affected? Everyone should feel welcome in the system.

A recurring paradox is experienced by planners between meeting "unequal travel needs" and changing the transport system. It is largely based on the existing process when one considers and discusses how gender equality can be considered in the planning. Sometimes when gender equality and diversity were described as "new" elements in regional transport infrastructure planning, at least compared to issues such as accessibility, economic growth and environment. it was obvious that deep-rooted planning traditions impinged on the development process and that a lack of skills in how to integrate gender in the planning practice may preclude data and complementary analysis that would help shape a future that is consistent with the sustainable development goals.

6 Discussion

The research cited here looked at the prospective regional planning for new transport and infrastructure in the perspective until 2035. The first planning processes followed were the regional plans for the period of 2014–2025; and then a follow up study about the planning for the period of 2018–2029 and the metrobus planning aiming for the year of 2035.

The examples were taken from transdisciplinary studies where researchers came into a regional transport planning and introduced an idea about more gender smart planning but were met with a planning tradition with difficulty in embracing a development process in how to integrate gender into the planning practice. This was still part of the problem reported in the follow up study on Region Dalarna.

Working systematically on gender mainstreaming in transport infrastructure entails implementing a gender perspective in all stages of decision making, planning, and execution (Christensen et al. 2017; Levin and Faith-Ell 2019; Woodward 2003). None of these regions had hitherto succeeded with such an effort. Many of the regions in Sweden have an aim to work with gender equality and to some extent also diversity mainstreaming in the transport and infrastructure plans for the upcoming twelve years. The regional plans will fit into the national transport and infrastructure plan driven with a new political paradigm, aiming at more sustainable "smart" mobilities. Nevertheless, the idea of the future is somewhat in line with what Freudental-Pedersen et al. (2019) call an "evocative slogan". The core idea is in an expectation of consumers' demand for high speed and to increase the travel distances i.e. a travel practice dominated by men. The idea of new tecno-centric mobility ends up in faster mobility and the enlargement of the commuter space, which in many cases seems to lack connections with other central aspects of mobility planning and governance. There were a few utterances in the focus groups problematising the idea of increased mobility from the perspective of gender equality. The ideal traveller seems to be a person who can choose transport mode, and even has the opportunity to choose distance work and thus overcome geographical distance by media technology instead of using the transport infrastructure. This approach however is gender-blind. Hospitals, primary schools, day care centres, elderly care, as well as grocery stores and hotels; still need workforce that is in place, and the majority of this labour consists of women.

User friendly, connected technology e.g. automated vehicles, integrated information and mobility services, intended to improve the transport system is part of a political process. Large transport and infrastructure projects have hitherto often failed to approach diverse citizen needs because they have tried to solve problems from top down, when they also need to take account of power struggles and lack of power, and the need to increase peoples agency, whether concerning matters of employment, market production, or household reproduction (Mouffe 2005, 2013; Sharp et al. 2003: cf. Levin and Faith-Ell 2019.) Still the planning must be done by skilled personnel who are capable to include knowledge on people's reality, i.e. in everyday life and everyday travel. Sweden is a leader when it comes to gender equality in many areas, but not in transport planning. Though there is a policy for gender mainstreaming, the planning shows shortcomings when it comes to implementing policy in practice. Both the national and regional transport plans are weak when it comes to gender and diversity. From this point the research has displayed problems to change a traditional planning perspective into a new paradigm including gender and an intersectional perspective early in the planning process on both national and regional levels (Levin et al. 2016; Levin and Faith-Ell 2019; Smidfelt Rosqvist 2019; Smidfelt Rosqvist and Wennberg 2012). The understanding of gender equality in transport planning – based on planning routines are not institutionalised. Noticeable are conflicting practices, while new thoughts about social sustainability are increasingly attached to the planning there are still beliefs that transport planning can be gender neutral and free from social impacts. Following Papa and Lauwers' (2015) definition of conventional mobility planning it often focuses on the physical dimensions and on traffic (and in particular on the car) rather than on people: it is large in scale, rather than local, it is forecasting traffic and it is based on economic evaluation and may fail to adequately consider wider impacts (Jeekel 2017). We here recognised the discourse centred on the conventional approach (cf. Banister 2008). It was only in the metrobus project in Gothenburg that an obvious social impact approach was noticeable. The present study showed that regional transport planning is far from the interpretation of "smart mobilities" as s "social" system and a system with new technologies that responds to a project that start from the bottom, from the real needs of the citizens. The challenge is to urbanise technologies to make them useful, technologies can empower people to act and co-construct their mobility and the mobile environment in what may be called civic empowerment. Smart mobility is an approach that promises the improvement of mobility and access opportunities for all. However, the smart transport system does not get smart until it can be used by a wide range of users without too much effort and prior knowledge. Planning for smart mobility has only started to recognise the importance of integrating knowledge on gender and diversity.

The present research will be further development in the new research and innovation action: TInnGO, Transport Innovation Gender Observatory. (EU Grant agreement #824349).

Acknowledgements. This paper was partly supported by the Swedish Governmental Agency for Innovation Systems Vinnova (#2013-02700), Region Dalarna and The Swedish Research Council for Environment, Agricultural Sciences and Spatial Planning, Formas (#2015-1140). Focus groups cited in this report have been discussed more thoroughly in a working paper written in Swedish by Karin Thoresson and Lena Levin.

References

Aldred, R., Woodcock, J., Goodman, A.: Does more cycling mean more diversity in cycling? Transp. Rev. (2015). https://doi.org/10.1080/01441647.2015.1014451
Alnebratt, K., Rönnblom, M.: Feminism som byråkrati. Leopard förlag, Stockholm (2016)
Angeles, L.C.: Transporting difference at work. taking gendered intersectionality seriously in climate change agendas. In: Griffin, C.M. (ed.) Climate Change and Gender in Rich Countries: Work, Public Policy and Action, pp. 103–114. Routledge, Abingdon (2017)
Banister, D.: The sustainable mobility paradigm. Transp. Policy **15**, 73–80 (2008)

Bencardino, M., Greco, I.: Smart communities. Social innovation at the service of the smart cities. TeMA. J. Land Use, Mobil. Environ. 39–51 (2014). https://doi.org/10.6092/1970-9870/2533

Benevolo, C., Dameri, R.P., D'Auria, B.: Smart mobility in smart city. In: Torre, T., Braccini, A. M., Spinelli, R. (eds.) Empowering Organizations. LNISO, vol. 11, pp. 13–28. Springer, Cham (2016). https://doi.org/10.1007/978-3-319-23784-8_2

Blumenberg, E.: Moving welfare participants to work: women, transportation, and welfare reform. AFFILIA 15(2), 259–276 (2000)

CIVITAS: Smart choices for cities Gender equality and mobility: mind the gap! (2014). http://www.eltis.org/resources/tools/civitas-policy-note-gender-equality-and-mobility-mind-gap. Accessed 01 Dec 2018

Codex: Rules and guidelines for research: informed consent. http://www.codex.vr.se/en/manniska2.shtml. Accessed 25 Jan 2019

Crenshaw, K.W.: Mapping the margins: intersectionality, identity politics, and violence against women of color. Stanf. Law Rev. 43(6), 1241–1299 (1991)

Christensen, H.R., Hjorth Oldrup, H., Poulsen, H: Malthesen, T.: Gender Mainstreaming European Transport Research and Policies – Building the Knowledge Base and Mapping Good Practices. (EU/FP 6, 2017) by Koordinationen for Kønsforskning, Københavns. Universitet. http://koensforskning.soc.ku.dk/projekter/transgen/eu-rapport-transgen.pdf. Accessed 25 Jan 2019

EIGE: European Institute for Gender Equality. Review of the Implementation in the EU area K of the Beijing Platform for Action: Women and the Environment. Gender Equality and Climate Change. (2012). http://eige.europa.eu/, Accessed 01 Dec 2018

EU: GAP, Gender Action Plan. https://ec.europa.eu/europeaid/sectors/human-rights-and-democratic-governance/gender-equality/gender-mainstreaming_en. Accessed 25 May 2018

EU: SDS, Sustainable Development Strategy. http://ec.europa.eu/eurostat/web/sdi/. Accessed 01 Dec 2018

Feldman, M., Orlikowski, W.: Theorizing practice and practicing theory. Organ. Sci. 22(5), 1240–1253 (2011)

Fairclough, N.: Discourse and Social Change. Polity, London (1992)

Faith-Ell, C., Levin, L.: Jämställdhet och genus i infrastrukturplanering – en studie av tillämpningen inom järnvägsplaneringen. Linköping and Stockholm: VTI Report 768 (2012)

Faith-Ell, C., Levin, L.: Kön i trafiken. Jämställdhet i kommunal transportplanering. [Gender Mainstreaming in Transport Planning. Guidance for Regional and Local Transport Planners.] SKL, Stockholm. (2013). http://webbutik.skl.se/sv/artiklar/samhallsbyggnad/kon-i-trafiken-jamstalldhet-i-kommunal-transportplanering.html. Accessed 01 June 2018

Freudendal-Pedersen, M., Kesselring, S., Servou, E.: What is smart for the future city? mobilities automation. Sustain. 11(221), 1–21 (2019)

Garfinkel, H.: Studies in Ethnomethodology. Prentice-Hall, Englewood Cliffs (1967)

Greed, C.: Women and Planning. Creating Gendered Realities. Routledge, London (1994)

Greed, C.: Are we there yet? Women and transport revisited. In: Cresswell, T., Priya Uteng, T. (eds.) Gendered Mobilities, pp. 243–253. Ashgate, Aldershot (2008)

Greed, C.: Are we still not there yet? Moving further along the gender highway. In: Joelsson, T., Scholten, C. (eds.) From One Track to Many Tracks – Integrating Gender in Transport Politics, pp. 25–42. Palgrave MacMillan, London (2019)

Government Bill 2008/09:93. Transport Policy Objectives. https://www.trafa.se/en/commissions/transport-policy-objectives/. Accessed 24 Feb 2019

Göransson, A., (ed.): Maktens kön. Kvinnor och män i den svenska makteliten på 2000-talet. Nya doxa, Nora (2007)

Halling, J., Faith-Ell, C., Levin, L.: Transportplanering i förändring: En handbok om jämställdhetskonsekvensbedömning i transportplaneringen. [Transport planning in change: A handbook on gender impact assessment]. K2, Linköping/ Lund/Stockholm (2016)

Hanson, S.: Gender and mobility: new approaches for informing sustainability. Gend. Place Cult. J. Fem. Geogr. **17**(1), 5–23 (2010)

Hjorthol, R.: Same city—different options: an analysis of the work trips of married couples in the metropolitan area of Oslo. J. Transp. Geogr. **8**(3), 213–220 (2000)

Hjorthol, R.: Gendered aspects of time related to everyday journeys. Acta Sociol. **44**(1), 37–49 (2001)

Hjorthol, R.J., Levin, L., Siren, A.: Mobility in different generations of older persons: the development of daily travel in different cohorts in Denmark, Norway and Sweden. J. Transp. Geogr. **18**, 624–663 (2010)

Jeekel, J.F.: Social sustainability and smart mobility: exploring the relationship. Transp. Res. Procedia **25**, 4296–4310 (2017)

Joelsson, T., Scholten, C. (eds.): From One Track to Many Tracks – Integrating Gender in Transport Politics. Palgrave MacMillan, London (2019)

Jones, A., Goodman, A., Roberts, H., Steinbach, R., Green, J.: Entitlement to concessionary public transport and wellbeing: a qualitative study of young people and older citizens in London, UK. Soc. Sci. Med. **91**, 202–209 (2013)

Kaufmann, V., Bergman, M.M., Joye, D.: Motility: mobility as capital. Int. J. Urban Reg. Res. **28**(4), 745–756 (2004)

Law, R.: Beyond 'women and transport': towards new geographies of gender and daily mobility. Prog. Hum. Geogr. **23**(4), 567–588 (1999)

Levin, L.: How may public transport influence the practice of everyday life among younger and older people and how may their practices influence public transport? Soc. Sci. **8**(3), 96 (2019). https://doi.org/10.3390/socsci8030096

Levin, L., Faith-Ell, C.: Women and Men in Public Consultation of Road Building Projects. Publication of selected articles from the Fourth International Conference on Women's Issues in Transportation. TRB, Washington (2011)

Levin, L., Faith-Ell, C.: Methods and tools for gender mainstreaming in Swedish transport planning. In: Proceedings from the 5th International Conference on Women's Issues in Transportation - Bridging the Gap, 14–16 April 2014. TRB, Paris, Marne-la-Vallée, France, pp. 215–223 (2014)

Levin, L., Faith-Ell, C.: How to apply gender equality goals in transport and infrastructure planning. In: Lindkvist Scholten, C., Joelsson, T. (eds.) Integrating Gender into Transport Planning: From One to Many Tracks, pp. 89–118. Palgrave Macmillan, London (2019)

Levin, L., Faith-Ell, C., Scholten, C., Aretun, Å., Halling, J., Thoresson, K.: Jämställdhetskonsekvensbedömning i regional transportplanering. [Gender impact assessment in regional transport planning]. K2 Research 2016:1, Lund (2016)

Levin, L., Gil Solá, A.: Integrating social impact assessment in local and regional transport planning (2019, ongoing project)

Levy, C.: Travel choice reframed: "deep distribution" and gender in urban transport. Environ. Urban. **25**(1), 47–63 (2013)

Litman, T., Burwell, D.: Issues in sustainable transportation. Int. J. Glob. Environ. Issues **6**(4), 331–347 (2006). http://www.vtpi.org/sus_iss.pdf

Metz, D.: Mobility of older people and their quality of life. Transp. Policy **7**, 149–152 (2000)

Mollenkopf, H., Marcellini, F., Ruoppila, I., Tacken, M.: What does it mean to get old and more immobile – and what can be improved? In: Mollenkopf, H., Marcellini, F., Ruoppila, I., Tacken, M. (eds.) Ageing and Outdoor Mobility: A European Study. IOS Press, Amsterdam (2004)

Mouffe, C.: The Return of the Political. Verso Books, London (2005)

Mouffe, C.: Agonistics. Thinking the World Politically. Verso Books, London (2013)

Papa, E., Laouwers, D.: Smart mobility: opportunity or threat to innovate places and cities (2015). http://www.westminster.ac.uk/westminsterresearch. Accessed 10 Nov 2017

Polk, M.: Gender mainstreaming in Swedish Transport Policy. In: Cresswell, T., Priya Uteng, T. (eds.) Gendered Mobilities, pp. 229–241. Ashgate, Aldershot (2008)

Rosenbloom, S.: Understanding women's and men's travel patterns: the research challenge. In: Research on Women's Issues in Transportation: Volume 1, Conference Overview and Plenary Papers, Conference Proceedings, vol. 35, pp. 7–28. National Research Council, Washington, DC (2006)

Rönnblom, M.: What's the problem? Constructions of gender equality in Swedish politics. Tidskrift för genusvetenskap 2–3, 33–55 (2011)

Rönnblom, M.: Analysing power at play: (re-)doing an analytics of the political in an era of governance. In: Hudson, C.B., Rönnblom, M., Teghtsoonian, K. (eds.) Gender, Governance and Feminist Post-Structuralist Analysis: Missing in Action?, pp 162–180. Routledge, Oxford (2017)

Schatzki, T.R.: Introduction: practice theory. In: Schatzki, T.R., Cetina, K., von Savigny, E. (eds.) The Practice Turn in Contemporary Theory, pp. 1–14. Routledge, London (2001)

Scheiner, J., Holz-Rau, C.: Gendered travel mode choice: a focus on car deficient households. J. Och Transp. Geogr. 24, 250–261 (2012)

Sharp, J., Briggs, J., Yacoub, H., Hamed, N.: Doing gender and development: understanding empowerment and local gender relations. Trans. Inst. Br. Geogr. 28, 281–295 (2003)

Sheller, M.: Automotive Emotions. Feeling the Car in Automobilities. Theory Cult. Soc. 21(4–5), 221–242 (2004)

Sheller, M.: The new mobilities paradigm for a live sociology. Curr. Sociol. Rev. 62(6), 89–811 (2014)

Shove, E., Pantzar, M., Watson, M.: The Dynamics of Social Practice: Everyday Life and How it Changes. Sage, London (2012)

Shove, E., Watson, M., Spurling, N.: Conceptualizing connections: energy demand, infrastructures and social practices. Eur. J. Soc. Theory 18(3), 274–287 (2015)

Smidfelt Rosqvist, L.: Gendered perspectives on Swedish transport policy-making: an issue for gendered sustainability too. In: Joelsson, T., Scholten, C. (eds.) From One Track to Many Tracks – Integrating Gender in Transport Politics, pp. 69–87. Palgrave MacMillan, London (2019)

Smidfelt Rosqvist, L., Wennberg, H.: Harmonizing the planning process with the national visions and plans on sustainable transport: the case of Sweden. Procedia - Soc. Behav. Sci. 48, 2374–2384 (2012)

Thoresson, K., Levin, L.: Analysis of focus group talk on gender equality, planning and transportation. Working paper. VTI and K2, Lund and Linköping (2014)

Trafikanalys.: Uppföljning av de transportpolitiska målen. [Traffic Analysis: Follow-up of the transport policy objectives]. Report 2017:7, Stockholm (2017)

UN. SDG, Sustainable Development Goals. https://sustainabledevelopment.un.org/sdgs. Accessed 01 Nov 2018

Vagland, Å.: Gender equality as a subsidiary objective of Swedish transport policy. In: Paper for the Conference for Research on Women's Transportation Issues in Chicago, 18–20 November 2004. SIKA Document 2004:2, Stockholm (2004)

Vagland, Å.: Gender equality. a subsidiary objective of Swedish transport policy. In: TRA-Transport Research Arena Europe, Göteborg, Sweden, 12th–15th June 2006. Greener, safer and smarter road transport for Europe (2006)

Watson, M.: How theories of practice can inform transition to a decarbonised transport system. J. Transp. Geogr. **24**, 488–496 (2012)

Wittbom, E.: Att spränga normer – om målstyrningsprocesser för jämställdhetsintegrering. Diss. Stockholm University (2009). http://jamda.ub.gu.se/bitstream/1/91/1/wittbom.pdf. Accessed 25 May 2018

Wittbom, E.: Jämställdhetsintegrering med styrning bortom new public management. Tidskrift för genusvetenskap **39**(2–3), 93–114 (2018)

Woodward, A.: European gender mainstreaming: promises and pitfalls of transformative policy. Rev. Policy Res. **20**(1) (2003)

Yamini J. S.: Is smart mobility also gender smart? NOS–HS workshop: Gendering Smart Mobilities in the Nordic region, Oslo, August 2017 (2017, Unpublished paper)

User Centred Design of a Knowledge Repository to Support Gender Smart Mobility

Cathleen Schöne[(⊠)], Sebastian Spundflasch, and Heidi Krömker

Technische Universität Ilmenau, Postfach 10 05 65, 98684 Ilmenau, Germany
cathleen.schoene@tu-ilmenau.de

Abstract. Gender and diversity issues are becoming increasingly important, but are still neglected in the mobility field. In order to counteract this situation, a knowledge repository is being developed with the aim of bundling and making available international gender- and diversity-specific mobility knowledge. As a result, stakeholders, such as transport planners, can call up this knowledge, incorporate it into their planning processes and implement it within the scope of their tasks. For this, however, the needs of the stakeholders must first be identified and a user-centred and task-oriented preparation of the knowledge must be conceived. This article therefore shows how internationally distributed knowledge can be made available in a user- and task-oriented manner and how an interactive and transparent flow of knowledge can be supported in the long term.

Keywords: Gender smart mobility · Knowledge repository ·
Information platform

1 Introduction

The aim of the development of the Transport Innovation Gender Observatory is to prepare internationally scattered knowledge for the consideration of both gender and diversity issues in a user-centered way for stakeholders in the field of mobility.

The term *gender* is more than just the mere allocation of biological sex, it is rather "[…] a matter of the socially and culturally shaped gender roles of women and men, which (in contrast to biological sex) are learned and thus changeable" [1]. The term *diversity*, on the other hand, refers to social diversity and different life situations. In order to be able to anchor the entire scope of the concepts of gender and diversity in the field of mobility, the specific concerns of women, for example with regard to their safety, but also of persons with special contexts of use, such as age, religion, ethnic origin and restrictions due to special needs, must be taken into account.

In some countries the turning away from thinking in exclusively traditional gender roles progresses faster than in others. Nevertheless, it is still mainly the women, who are responsible for looking after the children or caring for relatives. However, it is not only important that these particular contexts of use are taken into account, but also that women are employed in the companies themselves. This applies both at the operational level and at the decision-maker level. Only then it can be ensured in the long term that the special contexts of use can also be anchored in the mindset of the company.

© Springer Nature Switzerland AG 2019
H. Krömker (Ed.): HCII 2019, LNCS 11596, pp. 413–429, 2019.
https://doi.org/10.1007/978-3-030-22666-4_30

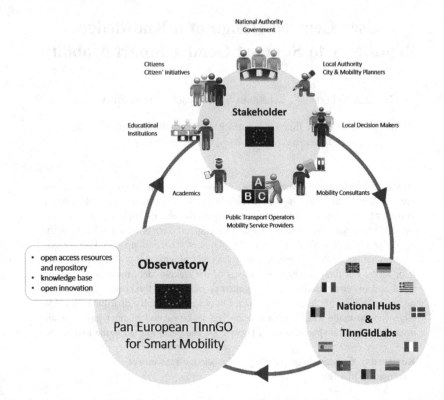

Fig. 1. Architecture of the knowledge repository with national hubs and an observatory for international stakeholder.

So they might have particular needs as mobility users, for example when travelling with a pram, but also when they are dependent on flexible working hours as (future) employees in the mobility sector.

Consequently, "the implementation of gender mainstreaming in transport means that every decision (on the planning, dimensioning and design of transport systems and facilities, on transport infrastructure, etc.) is examined to see whether it hinders or promotes equal opportunities for women and men" [1]. In order to advance the implementation, the international knowledge on gender and diversity in the field of mobility is to be collected, processed and made available in the knowledge repository.

This knowledge management takes place within "[…] a knowledge repository which handles the organization's meta-knowledge describing knowledge elements that come from a variety of sources with the help of meta-data for a number of dimensions, e.g., person, time, topic, location, process, type. A taxonomy, a knowledge structure or an ontology help to meaningfully organize and link the knowledge elements and are used to analyse the semantics of the organizational knowledge base" [2]. Typical functions are inherent in this knowledge management:

- knowledge search
- knowledge presentation

- knowledge publication, structuring and linking
- knowledge acquisition
- knowledge communication and cooperation
- [...] training and [...] learning [2]

Typical stakeholders, as shown in Fig. 1, who can influence the improvement within the mobility sector in this respect are:

- *National Authority/Government* who are responsible for the strategic planning of infrastructure at a national level.
- *Local Authority – City and Mobility Planners* who are responsible for the strategic planning of local infrastructure.
- *Local Decision Makers* who represent the interests of the citizens and mould the planning process regarding e.g. public transport.
- *Mobility Consultants* who organize calls for tenders and consult other stakeholders on how to exploit the potential of new forms of mobility.
- *Public Transport Operators and Mobility Service Providers* who develop and improve their services to satisfy the user's needs.
- *Academics* who teach and conduct research in the field of mobility.
- *Educational Institutions* that train and further educate staff, actors and interested parties.
- *Citizens/Citizen' Initiatives* who make use of their democratic rights and address specific issues.

Each stakeholder pursues a specific task, such as planning a smart station or designing mobility services, which must incorporate the knowledge required to take gender and diversity into account. Therefor the required knowledge needs to be collected, prepared and distributed through the repository.

As it is shown in Fig. 1 this process is being promoted by the stakeholders already mentioned. They collect knowledge in the tension field of gender and mobility from their countries, which would otherwise remain invisible to other stakeholders from different countries. This knowledge forms the basis for the work of the national Hubs and TInnGIdLabs (Transport Innovation Gender Idea Labs). The Hubs analyse the state of the art as well as the accumulated knowledge. They apply qualitative, quantitative and design research methods, combining hands-on knowledge, concrete actions and best practices. Based on this information the associated TInnGIdLabs will engage in training, dissemination and citizen engagement activities (such as smart labs, ideas factories and living labs) using multimedia resources and participatory design techniques to develop gender and diversity sensitive smart mobility and solutions.

The presented knowledge is maintained in the observatory at everyone's disposal. This is also the great challenge for human-technology interaction. The Observatory must be designed with a view to the various stakeholders and their associated tasks. Regarding professional and general interest, the knowledge must be explorable through different access points. However, in order to guarantee the sustainability of the platform, the process does not end with the provision of knowledge. Rather, it is a cycle, in which the stakeholders maintain and update the collected knowledge and disseminate the platform beyond the project time and boundaries, so that new knowledge always finds its way into the platform.

The research question therefore is:

How can existing internationally scattered knowledge be prepared for the different profiles of stakeholders and their individual tasks and information interests?

The knowledge structuring and presentation on the user interface should be conducted user-centered from the beginning. "A human-centered design process is one that involves human feedback at all stages of the design process starting from the identification of the need for a DSS and the gathering of user requirements, all the way though testing and evaluation of the final products. Human-centered design processes take into account human skills, needs and limitations" [3].

In the context of this user-centered development process, methods from various disciplines must be used and supplemented by aspects of knowledge management.

1. For the definition of user profiles and tasks, the established methods of the HCI can be used.
2. For the knowledge structuring an appropriate model for knowledge structuring must be found. Different access points and links between the knowledge elements as well as a balance between overview and details must be determined, so that the stakeholder-specific gender action goals can be served.
3. For the visualization and interaction with this knowledge structure the access points in connection with the stakeholder-specific gender action goals must be prepared within the framework of the platform, so that the highest possible utility, usability and a good user experience (UX) can be guaranteed.
4. For the sustainability the UX is particularly important so that individual users remain motivated to use the platform and provide knowledge, throughout the already mentioned desired knowledge cycle, in which not only new knowledge is generated, but also information is constantly updated, networked, shared and used.

Within the scope of this empirical study, HCI methods and knowledge management methods were combined to prepare knowledge. This enabled a definition of user profiles and tasks as well as a determination of a basic knowledge structure, as stated in the results.

2 Background

User-centric development processes exist, e.g. the model of the interdependence of human-centered design activities as it is shown in DIN EN ISO 9241-210 [4], but they do not address the specific challenges of access points to the knowledge repository.

Knowledge repositories, on the other hand, are primarily discussed in the context of knowledge management. Figure 2 shows typical modules of knowledge management, but here the access points must be defined in a user- and task-oriented way in interaction with the user-centered development process.

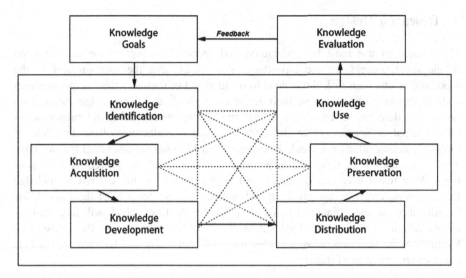

Fig. 2. Modules of knowledge management (translated in accordance with Probst 2012) [5]

The definition of the modules can be integrated into the user-centered development process in the following way:

- Knowledge Goals can be derived from the specific taskflows of the stakeholders. With which information interest do they access the platform?
- Knowledge identification is closely related to transparency. The knowledge environment regarding gender and diversity in the field of mobility research must be analysed. This creates the references to the content preparation and design of the user interface.
- Knowledge acquisition comprises the user-centred preparation of the required knowledge, which can only be acquired through international and/or cross-target group exchange.
- Knowledge Development implies the conception of new knowledge as well as the handling of new knowledge.
- Knowledge Distribution deals with the interest-driven transfer of knowledge within and across stakeholder groups. "Not everything has to be known by everyone" [5].
- Knowledge Preservation ensures the sustainability of the platform. How can knowledge be stored, updated and maintained?
- Knowledge Use describes the productive use of the knowledge gained within stakeholder task flows.
- Knowledge Evaluation is closely related to the typical cycles of action in the field of HCI, in which the result is always evaluated with regard to the goal.

A holistic strategy must be found to integrate the elements of knowledge management into the user-centered development process. At the same time, the boundaries of the individual companies and institutions must be broken down and the needs of all stakeholders involved must be taken into account when developing the knowledge repository.

3   Research Design

The context of use must be understood and specified as part of the user-centered development process for the repository. This means that the user profiles of the stakeholders and their task flows need to be analyzed in regard to their responsibilities and competences. Furthermore their Knowledge Goals and Knowledge Acquisition needs to be determined. To do so a workshop with experts consisting of representatives of municipal authorities, research institutions and manufacturers from the field of mobility, among others was held. 30 experts from 13 countries discussed for two days which responsibilities, tasks and information interests should be served by the repository. As a result user and organizational requirements, i.e. the desired Knowledge Development, could be specified. In addition, during the conduct of the Knowledge Identification, access points to knowledge could be identified which will flow into the design solution process (see Tables 2, 3, 4, 5, 6, 7, 8 and 9). For the Knowledge Evaluation, the Knowledge Use or rather the fulfilment of the user Knowledge Goals is the decisive criterion (Table 1).

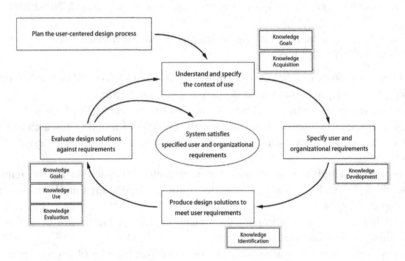

Fig. 3. The modules of knowledge management and their assignment in the user-centric development process for the repository [based on 4]

Table 1. The modules of knowledge management and their assignment to the phases of the user-centric development process for the repository

Phase	Requirements
Context of use	User profiles (Stakeholder) and Task flows Knowledge Goals and Acquisition
Organizational requirements	Knowledge Development
Design solutions	Knowledge Identification
Evaluation	Knowledge Goals, Use and Evaluation

Table 2. Context of use of the stakeholder *National Authority/Government*

	NATIONAL AUTHORITY/ GOVERNMENT
CONTEXT OF USE RESPONSIBILITIES & COMPETENCES	• strategic planning and coordination of mobility policies at national level • cooperation with government leaders, ministries, associations and institutions at international level • ability to grasp complex knowledge quickly and to map it to one's own problem • high level of knowledge based on a high level of education Regarding gender issues, this means e.g. → integrating gender issues, from the highest level, into national legislation or policies → creating awareness within society
TASKFLOW	• call for tenders for funding projects • allocation of funds • shaping of law and guidelines • preparation and dissemination of information material • responsibility in their function as employer (e.g. personnel development, workplace design)
KNOWLEDGE GOALS	Regarding to the repository, they pursue the following knowledge goals: → basic theoretical concept of gender and diversity → comprehensive overview of fields of action and concrete activities → research findings TK → studies, field reports EK → legal framework conditions in other countries RC

Table 3. Context of use of the stakeholder *Local Authority (City & Mobility Planners)*

	LOCAL AUTHORITY (CITY & MOBILITY PLANNERS)
CONTEXT OF USE RESPONSIBILITIES & COMPETENCES	• strategic planning and coordination of mobility policies at local level • cooperation with citizens, local councils, mayors and local politicians • ability to grasp complex knowledge quickly and to map it to one's own problem • high level of knowledge based on a high level of education Regarding gender issues, this means e.g. → implementing government policies at local level → creating awareness within society
TASKFLOW	• preparation of local gender neutral policies and strategies for mobility (bicycle plan, public transport plan etc.) • allocation of contracts, studies, expert opinions, consideration of gender issues • preparation of Roadmaps • provision of data / statistics (data collection methods, tools, data sets – how to collect gender diversity data) • role model in their function as an employer • planning and implementation of strategies and concrete measures in various fields of action (e.g. set up measurable targets, ensure equal access to transport regarding employment, data collection and monitoring)
KNOWLEDGE GOALS	Regarding to the repository, they pursue the following knowledge goals: → application-ready knowledge and field reports → ideas and Good Practice examples and thoughts on transferability → facts on concrete problems in a concrete field of action → practical tool EK
	→ legal framework conditions in the own country → repository of laws & regulation RC

Table 4. Context of use of the stakeholder *Local Decision Makers (City Council)*

	LOCAL DECISION MAKERS (CITY COUNCIL)
CONTEXT OF USE RESPONSIBILITIES & COMPETENCES	• regional planning, development planning, general traffic planning, development plans, building regulations, building construction, civil engineering, street lighting, green areas, road traffic issues, traffic promotion, urban redevelopment, public transport • representation of the interests of the citizen in an honorary capacity • few resources to deeply familiarize themselves with the technical side of the business • different levels of education (from master craftsman to computer scientist with a doctorate) Regarding gender issues, this means e.g. → representing the interests of society/ citizens towards the local authority → implementing measures and strategies requires the approval of the City Council
TASKFLOW	• quick understanding and evaluation of problems and problem areas in order to meet decision on appropriateness of strategies and concrete measures • request for expert opinions and set up advisory committees • establishment of basic requirements for the consideration of gender and diversity aspects
KNOWLEDGE GOALS	Regarding to the repository, they pursue the following knowledge goals: → short and comprehensible information on specific topics → summaries and key messages → reports of experience `TK` → relevant Good Practice examples `EK` → contact information of experts `CI`

Table 5. Context of use of the stakeholder *Mobility Consultants*

	MOBILITY CONSULTANTS
CONTEXT OF USE RESPONSIBILITIES & COMPETENCES	• development of concepts on how national, local authorities and local decision makers can exploit the potential of new forms of mobility (e.g. Connected Car) on the basis of the latest knowledge • ability to quickly grasp complex knowledge and map it to one's own problem • high level of knowledge based on a high level of training Regarding gender issues, this means e.g. → consulting for local authorities, providers of mobility services, public transport operators and companies on mobility questions → raising awareness of gender and diversity and consider it in proposed solutions and planning processes
TASKFLOW	• conduction of studies and investigations (feasibility studies, cost-benefit analysis, social impact assessment) • utterance of policy recommendations
KNOWLEDGE GOALS	Regarding to the repository, they pursue the following knowledge goals:
	→ ideas and Good Practice examples
	→ experience reports
	→ concentrated knowledge about research findings
	→ Guidelines `TK`
	→ legal framework conditions `RC`
	→ contact information of experts `CI`

Table 6. Context of use of the stakeholder *Public Transport Operators & Mobility Service Providers*

	PUBLIC TRANSPORT OPERATORS & MOBILITY SERVICE PROVIDERS
CONTEXT OF USE RESPONSIBILITIES & COMPETENCES	• preparation of and participation in strategic planning concepts internally and externally • network expansion and line network design, updating of urban framework planning such as land use plan, urban development plan for transport • route identification and optimisation, determination of stop locations, securing trafficability • ability to quickly grasp complex knowledge and map it to one's own problem • a high level of knowledge based on a high level of education in the field of transport and civil engineering Regarding gender issues, this means e.g. → implementing gender-sensitive measures in transport companies
TASKFLOW	• user-centered design and operation of mobility and service offerings • implementation of safety and security standards • further development of mobility services with a focus on accessibility • responsibility in the role as employer, improve working conditions, design of workplaces and work tasks • participation in tenders • customer acquisition
KNOWLEDGE GOALS	Regarding to the repository, they pursue the following knowledge goals:
	→ studies, data, numbers
	→ knowledge of the diversity of users (Personas und typical Scenarios) to better address their services TK
	→ relevant Good Practice examples
	→ work experiences from other providers and operators EK
	→ regulatory framework RC

Table 7. Context of use of the stakeholder *Academics*

	ACADEMICS
CONTEXT OF USE RESPONSIBILITIES & COMPETENCES	• transport economics, safety aspects • statistical recording and evaluation of traffic data • ability to generate new knowledge on the topic of gender in the context of research • high expertise Regarding gender issues, this means e.g. → raising awareness of gender and diversity within the academic field and society → disseminating gender and diversity concepts → researching and developing
TASKFLOW	• application for research projects • work on various research projects in the field of mobility • conduction of studies • addressing gender aspects in university teaching
KNOWLEDGE GOALS	Regarding to the repository, they pursue the following knowledge goals:
	→ gender assessment methods, surveys, checklists
	→ theoretical background knowledge, models
	→ guidelines TK
	→ ideas and Good Practice examples
	→ experience reports
	→ Good Practice examples EK
	→ legal framework conditions RC
	→ contact information of experts CI
	→ daily news and information on current activities
	→ events and conferences UD

Table 8. Context of use of the stakeholder *Educational Institutions*

	EDUCATIONAL INSTITUTIONS
CONTEXT OF USE **RESPONSIBILITIES & COMPETENCES**	• teachers from the training and further education sector in the transport and logistics sector • ability to professionally integrate new knowledge into training plans • high level of knowledge based on training and further education concepts Regarding gender issues, this means e.g. → creating awareness of gender and diversity issues within society (key role)
TASKFLOW	• management of educational campaigns • training of professionals • influence on the vocational orientation of young people • shaping educational environment (accessibility, safety, services) • knowledge teaching with consideration of the gender perspective • transport from/to school • gender-appropriate design of teaching materials
KNOWLEDGE GOALS	Regarding to the repository, they pursue the following knowledge goals:
	→ theoretical gender concepts
	→ guidelines
	→ information and materials for trainers and speakers
	→ material supporting gender and transport education TK
	→ Good Practice examples EK

Table 9. Context of use of the stakeholder *Citizens/Citizen's Initiatives*

	CITIZENS/ CITIZEN`S INIATIVES
CONTEXT OF USE RESPONSIBILITIES & COMPETENCES	• political commitment of citizens is an important component in a democratic state system • represent their own interests on their own initiative • few resources to deeply familiarize themselves with the technical side of the business • different education level Regarding gender issues, this means e.g. → pointing out gender and diversity issues within society → pushing forward the integration process → influencing public opinion, state institutions, political parties or other social groups in order to counteract a planning that develops past the user
TASKFLOW	• user of public transport – planning, departure, entry, exit and transfer, ride, arrival • communication with service and policy providers
KNOWLEDGE GOALS	Regarding to the repository, they pursue the following knowledge goals: → access to practical info → general, short and comprehensible information material → material and information on how to influence policy TK → communication channels, citizen panels CI

Within the framework of the analysis of the context of use, the knowledge goals and access points of the international users for knowledge identification could be localised from the task flows.

The access points can be classified into the following categories:

Theoretical knowledge (TK)	ranging from quick overview information to well-founded theoretical concepts. Guidelines and checklists were also in demand for the rapid operationalisation of research knowledge.
Experience knowledge (EK)	which can be drawn from the large number of existing projects, was attributed special relevance. Extraction of Good Practice was particularly in demand, as were experience reports from the wealth and variety of projects.
Country-specific regulatory conditions (RC)	that play an important role in the implementation of gender issues.
Communication information (CI)	such as the exchange of contact data for personal discussions with experts but also with citizen panels are requested.
Up-to-date information (UD)	is requested, especially with regard to ongoing projects and planned events and conferences.

As Table 10 shows, there is an interest in theoretical knowledge across almost all stakeholder groups. This includes above all information material, guidelines, surveys or similar sources, which are intended to contribute to a better understanding of the underlying gender and diversity concept. Furthermore, this knowledge offers the possibility to derive the public perception and to identify implementation potentials. There is also a demand for experience knowledge, which primarily includes good practice examples. An exchange of experience across target groups and national borders should open up new perspectives and have a groundbreaking effect. For both categories, the restriction that this knowledge is mostly held within one's own ranks and therefore remains invisible to other mobility actors and/or countries has applied so far. Country-specific regulatory conditions e.g. legal frameworks are interesting to the extent that they reveal limitations, but also adaptation points, which would hinder or facilitate the implementation of gender mainstreaming in the field of mobility. In addition, some countries can learn from the progressiveness of other countries. Communication information includes above all contact data which, if made available in a

Table 10. Information interest of stakeholder groups

Stakeholder - Groups	TK	EK	RC	CI	UD
National Authority/ Government					
Local Authority (City & Mobility Planners)					
Local Decision Makers (City Council)					
Mobility Consultants					
Public Transport Operators & Mobility Service Providers					
Academics					
Educational Institutions					
Citizens/ Citizen's Initiatives					

lively manner, is representative of a well-functioning exchange of knowledge. The question of who can make a certain knowledge available within which research or work area is essential. In the course of this, the demand for up-to-date information naturally also plays an important role. This involves the communication of dates such as planned events, conferences, but also project-related start and end times or milestones.

Even if a stakeholder has not explicitly stated a need for information in all extracted areas, it can nevertheless be assumed that there is also a fundamental interest in these areas, but by no means the same weighting. Rather, it can be assumed that there is a hierarchy of interests, which is mostly based on their task flows.

Thanks to these evaluated knowledge goals, access points could be extracted that can serve as the basis for organizing access to the repository.

4 Results

The main objective of the investigation was to ensure the utility of the repository by defining knowledge goals and access points for knowledge identification.

For the development of the repository, the method of user-centered design was extended in such a way that modules of knowledge management were assigned to the individual phases.

In the analysis of the usage context, particular emphasis was placed on the target-group-specific collection of the task flow with regard to knowledge goals and access points to the knowledge for gender-specific questions.

This made it possible to extract the essential categories of theoretical knowledge, empirical knowledge, regulatory conditions, communication information and current information for the preparation of content.

In further steps, it must be determined how knowledge can be acquired, developed, distributed and preserved in the context of the repository.

For the concrete use of knowledge, the access points must be refined in content, networked and visually prepared depending on the workflow.

References

1. Krause, J.: Gender Mainstreaming im Verkehrswesen - Einführung, Folie 2 "gender" Folie 3 "gender mainstreaming". AG 1 Verkehrsplanung, Forschungsgesellschaft für Straßen- und Verkehrswesen e. V. (FGSV). https://www.fgsv.de/fileadmin/road_maps/GM_Einfuehrung. pdf. Accessed 01 Mar 2019
2. Maier, R.: Knowledge Management Systems: Information and Communication Technologies for Knowledge Management, 2nd edn. Springer, Heidelberg (2004). https://doi.org/10.1007/ 978-3-540-24779-1
3. Jacko, J.: The Human-Computer Interaction Handbook: Fundamentals, Evolving Technologies and Emerging Applications, 3rd edn. CRC Press, Boca Raton (2012)
4. DIN EN ISO 9241-210:2010, Ergonomics of human-system interaction – Part 210: Human-centered design for interactive systems (2010)
5. Probst, G.: Wissen managen: Wie Unternehmen ihre wertvollste Ressource optimal nutzen, 7th edn. Springer Gabler, Wiesbaden (2012). https://doi.org/10.1007/978-3-8349-4563-1

Towards an Integrated Mobility Service Network

Cindy Mayas[⊠], Tobias Steinert, and Heidi Krömker

Technische Universität Ilmenau, Ilmenau, Germany
{cindy.mayas,tobias.steinert,
heidi.kroemker}@tu-ilmenau.de

Abstract. The approach of a linear mobility chain changes to a network-based mobility. Mobility not only consists of interactions with transport services. The transport services are enhanced by a variety of services concerning ticketing, information, refreshments, or entertainment. These services are used in parallel with transport and form an individual service network of travelers' journeys. This paper introduces the method of mobility service diaries to collect data about the usage of diverse mobility services. Based on the data analysis, the results present a classification of mobility services, typical patterns of the interplay of mobility services, and a framework of an integrated mobility service network. The aim of this research is to support the user-oriented development of human-computer interactions for mobility services.

Keywords: Mobility service · Mobility experience · Mobility service diary

1 Introduction

A major trend in the mobility market is the shift from private ownership of mobility vehicles to the use of mobility services. Besides to the rather traditional sector of public transport sector, other mobility services from multiple mobility providers are available, such as sharing services for cars and bikes, long distance bus services or self-driving car services [1]. This development changes the approach of a linear mobility chain of transport services to a network-based mobility, which creates a holistic mobility experience of the traveler [2]. The state of the art of these current results in mobility research are presented in Sect. 2. Based on these trends, the aim of this paper is to systematically describe this network-based mobility. Therefore, the method of mobility diaries is adapted to mobility service diaries and tested in an empirical pre-study, which is described in Sect. 3. Resulting from this data, the derived framework is presented in Sect. 4. The framework consists of a suggestion of standardized mobility service categories and touchpoints, typical mobility service pattern and a holistic description of an integrated mobility service network.

© Springer Nature Switzerland AG 2019
H. Krömker (Ed.): HCII 2019, LNCS 11596, pp. 430–440, 2019.
https://doi.org/10.1007/978-3-030-22666-4_31

2 State of the Art

2.1 Mobility Services Chain

According to the European Standard "Public Transport Reference Data Model" [3] a trip consists of the "complete movement of a passenger (or another person, e.g. driver) from one place of any sort to another". This approach focuses on the traveler's change of location. This change of location is caused by a typical sequence of traveler's actions, which are represented in the travel chain [4]. In the travel chain, the tasks are structured in terms of content and place from the user's point of view. Figure 1 represents the linear travel chain starting with the preparation activities such as planning and booking. Furthermore the travel chain includes the ways to the stopping point as well as the orientation and waiting times at the stopping point. After travelling with the public vehicle and alighting, the journey might include either a further public transport trip or ends with the arrival at the destination.

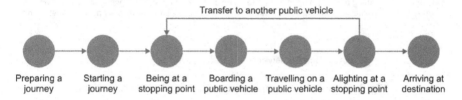

Fig. 1. Linear travel chain in public transport [5]

The travel chain for public transport can be transferred into a travel chain for mobility services in general (see Fig. 2) by replacing public vehicles to mobility services and replacing the stopping points by mobility points, which are any locations, where the use of a mobility service can be started or terminated. Therefore, mobility points include stops for several mobility providers, for example bus stops, train stations, sharing stations and any parking places for free-floating services. Mobility points may also consist of combinations of more than one mobility provider.

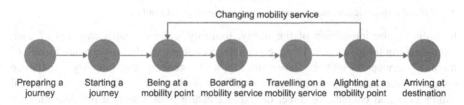

Fig. 2. Linear travel chain for transport mobility services [2]

The term of "mobility service" can be even applied more general. When providing a service, the customer is at the same time a part of the service provision [6]. This is not only the case for the service of transport, but also for other services along a journey, e.g. information on an individual travel request on a website is also a service. Other services that are often used during a journey include the sale of meals or drinks as well entertainment services. Furthermore, a lot of services have evolved, which support the trip with private-owned vehicles, e.g. parking service or e-mobility infrastructure. From that point of view, all mobility services, both transport services and extended services, support the tasks of the users and are orchestrated along the travel chain, as shown in Fig. 3. Some of the extended mobility services are independent from specific locations at the journey, such as information and navigation providing websites and mobile applications. Instead of location, time and individual restrictions became further criteria of provided mobility service touchpoints.

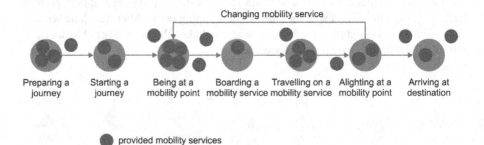

Fig. 3. Travel chain for transport and extended mobility services

2.2 Mobility Experience

Customer experience within the mobility context can be analyzed in different terms, such as travel experience [7] or mobility experience [2].

Based on a literature review Wienken [2] mentions the following key elements mobility experience:

- Mobility experience is the sum of all experiences of the user with the provider and his mobility services.
- Mobility experience is generated directly and indirectly at all touchpoints between provider and user and over the entire duration of the relationship.
- Mobility experience has a holistic and subjective character.

In addition, the interaction of the utility, usability and user experience [8] of each individual mobility service is an essential component of the conceptual model of mobility services, which is shown in Fig. 4. In general, the mobility experience comprises the perceptions and reactions of a mobility user resulting from all actual and/or expected contacts of direct or indirect nature with a service provider along a journey.

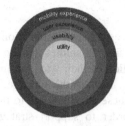

Fig. 4. Conceptual model of mobility experience [2]

From this point of view, each mobility service presents a possible touchpoint for mobility experience during a journey (see Fig. 5). These touchpoints include not only transport services but also extended mobility services, which are supposed to influence the mobility experience in addition to the transport services [2].

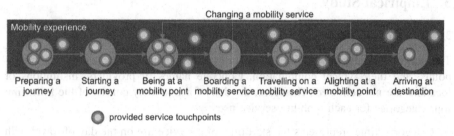

Fig. 5. Mobility services as mobility experience touchpoints along the travel chain

This variety of mobility services creates an unmanageable number of possibilities for the individual combination of mobility services, within the network of mobility service. Travelers select only some mobility services, according to location, time and personal restrictions, along an individual journey, which creates the mobility experience, shown in Fig. 6.

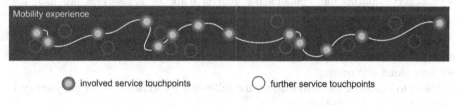

Fig. 6. Individual mobility service chain

Therefore, the challenges of the user-oriented development of mobility services are the integration of diverse mobility services into an individual journey and the conception of user-oriented supporting tools. In order to address these challenges, an empirical data base about the usage of mobility services is required.

2.3 Mobility Diaries

National travel surveys are an established international method for the analysis of mobility [9–13]. Next to an inquiry, all studies include the method of mobility diaries for the participants' documentation of their ways of one or more days. The key categories of these mobility diaries include the purpose, the length and the means of transport per way. The aim of these studies refers to the long-term quantification and development of mobility, in order to support strategic decisions for the mobility infrastructure and mobility market. Therefore, the modal split, which represents the share of each means of transport by volume of the national mobility, is a key result of mobility diaries. In relation to the demographic data about users and household even mobility user types may be derived [14]. In contrast, data about the use of extended mobility services is not included. Hence, there is a lack of data basis for a systematically analysis of the interdependencies of the mobility service network.

3 Empirical Study

3.1 Method of Mobility Service Diaries

The aim of this mobility service diary study is to identify service categories, touchpoints, and their typical combination. Concrete names of providers, institutions or locations are not collected. Therefore, the diary for a journey consist of the following four categories for each mobility service use:

- Category "time" represents the start time of a service use on the day of travel with an accuracy of about 5 min. Services which are related to the journey and were used before or after the day of travel, e.g. travel planning or billing, are marked as "before the day of travel" or "after the day of travel".
- Category "location type" documents the generalized description of the place of service use e.g. at home, bus stop, train station, airport, sharing station, etc. without specifying the place.
- Category "service type" means the general extent of the service used e.g. information, transport, food, drinks, hygiene, or entertainment without further details.
- Category "touchpoint" contains the general name of the service provider, e.g. restaurant, bus, sanitary facilities or websites.

According to the explorative character of the study the category input is not a standardized selection but open input fields, in order to find new indicators for mobility services along a journey.

Next to demographic information, the following information about the journey is collected in standardized items:

- "Length": less than 15 min/15 min up to less than 60 min/1 h up to less than 3 h/3 h up to less than 6 h/6 h and more
- "Motivation" (multiple responses possible): work/shopping and errands/free time/vacation/other

- "Companions" (multiple responses possible): one or two persons (older than 12 years)/more than two persons (older than 12 years)/children (less than 12 years)/ none
- "Familiarity": I'm familiar with the whole journey./I'm familiar with the majority of the journey./I'm not familiar with the majority of the journey./I'm not familiar with the whole journey.

3.2 Conducting the Survey

The participation at the study is voluntary and is not related to any financial or material rewards. After the acquisition, the mobility service diaries are handed over to 10 participants. Finally, 6 participants took part in the survey. 4 participants are male, 2 participants are female, and all of them are in the age between 20 and 40 years. Each participant documented one up to three journeys in a paper-based form within a range of one week.

The collected data of these participants contains 135 uses of mobility services within 12 journeys. The documented journeys include short- and long-distance journeys, journey with and without companions, journeys for work, free time, and vacation, as well as familiar and unfamiliar journeys.

4   Results

4.1   Classification of Mobility Services

In a first step, the data of "service type" is grouped by similar terms. This grouping revealed 15 different service types. In a second step, these service types are categorized by their relation to a journey. This systematization revealed the following five general categories of service types:

- "Transport services" intent the movement of a person from one place to another, e.g. bus, train, plane, or ship as well as elevators and escalators.
- "Transport-related services" intent the movement of luggage, additional help for personal movement, or preparing private or shared means of transport, e.g. refueling.
- "Ticketing services" include the sale, validation and further management of reservations and tickets.
- "Information services" provide requested individualized data about the planned or current itinerary, e.g. mobile applications or information desks.
- "Comfort services" cover all services to ensure viability and well-being of the traveler, e.g. food and drinks, hygiene, or entertainment services.

As shown in Fig. 7, the information service is the most used mobility service. Due to the fact, that individual travel information services are ubiquitous via mobile applications, this service becomes a high relevance for all travelers. The next important services are transport services and comfort services, which are used often especially in long-term journeys.

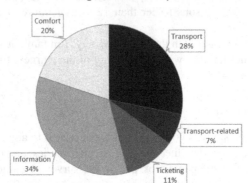

Fig. 7. Share of mobility service categories in the mobility service diaries

Furthermore, the data about "location types" is also analyzed by grouping and systematization. The 14 grouped locations are related to the following three categories of location types:

- "Whereabouts before or after a journey", e.g. home, office, hotel, or other destinations.
- "Mobility points", e.g. stop, station, airport, harbor, or parking places.
- "Means of transport", e.g. bus, train, plane, or ship.

4.2 Mobility Service Pattern

According to the analyzed categories, the following analysis assigns each documented mobility service to a time-location-map, grouped by service type. The map, which is shown in Fig. 8, reveals the following five pattern of mobility services:

- Pattern A - "Planning": Before the start of the journey, there is a close relationship between information and ticketing services. Both service types are often used in a direct sequence. Nevertheless, due to technical restrictions, many participants had to use several touchpoints for the information and ticketing process.
- Pattern B – "Starting": The participants documented different services at the start of their journey. Using a transport service, using a service at a mobility station, and using information services at the whereabouts before the journey start. This information service is the most documented starting service of journeys, which correlates to the high information needs of travelers especially at the start of unfamiliar journeys.
- Pattern C – "Waiting": When changing the means of transport, travelers often have to wait for a while at mobility stations for public transport. The data shows, that especially after the first trip section many travelers use several comfort services at the mobility station. This use of a comfort service is often in a direct sequence to the use of an information, probably to check, if the remaining waiting time is sufficient for the comfort use.

- Pattern D "Long-distance traveling": When passengers travel with one means of transport for hours, the use of comfort services rises. In contrast to pattern C, this use is not that much related to further information services, because the duration of the travel time in a vehicle is less dynamic.
- Pattern E "Continuous information": The need for individualized information remains high along the whole journey, independently from the duration of the travel. A continuous information supply therefore might become a key element for mobility experience.

These analyzed categories and pattern provide important indicators for the user-oriented development and improvement of new and existing mobility services.

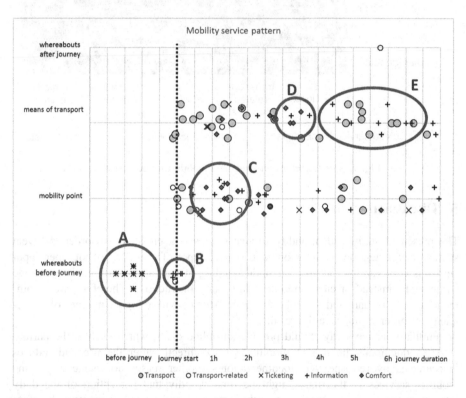

Fig. 8. Overview of mobility services according to location type, service type, and travel time

4.3 Mobility Service Network

The approach of mobility experience requires a holistic view of interplay of touch-points and mobility services. The analysis shows, that typical combinations of service types build the basis of the traveler interactions. Nevertheless, all service types should be considered at the different location types and be provided to the traveler, in order to improve the travelers' individual mobility experience. Figure 9 presents an overview of the derived mobility service network with the focus on creating individual mobility experience.

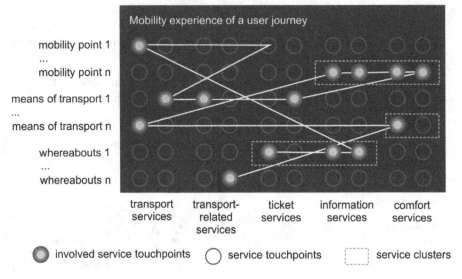

Fig. 9. Framework of an integrated mobility service network

5   Discussion

The presented study with mobility service diaries is a pre-study, in order to reveal standardized categories of locations and service types. The quality of the open-input entries is very varying, because the understanding of the variables differs between the participants, instead of clear and exemplary diary instructions. Therefore, we recommend to use standardized variables in further studies with a higher number of participants, in order to improve the results.

Furthermore the study should provide an online diary, which enables the participants to document the use of mobility service directly after the use and reduces inaccuracies by retrospective documentation. Another reason for neglecting some mobility services in the diary might be limited individual perception of the daily interactions as a mobility service. A further sensitization of the participants is therefore recommended.

The further research of mobility service should also analyze the perceived mobility experience of the documented journeys. Correlation between the use of mobility service types and the mobility experience might reveal influence factors for improvement by the service providers.

6 Conclusion

The presented method of mobility service diaries is appropriate to detect categories of mobility services and locations. The analyzed mobility service pattern reveal challenges for the user-oriented combination of mobility services, such as individual information and ticketing above all providers of a journey. The results will be integrated in research projects for development of a service-oriented mobility platform, e.g. DIMO-OMP[1], and consider the revealed typical mobility service pattern.

Further studies with mobility service diaries and a higher number of participants can be used, in order to detect lacks of mobility services and improve the mobility infrastructure in general.

Acknowledgements. Part of this work was funded by the German Federal Ministry of Transport and Digital Infrastructure (BMVI) grant number 19E16007F within the project DIMO-OMP.

References

1. Verband Deutscher Verkehrsunternehmen e. V. (eds.): Mobilitätsdienstleistungen. In: Mobi-Wissen. Busse und Bahnen von A-Z. http://www.mobi-wissen.de/Verkehr/Mo-bilit%C3% A4tsdienstleistungen. Accessed 01 Mar 2019
2. Wienken, T., Krömker, H.: Designing for mobility experience - towards an understanding. In: Stopka, U. (eds.) Mobilität & Kommunikation. Winterwork, Borsdorf (2018)
3. CEN EN 12896-1:2016: Public transport - Reference data model - Part 1: Common concepts (2016)
4. Verband Deutscher Verkehrsunternehmen (VDV), VDV-Förderkreis (Hrsg.): Telematik im ÖPNV in Deutschland. Alba Fachverlag, Düsseldorf (2001)
5. Hörold, S.: Instrumentarium zur Qualitätsevaluation von Mobilitätsinformationen. Springer, Wiesbaden (2016). https://doi.org/10.1007/978-3-658-15458-5
6. DIN EN ISO 9000:2005 Qualitätsmanagementsysteme – Grundlagen und Begriffe (2005)
7. Carreira, R., Patrício, L., Jorge, R.N., Magee, C., Hommes, Q.V.E.: Towards a holistic approach to the travel experience: a qualitative study of bus transportation. Transp. Policy **25**, 233–243 (2013)
8. DIN EN ISO 9241-210:2010: Ergonomics of human-system interaction – Part 210: Human-centered design for interactive systems (2010)
9. Department for Transport (eds.): Statistical release: National Travel survey England. https:// assets.publishing.service.gov.uk/government/uploads/system/uploads/attachment_data/file/ 729521/national-travel-survey-2017.pdf. Accessed 01 Mar 2019
10. Institut national de la statistique et des études économiques (eds.): Enquête mobilité des personnes : une enquête sur les déplacements des personnes et leurs modes de transport. https://www.insee.fr/fr/information/3365007. Accessed 01 Mar 2019
11. Ecke, L., Chlond, B., Magdolen, M., Eisenmann, C., Hilgert, T., Vortisch, P.: Deutsches Mobilitätspanel (MOP) – Wissenschaftliche Begleitung und Auswertungen, Bericht 2017/2018: Alltagsmobilität und Fahrleistung. https://daten.clearingstelle-verkehr.de/192/ 162/Bericht_MOP_17_18.pdf. Accessed 01 Mar 2019

---

[1] www.dimo-omp.de.

12. Infas Institut für angewandte Sozialwissenschaft GmbH (eds.): Mobilität in Deutschland. Kurzreport. http://www.mobilitaet-in-deutschland.de/pdf/infas_Mobilitaet_in_Deutschland_2017_Kurzreport.pdf. Accessed 01 Mar 2019
13. Federal Highway Administration: National Household Travel Survey. https://nhts.ornl.gov/. Accessed 01 Mar 2019
14. Kurosu, M. (ed.): HCI 2016. LNCS, vol. 9733. Springer, Cham (2016). https://doi.org/10.1007/978-3-319-39513-5

Mobility-as-a-Service (MaaS) Testbed as an Integrated Approach for New Mobility - A Living Lab Case Study in Singapore

Zhanhe Ryan Jin[1(✉)] and Anna Zhi Qiu[2(✉)]

[1] Energy Research Institute at NTU (ERI@N), Nanyang Technological University, Singapore 637553, Singapore
RyanJin@ntu.edu.sg
[2] mobilityX Pte. Ltd., 61 Ubi Ave 1, Singapore 408941, Singapore
Qiu@mobility-x.com

Abstract. Transportation plays a key role in our daily life. General public perception of freedom of mobility has been around ownership of private cars, resulting to traffic congestion and air pollution in many cities. In addition, with increasing demands for personalised transportation services, the traditional transport services provider are not able to cater to this demands citing challenges in market fragmentation, payment and integration technology.

The jalan[2] app is an integrated Mobility-as-a-Service (MaaS) solution developed by NTU, JTC and SMRT, to improve commute experience by seamlessly integrating public and private transport services with next-generation transport options including autonomous vehicles, on-demand ride sharing and Personal Mobility Devices sharing.

NTU as a Living Lab ecosystem, the jalan[2] app allow users to personalise their transport experience on NTU campus based on preference such as cost, time, transport modes or carbon footprint. The app will also crowd-sourced data for transportation modelling, smoothening of peak demand and predictive scheduling to improve NTU campus operation and supporting Singapore vision towards Car-lite.

Keywords: MaaS · Living lab · Stakeholder engagement · Project management

1 Introduction

1.1 Challenges in Urban Mobility and Rise of Smart ICT for Mobility

There are many challenges in urban mobility, ranging from traffic congestion and parking difficulties, public transport inadequacy, high infrastructure maintenance costs, to environmental impact and energy consumption. Traffic congestion and parking difficulties is one of the most dominant transportation problems in bigger urban cities. This is because although there is growth in mobility, parking space remain very limited. The other reason is high parking fares. As a result, many delivery vehicles tend to park illegally at the closest spot to unload goods without paying for parking [1].

© Springer Nature Switzerland AG 2019
H. Krömker (Ed.): HCII 2019, LNCS 11596, pp. 441–458, 2019.
https://doi.org/10.1007/978-3-030-22666-4_32

Much of the Public Transport System is Either Over- or Under-Used. During peak hours, there is a sudden spike in demand for public transport and the crowded conditions cause discomfort. This will be a huge issue for suburban areas due to finances. During high ridership periods, it will impossible to meet the demand and yet during low ridership periods, many services are financially unstainable. With the fast-growing urban mobility landscape, the need for infrastructure maintenance is also high. Thus, this causes a financial burden. Urban mobility also causes air and noise pollution. This environmental pollution will impact the quality of life of living organisms. The energy consumption of urban mobility has also increased, especially with the use of petroleum which increased the energy price [1].

With technology making it possible for the private sector to get involved and create new mobility solutions, the challenge impeding the rise of smart mobility is the lack of government policies. While government support lag, mobility solutions are created without following any legal framework or requirement. Since most mobility solutions are still at their infancy and have only been in the market for a few years, no exact goals have been defined. Hence, the markets need to be more flexible to adapt [2].

1.2   NTU-JTC-SMRT MaaS Testbed: Purpose and Intent

The Mobility-as-a-Service (MaaS) testbed is a joint endeavor between Nanyang Technological University (NTU) Singapore, JTC Corporation and SMRT Corporation to deploy and demonstrate an integrated mobility solution. The NTU-JTC-SMRT MaaS Testbed integrates existing transport services (public transport, campus shuttle buses) together with new mobility services such as Personal Mobility Devices (PMDs) sharing and Autonomous Vehicles. Some examples of PMDs include e-scooters, e-bicycles and bicycles. This is done via a single mobile application platform called *jalan*[2] (read as jalan-jalan), providing a one-stop journey planning, booking and payment solution. The Testbed brings together the strengths of the partners in Research & Development and Living Lab environment (NTU), industrial land planning and estate management (JTC), and multi-modal transport operations (SMRT).

2   Literature Review

2.1   Concept: Mobility-as-a-Service (MaaS)

MaaS refers to the combination of different transport modes that offers tailored mobility packages. This bundling of mobility modes allows a shift away from existing ownership-based transport system toward an access-based one. This concept offers a tailored hyper-convenient mobility solution. MaaS is seen as a mobility distribution model that deliver users' transport needs through a single mobile application.

The MaaS Operator uses the IT structure called the MaaS Platform to provide integrated mobility service to the end-users. This IT structure is comprise of a consumer-facing mobile application and/or web application, as well as a complex backend support system which delivers integral functions such as data import, data storage, journey planning, optimisation, ticketing, payment and communication [3].

Multi-stakeholders are popularising and accelerating the MaaS adoption in urban mobility settings. Governments intend to provide easy access to public transportation and other infrastructure provision for all urban dwellers. Traditional transport manufacturers are establishing new value propositions and transforming into service providers. Tech savvy consumers demand convenience, flexibility and more options for their daily commute. Existing transport operators intend to improve operation efficiency and better travel experience for customers [4].

2.2 Role of Universities and Its Importance

Universities are important drivers of knowledge creation, innovation, local economics, and more recently sustainability [5]. Universities are able to accelerate end-to-end process of knowledge creation, translation application and commercialization, while actively involving end users in the process. Besides, products and services development cycles tend to be long and are subjected to corporate bureaucracy or politics. The demand-pull development cycle is at risk of market failure even after launching [6].

However, universities haves been criticized for conducting "ivory tower research" whereby academia is disconnected from the "real-world" and focusing on publishing academic articles that are incomprehensible [7]. Universities are also not spared from the long and daunting red tape and lack the agility for rapid innovation. Despite having the resources and knowledge, the agility and structure of corporates and universities are insufficient in driving the process fast enough [8].

2.3 Adoption of Living Lab Approach

An increasingly common trend is the adoption of the Living Lab approach to develop an iterative, multi-stakeholders, concurrent research and open innovation ecosystem. Both universities and corporations use the Living Lab approach for agility, advancement and risk mitigation in the development and adoption of new technology or innovation. Living Lab is an open-innovation process that focuses on co-creation of innovations in real-world contexts. Not only does this process enable real-life monitoring of social settings, it also involves end-users [9].

As a user-centric approach, the Living Lab approach also entails working closely and frequently with multiple stakeholders to facilitate experimentation on innovative and sustainability Solutions. Through a Public-Private-People Partnership (4Ps), the Living Lab approach includes the following key features:

1. User involvement
2. Service Creation
3. Infrastructure
4. Governance and organization
5. Innovation Outcomes
6. Methods & tools

Organising Living Lab projects require three different levels of analysis: the organizational level (macro), project level (meso) and user activity level (micro). The macro level is whereby the Living Lab approach enables and fosters innovation, typically in certain area. Such innovation is usually enabled by Public-Private-People Partnerships. The meso level refers to Living Lab projects that take place following an organization-specific methodology of fostering innovation. Lastly, the micro level refers to the involvement of users and/or stakeholders. Multi-stakeholder participation often includes representatives from public sector, private sector, academia and general populace. However, the key to success lies in involving and engaging end-users to ensure the products meets their needs [10].

Seeking behavioural change is one of the major challenges the world is facing. With this Living Lab concept, this approach not only uses a lab-environment but also tests behavioral insight in a multi-disciplinary way and in a 'real-life' situation. The living Lab approach can adequately handle complex problems and environments [11]. The Living Lab project can function as a platform to research on mutual effects and the resulting lifestyles of people. This is because varying consumption patterns may be connected to different recognizable consumption decisions [12].

Given the benefits of the Living Lab approach, the NTU EcoCampus initiative was launched in 2014. The $20 million Living Lab initiative undertook a board range of projects in innovation and sustainability. The NTU EcoCampus initiative received support from the Singapore government and several multi-national corporations to transform the university campus to a living testbed for innovative, sustainable urban solutions.

2.4   User Engagement: Method, Interest and Engagement

MaaS is about putting commuters at heart and bringing together a personalized transportation experience at the system level. By acknowledging the close relationship between end-user satisfaction and project success, this gives a new spin to the traditional project management methodologies, which makes the Living Lab approach a great fit. The Living Lab approach internalizes end-user feedback and continues to optimize the implemented solutions to bring the enhanced outcome for end-users. By adopting a more agile approach, the project team can engage end-users at a relatively frequent period. In order to bring end-user insights into system design and daily operations, their feedback is collected in a variety of methods, such as through questionnaires, focus group discussions and one-on-one interviews. Although incorporating end-user feedback introduces challenges such as change management and stakeholder management, end-user feedback is still of increasing importance in the MaaS testbed. This is because considering end-user inputs improves the chances of user adoption and potential commercial success.

Throughout this one-year testbed, successive mobile application versions were released as part of the effort to improve user experiences. Some of the updates include operation scope, and refreshed hours and locations to better serve commute demands. The pricing schemes were also reviewed to find out end-user perception of different price points and the suitable business model to adopt. All aspects of the project were continuously updated based on the valuable insights and data derived from the testbed users. However, the complexity of user requirements was also recognized. For example, there are implications for releasing an in-app user feature in operations and business. Downstream effects should be thoroughly thought to avoid having too many iterations that could be a waste of the project team's resources.

2.5 Behavioral Change/Consumer Pattern Change

MaaS introduces a paradigm shift in the traditional transportation industry, re-focusing transportation services from being infrastructure-focused to being user-centric [13].

Traditionally, transportation services are separately run by individual transportation operators. This means that commuters have to go to the specific train station or bus-stop to enjoy the public transportation service. Commuters have to switch between multiple ride-hailing mobile applications to enjoy the best rate or secure the fastest ride available. Commuters also utilize multiple payment tokens and mechanisms throughout one day of travelling. With a single MaaS platform, various transportation services are brought together in a more integrated way than ever, providing a full suite of transportation options for end-users [14].

Many a times, users might not immediately understand the concept when they first encounter the idea in such innovative testbed projects. Nevertheless, MaaS is considered a concept which is relatively easy to communicate to users. This is because popular platforms such as Amazon and taobao also utilizes a similar concept of Platform-as-a-Services. According to MaaS Global, a Finnish company which provides MaaS subscription services in a few cities, MaaS is deemed the "Netflix of transportation" [13]. With MaaS, transportation services of enhanced optimisation and customisations can be offered to end-users on a single platform. Regardless, it is still a novel concept with a predictable learning curve for the masses. This might be the case for certain functions, such as the all-in-one mobile application usage, new form of transportation payment and pricing model for an integrated transportation service. It is the testbed's intention to anticipate, create and simulate the end-user adoption and response of a commercial MaaS product. During this process, the project team is expected to socialise the MaaS concept, engage and educate users on the above-mentioned purpose [14].

3 Case Study: NTU-JTC-SMRT MaaS Testbed

3.1 MaaS Testbed Vision and Activities

The Mobility-as-a-Service (MaaS) Testbed aims to reduce the usage of privately-owned transportation through an integrated mobility solution that meet the changing commuter needs and demands and to fulfil Singapore's vision of being a car-lite society. The Testbed is in close alignment with the whole-of-government Smart Nation vision [15] and the Sustainable Singapore Blueprint [16]. The Testbed uses NTU Singapore and the adjacent JTC CleanTech Park as a Living Lab.

The MaaS Testbed began by digitalizing existing transport modes such as campus shuttle buses with real-time arrival timings and crowdedness level Internet-of-Things (IoT) sensors. The Testbed also introduced several new greener mobility services such as autonomous vehicles, electric vehicles and PMD-sharing. The integration of both traditional and new mobility services through the MaaS mobile application *jalan*^{2} enables new connectivity that improves commuters' travelling experience.

3.2 Project Deployment

Mobility Partners and Services Deployed
The testbed is designed to be an open platform that connects and integrates multi-modal transport services. Some of the mobility services deployed include PMD-sharing, shuttle buses, autonomous vehicles, on-demand buses and public transport. Below is a summary of mobility partners and their respective involvement:

1. **Continental,** a leading German manufacturing company specialising in various automotive components. For the Testbed, Continental supplied GPS sensors and people counters to be installed onto existing campus shuttle buses. Real-time location and crowdedness level data are captured and analysed to provide recommendations on schedule and route improvement.
2. **Telepod,** a local start-up providing e-scooter-sharing services to achieve sustainable urban mobility and last mile solutions. For the Testbed, Telepod supplied 100 e-scooters, which can be booked and unlocked through the *jalan*^{2} app.
3. **oBike,** Singapore's first home-grown station-less smart bicycle-sharing company, uses technology to allow commuters to travel during one-way first and last mile commuting. For the Testbed, oBike supplied 100 third generation oBike. Testbed users are able to book and unlock an oBike through the *jalan*^{2} app.
4. **SWAT,** a local start-up focusing on dynamic routing technology. The project team adopted their technology to operate a handful of on-demand bus services routes in and between the Testbed and public housing towns.
5. **Airbike,** a local start-up providing e-bicycle sharing service. The e-bike sharing service enable users to book and unlock an e-bicycle through the *jalan*^{2} app.
6. **2getthere,** an autonomous shuttle service that was set up for a trial route between 3 NTU residential halls. Testbed users can use the *jalan*^{2} app to check the schedule of the autonomous shuttle service.

Release of *jalan*^{2} app
A mobile application called *jalan*^{2} was developed for the purpose of the Testbed. The *jalan*^{2} app offers customized mobility service options for commuters to travel from point A to point B. The app is named after the Malay phrase "jalan-jalan", which means "a care-free stroll". It is the project team's hope that this new mobility app could bring the simplicity and convenience of travel to the Testbed. Through the Living Lab approach, we observed growth of user base every time a new feature was introduced. At the end of the Testbed, we see the number of downloads of the *jalan*^{2} app arrived at more than 12,000 times. The download number served as an indication that the value proposition of testbed was well received by users.

Here are a short list of key functionalities and some app screens that demonstrates features implemented in the app (Fig. 1).

1. Location picker page that is customized to user's past journey history
2. A wide range of options that is sorted based on user's preference

3. A map overview of the journey option selected by the user
4. Full integration of all mobility services
5. Crowdedness level of campus shuttle bus
6. Arrival time of campus shuttle bus

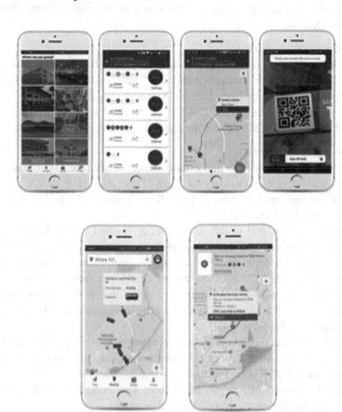

Fig. 1. Sample screenshots from jalan[2] app

3.3 Key Project Dimensions

Stakeholders' Management

Using a Living Lab approach, the MaaS testbed brings together a large number of stakeholders ranging from the core Consortium partners (NTU Singapore, JTC and SMRT), over 20 internal stakeholders (Functional, Operational, Schools/Colleges and subsidiary Business Unit) within the university, to many more external partners. With such a massive group of stakeholders involved, each with their unique concerns and circumstances, the MaaS Testbed could be compared to a huge ship with many paddles.

Consortium Partners

NTU as a Living Lab is well-known for its deep R&D expertise in engineering and low-carbon transportation solutions, the development of a more sustainable and eco-friendly vehicles will be a great advantage toward the MaaS Testbed. JTC is a public sector developer that is responsible for supporting and catalysing the growth of industries and enterprises in Singapore. Hence, this project would able to improve and transform commuter experience. SMRT is Singapore's premier multi-modal land transport provider that is responsible for the needs of the user by providing a better transportation solution.

Internal Stakeholders or Partners
As a large public education institution, the university has a higher moral obligation and duty towards its' community (students and staff). There are many internal stakeholders within the university ranging from Facility Management, Students' Union, Housing and Auxiliary Services, Student Affairs Office, Student Life, Committee on Space Planning and Allocation (CSPA), and the individual schools and colleges.

Other Stakeholders or Partners
Other stakeholders include tenants and estate management of the JTC CleanTech Park, BusPlus and AutoService (subsidiary of SMRT) for providing jalanX services and installation of sensors, and 2getthere Asia (Joint Venture of SMRT) for providing autonomous shuttle bus services.

A project as complex and visible as the MaaS Testbed would definitely raise the attention of the community stakeholders, especially since it involves the installation of mobility stations and the deployment of higher risk equipment such as electric scooter and autonomous vehicles. Some of the earlier challenges faced by the project team include stakeholders' concerns for community safety due to the introduction of new mobility service such as autonomous buses and the increased use of PMDs on campus as we deployed more for sharing.

Concerns were raised on the operational disruption as the introduction of mobility services involved the closure of carpark for autonomous shuttle bus operations and the high utilisation rate of PMDs on the limited space within NTU campus.

The respective operational departments were also concerned with the increased manpower constraint due to the MaaS Testbed activities.

Mobility Services Integration
As a service aggregation app, the number of mobility service partners onboarding this platform is crucial. How partners' mobility services are integrated with this MaaS platform technically and operationally would greatly impact user experiences.

Platform Integration Through Application Program Interface (API)
At the start of project, two methods of integration were examined and compared, namely API integration and deep link through URL scheme. API integration enables real-time data sharing between application platforms and allows the MaaS app to be the only user-facing mobile app required for the Testbed. It would take 2 to 3 months' integration effort in general, as there might be customization, co-development and tuning of API from the parties involved. URL scheme requires much lighter integration effort. However, this means that Testbed users will need to install various mobility service apps and to switch from the MaaS app to individual mobility service app so that their travel needs could be served.

With the MaaS objectives to provide homogeneous and complete user experiences, and to capture sufficient real-time data for further analytics for transport provision enhancement, the project team carefully decided to adhere to API integration for all mobility service providers in this Testbed as much as possible.

Operation Integration with Partners
Functioning as a MaaS operator within this testbed, the project team centrally coordinates first-last-mile transport resources, for example, shared e-scooters and shared bikes, with mobility service providers. This is to supplement and better improve the connectivity within the campus that is mostly supported by 2 campus shuttle bus routes and 2 public bus routes.

As each mobility service faces unique operational challenges, our project team dived into continuous in-depth discussions with individual mobility partners on the platform to formulate customized operation arrangement that works to best leverage on existing operation resources from each team and to optimize the operation results.

- **Telepod.** Telepod was still in the process of building operational teams during the project. Yet, the performance of an e-scooter-sharing service is dependent on its operations; specifically relocation, replenishment and recharging (3Rs) management. The 3Rs have to be taken care of so that the supply matches the Testbed users' demand. After deliberations, project team and Telepod worked out two measures to ensure the 3Rs. Firstly, the two teams worked together on a shift arrangement to support an extended operation hour from 7 am to 2 am since active e-scooter use on campus was identified to happen post mid-night. Secondly, a major charging station was set up upon reviewing the existing charging infrastructure. This new charging station has a high capacity and easy access, which makes the 3Rs much more effective.
- **OBike.** Obike's operation team is well-staffed and experienced about redistribution and replenishment (2Rs). Its operations require relatively lighter involvement from the project team as compared to managing the bicycle fleet. Hence, the project team established communication protocols for the 2Rs, while OBike took instructions from the project team to perform 2Rs on the ground.

Continuous User Engagement: Pre, During and Post Testbed
As part of any Living Lab project, users' involvement and co-creation is fundamental to the project development. To encourage user involvement in the early stage of the MaaS Testbed, a design thinking workshop was conducted with roughly 20 NTU students to gather their feedback on the gaps and consolidate a vision for the campus transportation landscape.

Several publicity roadshows and marketing efforts were done to engage with end-users and increase users' adoption. During the marketing activities, awareness for safety education and usage of the mobility services like PMD-sharing and Autonomous Vehicle were emphasised, and users were welcomed to provide feedback if any.

There are also on-going surveys conducted over email, in-app and in-person at different stages of the Testbed to consolidate a better understanding of users' experience. At the end of each journey, there will be an in-app survey for users to rate their satisfaction level and provide feedback if any. Through the series of on-going users' engagement efforts, it enables an iterative and continual improvement to the MaaS provisioning in a 'real-life' scenario.

Post-Testbed surveys were carried out with 196 testbed users to assess their awareness and perception of $jalan^2$ services for future improvement. According to the survey, 136 respondents out of 196 were aware of the app, and 65 of them had the app on their phone. This means that $jalan^2$ has captured 33.2% of the entire market based on the sample size by the end of the Testbed. This indicates a significant behaviour shift in how people travel within campus. One-third of Testbed users have picked up the convenience value of having PMDs to connect their first-last-mile journeys any-time, anywhere. It is also interesting for us to learn that 43% of the Testbed users heard of the app through word-of-mouth, either positive or negative. Hence, creating a positive user experience is the key to a positive word-of mouth. Here are the top four concerns that hindered positive word-of-mouth, all of which were then taken into consideration to guide future app development and service deployment.

1. Smartness of the app, in terms of routing and recommendations
2. The necessity to have deposit to access mobility services
3. Differentiation from alternative transport apps
4. Safety concern of riding e-scooters

4   Insights from the NTU-JTC-SMRT MaaS Testbed

The NTU-JTC-SMRT MaaS Testbed showed the possibility of rapid deployment of the MaaS concept within a Living Lab environment such as the NTU campus. Data obtained from users and utilisation data from the respective transportation modes show high adoption rate.

4.1   Understanding the Roles and Responsibilities of a MaaS Operator

In this testbed, a deep involvement in individual transport service was committed by the project team as the designated MaaS operator. This unique arrangement was made as the project team was also the central party bringing various mobility service providers into the testbed.

With a hands-on approach, the project team was able to centrally coordinate transport resources in order to achieve high usage rate of PMDs. In particular, e-scooter-sharing services has achieved an average utilisation rate of 15 trips per e-scooter per day throughout Semester 2. On 19th March 2018, a record high usage was observed at 23 trips per e-scooter per day. However, it is worth highlighting that the manpower cost involved for such hands-on central management in a semi-confined estate shall not be overlooked.

At its peak, we have a 5-man operations team centrally managing 7 different transport modes. They would:

- Relocate, replenish and recharge for a fleet of 100 shared e-scooters.
- Plan, operate and provide customer support for a fleet of 5 on-demand buses.
- Monitor and coordinate for the operation of the 100 shared bicycles.
- Set up and maintain 16 *jalan*² integrated mobility stations.
- Set up and maintain 2 major PMD charging stations.
- Deploy and operate 1 autonomous vehicle along a trial route.

In addition, a track with one driver and one worker and a few student ambassadors are employed during the testbed to supplement project team's operation requirements.

One of the objectives for various mobility service providers to join the Testbed is that they could leverage on the learnings to improve their latest hardware and software. There are times that some features were offline for upgrading and/or the usability of certain function was not ideal. For example, this Testbed implemented the first GPS-based geo-fence for parking at integrated PMD stations in Singapore. However, drifting of GPS signal have always imposed challenges like indiscriminate PMD parking. That is why besides centrally managing various transport services in the Testbed, the project team also took on the responsibility to provide customer support to Testbed users who encountered difficulties using the app to access services.

Lastly but importantly, taking care of the riders' safety and partner assets' security when it comes to shared PMD services require the operation team's constant monitoring throughout the day. During this testbed, there were few theft cases and minor user injury cases reported. All cases were investigated thoroughly, during which the MaaS operator spent considerable amount of effort working with the police and the campus security to close the cases.

4.2   Review of Individual Mobility Services

Autonomous Vehicle

Autonomous vehicle was the most innovative concept that was tested in this project. Although the operating route was deliberately chosen to have a somewhat controlled environment for trial, the project team were extremely concerned about how Testbed users would react to such cutting-edge technology, especially when interactions between Testbed users and autonomous vehicle are unpredictable. Safety marshals were deployed at the very beginning of the Testbed period whenever the autonomous vehicle was in operation.

Upon the launch of the autonomous vehicle service, we received lots of attention from students, faculty member and staff. People travelled from other parts of the campus to try out what it was like to ride on a driverless shuttle. A few months in, we observed Testbed users becoming rather comfortable with having an autonomous vehicle moving in a live environment. Autonomous vehicle was perceived as just one of the transport options to get people from one point to another.

PMD-Sharing: E-scooters, Bicycles, E-bikes
Shared E-mobility services such as e-scooters and e-bikes have different sets of operational emphasis on charging and redistribution as compared to already popularised bike-sharing services. This naturally results in higher investment in operating manpower and operating assets. Meanwhile, users are observed to have reservations about safety when riding an e-scooter. This creates a high barrier for behavioural shift that requires constant user engagement and safety riding education.

One challenge that all PMD-sharing service providers have in common is the lack of adequate infrastructure provisions such as parking locations, shared path and charging stations. During this Testbed, our project team took the lead to improve the infrastructure facilities, which largely benefited the mobility service partners on board.

By participating in the Testbed, PMD-sharing service providers gained experiences and insights that prepared them for mass service deployment in the future. For instance, introduction of design-for-operations concept, implementation of maintenance practices for shorter operational turnover rate and efficient operation protocols between the MaaS operator and platform partners were recommended by our project team as the MaaS operator. It was a process of co-learning and co-development between PMD-sharing service providers and the MaaS operator.

On-Demand Buses
On-demand bus services leverage on data analytics and dynamic routing technology to optimise the fulfilment rate of shuttle bus services.

From 109 survey results obtained, Testbed users consider on-demand bus as an innovative transport options that could potentially bring convenience to their daily commute. Among the survey group, 16% has tried on-demand bus service. A few factors that may have hindered a wider adoption of such services were identified:

- On-demand bus routes were available only in a handful of operation zone.
- Frequency and window of operation were undesired due to the limited number of buses being operated.
- On-demand mobility is a new concept to the Testbed, whose service differentiation required a lot more communication to Testbed users.
- Fierce competition was seen from existing transportation modes regarding comfort level, pricing strategy and travel duration.

Throughout the project, the project team continued to configure parameters like operation zone, operation window, bus capacity to capture most demand, to provide best user experience and to achieve high efficiency. However, it remained challenging to have willing customers to subscribe to this service. It is believed that with a wider user base or under other user cases such as optimisation of existing service route during off peak hours, on-demand bus service would yield a higher chance to succeed.

Digitalization of "traditional" Transport
Installation of people counters and location sensors transformed the traditional campus shuttles from an information black box into a moving data collection channel. The challenges during the installation are the calibration of such devices to minimise double counting and differentiate boarding and alighting passengers. The installed sensory system yielded a 92% accuracy rate according to the on-site survey.

As campus shuttle bus accounted for 30% of intra-campus travel, this newly-enabled information availability supplied much insight into how Testbed users commute every day. We are able to assess real-time crowdedness of each individual bus and station, capacity and supply of the bus fleet, and efficiency of every bus routes. Future improvement of campus shuttle bus service shall be supported with data and insights.

The Fig. 2 below illustrated the overcrowding situation of one campus shuttle route throughout a specific school day at different stations.

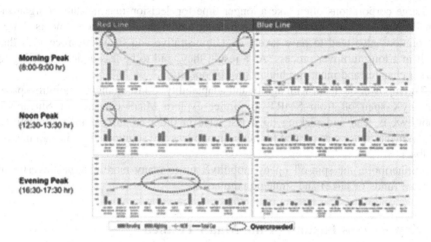

Fig. 2. Campus shuttle bus overcrowding study

4.3 Harnessing the Power of MaaS

MaaS platform aims to provide an end-to-end functionality from planning a trip, reservation and booking of transport services, to payment for transport services. When a wide range of mobility service providers and ticketing/payment service providers are fully integrated within the MaaS platform, it truly reflects the MaaS value of providing convenience and choices to daily commuters. At the same time, the MaaS platform captures all trips data of daily commutes, which are eventually channeled to a central data lake for future transport planning studies and improvement of current MaaS service.

In this testbed, great insights for the MaaS operator and the mobility service providers were gained in the following areas:

- Understand the current transport demands, patterns and provisions based on different transport modes,
- Conduct quantitative transport accessibility analyses to highlight the gaps of current transport network,
- Identify the existing transport issues and potential opportunities for improvements,
- Estimate student and staff's price elasticity and their behavioral reactions to changes in transport network, mode and services.

By leveraging on mobility service providers' existing operation networks across Singapore, the MaaS platform could extend beyond the Testbed for further learnings and insight.

4.4 Spin-Off of mobilityX for MaaS Commercialization

Over the Testbed period, we observed some limitations of a large corporation and university like SMRT and NTU Singapore in accelerating rapid deployment in a dynamic environment.

Large corporations often take a longer time for decision-making due to organizational complexity and several decision-making authorities across departments. Large corporations also tend to have layers of administrative and financial procedures that result in a long turnover process. As a result, they tend to be less agile in reacting to market changes and incurring higher cost of operation.

To empower the team with agility and adaptability in fast-changing mobility space, mobilityX span off from SMRT to further explore MaaS concept for Singapore. mobilityX is lean and picking up a fast pace in order to survive in a competitive environment. Space was created for iterating on ideas and solutions from instant market feedback.

Furthermore, the spin-off opens mobilityX up to many more external investment (venture funds or alike) and strategic partnership.

5 Key Success Factors of the MaaS Testbed

Taking account the individual stakeholders' expectations and strategic objectives, the MaaS Testbed is closing the gaps for the campus (and estate) and campus mobility needs (both from NTU requirement, as well as individual users and mobility service providers). Some objectives include enhancing life of the community, sustainability and mobility. Clearly, there are different key factors that would be considered significant for other stakeholders from their respective perspective.

However, from a project perspective, the Key Success Factors for the NTU-JTC-SMRT MaaS Testbed could be attributed to 3 key areas: **Smart Cities readiness**, **Regulatory and Market readiness**, and **Living Lab approach**.

Smart Cities Readiness

- Cyber and Data security
- Smart Phone adoption and Digitalisation
- Integration and real time sharing with Government and transport operator data

Regulatory and Market Readiness

- Transport eco-system "sandbox" for new entry startup– to avoid overly restrictive policies
- Platform efficiency
- Openness in new business model

Living Lab Approach

- 4P approach
- Users' centric
- Iterative approach
- Risk acceptance

5.1 Smart City Readiness: IOT, Openness of Data, IT, Smart Mobility, Smart Phone Adoption Rate

Smart City readiness is a combination of factors including but not limited to Smart Phone adoption rate in the population, availability and openness of transport and commuter data to the MaaS platform and other transport operators in optimising transport planning and demand aggregation. In order for transport operators to have better overview and visibility of their assets in real time, smart IoT sensors should communicate vehicular and traffic data with analytics function, allowing transport operators to predict and activate idling assets to meet demands over high-speed low-latency cellular mobile communications.

On the Users' End, Technology Disruption Would Inevitably Result in Lifestyle Changes. While users should be adaptable in embracing new technology and trends, the technology disrupter should consider user-centric designs (User Experience Design, and User Interface design) for a seamless experience. Compatibility of new technology with existing devices should also be deliberated in allowing users to switch with low barriers (cost and convenience).

5.2 Regulatory and Market Readiness

Since 2014, Singapore has embarked on the journey of supporting better living using technology through Smart Nation initiative. Digitisation of the transport sector is one of the six pillars under this initiative. To catch up with recent rapid transformation in transportation industry, the Land Transportation Authority of Singapore worked closely with mobility service providers to establish regulation framework, licensing scheme and trial projects. To name a few, on-demand public bus trial, license to operate for bike-sharing companies and sandbox for e-scooter sharing services are all in action at this moment.

In order to strengthen platform offerings to everyday commuters and to harness the power of data aggregation, the MaaS platform requires a full range of mobility service partners with decent services level. While respecting the spirit of market competition, splitting urban mobility market among various MaaS operators in one city will not benefit commuters. In addition, being an aggregator platform, the MaaS operator has a more complex system construct as compared to a single mode operator. Given the platform nature, the resources involved and the business complexity, it is recommended for government to work with one MaaS operator to digitalise end-to-end commuter experiences. Sandbox can serve as a means to select a MaaS operator. By opening up to shortlisted MaaS operators, government regulators could set up an overall framework that allows innovation and creativity from individual Sandbox participants. Some guiding principles could be:

- Encouragement of Public Transport and existing infrastructure usage
- Enhancement of travel experience
- Transport insight through digitisation and big data analytics.

5.3 Project Management Approach: Living Lab

Designated as a Living Lab environment, the entire NTU campus and JTC CleanTech Park are accustomed to iterative deployment or Testbed of Sustainable Urban Solutions. As a whole, the university, including the respective operational and functional stakeholders, are aware of the iterative nature of Living Lab projects in Innovations and RD&D (Research, Development and Demonstration) projects.

Similar to a sandbox for technology, we are able to deploy, demonstrate and iterate the technological solutions to meet the users' actual requirements, using a 4P approach. Bringing in partners from the different sector and domain, we are able to tapped on the strength and resources of the respective area. For future nationwide deployment, the same approach should be adopted.

6 Conclusion and Future

Digitalisation and technological advancement has cause disruption to the transportation sectors and our daily life. Users are expecting a more seamless, point-to-point transportation experience with an integrated approach throughout the journey planning, booking and payment process. Rapid development in new transportation trends and technologies such as ride-sharing (Uber/Grab), Electric/Autonomous Vehicles, PMD-sharing, and Intelligent Transport Systems enable new and more efficient services to be deployed. However, traditional transportation operators have been slow in the transformation due to high inertial and risk in large scale adoption.

MaaS requires large scale deployment to gain the economies of scale in matching the users and transportation services providers. In a commercial deployment setting, it means connecting the public and private transportation service providers in a single platform, and at the same time distributing the profit margin and risk accordingly.

While innovation development offers opportunities to transform the entire industry, it also inherently consists of other associated challenges including business and financial risks. The innovative eco-system of the Living Lab approach could provide an avenue in deployment of such solutions.

The MaaS Testbed has been successful through the strong Living Lab approach. The Living Lab approach is being tested in NTU Singapore through the EcoCampus initiative. On the Future Mobility front, the MaaS Testbed brings together multi-stakeholders and end-users in an iterative process throughout the development and deployment stage. The strong multi-partnership consortium brings in different experiences and strengths to make the Testbed possible, with the strong support from the functional and operational stakeholders and partnership. Several channels of engagement were used to continuously iterating and improving the MaaS experience on campus. Particular focus is on engagement with the end-users. The Living Lab approach also accelerated the process of innovation and functionality deployment by breaking down the key priorities

With the strong grounding of the MaaS Testbed team, we are able to deep-dive into the operational and business challenges of both MaaS operators and mobility service partners. The demonstration of the MaaS concept in a fairly large campus is subjected to the same level of regulatory and users' scrutiny in the real world environment. The testbed experience enables the project team to develop processes, procedures and build up case experience in handling different dimensions of MaaS deployment. The SMRT team also developed experience in identifying suitable and ready mobility service partners which are often start-ups that are financially lean but lack professional experience.

The utilisation data from the MaaS Testbed shows strong user acceptance in the value of MaaS solutions. Usage of *jalan*² has been high and some mobility services enjoyed higher utilisation rate than the average rate out of the Testbed. User satisfaction survey shows high acceptance alignment and that users are likely to embrace MaaS as a lifestyle option in the near future.

The Living Lab Approach also helped the mobilityX team to gain confidence in future commercialised deployment in Singapore and the wider region. It enables a rapid transformation of the team, products and servicing offerings to evolve according to the desire of the users and market, hence achieving a product-market fit with a reasonable business case.

Acknowledgements. The Mobility-as-a-Service Testbed is jointly supported by Nanyang Technological University, Singapore together with JTC Corporation (JTC) and SMRT Services Pte Ltd. (SMRT), within the University Living Lab Program "EcoCampus Initiative – Sustainable Urban Transportation" and hosted under Energy Research Institute @NTU and Sustainable Earth Office at NTU.

References

1. Rodrigue, J.: The Geography of Transport Systems, 4th edn. Routledge, New York (2017)
2. May, A., Boehler-Baedeker, S., Delgado, L., Durlin, T., Enache, M., van der Pas, J.: Appropriate national policy frameworks for sustainable urban mobility plans. Eur. Transp. Res. Rev. **9**(1) (2017). Article 7
3. MaaS Lab: The MaaS dictionary. https://docs.wixstatic.com/ugd/a2135d_d6ffa2fee2834782b4ec9a75c1957f55.pdf. Accessed 27 Feb 2019
4. Falconer, R., Zhou, T., Felder, M.: Mobility-as-a-service: the value proposition for the public and our urban systems. Arup (2018)
5. Trencher, G., Yarime, M., Kharrazi, A.: Co-creating sustainability: cross-sector university collaborations for driving sustainable urban transformations. J. Clean. Prod. **50**, 40–55 (2013)
6. National Academy of Sciences, National Academy of Engineering, and Institute of Medicine.: The Roles of Universities. In: Trends in the Innovation Ecosystem: Can Past Successes Help Inform Future Strategies? Summary of Two Workshops, pp. 17–24. The National Academies Press, Washington DC (2013)

7. Perumal, R.: The Development of Universities of Technology in the Higher Education Landscape in South Africa (2010)
8. Innovate UK: Turning red tape into a golden thread of innovation, https://innovateuk.blog. gov.uk/2018/10/22/turning-red-tape-into-a-golden-thread-of-innovation/. Accessed 27 Feb 2019
9. Bergvall-Kåreborn, B., Ståhlbröst, A.: Living lab: an open and citizen-centric approach for innovation. Int. J. Innov. Reg. Dev. 1, 356–370 (2009). https://doi.org/10.1504/IJIRD.2009. 022727
10. Schuurman, D., De Marez, L., Ballon, P.: Living labs – a structured approach for implementing open and user innovation. In: Proceedings of the 13th Annual Open and User Innovation Conference (2015)
11. Mavridis, A., Molinari, F., Vontas, A., Crehan, P.: A practical model for the study of living labs complex environment (2009)
12. Welfens, M., Liedtke, C., Rohn, H., Nordmann, J.: Living lab: research and development of sustainable products and services through user-driven innovation in experimental-oriented environments. Delft (2010)
13. Business Insider Nordic: 'Netflix of Transportation' is a trillion-dollar market by 2030 – and this Toyota-backed Finnish startup is in pole position to seize it. https://nordic. businessinsider.com/this-finnish-startup-aims-to-seize-a-trillion-dollar-market-with-netflix-of-transportation–and-toyota-just-bought-into-it-with-10-million-2017-7/. Accessed 27 Feb 2019
14. Deloitte Insights: The rise of mobility as a service, https://www2.deloitte.com/insights/us/en/deloitte-review/issue-20/smart-transportation-technology-mobility-as-a-service.html. Accessed 27 Feb 2019
15. Smart Nation Singapore: Strategic national projects to build a smart nation. https://www. smartnation.sg/whats-new/press-releases/strategic-national-projects-to-build-a-smart-nation. Accessed 27 Feb 2019
16. Sustainable Singapore Blueprint. https://www.clc.gov.sg/docs/default-source/books/ssbcombined-cover-text.pdf. Accessed 27 Feb 2019

Business and Billing Models for Mobile Services Using Secure Identities

Ulrike Stopka, Gertraud Schäfer, and Andreas Kreisel[✉]

Technische Universität Dresden, Dresden, Germany
{ulrike.stopka,gertraud.schaefer,andreas.kreisel}@tu-dresden.de

Abstract. Worldwide, strongly increasing trends towards the use of mobile services can be seen in almost all areas of society, economy and administration. A secure eID service is the user's key to mobile online services. Many branches are recognizing this. The research project OPTIMOS 2.0 develops an open SE-based ecosystem for mobile services. This supports a straight forward approach for mobile eID apps that covers all requirements and recommendations given by the EU's eIDAS directive. Different usage scenarios for secure mobile services are possible. The number and roles of stakeholders are very complex – especially by addressing several markets. The paper presents the generic approach to establish business and billing models for the involved stakeholders on a multi-sided platform. This includes the Platform Business Model Canvas, the more comprehensive Business Model Canvas and a Morphological Matrix.

Keywords: Multi-Sided platform · Secure electronic identity ·
Open ecosystem · Mobile application · Business modeling ·
Billing model · Morphological analysis

1 Motivation

Since 2017, websites have been accessed more often by mobile devices than by desktop PCs. In the third quarter of 2018, 52% of global website visits came from mobile devices [16]. Globally, the trend towards mobile online services offers great opportunities, but also a lot of challenge for service and technology providers from all market sectors.

In the future a secure Electronic Identity (eID) will be the user's key to a wide variety of (mobile) online services. A non-discriminatory access and universal applicability is necessary for participation in the digital world. Many branches, companies, administration and government have recognized the relevance of eID.

Already today many mobile devices have been equipped with Embedded Secure Elements (eSEs) or Embedded Universal Integrated Circuit Cards (eUICCs). Therefore, many things are conceivable:

© Springer Nature Switzerland AG 2019
H. Krömker (Ed.): HCII 2019, LNCS 11596, pp. 459–476, 2019.
https://doi.org/10.1007/978-3-030-22666-4_33

Mobile ticketing in public transport: eID apps for customer onboarding by storing high-priced tickets on a Secure Element (SE), online payment

Carsharing: registration, payment and storage of the electronic car key in the SE, onboarding via eID app

Air traveling: SE used for boarding pass, on-board-eID and payment

Hotelling: Mobile check-in, storing the room key on SE, access to different hotel facilities

But there is still no infrastructure for these kind of mobile services that meets the reliability requirements (substantial) given by the European Union (EU)'s Electronic Identification, Authentication and Trust Services (eIDAS) directive for electronic transactions in European markets.

In 2018 the research project OPTIMOS 2.0 started with the target to define an open, practical ecosystem using secure identities on mobile services. The project wants to demonstrate scalable eID applications utilized for secure identification

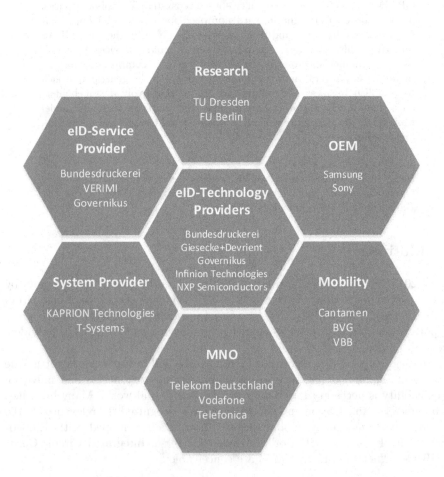

Fig. 1. Stakeholders of the ecosystem offering eID secured mobile services

and authentication in Internet of Things (IoT), mobility sector as well as mobile networks and services. The German Federal Ministry of Economic Affairs and Energy funds the project and made sure that all relevant stakeholders and associated partners from all parts of the ecosystem were present (see Fig. 1).

The number and roles of stakeholders of the SE-based ecosystem for mobile services is very complex – especially when addressing several markets. By using the first results of the OPTIMOS 2.0 project, this paper wants to introduce a sequence of steps leading to business an billing models for Multi-Sided Platforms (MSPs) for the involved stakeholders such as platform owners, eID service and technology providers, mobile network operators, mobility service providers, system providers and Original Equipment Manufacturers (OEMs).

Market-dominating international corporations (e.g. Facebook, Google, etc.) use their enormous customer base to establish their own ecosystems, to keep customers in them (lock-in-effect) and to collect far-reaching data about the user's behavior. Service or technology providers may face barriers to access and unreasonable obligations towards the operators of such closed systems, hamper competition and undermine the implementation of national and European objectives on IT security and data protection. As an alternative, OPTIMOS 2.0 aims to create an infrastructure for mobile services that fully meets the criteria of an open ecosystem:

- Accessible to service providers and technology suppliers on a non-discriminatory basis,
- Customers should be able to choose between as many service providers as possible, and
- Comprehensive support of the national and European requirements on IT security and data protection.

The task of the Technische Universität Dresden is to investigate and evaluate efficient business and billing models for all partners involved in an open ecosystem offering mobile services by using secure identities. One of the main questions is the pricing of the different eID mobile services and the revenue sharing among the various partners. The major goal is to increase the attractiveness of the platform for all user groups.

2   Project OPTIMOS 2.0

2.1   Main Innovations

The main innovations provided by OPTIMOS 2.0 will be the following:

- **Specification and development of an open trusted service management system for all types of secure elements.**
  The main challenge with SE-based architectures for mobile services is to provide all interested service partners with a non-discriminatory place on a SE in the customer's mobile device, to implement the API and to manage it.

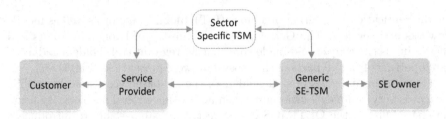

Fig. 2. General architecture of the SE-based ecosystem.

Therefore, OPTIMOS 2.0 specifies and integrates a Trusted Service Manager (TSM) that provides service providers from different markets with a defined technical and commercial interface for access to the secure platforms of mobile device manufacturers (eSE) and mobile network operators (eUICC) in the sense of one-stop shopping (see Fig. 2).

– **Specification and development of a TSM API for developers of mobile applications.**
Service providers should be able to use the SE-based ecosystem for mobile services without having to acquire a detailed knowledge neither of the architecture of mobile devices with SE nor the existing Application Programming Interface (API) for accessing SE.

– **Minimization of access barriers for small and medium enterprises and new entrants – research, specification and development of a *Secure Wallet***
The *Secure Wallet* work package explores, specifies and demonstrates a concept that allows multiple service providers to share a storage space on the SE, ensuring that data from different applications can coexist in a single storage space. This offers commercially viable options for companies that have no expertise with SE or whose services are only used occasionally or by a limited number of customers for a limited time.

– **Definition of new open interfaces and direct cooperation in standardization**
The innovations resulting from the activities described above are to be introduced into the responsible international standardization committees such as the International Organization for Standardization (ISO), the Near Field Communication (NFC) Forum, the GSM Association and the Global Certification Forum (GCF) for Mobile Devices.

2.2   Fields of Application and Use Cases

The basis for numerous eID mobile application fields e.g. in transport, mobility, government or in the hotel industry is a trustworthy identification with the help of an NFC-enabled mobile device and the eID function of the identity card.

As shown in Fig. 3, the identification data is taken from the identity card using an online platform of the operator of the secure eID app. Based on the

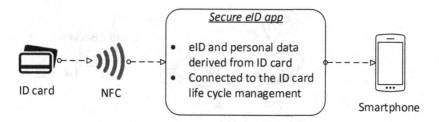

Fig. 3. Data capturing by the eID secure app

requirements of the OPTIMOS 2.0 deployment scenarios, additional data may be collected and managed.

Use Case Carsharing. The use case Carsharing demonstrates the flexibility and wide range of the possibilities planned in OPTIMOS 2.0:

1. Creation of an online account and login using the *Secure eID App*, i.e. customer onboarding, registration, and payment is executed/handled via the app
2. Storing the electronic vehicle key in the secure element
3. Using the vehicle key via the NFC or Bluetooth Low Energy (BLE) interface also when the mobile device is offline or in *battery low* mode

The demonstration will be based on the *Secure Wallet*. The carsharing provider can use the tools and the protected area of the *Secure Wallet* in the SE for his own app development. It is not necessary to rent a separate area on SE via the TSM.

Use Case Onboarding and Management of IoT-Devices. IoT devices can be securely assigned to a responsible person using mobile eID applications by introducing secure authentication for device management (see Fig. 4). One field of utilization will be the customer on-boarding for eUICC devices of the preferred mobile network operator. Manufacturers of SEs for IoT devices will work together on open standards to define a method that can be used across manufacturers. Similar fields of application can be identified:

- In air traffic for passenger check-in (customer onboarding by eID app, SE used for boarding pass, on-board-eID and -payment)
- For ticketing in local public transport (same security level as chip card, supporting all ticket products, interoperable with existing public transport readers, cards, etc.)

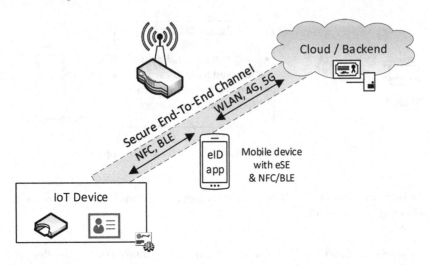

Fig. 4. On-boarding by the eID app online or at the point of sale and authentication for the device maintenance

- For check-in, registration and payment in hotels key management for hotel rooms, access to hotel facilities of any kind and checkout

The services can be used in any case (offline or in low-battery-status).

3   Concept for Creating Business Model for Multi-Sided Platforms

3.1   Multi-Sided Platforms

MSPs enable interactions between several market players. In traditional markets they are already exist, e.g. broadcasting, real-estate agents, flea markets or shopping malls. Typically, those platforms bring together suppliers and consumers, whereby a large number of suppliers attracts a large number of customers and vice versa. Additionally they can add values to both sides as well as attract third party actor for entering on such a market place (like advertisement in broadcast media or newspapers).

Characteristics of Electronic Platforms. Platforms in electronic markets have an even much wider focus than in traditional markets. With their digital algorithms, they are able to extend existing processes, emerge new value creation structures and generate additional value for all platform partners [7].

Subsuming the results of literature research [2, 3, 5, 14] successful digital MSPs

- have a high scalability and wide reach, because computing capacity can be adapted quickly and flexible,

- facilitate transaction between different market partners anytime and anywhere even without knowing each other before,
- create the basis for complementary services,
- accelerate processes on provider as well as on demand side,
- make transactions cheaper and simpler[1].

3.2   Platform Types and Characteristics in the OPTIMOS 2.0 Project

Different electronic MSPs cooperate on different value-added stages within the OPTIMOS 2.0 ecosystem.

The Generic SE-TSM (platform type I) is the core of the ecosystem. It is operating a trusted service managing platform pooling and brokering SE memory space between all kind of SE owner and sector specific TSMs (see Fig. 5). It enables mobile eID services on a substantial secure level. Service providers (platform type III) with customer contact can integrate them for

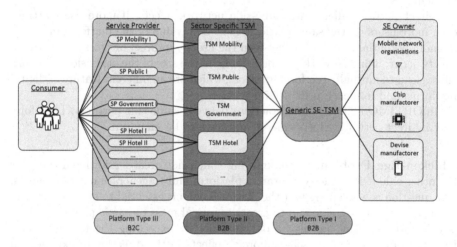

Fig. 5. Identified platform types of the OPTIMOS 2.0 ecosystem

- providing one-step electronic services to their customers,
- simplifying their operations or sale processes or
- adding customer value to a wide variety of service applications.

---

[1] Hagui stated "At the most fundamental level there are two types of basic functions that MSPs can perform: reducing search costs, incurred by the MSP's multiple constituents before transacting, and reducing shared costs, incurred during the transactions themselves. Any feature or functionality of an MSP falls into either of these two fundamental types." [5, p. 5].

Furthermore, there is a possibility for new players (platform type II) in the ecosystem bundling the demands of a specific market field (e.g. public transport, government, banking, etc.). All these platforms "have two vital goals: increase their service base and sustain their service offerers" [4, p. 2]. They all look for sustainable business models. Therefore, we are discussing the features and principles of MSPs as well as methods for developing business models taking into account design features and characteristics.

Regarding to OPTIMOS 2.0 we expect simpler and cheaper transactions for services requesting a substantial secure identification by using the eID instead the original identity card. Service providers can optimize their operating processes for registration, using already verified electronic data, and offer services anywhere and anytime.

Network Effects. Managing a MSP, we have to consider the fundamental features of network effects. We can distinguish between direct and indirect network effects. They can have positive but also negative influence on the platform participants.

Direct network effects are same-side network effects, if more users attract even more users on the same platform side. The value of the platform increases by more potential interactions between the users[2]. Indirect or cross side network effects occur, if one network side benefits from the size of the other side. Network effects lead to mutually reinforcing effects, which are an inherent feature of digital platforms and influence price setting.

In OPTIMOS 2.0 are no interactions between same-side partners. Therefore, only indirect network effects emerge and have to be managed.

Chicken-Egg-Problem. Even knowing which side creates additional value for the other side it is not easy to start a platform and to generate network effects. The question is who enters first the platform, who is essential to attract partners of the other platform side and to create a critical mass of participants too. This is called the chicken and egg problem. Only by exceeding the tipping point, the platform can grow by self-reinforcing effects in the direction of exponential network effects [6, p. 81].

Establishing a platform in an ecosystem like OPTIMOS 2.0 the question is fundamental which group of participants has to enter first in order to attract other participants.

To solve the *chicken-egg-problem* Parker et al. [12, pp. 89–99] determined eight different strategies applied by successful platforms:

Follow the rabbit: The functionality of the business model is proved initially on an one-sided market (Practical example: Amazon Marketplace).
Piggybacking: The new platform takes over the user base of another platform (Practical example: PayPal payment as a solution for eBay).

---

[2] Prominent example are social media services like Facebook or Twitter. Their main business is the interactive communication between users.

Seeding: The platform creates value in advance in order to convince users and to push the market. (Practical example: Innovations in the Google Play Store)

Marquee: The platform incentivizes and motivates important and well-known users (Practical example: Influencers on LinkedIn).

Single-side: The platform acquires first market group (Practical example: reservation system at OpenTable).

Producer evangelism: The supply side introduces its own customer base in to the platform (Practical example: auction process at Mercateo).

Big bang: The platform attracts attention to its launch with maximum marketing effort. (Practical example: Twitter party on SXSW festival).

Micromarket: The platform initially establishes itself in an already existing community. (Practical example: Facebook at the Harvard University).

In OPTIMOS 2.0, it is important to get partners early on board who are sufficient to reach the critical mass. Therefore, it is most essential to commit partners offering the data spaces on SEs. This is necessary for attracting the service providers with their large number of potential end users benefiting from one-step-services supported by a secure mobile eID. Since the offered eID-service still is new on the market and end users are not used to such kind of identification a fast market penetration can build user acceptance necessary to reach the critical mass.

The *piggybacking strategy* is the most promising one. Already established provider platforms will implement mobile eID-services to add value to their core business. It is essential to take on board suitable and promising business fields and to run first demonstrators to show the value of the new implemented service in the early project phase.

The *seeding strategy* can also be important. The new implemented security architecture and features for secure digital identification can attract end users to ask for such services even in other business fields.

3.3   Developing Business Model

Business models describe how businesses are running. They consider combinations of production factors, competitive strategies together with functions and rules of all actors.

The *Integrated Business Model* describes already established businesses and gives an idea how a specific business runs and which success factors are essential. It focuses on different sub-models like market model with competitors and the demand side, procurement model, performance model, model of service offer, distribution model and capital model focusing on financing and revenue [17].

Starting a new MSP like OPTIMOS 2.0 we are facing the task to develop a successful business design attracting all partners in the ecosystem as well as investors. Therefore, we approach the process of business modeling as "the managerial equivalent of the scientific method – you start with a hypothesis, which you then test in action and revise when necessary" [9]. It will be an ongoing process focusing to the following core questions [11,13]:

- Who are the customers and who are potential partners at the case of MSPs?
- What is the specific value for customers and partners? What is the Unique Selling Proposition (USP)?
- What is the economic logic delivering value to customers?
- How can we earn money? Who is paying for what?
- What are the appropriate costs?

We designed a schedule to answer these questions illustrated in Fig. 6. The design process starts by using the *Platform Business Model Canvas*[3] to detect the core elements of a business model: value propositions, value transactions and key platform components for all partners. Based on these results the *Business Model Canvas*[4] helps to become deeper insights to all related business fields like customer segments, channels of distribution and interaction, customer relationship, cost structure alongside with key resources, partners and activities. Afterwards, we can formulate complete business models for a participants. The *Morphological Matrix* helps to describe, structure and analyse the developed models. Finally, we can design new, so far not noticed combinations of business models and present all partners their realistic opportunities.

Platform Business Model Canvas. The main idea behind this model is to take the perspective of each platform participant. This includes the supplier or producer side, the consumer or demand side as well as the platform itself, which enables the core service between two or more market sides by providing their infrastructure. In addition, third parties can offer services via the platform to both suppliers and customers. The three main perspectives – value proposition or USP, transaction activities including billing approaches and key components – are investigated and developed in a design process. The process is carried out iteratively.

In the following we will present some first results focusing on *Value Transaction* for all partners of the Generic SE-TSM platform. Through its services, the platform owner facilitates interactions between the SE owners (supplier side) and Sector Specific TSMs (customer side).

Value Transactions. On the first step, we have to think about the transactions the participants of OPTIMOS 2.0 want to carry out, esp. what kind of transactions take place and who will pay for which transaction. Here we start with our first drafts:

- SE owner transactions:
  - Allocation of memory space to Generic TSM platform
  - Generating revenue for providing memory space
- Generic SE-TSM platform transactions:

---

[3] The *Platform Business Model Canvas* was developed by WALTER to fit the Business Model Canvas for platforms markets [15].

[4] The *Business Model Canvas* was created by OSTERWALDER and PIGNEUR in 2008 [10].

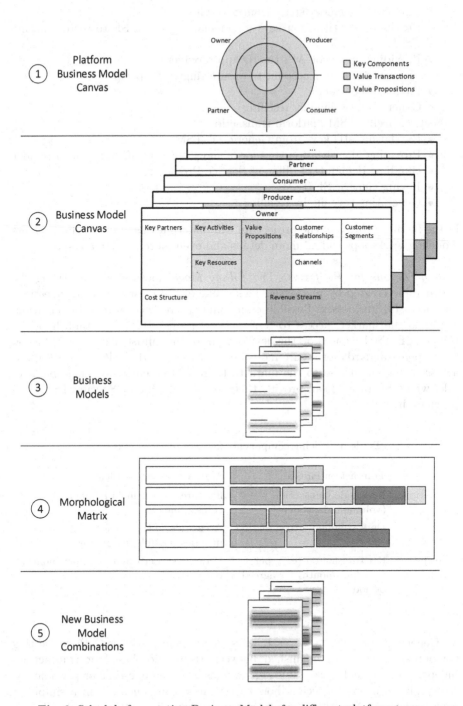

Fig. 6. Schedule for creating Business Models for different platform types

- Purchasing memory space from SE owner
- Providing identification keys for memory space of SE to sector specific TSM
- Providing applet or API to eID app provider
- Managing end user account to ensure singly eID per person
- Paying SE owner
- Generating revenue by reselling
- Sector Specific TSM platform transactions:
  - Purchasing eID keys from Generic SE-TSM
  - Providing identification keys for memory space of SE to Service Provider
  - Bundling demands of different Service Providers
  - Paying Generic SE-TSM
  - Generating revenue by reselling

In the further process of the project we will underpin this approach by business modeling workshops to find more details and even more reliable answers.

Billing Options for Platforms. The *Billing Model* focuses on monetizing the added value created by platforms. This is one of the most important questions for successful businesses. Possible opportunities are "access to value creation, access to the market, access to tool, and curations" [12]. By establishing the Generic SE-TSM – based on the development of the infrastructure as well as on their open standards for secure mobile eID services – OPTIMOS 2.0 will grant access to tools and also to the market. Thereby, this creates a single point of sale, where SE owners can provide their service and Sector Specific TSMs can purchase it.

Table 1. Pricing components for core platform business

	Transaction-based	Transaction-independent
One-time	Per single transaction (volume-independent)	flat rate for a defined number of transactions
	Data volume per transaction	Access fee to the platform (this also includes possible license fees)
Per period	Per transaction (e. g. per year, per month) for agreed period	Periodic subscription (e. g. per month, per year)

Generally, there are several ways for pricing services. Price setting components can be distinguished between transaction-based or transaction-independent elements (see Table 1) as well as charging one-time or periodically. Every platform has to decide, how to use these components – in a simple or combined form. Also they have to consider, how price setting could influence network effects.

Nevertheless, depending on the business model platforms have the possibility to create value out of data collected on their platform by generating indirect revenues out of aggregated information, trend analysis, selling contacts, etc. They also can integrate other actors, e.g. advertising partner and suitable additional services, to improve their own services or by pricing the access to data (see Table 2).

The pricing mechanism can also be distinguished in static and dynamic pricing as shown in Table 3.

Looking into the OPTIMOS 2.0 ecosystem the integration of third party partners seems only reasonable for service provider platforms acting in Business to Consumer (B2C) markets and not for the TSMs, since they operate in Business to Business (B2B) markets.

With respect to OPTIMOS 2.0 we consider four billing models for the different transactions between the integrated partners (see also Fig. 5).

Billing Model 1 (B2B) between the Generic SE-TSM and the SE owner (e.g. Mobile Network Operator (MNO)): The Generic SE-TSM pays the owner of the eSE or eUICC for provisioning certain memory space on the SE. The most reasonable way is paying per transaction only one-time for a certain period or periodically over an agreed period of validity.

Billing Model 2 (B2B) between Generic SE-TSM and Sector Specific TSM (e.g. Mobility TSM): The Generic SE-TSM could use all pricing components of Table 1 to generate revenues for his services as a broker providing access to secure elements in smartphones from a single source. It seems to be most useful to set prices for transactions. Depending on the market situation and the acceptance of the SE-TSM integrating transaction-independent pricing elements are possible.

Billing Model 3 (B2B) between Sector Specific TSM (e.g. mobility TSM) and Service Provider (e.g. car sharing provider): All options according to Table 1 are possible. Service providers pay for the trusted services provided

Table 2. Pricing components for additional services on the platform by integration third party partners

	Transaction-based	Transaction-independent
Direct from third party partner	Provision	Advertisement (pay per view or click)
	Advertisement (pay per sale, per lead)	Access fee to the platform
		Periodic subscription (e.g. per month, per year)
Indirect		Selling contacts
		Selling information out of data mining results

Table 3. Pricing mechanisms [10, p. 33]

| Fixed Menu Pricing | Dynamic Pricing |
| *Predefined prices are based on static variables* | *Prices change based on market conditions* |

	Fixed Menu Pricing		Dynamic Pricing	
List price	Fixed prices for individual products, services or other value propositions	Negotiation	Price negotiated between two or more partners
Product feature dependent	Price depends on the number or quality of value proposition features	Yield management	Price depends on inventory and time of purchase
Customer segment dependent	Price depends on the type and characteristic of an customer segment	Real-time-market	Price is established dynamically based on supply and demand
Volume dependent	Price as a function of the quantity purchased	Auctions	Price determined by outcome of competitive bidding

by the Sector Specific TSM. He decides which pricing components has to be included in his pricing model.

Billing Model 4 (B2C) between end customer (e.g. carsharing user) and service provider (e.g. carsharing provider): All options shown in Tables 1 and 2 are possible to use and to combine. Service Providers have to decide how to attract their customers and third party partners. It is possible that one group subsidizes the others. The service provider has to create acceptance for the billing model, depending on the core services, the customers' willingness to pay, the possibility and necessity for fast roll out.

In the next step we are clarifying which monetization strategies and models should be considered in order to architect the platform "in a manner that affords its control over possible sources of monetization" [12]. Considering network effects we have to decide who should be charged.

Business Model Canvas. Based on the result of the Platform Business Model Canvas for all participant groups we can get comprehensive insights in the business modeling process by deploying the *Business Model Canvas* (see Fig. 6). It is a tool for describing, analyzing and designing business models and gives a good structure especially for understanding and evaluation the platform owner's business. The model defines "nine basic building blocks that show the logic of how a company intends to make money" covering "the four main areas of a business: customers, offer, infrastructure, and financial viability" [10, p. 15] as follows:

Customer Segments: This block focusses on the question for which customer groups a platform will create value by facilitating interaction between two sides. They will realize transactions via the platform in case it is more convenient and cheaper than direct transactions.

Value Proposition: The OPTIMOS 2.0 ecosystem enables mobile services using identification on an eIDAS substantial level to ensure high confidence.

Channels to reach the different customer segments: Via service provider applications end users request eID services for identification.

Customer Relation: Customer Relation bases on an automated process in form of customer self-service mobile apps with access to customized features and characteristics.

Revenue streams: Based on the different options for setting billing models or pricing schemes (see billing models) the more detailed investigation focuses on the willingness to pay based on competition situation.

Key Ressources: For running a business essential physical, intellectual, human and financial resources have to be identified.

Key Activity: In case of OPTIMOS 2.0 all main processes to enable the automated services belong to the key activities.

Key Partners: Since OPTIMOS 2.0 will create an ecosystem, all partners have to be identified, needed to run the business, to optimize the business, to reduce risks, or to acquire financial resources.

Cost Structure: General we can distinguish between the two extremes of cost structure: cost driven or value driven business model. Generally, we often find approaches between those two extremes. For getting an understanding and later a more realistic calculation we have to identify fix and variable costs and determine how they will be effected by economies of scale or economies of scope.

Based on that structure possible business models for the OPTIMOS 2.0 partners will be described and analyzed. In case the result does not show a sustainable business prospect we will have to adjust the business models and start a new modelling process.

Morphological Matrix. A *Morphological Matrix* allows a qualitative description of pattern or features of a specific business model. It is a tool to examine different combinations and variations of electronic businesses [1, p. 5]. It can be used for business model creation "so that proven ideas in one business could be transplanted in another" [8, p. 4].

For OPTIMOS 2.0 we want to deploy this tool for evaluation the business models designed by the modeling process with the canvas methods. Based on the work of Clement et al. [1, p. 6] and Lee et al. [8, p. 4] we have developed a first draft for a *Morphological Matrix* (see Fig. 7) taking into account various main and sub design fields and their possible characteristics useful to describe potential business models for OPTIMOS 2.0 partners. By implementing the demonstrator system further research and project work will lead to more sophisticated perceptions to complete this approach.

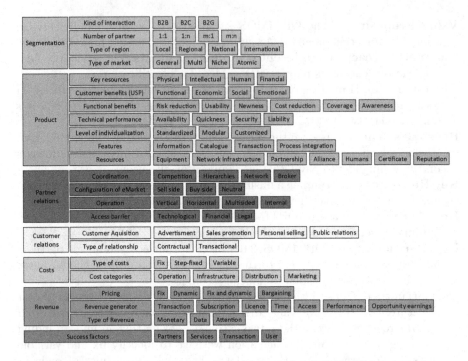

Fig. 7. First draft of the Morphological Matrix for OPTIMOS 2.0 project

4   Outlook

Based on the introduced methodology we will carry out various workshops for defining and validating business models for all OPTIMOS 2.0 partners. Furthermore, we are identifying legal and economical conditions for specific use cases.

Acronyms

Notation	Description	Page List
eIDAS	Electronic Identification, Authentication and Trust Services	459, 460, 473
API	Application Programming Interface	462, 470
B2B	Business to Business	465, 471, 474
B2C	Business to Consumer	465, 471, 472, 474
BLE	Bluetooth Low Energy	463, 464
eID	Electronic Identity	459–467, 470, 473
eSE	Embedded Secure Element	459, 462, 471
EU	European Union	459, 460
eUICC	Embedded Universal Integrated Circuit Card	459, 462, 463, 471
IoT	Internet of Things	461, 463, 464
ISO	International Organization for Standardization	462
MNO	Mobile Network Operator	460, 471
MSP	Multi-Sided Platform	461, 464–468
NFC	Near Field Communication	462–464
OEM	Original Equipment Manufacturer	460, 461
SE	Secure Element	459–463, 465, 467, 468, 470, 471
TSM	Trusted Service Manager	462, 463, 465, 468, 470–472
USP	Unique Selling Proposition	468, 474

References

1. Clement, R., Schreiber, D.: Internet-Ökonomie. Springer, Heidelberg (2016). https://doi.org/10.1007/978-3-662-49047-1
2. von Engelhardt, S., Wangler, L., Wischmann, S.: Eigenschaften und Erfolgsfaktoren digitaler Plattformen. iit-Institut für Innovation und Technik in der VDI/VDE Innovation + Technik GmbH (2017)
3. Evans, P., Gawer, A.: The rise of the platform enterprise: a global survey. The Center for Global Enterprise (2016)
4. Ghazawneh, A., Mansour, O.: Value creation in digital service platforms. In: 28th Australasian Conference on Information Systems (2017)
5. Hagiu, A.: Multi-sided platforms: from microfoundations to design and expansion strategies (2009)
6. Jaekel, M.: Die Macht der digitalen Plattformen. Springer, Wiesbaden (2017). https://doi.org/10.1007/978-3-658-19178-8
7. Kollmann, T.: E-Business: Grundlagen elektronischer Geschäftsprozesse in der Net Economy (German Edition), 5th edn. Springer, Wiesbaden (2013). https://doi.org/10.1007/978-3-658-26143-6
8. Lee, J.H., Hong, Y.S.: A morphological approach to business model creation using case-based reasoning. In: Proceedings of the 18th International Conference on Engineering Design (ICED 2011), pp. 165– 175 (2011)
9. Magretta, J.: Why business models matter. Harv. Bus. Rev. 5 (2002)
10. Osterwalder, A., Pigneur, Y.: Business Model Generation. Wiley, Hoboken (2010)
11. Ovans, A.: What is a business model. Harv. Bus. Rev. 5, 2016 (2015)
12. Parker, G., van Alstyne, M.W., Choudary, S.P.: Platform Revolution, 1st edn. Norton & Company, New York (2016)
13. Stähler, P.: Geschäftsmodelle in der digitalen ökonomie. Merkmale, Strategien und Auswirkungen. Josef Eul Verlag (2001)
14. Uenlue, M.: The Complete Guide to the Revolutionary Platform Business Model (2017). https://www.innovationtactics.com/platformbusiness-model-complete-guide/. Accessed 2 Nov 2019
15. Walter, M.L.: Endlich ein Canvas für Plattform-Geschüftsmodelle (2016). Fleing, E. (ed.). https://www.deutsche-startups.de/2016/04/05/endlich-ein-canvas-fuer-plattform-geschaeftsmodelle/. Accessed 2 Dec 2019
16. WeAreSocial: Global Digital Report 2018 (2019). https://wearesocial.com/blog/2018/01/global-digital-report-2018. Accessed 2 Nov 2019
17. Wirtz, B.W.: Electronic Business. Gabler Verlag, Wiesbaden (2001)

Study on Evaluation of Subway Passenger Wayfinding Systems Based on SEM

Chuan-yu Zou[1], Guangxin Wang[2(✉)], and Yongquan Chen[1]

[1] Research Group of Wayfinding, China National Institute of Standardization,
Beijing 100191, China
[2] Department of Psychology, School of Humanities, Beijing Forestry University,
Beijing 100083, China
wgx8868@163.com

Abstract. It is a major issue affecting the operations of subway whether the passenger wayfinding system is scientific, standardized and reasonable, and it is also one of the important indicators when evaluating subway satisfaction. Based on the expert argumentation, customer interviews and manager's argumentation, the satisfaction assessment questionnaire for subway passenger wayfinding system was compiled. When evaluating the satisfaction of subway wayfinding system, the objective structural equation model was used to give weights. The research shows that the assessment questionnaire consists of three dimensions: the safety signs, the traffic signs, and the continuity and rationality of the signs. The evaluation scores of subway wayfinding system of different subway lines can be divided into three dimensions. Line 1, line 15, line 9 obtained higher scores on three dimensions, while line 2, line 8 obtained lower ones at the bottom of the scores. In the future, further researches should be carried out to verify the way in which structural equation modeling is established, and to evaluate the evaluation effectiveness of subway wayfinding system.

Keywords: Structural equation · Subway wayfinding system · Beijing subway

1 Introduction

In today's modern public transportation environment, the lack of a systematic and standardized public information guidance system is not only an obstacle to cognitive communication, but also an obstacle to life behavior, or a psychological and spiritual obstacle. It could lead to the weakening and loss of function of public transportation facilities.

The common elements of the wayfinding systems are divided into two categories: public information guiding elements and safety information elements. Since the publication of the first national standard GB 3818-1983 "Public Information Graphic Symbol" in 1983, after more than 20 years of research, promotion, publicity and implementation, the national standard system of public information guidance system has been initially established. It has played a guiding role in the construction of Chinese public information wayfinding systems [1].

© Springer Nature Switzerland AG 2019
H. Krömker (Ed.): HCII 2019, LNCS 11596, pp. 477–486, 2019.
https://doi.org/10.1007/978-3-030-22666-4_34

Now, the National Standardization Technical Committee on Graphical Symbols (SAC/TC59) has basically completed the national standard system of public information wayfinding systems in China through more than 20 years of efforts. The GB/T 15566 "Public Information Guidance Systems- Setting Principles and Requirements" series of standards is the basis for implementation [2], and the GB/T 20501 "Public Information Guidance Systems - Design Principles and Requirements for Elements" is the implementation method [3], GB/T 10001 "Public Information Graphical Symbols" is the basis [4]. China has established a perfect national standardization system for public information guidance systems. The system has played a guiding role in the establishment of Chinese guidance system standard system by standardizing the corresponding guiding elements, clarifying the requirements of elements, system design and setting.

Internationally, ISO/TC 145, the technical committee specializing in the standardization of graphical symbols in the International Organization for Standardization (ISO), has begun to focus on the development of international standards related to public information wayfinding systems. The specific work is undertaken by the Working Group ISO/TC 145/SC 1/WG 5 "Public Information Guidance Systems". The working group was convened by the Chinese expert, and in November 2010, the first international standard for public information guidance systems, ISO 28564-1, "Public information guidance systems - Part 1: Design principles and element requirements for location plans, maps and diagrams" was officially released [5]. In addition, other countries in the world pay more attention to the research and establishment of urban wayfinding systems, such as the Philadelphia wayfinding project launched by Philadelphia in the United States in 1992.

Although most public information wayfinding systems are built on the basis of national standards, in practical applications, there are still problems such as disjointed planning and operation, user complaints, and media exposure. The reason is mainly including: planners' different understandings of national standards, user's different hard conditions (such as vision, height, physical activity and other physical conditions) and soft conditions (such as cultural education background, understanding ability, etc.), and, poor management. The research on the evaluation method of planning, design, construction and operation and maintenance level of public information guidance system is still a blank, and it has become an urgent problem to be solved. It requires us to pay attention on its importance with a scientific and systematic vision.

Ji et al. established fairness structural equation of public transportation system, and the questionnaires were collected by Kunming public transportation as an example. The main parameters of the structural equation model were programmed and evaluated by LISREL software [6]. Zhang et al. combined the characteristics of urban subway to establish a post-evaluation index system for urban subway construction projects. Based on this, an empirical study was carried out on the evaluation model using structural equations [7]. Chen et al. constructed a structural equation model of urban subway passenger satisfaction group, constructed a multiple-group analysis model of urban subway passenger satisfaction, quantitatively analyzed the path coefficient between each latent variable, and finally based on the sample characteristics did group research [8].

Although the national standards and subway wayfinding system in line with national standards have been established, the systematic evaluation of subway signs has rarely been studied at home and abroad. Therefore, our thinking is to construct a scientific r subway sign evaluation system, and further improve the basis for the scientific, standardized, rational, and humanized subway signs.

2 Objects and Methods

2.1 Objects

The study was conducted in July 2016 and selected 8 subway lines in Beijing for field investigation. The participants were randomly selected from those 8 subway lines. In the selection of subway lines, taking into account the new subway lines and old lines, and considering the location of the selected points, For example east, west, north and south in Beijing. Based on the above considerations, it was determined that 10 subway stations were selected for testing. 30–50 copies of the questionnaire were distributed at each station. The questionnaires were randomly distributed and then collected on the spot. 470 questionnaires were distributed, and 468 questionnaires were returned, with a recovery rate of 99%. In the survey sample, there were 247 males and 221 females. The average age of the participants was 26.79 years old. The average number of subway rides per week was 6.45.

2.2 Tools

From national standards, industry standards, local standards and various related regulations and documents, the factors affecting the construction level of public transport passenger wayfinding system were selected, experts' opinion surveys and public surveys were conducted, and opinions and suggestions from different stakeholders were collected to form a subway sign assessment questionnaire.

The assessment questionnaire consists of three dimensions: the continuity and rationality of the safety signs, the wayfinding signs and the public information signs. The safety signs were divided into two sub-dimensions: emergency signs and common safety signs. The wayfinding signs were divided into four sub-dimensions: traffic route maps, toilet direction signs, block guide maps, and station space diagrams. The continuity and rationality are divided into two dimensions: the continuity of the direction signs and the rationality of setting. The meaning of each dimension of the questionnaire is as follows (Table 1):

The conceptual model of the confirmatory factor analysis of "Subway passenger wayfinding systems questionnaire" is as follows. The dimension of each question is in line with the exploratory factor analysis result. The standardized path diagram of the model is shown in Fig. 1.

The fitting test of subway passenger wayfinding system model is shown in Table 2. There are many measurement criteria for determining the overall goodness of the fit of the model. Commonly used indexes are: gauge fitting index (NFI), comparison fitting index (CFI), incremental fitting index (IFI), goodness of fit index (GFI), adjusted

Table 1. Dimensions of subway wayfinding system evaluation and their meanings.

	Dimensions	Meaning
Safety signs	Emergency signs	Signs of escape routes, fire equipment, emergency exits, etc. in the subway
	Common safety signs	Signs giving information to passengers to travel safely in the subway
Wayfinding signs	Traffic route maps	Orientation, comprehension and convenience of traffic route maps
	Toilet direction signs	Direction, distance and convenience of the toilet that passengers value very much
	Block guide maps	Direction information, location information, important neighborhoods and landmark information when passengers leave the station
	Station space diagrams	Overall layout of the space inside the station
Public information signs	Continuity of direction signs	Continuity of direction signs for passengers entering, transferring, exiting the station
	Rationality of setting	Overall evaluation signs' location, font, height, applicability

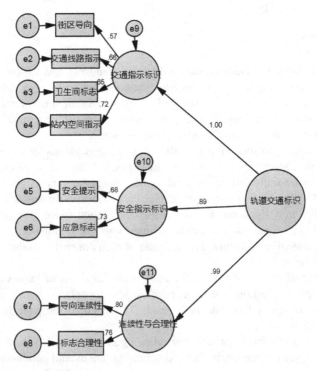

Fig. 1. Second-order confirmatory factor analysis model for subway passenger wayfinding systems questionnaire

goodness of fit index (AGFl)), relative fit index (RFl), root mean square residual (RMR), approximate root mean square residual (RMSEA), and so on. It is generally accepted in the academic community that in large sample cases, NFI, CFI, IFI, GFI, AG-FI, RFI are greater than 0.9, RMR is less than 0.05, and RM-SEA value is less than 0.08, indicating that the model and data fit well. The model fit index for this study was CFI = 0.950 > 0.90, IFI = 0.950 > 0.90, TLI = 0.926, RM-SEA = 0.091. The display model has a good overall fit and can fit the sample data very well.

Table 2. Goodness of fit test of subway passenger wayfinding system

Indexes	CFI	RMSEA	NFI	RFI	IFI	TLI
Fitted value	.950	.091	.938	.909	.950	.926

3 Beijing Subway Passenger Wayfinding System Index Weight Assignment

The three dimensions of the subway passenger wayfinding systems and the weight of each sub-dimension index are the first considerations when conducting comprehensive evaluation. Weight determination generally has two methods, e.g. Delphi method and AHP method. It is a common method for Delphi to obtain the weight of evaluation index. The AHP method calculates the total evaluation value by combining the evaluation of both experts and respondents. In this test, the structural equation method is used to determine the weight of the satisfaction index in the evaluation of subway passenger wayfinding system. The structural equation model is a comprehensive statistical method. It is based on many traditional statistical methods and is a comprehensive application and improvement of statistical methods such as multivariate homogeneity, principal component analysis, path analysis and simultaneous equations. The structural equation model enables researchers to process measurement errors in the analysis and analyze the structural relationships between latent variables. It is obtained by summarizing the weighted values of each sub-indicator in the questionnaire. Its calculation formula is as follows:

$$\text{Subway wayfinding} = T_i \left(\sum_I^4 (W_i \times O_i) + \sum_I^2 (Y_i \times P_i) + \sum_I^2 (Z_i \times Q_i) \right) \quad (1)$$

W_i identifies the weight of each test dimension i index of "wayfinding signs" index. The scores are derived from the normalized estimates of the structural equations (see Table 3).

O_i is the initial score for each test dimension of "wayfinding signs" index.

Y_i identifies the weight of each test dimension for "safety signs".

P_i identifies the initial score for each test dimension for "safety signs".

Z_i identifies the weight of each test dimension for "public information signs".

Q_i identifies the initial scores for each test dimension for "public information signs".

T$_i$ is the weight of the total score of subway wayfinding systems in the three dimensions of safety signs, wayfinding signs, and public information signs. The weighting factors of each dimension and sub-dimension are shown in Table 3.

Table 3. Subway wayfinding systems scoring weight

Dimension		Dimension	Weight
Wayfinding signs	←	Traffic route maps	.57
Wayfinding signs	←	Toilet direction signs	.66
Wayfinding signs	←	Block guide maps	.65
Wayfinding signs	←	Station space diagrams	.72
Safety signs	←	Emergency signs	.68
Safety signs	←	Common safety signs	.73
Public information signs	←	Continuity of direction signs	.80
Public information signs	←	Rationality of setting	.76
Total score	←	Wayfinding signs	1.00
Total score	←	Safety signs	.89
Total score	←	Public information signs	.99

4 Research Result

According to the weight coefficient, the data is summarized, and the overall situation of Beijing subway wayfinding systems is obtained. The total score is: 91.13 ± 14.62.

The evaluation scores of those Subway Passenger Wayfinding Systems fell into three different categories. The high level included subway Line 1, Line 9, Line 10 and Line 15. The medium level included Line 5, Line 7 and Line 13, while the low-level included Line 2 and Line 8. This trend and discipline were reflected in all 8 dimensions.

In the overall evaluation of Beijing subway wayfinding system, subway Lines 9, 10 and 15 are among the best, which is closely related to the overall construction planning of Beijing subway. Beijing subway Line 15 was put into use in 2016, and the southern section of subway Line 9 (except Fengtai East Street) was put into use on December 31, 2011. The northern section (except the Military Museum) was put into use on December 30, 2012. Most of the second phase of subway Line 10 (Bagou Station - Xizhan Raiway Station, Shou Jingmao Station - Jinsong Station) was put into use on December 30, 2012 (Jiaomen East Station was temporarily suspended), until December 1, 2017, the line 10 was put into loop operation. Compared with other lines, subway Lines 9, 10 and 15 are new, and the planning and design is more reasonable. The overall evaluation of Beijing subway wayfinding systems is among the best, and it is also reasonable. Subway Line 1 is the earliest one in Beijing and also the earliest subway line in China.

What are the differences between the continuity and rationality of safety signs, wayfinding signs, and public information signs for each line of Beijing subway wayfinding systems? The specific indexes are shown in Table 4.

Table 4. Differences between the continuity and rationality of safety signs, wayfinding signs, and public information signs

Subway line	Wayfinding signs	Safety signs	Public information signs	Total score
1	31.07 ± 4.55	34.64 ± 5.82	32.34 ± 4.24	93.92 ± 12.43
2	28.17 ± 7.62	31.46 ± 8.07	29.25 ± 7.90	85.14 ± 21.49
5	29.97 ± 5.23	34.15 ± 6.73	30.51 ± 5.27	90.58 ± 12.90
7	29.80 ± 5.26	33.79 ± 5.59	30.83 ± 4.99	90.40 ± 13.44
8	29.12 ± 3.89	32.40 ± 4.01	29.49 ± 5.01	87.18 ± 9.73
9	31.39 ± 5.70	34.29 ± 4.27	31.22 ± 6.43	92.82 ± 15.79
10	31.50 ± 5.12	35.49 ± 4.27	32.30 ± 5.16	95.07 ± 11.81
13	29.95 ± 6.62	33.30 ± 5.67	31.23 ± 6.34	90.51 ± 16.49
15	30.31 ± 5.59	34.34 ± 6.62	31.87 ± 5.96	93.66 ± 14.64

Table 4 shows that different subway lines in Beijing vary in the three dimensions of wayfinding signs, safety signs, and public information signs. However, those three dimensions show the same change rules and trends. Table 4 shows: Lines 1, 15, and 9 score higher on the three dimensions. Subway Lines 2 and 8 have lower scores in the three dimensions, and their scores are in the bottom of the score valley. Also, the peaks and valleys of each line have the same trend. Therefore, it can reflect the basic situation of the subway wayfinding system (Fig. 2).

Taking the total score of subway wayfinding systems as the dependent variable, the subway line as the grouping variable, the one-way analysis of variance (ANOVA) results are as follows: on the total score of subway wayfinding systems, $F_{(8, 460)} = 2.172$, $p = 0.028 < 0.05$, showing a significant difference between the evaluation total scores of subway lines. Further multiple comparisons found that the total score of subway Line 2 was significantly lower than that of Line 1 ($p = 0.009 < 0.05$), and the total score of Line 2 was significantly lower than that of Line 9 ($P = 0.018 < 0.05$), Line 10 ($P = 0.002 < 0.05$), and Line 15 ($P = 0.003 < 0.05$). The total score of subway Line 8 is significantly lower compared with Line 15 ($P = 0.026 < 0.05$).

Taking "public information signs" as the dependent variable and subway lines as a grouping variable, the results of one-way analysis of variance (ANOVA) are as follows: Line 10, Line 15 and Line 1 score higher, showing Line 10, Lines 15 and 1 are better at the location of the signs, the font and height of texts, and the suitability of the signs. The "Rationality" score of the subway Line 2 is low indicating that the sign of Line 2 is unreasonable. Rationality score of Line 2 is significantly lower than that of Line 1 ($p = 0.018 < 0.05$), and Rationality score of Line 2 is significantly lower than that of Line 10 ($P = 0.016 < 0.05$), and Line 15 ($P = 0.017 < 0.05$).

Taking "Wayfinding signs" as the dependent variable and the subway line as the grouping variable, the one-way analysis of variance (ANOVA) results are as follows: on the "Wayfinding signs" score, $F_{(8,460)} = 2.09$, $p = 0.035 < 0.05$, it shows a significant difference between the lines on this index. Further multiple comparisons found that "Wayfinding signs" scores of Line 2 were significantly lower than Line 1

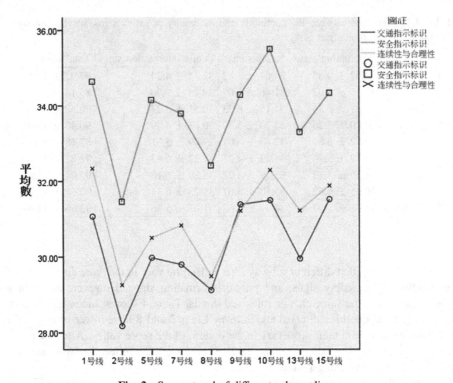

Fig. 2. Scores trend of different subway lines

(p = 0.024 < 0.05), and "Wayfinding signs" scores of Line 2 were significantly lower than Line 9 (p = 0.01 < 0.05), Line 10 (P = 0.008 < 0.05), Line 15 (P = 0.002 < 0.05). "Wayfinding signs" of Line 15 were significantly higher than Line 8 (P = 0.026 < 0.05).

"Safety signs" is used as the dependent variable, and the subway line is used as the grouping variable. The one-way analysis of variance (ANOVA) results are as follows: on "Wayfinding signs", F(8,460) = 1.628, p = 114 > 0.05. Further multiple comparisons found that "Wayfinding signs" scores of Line 2 were significantly lower than Line 1 (p = 0.025 < 0.05), and "Wayfinding signs" scores of Line 2 were significantly lower than Line 5 (p = 0.024 < 0.05), Line 10 (P = 0.003 < 0.05), and Line 15 (P = 0.015 < 0.05).

5   General Conclusion

The scores of passenger wayfinding system for each subway lines can be divided into three levels. The high level included Line 1, Line 9, Line 10 and Line 15. Line 5, Line 7, and Line 13 were located in the middle level. Line 2 and Line 8 were at the low level. This trend is reflected in several dimensions of "Safety signs", "Wayfinding signs" and "Public information signs".

6 Analysis and Discussion

6.1 Time of Design and Use for Wayfinding Systems

According to the results, the score of subway Line 2 fell into low lever. This is basically consistent with researcher's theoretical expectation. Line 2 is a loop-line. It is the first loop line in China. Due to the long operation time of Line 2, the wayfinding system is out of date and cannot meet the requirements of the modern subway transportation. Therefore, it got a low score, not only in total score, but also in all 8 sub-dimensions. Therefore, when the subway line renovation project is costly and difficult, the upgrade of the wayfinding system of Line 2 can be done and done well, which is helpful for improving passenger satisfaction. Line 1 is the earliest subway line in Beijing and also the earliest subway line in China. This time, it is also in the forefront of the overall evaluation of Beijing rail transit signs. It is also a subway line with a long planning and long operation history. Line 1 is among the high level, which may be related to the characteristics of the line. Line 1 is East-West line and Line 2 is a loop line. The passengers' requirements for loop line are clearer and more accurate, while East-West line is relatively easy. This may be the reason why Line 1 scored higher in this evaluation.

6.2 Improve the Design of Wayfinding Systems

In order to host the 29th Beijing Olympic Games in 2008, the first phase of the Olympic branch of Line 8 was built in advance, and there were 5 interchange stations, which were transferred to Changping Line, Line 13, Line 10, Line 2 and Line 6. This branch Line 8 extended the central axis Line running through Beijing's north and south again. As a project of the Beijing Olympics, Line 8 is impeccable in terms of construction quality. However, the evaluation score of the Line 8 Subway Passenger Wayfinding Systems shows that the 7 dimensions of line 8 fell into low level among8 dimensions, i.e., safety sign, traffic route map, emergency sign, direction sign of toilet, location map, internal space map of the station, and the continuity of direction signs. The score of the rationality of signs' setting, including the location of signs, font, installation height and applicability, was acceptable.

6.3 Insufficient of Study and Future Prospects

In this study, the "Subway Passenger Wayfinding System Questionnaire" was compiled and the dimensions of the wayfinding system were established. The structural equation modeling method established the weight of each dimension and was a methodological innovation for the evaluation of subway passenger wayfinding system. It has certain practical and theoretical significance. At the same time, there are some shortcomings in this study. First, the sample size should be further expanded. In this assessment, the average sample size per subway line is 35, which constitutes a small sample. Further scientific assessments also need to expand the sample size. Second, the selection of subway stations requires further clarification of sampling rules. The staff intensity on site and whether it is a transfer site will have an impact on the assessment. In the future, further research should be carried out to further verify the way in which structural equation modeling is established, and to evaluate the effectiveness of the rail transit passenger marking system.

Acknowledgments. This research was supported by National Key R&D Program of China (2016YFF0201700, 2016YFF0202806).

References

1. GB 3818-1983 Public information graphical symbols
2. GB/T 15566 (All parts) Public Information Guidance Systems- Setting Principles and Requirements
3. GB/T 20501 (All parts) Public Information Guidance Systems - Design Principles and Requirements for Elements
4. GB/T 10001 (All Parts) Public Information Graphical Symbols
5. ISO 28564-1:2010 Public information guidance systems—Part 1: Design principles and element requirements for location plans, maps and diagrams
6. Ji, X., Wei, X., Chen, F.: Evaluation of public transport system equity based on structure equation model: taking kunming as an example. J. Highw. Transp. Res. Develop. **30**(1), 126–132 (2013)
7. Zhang, B., Wang, Y., Zhao, H.: A research on the post evaluation of urban mass transit construction project based on SEM. J. Math. Pract. Theory **42**(17), 47–54 (2012)
8. Chen, J., Tang, W., Cai, X., Duan, L.: Multiple-group structural equation model of passenger satisfaction in urban rail transit. J. Transp. Syst. Eng. Inf. Technol. **18**(1), 173–178 (2018)

User Journey with Intelligent Public Transportation System in Shanghai

Yuqian Ni, Fan Yi$^{(\boxtimes)}$, Shaolin Ma, and Yanjuan Sun

ISAR User Interface, 127 Guotong Rd, Shanghai, China
yuqian_ni@isaruid.com, fan_yi@isaruid.com

Abstract. Intelligent Public Transportation Systems (IPTS) are making contribution to better transport efficiency and energy user, as well as assisting travelers to have better experience while taking public transportation services. Existing literature paid more attention to the technical aspect around this topic, while left the user side not being examined thoroughly. The current study aimed to reconstruct user travel chains and to build a scenario-based user journey map for IPTS usage in the city of Shanghai. We approached this objective by conducting a combination of methods, including field observation, diary entries and in-depth interviews. Travelers' behaviors, perceptions and needs were analyzed at each stage of the trip. Findings showed that IPTS are making it easier for people to travel with public transportation services, but meanwhile there are problems unsolved. Travelers want to have more control and certainty about their trips, but these needs cannot be guaranteed via travel assisting tools, especially when the trip is complicated with multiple transfers or even multiple means of transportation. Besides, people's dependency on mobile trip planner and navigator seem to be high, which could lead to anxiety about not having enough mobile phone battery.

Keywords: Intelligent Public Transportation · User experience · Journey map

1 Introduction

Intelligent Public Transportation Systems (IPTS) have been developed for decades in urban areas worldwide. The purpose of building IPTS is two-fold: (1) to assist transportation staffs to manage service and maintain the performance of transportation networks; (2) to help users plan and execute trips with ease and satisfaction [1, 2]. Paying special attention to the user side of IPTS, this paper examined Shanghai travelers' behavior journey while taking public transportations. We adopted the approach of scenario-based modeling [3] and investigated users' behaviors and perceptions in each phase of the trip, i.e., before/during/after the trip. We aimed to build a comprehensive user journey map for Shanghai's IPTS, and to distill some key factors which affect users' behavior patterns and perceptions. Findings of the study can assist IPTS developers and designers to improve Shanghai IPTS's facilities so that better user experience can be attained.

As a metropolitan, Shanghai has relatively advanced IPTS. Real-time schedules are accessible at all metro stations, and also available at increasing number of bus stations,

© Springer Nature Switzerland AG 2019
H. Krömker (Ed.): HCII 2019, LNCS 11596, pp. 487–499, 2019.
https://doi.org/10.1007/978-3-030-22666-4_35

as well as a few mobile apps. Mobile payment is widely supported, travelers are able to use smart phones to either purchase tickets or add balance at vendor machines. If people choose to go to the ticket-less way, they can even scan phones on ticketing machines directly, without bothering taking out physical tickets or transportation cards. Complementing bus and metro, vehicle sharing (bike and e-car) and ride sharing (Uber equivalence) are easily assessable, helping travelers to arrive the stations and to complete the "final mile". Moreover, trip planning tools, which support not only unimodal travels, but also multimodal ones, are assisting travelers to familiarize with their trips in advance. Some software can also assist travelers during the trip, such as showing which station they are at, and remind passengers not to miss transfer point or destination.

To better understand people's usage of IPTS in the city of Shanghai, we paid attention to not only the primary public transportation modals, i.e. bus and metro, but also any complementary modals which fulfill the trips, such as vehicle sharing and ride sharing services. In addition, we examined trips with familiar routes and unfamiliar routes as distinct scenarios. For unfamiliar routes, travelers require to allocate more planning time in advance. Though trip planning is not a novice topic, most studies approach this topic from technical aspect [4, 5], while left the user experience side not getting sufficient attention. For familiar routes, the planning phase is less intense, yet travelers still need to manage their trips properly so as to avoid unexpected frustrations. For both route types, IPTS can provide helpful services and tools, but the question is how people are making use of these services and to what extent are these intelligent services really solving problems. To answer such questions, the current study adopted combined research methods, including field observation, diary entries and in-depth interviews, to investigate users journeys while using IPTS in Shanghai.

2 Literature Review

Intelligent public transportation has been studied for decades. Existing literature on this topic can be generally divided into three categories: (1) infrastructure oriented, (2) integrating user aspect, but still more technology related, and (3) user-centric. The first category deals with the foundation of intelligent public transportation system. It reviews the architecture and technologies which enable IPTS [1, 6]. The possibility of applying new technologies to enhance the systems are also discussed in these literatures, for example using Zigbee and NFC for better data transit [7, 8], and adopting radio-frequency identification technology or mobile communication network to track passenger flow [9, 10]. These studies mostly discuss IPTS from architecture-perspective, yet the user side is not covered.

The second category of relevant literature embraces the interaction between users and technologies, and covers various aspects throughout the journey. For example, during the trip planning phase, Földes and Csiszár proposed a personalized route planning model which takes users' personal preferences into account [11]. Optimizing solutions for multimodal journey planning are also discussed in prior studies [12, 13]. Besides, time estimation is a topic that has been studied intensively. Simeunović et al. worked on an algorithm that aim to yield better transfer time estimation [14]. Pagani et al. put forward a system which not only shows estimated arrival time, but also the cumulated probability for arriving on time [15]. Moreover, some other studies pay attention to the constant changing environment while travelling. Systems have been designed to recognize and reconstruct travelers' surrounding circumstances, and offer them with precise contextual instructions and suggestions [16, 17].

Nonetheless, though the second category of studies make contribution to the user experience of IPTS by investigating optimized algorithms and models, these studies mostly take the technical perspective, paying more attention to technical difficulties and corresponding solutions. Users' behaviors and perceptions about IPTS are barely mentioned in these studies, but are the main topic of the third category of relevant literature. Digmayer et al. analyzed travelers' activities by using a scenario-based approach [3]. They constructed personas and scenarios for unimodal and multimodal travel chains, and identified travelers' information needs at each phase of the journey. By identifying these user needs, travel information apps can be better designed with proper content and functions. Islam et al. examined how travelers are using, and what factors drive their usage of ubiquitous real-time passenger information tools [5]. They claim that better understanding user behavior of using such tools can help to clarify technological areas that are necessary for more investments, and assist to make IPTS more effective.

The current study will follow the direction of the third category of relevant literature, i.e. the user-centric approach. The user side of IPTS has not been studied thoroughly. More work need to be done in this area to assist people in relevant fields to better understand travelers' behaviors and perceptions around IPTS, which can be helpful to technical-oriented studies as well.

3   Methodology

We employed a combination of methods for the current study. Field observations, diary recordings and in-depth interviews were conducted consecutively, leading the researchers to gain a better understanding about user's behavior patterns and perceptions around IPTS related tools and services throughout their journeys.

3.1   Field Observation

Field observation was carried out as the first phase. This phase purpose to assist the researchers to familiarize with the current implementation of intelligent services in Shanghai's public transportation system, so as to construct a more accurate and up-to-date context for later examinations. Also, by doing field observation, the researchers were able to gain a general sense about how Shanghai citizens are using and thinking about their intelligent public transportation system.

We conducted observations at 5 bus stations and 5 metro stations, spread across 2 major districts in the city of Shanghai, i.e. YangPu District and HuangPu District. At each station, we spent 30 min, paying special attention to these stations' implementation of intelligent transportation services, including real-time schedule boards, mobile payment portals, availability of nearby sharing bikes and e-cars, as well as ride-sharing pick-up areas. We also focused on people's behaviors and perceptions around these services, taking notes about people's usage of intelligent tools and conducting random short interviews. Our field observation covered both rush and non-rush hours. All observed data were recorded properly and prepared for further analysis.

3.2 Diary Recording

Following field observation, 10 participants were recruited with snowball sampling technique to write diary entries. All participants take public transportation for daily commute in Shanghai, and have some experience of using intelligent transportation services. Participants aged from 20–40, as suggested by a prior study that people at this age group are the most likely to use intelligent public transportation tools [1]. Also, because significant gender difference about using intelligent public transportation service was not expected [1], the gender ratio of the current study was set to 1:1.

Participants were required to write diary entries for 4 days, including 2 workdays and a weekend. They needed to cover trips on both familiar and unfamiliar routes during these 4 days. Content of diary contain basic trip information, including date, purpose of trip, starting point, destination, departure and arrival time, as well as means of transport, see Fig. 1. Besides, usages of intelligent public transportation services were also recorded, see Fig. 2. Participants took notes about their behaviors and perception about using trip planner, travel assistant (assisting during the trip), real-time schedule board, mobile payment, biking-sharing, e-car sharing and ride-sharing, etc. Completed diary entries were sent to researchers on daily basis via the Internet.

3.3 In-Depth Interview

All the participants who wrote diaries were also invited to take part in in-depth interviews. The interviews were conducted either face-to-face or via video chat software, and typically lasted for 40–60 min. Though interview script was prepared in advance, when carrying out these sessions, we did them in semi-structured format and modified questions based on each individual's diary entries. Participants were encouraged to elaborate on their behaviors and feeling at different stages throughout the whole trip, to explain how and why they choose to use certain intelligent transportation services, and to propose where they think relevant tools can be added to improve their experience of the journeys. All the interviews were voice recorded and transcribed for further analysis.

3.4 Data Analysis

During the phase of data analysis, we adopted a scenario-based approach [3], which allows researchers to closely examine the activities of users through each step of their public transportation journeys. We divided the journeys into the following steps: before leaving, on the way to transport station, at the station, on vehicle, transfer, at the final station, leaving the final station. For each step, we analyzed travelers' behaviors, perceptions and needs, so as to reconstruct the travel chains, and to build a comprehensive scenario-based user journey map. Though the scenario and persona which represented in the journey map are hypothetical, they actually originated from coded qualitative data from our field observation, diary entries and in-depth interviews.

Trip Information			
Date	*5-Jan-19*	Purpose of Trip	*Visit friend*
Departure Time	*13:00*	Arrival Time	*15:20*
Starting Point	*Yangpu, Shanghai (Home)*	Destination	*Songjiang, Shanghai (Friend's Home)*
Means of Transport	*Walk - Metro Line 3 - Metro Line 9 - Riding-Sharing*		

Fig. 1. Diary sample – trip information section

Usage of Intelligent Public Transportation Services		
Trip Planner	Did you use it during the whole trip?	*Yes, GaoDe Map*
	When and where did you use it?	*Plan the trip the night before*
	Any satisfying experience worth mention?	*The routes are quite straightforward*
	Any unsatisfying experience worth mention?	*Can't remind me what time to leave home*
Trip Assistant for Whole Journey	Did you use it during the whole trip?	*Yes, HuaSheng Metro*
	When and where did you use it?	*When I was on Line 3, to check which station I'm at*
	Any satisfying experience worth mention?	*Easy to see which station I'm at, quite accurate*
	Any unsatisfying experience worth mention?	*Sometimes lost Internet connection*
Real-time Schedule Board	Did you use it during the whole trip?	*Yes, the real-time schedule at platforms*
	When and where did you use it?	*When I was waiting for the train, I always look at it*
	Any satisfying experience worth mention?	*Pretty clear*
	Any unsatisfying experience worth mention?	*Importation information are not big enough*
Mobile-payment	Did you use it during the whole trip?	*Yes, Apple Pay*
	When and where did you use it?	*When I enter and leave the station*
	Any satisfying experience worth mention?	*Easy to use, no need to unlock my phone*
	Any unsatisfying experience worth mention?	*No*
Bike-sharing	Did you use it during the whole trip?	*No*
	When and where did you use it?	*N/A*
	Any satisfying experience worth mention?	*N/A*
	Any unsatisfying experience worth mention?	*N/A*
Car-sharing (Renting)	Did you use it during the whole trip?	*Yes, DiDi*
	When and where did you use it?	*When I was at the final metro station, I requested a car*
	Any satisfying experience worth mention?	*I can share my location with friends*
	Any unsatisfying experience worth mention?	*No*
Ride-sharing (Uber Equivalence)	Did you use it during the whole trip?	*No*
	When and where did you use it?	*N/A*
	Any satisfying experience worth mention?	*N/A*
	Any unsatisfying experience worth mention?	*N/A*

Fig. 2. Diary sample – usage of IPT service section

4 Result

4.1 Persona

To construct scenario-based journey maps, persona first need to be created. By creating persona, it would be easier for researchers to build empathy and think from the perspective of travelers. Also, persona help the researchers to delve deeper into the meanings behind users' activities, and probe into travelers' needs throughout each stage of the trip. According to the profile of participants who took part in our study, we built a hypothetical female persona with age of 28, whose name is Lizhen Wang. Lizhen is a game operating manager based in Shanghai. She takes public transportation for daily commute and other in-city travels, and is a heavy smartphone user. Besides, Lizhen has very poor sense of direction.

4.2   Scenario 1. Taking Metro for Unfamiliar Trip

Lizhen was invited to a friend's apartment where she had never been. Her friend's place was at another district in the city of Shanghai. It would take around 1.5 h to go there with public transportation services. Lizhen took a metro-based route, and transferred metro lines once during the trip. A few intelligent transportation tools were assisting Lizhen to complete the trip.

Before Leaving. When planning the trip, Lizhen used a mobile trip planner. The planning activity was executed twice: first happened right after the invitation because she wanted to have a general sense about how to go there; Lizhen used trip planner again on the day before departure, which purposed to confirm the route. Among all suggested routes, Lizhen preferred metro-based ones, which in her mind were more reliable. Based on prior experience, Lizhen was quite confident about the estimated travel time for metro-based routes, yet she would still prepare extra 20 min in case unexpected situation happened. Lizhen hoped the trip planner could be more personalized and suggest routes according to her own preference. Lizhen barely choose to transfer between metro and bus, yet such routes were suggested to her quite often.

Way to the Initial Station. Lizhen did not want to spent too much time on the way to the initial station, so she rode a sharing bike for the "first mile". Since there were not plenty sharing bikes around Lizhen's apartment, she spent some time looking for one. When arrived metro station, she parked the bike at a specific area which is very close to the station. No navigation was needed at this stage.

At the Initial Station. Lizhen used mobile payment to enter the station. Before choosing a metro direction, she looked at her trip planner app for a double check. The metro platform has real-time schedule e-boards, but Lizhen did not look at it because she knew the metro would come every 3–5 min. She would look at the e-board when she had waited longer than the expected.

On Vehicle. In order to get information about where the train was at, Lizhen depended on the train's announcement as well as the electronic map. She did not use mobile travel assist for this information for the reason that her phone was occupied for entertaining activities, also because she concerned that mobile geo-location detection became inaccurate underground.

Transfer. Lizhen followed signs in the metro station to find the place where she would take the next train. To make sure taking the correct train, Lizhen looked at her trip planner again before choosing a train direction. Again, she did not look at the real-time schedule e-board, because she knew the next train would come soon.

At the Final Station. The final metro station was still relatively far away from Lizhen's destination, so Lizhen booked ride-sharing service when she arrived the final station. She met the driver at a specific station exit as they communicated via phone.

Leaving the Final Station. Lizhen shared her location with her friend when she was in the ride-sharing car, so that her friends could know where she was and when she would arrive.

4.3   Scenario 2. Taking Bus for Unfamiliar Trip

Lizhen needed to visit a client. The client's office was at an area that Lizhen had never been to, and that place could not be easily accessed via metro. Lizhen took bus for this trip. One transfer need to be made during the trip. Navigation tool played an important role assisting Lizhen to arrive on time.

Before Leaving. Lizhen always prefers metro because of its reliability. However, for this trip, taking metro would consume twice as much time as taking buses. With respect to time, Lizhen chose bus as the means of transport. Also, since Lizhen was not confident about the trip planner's estimated travel time for bus travels, she prepared extra 40 min.

Way to the Initial Station. Lizhen had never took that bus line before, so she used mobile navigation to help her look for the bus station. Though the navigator led Lizhen to the correct place, Lizhen did not recognize the bus station at the first place as that was a small station. Lizhen was hoping the navigator could show a picture of the bus station which might help people to recognize it.

At the Initial Station. Once arrived the station, Lizhen opened the bus schedule app to find out the estimated bus arrival time. Based on prior experience, the estimated arrival time is not always accurate, but it could gave Lizhen a general sense about how long she had to wait. Besides estimated time, this app also indicated how many stops remaining for the bus to come. This is a feature that Lizhen liked particularly because it added some certainty to bus travels.

On Vehicle. Because the announcement on bus could be inaccurate and misleading, when on the bus, Lizhen used navigator to track her own location. The navigator showed which station she was at, and how many stations remaining before getting off. Nonetheless, since phone was used for navigating, Lizhen could not do other entertaining activities such as watch streaming videos. Lizhen hoped the navigator could sent her a push notification few stops before getting off, so that she did not have to open the navigator for the whole time.

Transfer. Lizhen got off the first bus at station A, while she needed to take the second bus at station B. Once getting off the first bus, Lizhen used navigator to look for station B, and also opened bus schedule app to check the second bus's arrival time. Though this time Lizhen did not spend too much time on transfer, she actually could not know how long it would beforehand, neither did she trust the estimated transfer time from trip planner.

At the Final Station. Lizhen looked at the bus stop name to confirm that she arrived the correct station. Then she found a sharing-bike parking site near the station, so she took one and headed for destination.

Leaving the Final Station. Lizhen needed navigation for the "final mile", but she was riding a bike without a phone mount. Therefore, Lizhen had to stop occasionally to look at her phone to make sure that she was on the right way.

4.4   Scenario 3. Familiar Trip

For daily commute, Lizhen can either take bus or metro. When taking bus, there is no need for transfer, and usually takes less time. When taking metro, Lizhen needs to transfer once.

Before Leaving. Lizhen checks estimated bus arrival time via mobile app, and leave home when the bus is coming soon. No intelligent transportation tools are used if she choose to take metro to workplace.

Way to the Initial Station. Lizhen walks to bus/metro station from home. If she finds a sharing bike on the way, she will use it. However, she will not look for sharing bike purposefully, because looking for a well-functioning bike may consume more time than walking to the station.

At the Initial Station. If taking bus, Lizhen looks at estimated bus arrival time from phone once she arrives station. If taking metro, she uses transportation card to enter the station. Lizhen does not use mobile pay during rush hours because mobile pay is not reliable sometimes, which can cause frustration and delay for commute.

On Vehicle. Lizhen is very familiar with the daily commute route, so no travel assistants are needed during this stage.

Transfer. When taking metro for commute, Lizhen needs to transfer once. She will look at the estimated train arrival time on the platform to make sure that she can arrive office on time. The estimated arrival time for metro are usually accurate.

At the Final Station. Lizhen will try to find a sharing bike to reduce the time spent on the "final mile". Nonetheless, there are not always enough sharing bikes parked around the final bus/metro station, so Lizhen usually left home a bit earlier in case she has to walk the "final mile".

Leaving the Final Station. Lizhen is familiar with the route from the final station to office, so no mobile navigation is needed.

Fig. 3. User journey map for in-city public transport with intelligent services

5   Discussion

5.1   Travelers Heavily Rely on Intelligent Transportation Tools

With the existence of trip planner and navigator, travelers do not have to memorize unfamiliar routes and are able to refer to relevant mobile apps for confirmation at any time. As one participant pointed out in the interview that "I like to keep it (a mobile app merging the function of both trip planning and navigating) open for all the time. Every time I need to make a decision (such as choosing a side at the metro station, or deciding whether to get off a bus), I'll look at it. I just cannot remember those information, neither do I want to memorize it." Besides trip planning and navigating software, other intelligent transportation tools are also frequently used throughout various stages of the trip, for instance, scanning smartphone for mobile payment when entering/exiting metro stations or getting on/off buses, and unlocking sharing bikes for the trip's first/last mile. For participants in the current study, their usage of intelligent public transportation tools covers the timespan of the whole trip, and they have built high dependency on these tools.

These tools, on one hand, are assisting travelers to complete their trips with more ease, but on the other hand, are also creating an anxiety about ensuring their smart phones are functioning well throughout the whole trip. Battery is the prerequisite for a phone's well-functioning, and is also a factor that travelers concern a lot: "I barely used my phone on metro though I really wanted to, because I still need to use the phone to exit the metro station, and also use it to unlock a sharing bike. I was just hoping my phone do not die on the way." In addition to low phone battery, poor phone signal is another factor that can bother travelers. When travelling to an area where phone signals were barely receivable, it could be desperate. Travelers heavily rely on travel assisting tools, while these tools only work when the smartphones are functioning well. If foreseeing phones' incapability, participants in our study would choose alternate method in advance, such as using physical transportation cards for payment, or bring supplementing devices, for example portable power banks.

5.2   Travelers Want More Certainty During the Journey

Travelers tend to avoid transferring between metro and bus, or bus and bus, because of much uncertainty. As a participant mentioned in the interview "oftentimes I need to go to another bus station for the next part of my journey. This worries me because first I am not confident about my ability to find that bus station without too much effort, and second I do not know how long I need to wait for that bus." Neither are travel assisting tools dealing well with such uncertainty caused by transferring with buses. Though researchers are working on algorithms to estimate transfer time happened in public transportation [14], according to our participants, the performances of existing trip planners, with regard to estimating time for transfer with buses, are not satisfying. Participants never know how long it will take until they are actually doing the transfer. For such reasons, travelers are more likely to avoid paths with bus-transfer even tripper planners are indicating that these paths are taking the least time.

Another uncertain situation that travelers tend to avoid is using unreliable mobile payment methods during rush hours. Mobile payment methods are getting iterated and optimized gradually, however, at the current stage, the hardware and software that some traveler use do not perform well consistently. Problems that travelers encounter include slow loading time, QR code not being identified, no low balance alert, malfunction station gate, and etc. These problems can happen to travelers occasionally and make them late for commute or blocking other people who are waiting to enter the stations. For such reason, a more reliable alternate, such as physical transportation card, are still used by mobile payment users during rush hours. Nonetheless, participants also mentioned that unreliable issues are more likely to happen to QR code-based mobile payment methods, instead of NFC-based ones. Therefore, it can be expected that if NFC-based mobile payment become wider adopted or QR code-based methods resolve the reliability issue, mobile payment may play an important role during rush hours.

5.3   Travelers Trace for Smoothness While Travelling

The reason of travelers' avoiding uncertain situations is partly due to their pursuit of smoothness during the trip. Both unpredictable bus-transfer and unreliable QR code-based mobile payment can make travelers feel trapped in certain stage of the journey. On the contrary, transfer within metro stations and pay fare with reliable methods are less likely to yield unexpected troubles and afford smoother travel experience, therefore receive popularity. In addition to transfer and payment, another issue that closely relates to travel smoothness is the designated parking area of sharing bikes. Though sharing bikes are making contribution to resolving the problem of "first/last mile", assisting travelers to arrive destinations more efficiently, relevant experience is not always as satisfying as expected. As our participants mentioned, "sometimes I really have to spend a lot time looking for a bike, it makes me upset" and "a few times, I have to park the bike kind of far away from the station, and then walk there, it is like extra walking distance for me". It is reasonable to expect that the travel of "first/last mile" can be smoother if more sharing bike parking areas are designated around transportation stations, residential areas, as well as business districts. Nonetheless, such parking lot designation also need to ensure not causing other problems, such as blocking the pedestrian lane. Various stakeholders need to be involved to work out a solution that suits best for all kinds of travelers, as well as the city overall.

5.4   Travelers are Looking for More Human-Oriented Experience

Though existing trip planners and navigators allow users to modify paths by choosing fewer time/transfer/walking, or bus/subway first, participants in the current study are hoping the intelligent tools can consider even more experience-related factors, such as weather and condition of roads. For people taking public transportation, weather is a crucial factor to be taken into account, as different weather can dramatically affect travelers' choice of transport, the travel time, as well as the overall travel experience. However, to our best knowledge, current travel assisting tools have not taken weather into consideration, and make traveler feel that "it only plans routes based on some mysterious numbers while I would like it to consider more about my comfort." Some trip planners will suggest travelers to wait at an unsheltered bus station during a heavy raining day, which according to a participant "is very unhuman". Besides weather, road condition is another "human factor" that travelers care about. As a participant pointed out "I will not always follow its walking navigation, I'll look at the general direction and then choose my own path. The navigator sometimes points me to some unclean and narrow streets, which may take less time, but I definitely do not want to go that way". For travelers, they hope the travel assisting tools can regard them more as "human instead of data points", and give travelers' comfort and safety higher priority while planning routes for them.

6 Conclusion

Intelligent public transportation systems are providing helpful services for travelers to complete their journeys with more efficiency and better experience. It seems that some travelers have built very high dependency on intelligent travel assisting tools throughout various phases of their journeys. However, such high usage of relevant tools also causes anxiety about the well-functioning of smartphone. Besides, travelers are looking for journeys with more certainty and smoothness, while existing IPTS still have some incapability in addressing these objectives. Travelers also hope travel assisting tools can take more human factor, such as the degree comfort and safety, into consideration, so that people can regard them are more thoughtful and caring assistants, instead of tools without much warmth.

References

1. Elkosantini, S., Darmoul, S.: Intelligent public transportation systems: a review of architectures and enabling technologies. In: International Conference on Advanced Logistics and Transport (ICALT), pp. 233–238. IEEE (2013)
2. Esmaeili, L.: Rural intelligent public transportation system design: applying the design for re-engineering of transportation eCommerce system in Iran. Int. J. Inf. Technol. Syst. Approach (IJITSA) 8(1), 1–27 (2015)
3. Digmayer, C., Vogelsang, S., Jakobs, E.M.: Designing mobility apps to support intermodal travel chains. In Proceedings of the 33rd Annual International Conference on the Design of Communication, p. 44. ACM (2015)
4. Grotenhuis, J.W., Wiegmans, B.W., Rietveld, P.: The desired quality of integrated multimodal travel information in public transport: customer needs for time and effort savings. Transp. Policy 14(1), 27–38 (2007)
5. Islam, M.F., Fonzone, A., MacIver, A., Dickinson, K.: Modelling factors affecting the use of ubiquitous real-time bus passenger information. In: 5th IEEE International Conference ON Models and Technologies for Intelligent Transportation Systems (MT-ITS), pp. 827–832. IEEE, June 2017
6. Esmaeili, L., Hashemi, S.A.: Toward the design of rural intelligent public transportation system rural public transportation of Iran. In: 7th International Conference on e-Commerce in Developing Countries: With Focus on e-Security (ECDC), pp. 1–8. IEEE (2013)

7. Bian, J., Yu, X., Du, W.: Design of intelligent public transportation system based on ZigBee technology. Int. J. Performability Eng. **14**(3), 483 (2018)
8. Alan, U.D., Birant, D.: Server-based intelligent public transportation system with NFC. IEEE Intell. Transp. Syst. Mag. **10**(1), 30–46 (2018)
9. Oberli, C., Torres-Torriti, M., Landau, D.: Performance evaluation of UHF RFID technologies for real-time passenger recognition in intelligent public transportation systems. IEEE Trans. Intell. Transp. Syst. **11**(3), 748–753 (2010)
10. Pinelli, F., Nair, R., Calabrese, F., Berlingerio, M., Di, L.G., Sbodio, M.L.: Data-driven transit network design from mobile phone trajectories. IEEE Trans. Intell. Transp. Syst. **17**(6), 1724–1733 (2016)
11. Földes, D., Csiszár, C.: Route plan evaluation method for personalised passenger information service. Transport **30**(3), 273–285 (2015)
12. Talasila, P., Haldar, A., Pai, S.S., Goveas, N., Deshpande, B.M.: Multimodal transit scheduler: an actor-based concurrent approach. In: 20th International Conference on Intelligent Transportation Systems (ITSC), pp. 1–6. IEEE (2017)
13. Dalkılıç, F., Doğan, Y., Birant, D., Kut, R.A., Yılmaz, R.: A gradual approach for multimodel journey planning: a case study in Izmir. Turkey. J. Adv. Transp. **2017**, 14 (2017)
14. Simeunović, M., Pitka, P., Basarić, V., Simeunović, M.: Model and software-based solution for implementing coordinated timed-transfer passenger transport system. In: 5th IEEE International Conference on Models and Technologies for Intelligent Transportation Systems (MT-ITS), pp. 303–308. IEEE (2017)
15. Pagani, A., Bruschi, F., Rana, V.: Time of arrival cumulative probability in public transportation travel assistance. In: IEEE 20th International Conference on Intelligent Transportation Systems (ITSC), pp. 1–6. IEEE (2017)
16. Pagani, A., Bruschi, F., Rana, V., Restelli, M.: User context estimation for public travel assistance and intelligent service scheduling. In: IEEE 20th International Conference on Intelligent Transportation Systems (ITSC), pp. 1–8. IEEE (2017)
17. Stanciu, V.D., Dobre, C., Cristea, V.: Context-based service for intelligent public transportation systems. In: Eighth International Conference on Complex, Intelligent and Software Intensive Systems, pp. 353–358. IEEE (2014)

Investigating Users' Responses to Context-Aware Presentations on Large Displays in Public Transport

Romina Kühn[1]([⊠]), Diana Lemme[1]([⊠]), Juliane Pfeffer[1]([⊠]),
and Thomas Schlegel[2]([⊠])

[1] Software Technology Group, TU Dresden, Dresden, Germany
{romina.kuehn,diana.lemme,juliane.pfeffer}@tu-dresden.de
[2] Institute of Ubiquitous Mobility Systems,
Karlsruhe University of Applied Science, Karlsruhe, Germany
thomas.schlegel@hs-karlsruhe.de

Abstract. Public displays are increasingly used in public transport to present information such as departure and arrival times or network maps. Since this information is displayed generically, users often have problems to find the specific information they need. We propose context-aware visualizations on public displays to support passengers by improving personalized information access. Several visualizations for this domain were identified, for example, highlighting individual route or pricing information by fading out the background or increasing readability by font size adaptation. To investigate the influence of adapted content on the user we tested prototypical presentations that show personalized information concerning personal trips. In our user study with 20 participants we analyzed these visualizations to compare their efficiency in contrast to non-adaptive content by measuring time to perform specific tasks. This work presents the results of our user study. They show that especially highlighted information supports the user in finding personalized information faster.

Keywords: Public displays · Public transport ·
Personalized information · Visualization · Highlighting · User study ·
Context adaptation

1 Introduction

Public displays already found their way into public space and offer a wide range of information to users, e.g., latest news, advertisements [2,17] or information regarding a specific topic or location [19]. Displays offer various benefits over printed media to both information recipients and providers. On the one hand, content can be exchanged nearly automatically that simplifies updating. On the other hand, displays can present a wide range of media types (e.g., text, images, or videos). This variety enables to enhance information density compared to

© Springer Nature Switzerland AG 2019
H. Krömker (Ed.): HCII 2019, LNCS 11596, pp. 500–514, 2019.
https://doi.org/10.1007/978-3-030-22666-4_36

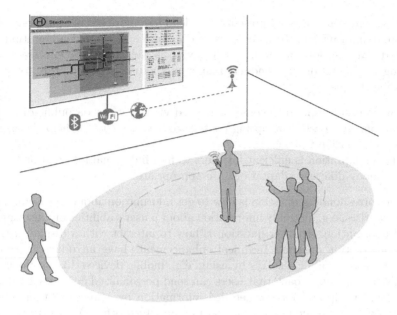

Fig. 1. Public display scenario with public transport content.

printed media. It extends visualization and consequently interaction possibilities and can provide better information at any time or situation.

We focus on public displays in the field of public transport (Fig. 1). Since passengers prefer displays at stopping points over printed information [9], displays are used to present, e.g., schedules, departure times, or platform information digitally. With the large amount of existing information in public transport, it can be challenging to find the specific information a user is searching for. For example, most passengers in public transport are only interested in a specific line or departure time to reach their destination. However, the needs for information differ from person to person [11] and depend on their respective context. Especially in case of homogeneous types of media or information, e.g., a lot of text or numbers, required information is difficult to find [15]. Since information is often displayed generically without user-specific individualization or preferences, users often have problems to find the specific information they need, especially in time-critical situations. Therefore, applying context-aware public displays to the public transport domain is comprehensible and useful.

We address these issues by providing personalized visualizations for public transport services from mobile phones, e.g., coming from route-planning applications, presented on public displays. Consequently, we aim at adapting the representation of a displays content to the users context. This is supposed to support the user to find personalized information faster. In this work, we investigate the approach of visualization adaptations in order to personalize for individual context and the efficiency of adapted in contrast to non-adaptive presentations.

We assume that adapted presentations lead to a more efficient understanding of display information. To investigate the influence of context adaptation and adapted visualization on the user's behavior as well as on the effectiveness of passenger information, we conducted an empirical user study in order to answer the following questions:

1. How do users respond to context-adapted visualizations on public displays?
2. Does our approach of adapting content to a user's context help the user to find information faster?
3. Which adaptation is efficient for users to find individual information?
4. How many different adaptations are appropriate?

One corresponding question is how to get information about the user context that comprises, e.g., not only information about a user's abilities but also specific travel information. Another question is how to interact with a display without direct interaction options. Technical enhancements have introduced means of interaction with public displays by using, e.g., mobile devices [13,14]. We utilize Kühn's et al. [14] approach that users can send personalized content,e.g., details of a planned trip and receive detailed information by connecting their mobile devices (with a specific mobile application and Bluetooth or Wi-Fi) to the public display. Personal settings of the mobile device, such as color or language preferences, can be used to adapt the presentation or travel details.

To clarify our approach this work is structured as follows: We present relevant *Related Work* in the field of context-aware public displays and personalized visualizations in public transport. Then, we describe the *User Study* we conducted to investigate the users' behavior during the usage of a public display and the corresponding *Results*. The paper concludes with a *Discussion* of the results and *Future Work*.

2   Related Work

To examine challenges and chances of context-aware presentations we give a brief overview of current public display approaches that also include context adaptation and personalization as well as visualizations in public transport.

2.1   (Context-Sensitive) Public Displays

Public displays are often located in areas with high pedestrian traffic, such as train stations, transport hubs, or public places and increasingly replace the usually analog road maps and timetables. As part of digital signage systems [14], they mostly show digital content, but passively. In contrast, active public displays enable the user to explore information but these kinds of displays are seldom in use so far. Possible interaction modes between users and displays are touch interaction [20] or distant interaction, for example, through gestures [4] or a user's mobile device [13,14,16]. The interactive public display UBI-Hotspot [19] offers three levels of interaction on a single physical display. These levels range from no interaction (only passive information display) over direct touch interaction to an interactive display that can be controlled by a personal mobile phone. Kühn et al. [14] described a concept for mobile interaction with a public display and implemented a prototypical application for a fictional public transport provider. By connecting their mobile devices to the display they receive detailed information, e.g., concerning their trip. Furthermore, the user can browse the displayed information on the screen using a mobile phone. In general, with the support of public displays in public transport, passengers feel more secure because of knowing the current departure time or time remaining until departure [8]. Providing various ways to interact with displays can also be an additional benefit for browsing information.

To adapt the visualization of a display to user preferences the users and their situation have to be known, so-called user context. According to Abowd et al. [1], there are four main types of context: location (where the user is), identity (who the user is), activity (what the user is doing) and time (when the user is doing this). A context-sensitive system uses context to provide relevant information and/or services to the user, whereby relevancy depends on the user's intention. Since collecting context data is not trivial there is a lot of research on this. For example, Cardoso and José [7] introduce a framework for designing context-aware displays. They describe the concept of digital footprints that can be used to dynamically characterize a place. Context-aware display systems can use this to adapt their content or interaction mode to the social environment of a place. Kühn et al. [14] propose to use personal settings from mobile phones as user context for adaptation.

2.2 Personalization

In general, personalization is defined as the adaptation of services to fit the needs and preferences of a single user or a group [12]. Personalization can be system- or user-initiated [6]. According to Tam and Ho [22] three types of personalization can be differentiated: user-driven personalization (user specifies relevant information and content), transaction-driven personalization (system generates personalized content) and context-driven personalization (adaption is related to current context information).

Fan and Pool [10] distinguish between three dimensions: what to personalize, whom to personalize and what does the personalization. Personalization can be performed by a system(explicit) or a user (implicit). Content functionality, channel or user interface can be personalized. Bus Catcher [5], a personalized and mobile context-sensitive system, enables users to access information of public transportation. GPS provides real-time data about bus stops, routes and schedules for passengers. With schedules from a server, the users get latest information. A map shows the current locations of users and buses. Another feature of Bus Catcher is an alerting mechanism for tourists: if a bus passes close to an attraction, the tourists, e.g., get information about that attraction including opening times. These approaches give some conceptual ideas how context can be identified and presented but they lack in propositions about efficiency of adapted content presentation.

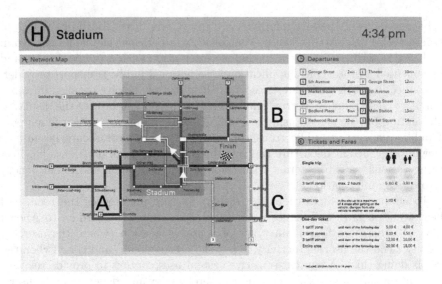

Fig. 2. Exemplary presentation of content in public transport with highlighted content. A: Shapes and colors are used for highlighting. B: Additional forms accentuate content. C: Blurring unimportant content. (Color figure online)

2.3   Visualization in Public Transport

One advantage of interactive visualizations is that information can be filtered or adapted to user preferences. Robinson [21] introduces various adapted presentations for public transport, such as color-based highlighting. Kühn et al. [15] differentiate their presentations in single and combined highlighting as well as levels of information complexity. The different existing kinds of accentuations, such as symbols, flashing or blurring gave us initial ideas for our highlight effects that are shown in Fig. 2. Color- and symbol-based highlighting is represented in Fig. 2A. Using a rectangle around a displayed location is shown in Fig. 2B. Figure 2C uses two different color techniques to emphasize important data and suppress unimportant data. The ticket rate is colored red to find it faster and other tickets and fares are blurred.

Lemme et al. [18] conducted a study to investigate the needs of personalized visualization in public transport. The participants of their user study stated that they would use a context-aware display for planning their trips with public transportation, obtaining information about tickets and fares, finding the shortest or fastest way from a current location to a destination, and having a look at different departure times. A survey proved advantages, such as a better overview of important information, time advantages by finding the correct information faster or "up-to-date" advantage by finding the current and correct information.

Our quantitative user study aims to validate the results of Lemme's et al. [18] user survey that proved that an adapted public display supports users in finding information more comfortably. We use their approach to examine the effects of context-aware presentations concerning efficiency.

3   User Study

To investigate the influence of context adaptation and adapted visualization on the user's behavior as well as on the effectiveness of passenger information, we conducted an empiric study. We set up a study to examine the potential of context-sensitive public displays, to analyze adapted visualizations regarding their efficiency as well as possible problems occurring during the usage of such presentations in public transport. Furthermore, the user study serves as part of a planned analysis of multi-user scenarios in public space. In this work, we focus on a single-user scenario to investigate the efficiency of presentation adaptations. We examine the aspect of supporting the planning of a user's trip by adapting a public display to the following contexts:

- Current location or destination by highlighting positions or routes using flashing,
- Daytime by inverting the colors of background and text - light background and dark text during the day, dark background and light text at night, and
- Distance between user and display by changed font sizes.

For this purpose a public transport display was adapted and divided in different information areas. This section reports on the study's *Test Set-Up, Participants, Procedure* as well as *Design and Tasks*.

3.1   Test Set-Up

In our user study, we investigated a scenario with one display and one person (1:1 scenario). The study was performed in a lab environment to focus on the reaction of the person without external distractions. The hardware set-up for our study consisted of a projector that presented the information and a laser pointer that was used by the test persons to select their answers on the projection. The participants were located in a distance of about 3 m in front of the projection (Fig. 3). To measure the response time of the users for each task the test leader used a stopwatch. For a better documentation and evaluation the tests were recorded with a video camera to capture the behavior of the participants and how they responded to the system.

3.2   Participants

For our user study we invited twenty participants by e-mail, aged between 21 and 45. The 9 male and 11 female participants all reported that they use public transport often. We divided our participants into two groups with 10 persons

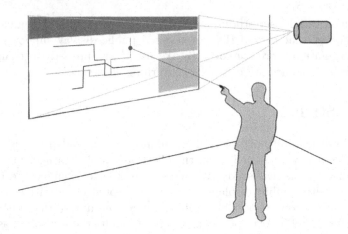

Fig. 3. Simplified view on test set-up.

each by a random selection process. Both groups had the same tasks but one group solved the given tasks with a context-aware public display with adapted information presentation (*context group*). The second group served as *reference group* and used a non-adaptive display. Each participant performed the study separately.

3.3   Procedure

For each participant we had the same test procedure: an introduction to the study and the performance of the given tasks. In the *introduction* the participants got a detailed description of the test scenario, the type of tasks that had to be performed and the test procedure. We used a similar example of the test display that was described in written form for explanation. Furthermore, the participants had the opportunity to ask questions concerning the user study. A pre-questionnaire gathered demographic data and included questions about their behavior in using the public transport.

During the *performance of the tasks* each participant had to complete several tasks. Starting with a black screen the tasks were provided in a random order. The test persons were asked to read the given task and to signal whether the test leader can start the test or has to answer questions concerning the respective task. In case of a positive signal the leader activated the projection and measured the time from the projector activation to a first answer (*reaction time*). To perform a task it was required that the test persons use the laser pointer to select their answer on the projection. Since the first answer was not necessarily correct the test leader also measured the time until the right answer was given (*task completion time*). Then, the test leader deactivated the projection and the participant had to perform the next task. This procedure was repeated for all questions. In the last part of the study the users were asked retrospectively about their impressions in using our public transport information system.

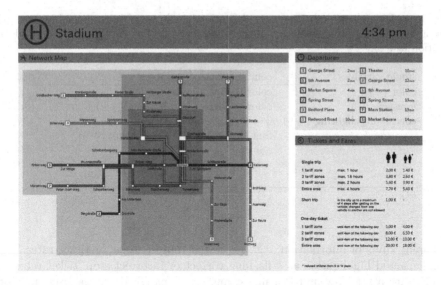

Fig. 4. Exemplary presentation of content in public transport without adaptation.

3.4 Design and Tasks

The users had to answer questions referring to a fictional trip. For this purpose, we illustrated the most important information on a large screen: a *network map* as used in many public transport networks, showing the different lines on a schematic map, the next *departures* from the station, the location the display is located at, the line numbers and destinations of the vehicles departing, as well as general ticket information for a fictional public transport provider. An exemplary presentation of this content is given in Fig. 4. The map is shown on the left side, departures and ticket rates are presented on the right side.

The participants had to perform two types of tasks with 11 tasks each. The first type included *simple tasks* where the participants had to find, for example, stops or routes on the network map. *Complex questions* were the second type of tasks that comprised combined information from different sections of the display representation. For example, the users were asked to find the cost for a specific trip. To answer this question, the participants had to find the route on the network map and then had to look up the correct ticket rate. For both types of tasks the test leader measured the time from the start of the task to the correct answer.

Since we wanted to investigate the influence of context on users we provided different types of context adapted visualizations. In a real world scenario these visualizations can be adapted to personal settings of a user's mobile phone, as mentioned previously, or environmental context. For example, to represent different day times we designed a night mode by *inverting the colors (c)*. The second context adaptation is the modification of the *font sizes (f)* to address the distance between user and display. The third type of adaptation is to highlight

Table 1. Different types of context adaptation: single variables and combinations of variables as well as the evaluation of their usefulness (++ = very useful to find information faster).

single variables	combinations of variables
color (+)	color + font size (+)
font size (+)	color + correct highlighting (++)
correct highlighting (++)	color + incorrect highlighting (- -)
incorrect highlighting (- -)	font size + correct highlighting (++)
	font size + incorrect highlighting (- -)
	color + font size + correct highlighting (++)
	color + font size + incorrect highlighting (- -)
(4)	(7)

different kinds of information, e.g., requested routes, stops, or ticket rates. We implemented highlighting as a flashing correct or incorrect stop, ticket rate, or route with a rotating circle for interchange stations. To prove the usefulness of highlighting we divided this type of adaptation into *correct (h)* and *incorrect highlighting (hi)* of information. With this approach we also wanted to test the distraction caused by wrong highlighting that could occur, if several persons would use the same display simultaneously and highlighting would be provided for only one of them.

We combined our four variables (color, font size, correct and incorrect highlighting) with each other and got 11 types of context adaptation (4 single variables and 7 combinations) (Table 1). Since there are simple and complex tasks each participant had to perform 22 tasks in total. For generating different public transport network maps for the single tasks for each test person, we implemented a prototypical software tool. The network map varied in its structure, so that the arrangement of stops and routes changed with each task in order to avoid learning effects.

4   Results

The user study provided a wide range of results. We got measured times for completing the tasks from each group and important feedback that helps to answer our research questions introduced in the first section.

4.1   Measured Completion Times and Task Correctness

The participants had to perform the tasks as fast as possible. We measured the time until the correct answer was given. In case of an incorrect answer (*error*) we measured both, the time until the first and the right answer. We extracted different types of data: *reaction time* and *task completion time*.

For simple tasks (Fig. 5), participants with context adaptation could give correct answers faster in five tasks (no. 5, 6, 9–11). The results for the complex tasks

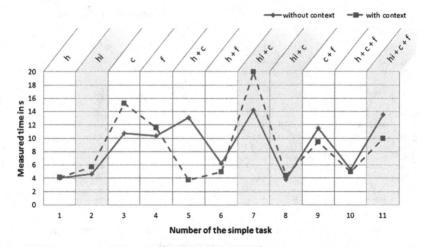

Fig. 5. Measured task completion time of simple tasks with and without context adaptation.

are shown in Fig. 6. Six tasks could be solved faster from the participants that used the adapted content (no. 12, 15–17, 20, 21). Comparing the tasks between both groups the test users completed 11 tasks faster with help of any kind of context adaptation. Even when comparing the average results with each other, persons with context information were considerably faster than those without context adaptation.

To answer the given tasks, the use of *correct highlighting* could be identified as a suitable visualization. Correct highlighting was applied in eight questions and influenced the task completion time positively seven times (no. 5, 6, 10, 12, 16, 17, 21). In these cases the correct answers were given faster compared to the reference group.

Color as well as *font size adaptation* could not achieve the desired effect. In particular, the color adaptation had no added value under laboratory conditions. Nevertheless, in combination with highlighting very good results were achieved. One explanation of this result could be the used flashing effect that is more intensive on a dark background (nos. 5, 16).

Incorrect highlighting seems to be confusing for the participants. In eight tasks incorrect highlighting was used (see grey columns in Figs. 5 and 6). Then, correct answers of seven questions were given substantially slower compared to the reference group.

The participants that used the adapted content presentation had 30 errors overall (see Table 2). In contrast to this, the reference group without any adaptions at all made 27 errors. Excluding the errors made due to incorrect highlighting a clearer picture emerges. Only 14 occurring errors remain in the context group. In both groups, complex tasks were prone to errors. Beside the incorrect answers given, we also distinguish invalid answers. In the context group, 13 (with incorrect highlighting) respectively 5 (without incorrect highlighting) tasks could

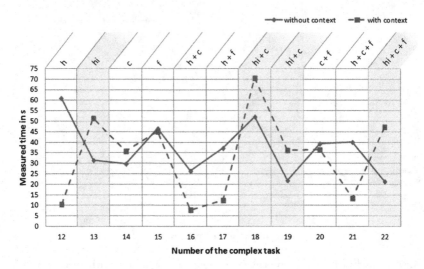

Fig. 6. Measured task completion time of complex tasks with and without context adaptation.

not be solved at all and therefore were invalid for the evaluation. Nevertheless, it was remarkable, that 8 out of 13 not valuable answers resulted from incorrect highlighting. The reference group only had two invalid answers.

Table 2. Overview of the number of errors that occurred in simple and complex tasks as well as invalid tasks.

	context group		reference group
	adaption excluding incorrect highlighting	adaption including incorrect highlighting	
errors in simple tasks	2	4	5
errors in complex tasks	12	26	22
$\sum$ errors	14	30	27
invalid in simple tasks	0	0	0
invalid in complex tasks	5	13	2
$\sum$ invalid	5	13	2

We found out that especially tasks that were a bit more difficult were often answered incorrectly. For task 13 *"How many stops do you pass between your current location and the stop Main Station?"* the participants had to know their current location and had to find the right destination to count the stops between them. The current location of the public display is shown on top of the display that is presented in Fig. 2. The name of the current stop is *Stadium*. Most of the test users of both groups did not find the name of their current location and, therefore, could not solve the task. Another example why our test users did not find the correct answer is that some participants did not know that short trip tickets are cheaper than single trip tickets and therefore more suitable for less than 4 stops. They often did not read the ticket fares properly and chose the wrong tickets. This is reflected in the results of task 18 again, that caused problems in both groups. Once more, these were tasks with incorrect highlighting that resulted in wrong or invalid answers. In summary, we can state that the test persons with correct highlighting were overall faster in solving a task and made fewer mistakes. In contrast, incorrect highlighting caused some problems. Highlighting false information confused and distracted the participants.

4.2 User Feedback

In general, our test users responded positively to context adaptation. Hence, they would use context-aware public displays in public transport to find the specific information they need. The reference group that performed the tests without adaptation on the display stated that context adaptation would be very helpful and that they would appreciate the option to get personalized information. Even though the context group had some difficulties during performing some tasks they agreed that correct context adaption is beneficial in finding specific information faster.

Using color and large font sizes was evaluated as a good way to adapt to daytime or distance between a user and a display. To highlight information with flashing was judged as a little distracting, especially in case of highlighting incorrect information. However, the correct content was found faster with flashing than without flashing.

As improvement suggestion, the test users stated that *highlighting with colors and/or shapes* would be more helpful. For example, a rectangle around a location (Fig. 2B) or a colored ticket rate (Fig. 2C) would be easy to find without distraction. To display *unimportant content in grey* while coloring important data (Fig. 2C) is another possibility to adapt information. The participants also stated that they do not want to exclude *flashing* as an option for highlighting. Instead, they would use it less to avoid an overload of accentuation.

Overall, the following aspects were emphasized considering content representation: The test users wanted more explanation to find important information such as detailed ticket information. Furthermore, they wanted a clearer presentation of their current location and suggestions for alternative routes in case they need to take another line.

5 Discussion

The positive user feedback as well as the measured task completion times suggests that our approach of an adaptive public display is useful to offer and to access personalized information for individual users. However, there are also some limitations of our approach, such as the variety of the used adapted visualizations as well as the exact number of different adaptations that are appropriate.

The test users of the context group performed their tasks faster with correct highlighting. In contrast, they were nearly three times slower in case of incorrect highlighting because they were distracted by incorrect information and also slower in comparison to no adaptation at all. However, in a real application context, incorrect highlighting is very likely because of different users catching up on public displays [3]. Additional investigations are required to find out how we can solve this problem.

In our user study, we investigated four types of adaptation: correct and incorrect highlighting, inverting colors, and changing font sizes (see Fig. 2). In particular, *correct highlighting* was rated very useful as well as a *combination* of correct highlighting and *inverting colors* (see Table 1). The combination of correct highlighting, color, and font size also led to good results. In contrast, incorrect highlighting was very confusing for the test users. The combination with other adaptations did not change this result that leads us to the assumption that finding information fast only depends on highlighting. This is a valuable approach for further investigations.

Generally, our approach of making a public display presentation context-aware by highlighting information and adapting to daytime and distance showed interesting results that may benefit users of public displays.

6   Conclusion and Future Work

In this paper, we presented an approach on supporting users in public transport by context-aware public displays. We carried out a user study to investigate different types of context adapted visualizations, their efficiency in finding information, and the users' responses to them. The results of this study varied with each context and task but showed that the adaptation to the context is generally useful and desirable. Especially, highlighted information supports users in finding personalized information faster.

Referring to our research questions in Sect. 1, we can conclude that our test users responded positively to adapted visualizations in general. The most efficient adaptation is correct highlighting, especially with additional adaptation such as color. We can also state that also three different adaptations were beneficial to the user. Overall, we found out that adapted presentations, as we used them, lead to find relevant information more efficiently.

In future work, a major challenge is to take into consideration that several viewers can use one display at the same time. Therefore, an important issue is an approach for multi-user visualization and interaction with large screens. According to our results with incorrect highlighting, one challenge is that personalization must not be too specific to both address security and privacy issues and enable everybody to use the public display simultaneously. The introduced context adaptations in this work represent examples of many diverse possible adaptations to user preferences. We propose that other adaptations have to be investigated as well, e.g., language adjustments or adaptation of the content colors according to the preferences of colorblind persons.

Acknowledgments. The European Social Fund (ESF) and the German Federal State of Saxony have funded this work within the project CyPhyMan (100268299). We would also like to thank Enrico Hinz for his work and support.

References

1. Abowd, G.D., Dey, A.K., Brown, P.J., Davies, N., Smith, M., Steggles, P.: Towards a better understanding of context and context-awareness. In: Gellersen, H.-W. (ed.) HUC 1999. LNCS, vol. 1707, pp. 304–307. Springer, Heidelberg (1999). https://doi.org/10.1007/3-540-48157-5_29
2. Alt, F., Müller, J., Schmidt, A.: Advertising on public display networks. Computer **45**(5), 50–56 (2012)
3. Alt, F., Schneegaß, S., Schmidt, A., Müller, J., Memarovic, N.: How to evaluate public displays. In: Proceedings of the 2012 International Symposium on Pervasive Displays, PerDis 2012, pp. 17:1–17:6. ACM, New York (2012)
4. Ardito, C., Buono, P., Costabile, M.F., Desolda, G.: Interaction with large displays: a survey. ACM Comput. Surv. (CSUR) **47**(3), 46:1–46:38 (2015)
5. Bertolotto, M., O'Hare, G.M.P., Strahan, R., Brophy, A., Martin, A., McLoughlin, E.: Bus catcher: a context sensitive prototype system for public transportation users. In: Proceedings of the Third International Conference on Web Information Systems Engineering (Workshops) - WISEw 2002, pp. 64–72 (2002)
6. Blom, J.: Personalization: a taxonomy. In: CHI 2000 Extended Abstracts on Human Factors in Computing Systems (CHI EA 2000), pp. 313–314 (2000)
7. Cardoso, J.C.S., José, R.: A framework for context-aware adaptation in public displays. In: Meersman, R., Herrero, P., Dillon, T. (eds.) OTM 2009. LNCS, vol. 5872, pp. 118–127. Springer, Heidelberg (2009). https://doi.org/10.1007/978-3-642-05290-3_21
8. Dziekan, K., Kottenhoff, K.: Dynamic at-stop real-time information displays for public transport: effects on customers. Transp. Res. Part A Policy Pract. **41**(6), 489–501 (2007)
9. Dziekan, K., Sedin, S.: Customer reactions to the implementation of a trunk bus network in Stockholm. In: Proceedings of UITP Conference (2005)
10. Fan, H., Poole, M.S.: What is personalization? Perspectives on the design and implementation of personalization in information systems. J. Organ. Comput. Electron. Commer. **16**(3–4), 179–202 (2006)
11. Grotenhuis, J.W., Wiegmans, B.W., Rietveld, P.: The desired quality of integrated multimodal travel information in public transport: customer needs for time and effort savings. Transp. Policy **14**(1), 27–38 (2007)
12. Jørstad, I., Thanh, D., Dustdar, S.: The personalization of mobile services. In: IEEE International Conference on Wireless and Mobile Computing, Networking And Communications - WiMob 2005, vol. 4, pp. 59–65 (2005)
13. Kaviani, N., Finke, M., Fels, S., Lea, R., Wang, H.: What goes where? Designing interactive large public display applications for mobile device interaction. In: Proceedings of the 1st International Conference on Internet Multimedia Computing and Service, pp. 129–138 (2009)
14. Kühn, R., Lemme, D., Schlegel, T.: An interaction concept for public displays and mobile devices in public transport. In: Kurosu, M. (ed.) HCI 2013. LNCS, vol. 8007, pp. 698–705. Springer, Heidelberg (2013). https://doi.org/10.1007/978-3-642-39330-3_75

15. Kühn, R., Moltkau, B., Schlegel, T.: Visual highlighting of information for public transport information systems. In: 8th Nordic Conference on Human-Computer Interaction Making Places: Visualization, Interaction and Experience in Urban Space (Workshop) (2014)
16. Kuikkaniemi, K., Jacucci, G., Turpeinen, M., Hoggan, E., Müller, J.: From space to stage: how interactive screens will change urban life. Computer **44**(6), 40–47 (2011)
17. Lasinger, P., Bauer, C.: Situationalization: the new road to adaptive digital-out-of-home advertising. In: Proceedings of IADIS International Conference e-Society, pp. 162–169 (2013)
18. Lemme, D., Kühn, R., Funke, A., Schlegel, T.: A feasibility study of context-adaptive visualizations in public transport information systems. In: Proceedings of the 13th International Conference on Mobile and Ubiquitous Multimedia - MUM 2014, pp. 262–263. ACM Press, New York (2014)
19. Ojala, T., et al.: Ubi-hotspot 1.0: large-scale long-term deployment of interactive public displays in a city center. In: Proceedings of the 2010 Fifth International Conference on Internet and Web Applications and Services - ICIW 2010, pp. 285–294. IEEE Computer Society (2010)
20. Peltonen, P., et al.: Extending large-scale event participation with user-created mobile media on a public display. In: Proceedings of the 6th International Conference on Mobile and Ubiquitous Multimedia, MUM 2007, pp. 131–138. ACM, New York (2007)
21. Robinson, A.C.: Highlighting techniques to support geovisualization. In: In Proceedings of the ICA Workshop on Geovisualization and Visual Analytics (2006)
22. Tam, K., Ho, S.: Understanding the impact of web personalization on user information processing and decision outcomes. MIS Q. **30**, 865–890 (2006)

Crossing the Street Across the Globe: A Study on the Effects of eHMI on Pedestrians in the US, Germany and China

Florian Weber[1]($\boxtimes$), Ronee Chadowitz[1], Kathrin Schmidt[2],
Julia Messerschmidt[2], and Tanja Fuest[3]

[1] BMW Group, Parkring 19, 85748 Garching, Germany
`florian.ww.weber@bmw.de`
[2] Spiegel Institut Mannheim GmbH & Co. KG,
Eastsite VI, 68163 Mannheim, Germany
[3] Chair of Ergonomics, Technical University of Munich,
Boltzmannstr. 15, 85748 Garching, Germany

Abstract. Automated vehicles (AVs) will be integrated into mixed traffic within the next few years. To replace human-human communication, different external human machine interfaces (eHMI) have been proposed. Interpretation of HMIs can be subject to cultural influences. To examine the potential of eHMIs across different cultures, as well as research the transferability of study results, three virtual reality studies were conducted in Germany, the United Stated and China. In all studies, participants had the role of pedestrians and had to recognize if an AV is yielding or passing the pedestrian. Two eHMIs, a lightband and a display were used to inform pedestrians of the AV's intention. When yielding to the pedestrian, eHMI showed benefits in terms of intention recognition in Germany and the United States. In China, however, eHMI did not show these effects. Results consistently show across all cultures, that eHMI deteriorates pedestrians' recognition of the AVs intention to pass and not yield. The authors conclude that eHMI should only be used in already safe states of the AV, such as when yielding. In addition, eHMIs should not be introduced into a different culture without considering necessary cultural adaptations. For testing eHMIs internationally, there is a need to carefully select comparable traffic scenarios. These scenarios have to take into account habitual behavioral patterns of the current traffic and therefore the learned behavior of road users as to what they expect to happen in such situations.

Keywords: Automated vehicles · Explicit communication ·
External human machine interface · Virtual reality · Vulnerable road users

1 Introduction

Within the next few years, automated vehicles (AV) will be integrated into today's traffic. From SAE level 4 onwards, no driver will be needed to steer the vehicle, and automation should handle nearly every situation independently [1]. AVs will communicate seamlessly with each other by means of new technologies such as Car2X and

© Springer Nature Switzerland AG 2019
H. Krömker (Ed.): HCII 2019, LNCS 11596, pp. 515–530, 2019.
https://doi.org/10.1007/978-3-030-22666-4_37

Car2Car [2]. AVs will, however, be integrated into manually driven traffic and must be able to communicate with human road users (HRU) in this complex socio technical system [3]. An insufficient integration might raise communication issues [4]. This can be particularly problematic when it comes to high risk traffic groups such as pedestrians [5]. The lack of communication between driver and pedestrian may decrease trust [6] and confidence [7]. Communication has been formed over time through social and cultural influences [8]. AVs therefore not only need to understand manual traffic, but have to adapt communication according to cultural specificities.

To communicate appropriately AVs can try to emulate human behavior. They can adapt their driving behavior and communicate intentions in an implicit way, such as by adapting speed and trajectories [9] or communicating directly [10] via already existing features, such as turn indicators or the horn [11]. Equipping AVs with novel external human machine interfaces (eHMI) which communicate additional messages might also be a beneficial approach [7, 12, 13] to replacing current driver behavior such as eye gaze, mimics, and head and hand gestures. In a video survey, it was found that pedestrians would cross the road more often when the AV was equipped with an eHMI [14].

Most eHMI solutions prevalent in showcars and academic research predominantly use technologies which rely on visuals, such as projection onto the street, externally legible displays showing either text or icons, or direct light from a light bar [15]. An overview of different eHMI approaches can be found in [16]. Replacing direct communication is a major challenge in the development of AVs, but also creates an opportunity to improve current interaction by establishing clear and consistent interaction patterns with eHMIs specifically designed for this purpose. Besides increasing the trust and acceptance in AVs, improvement of traffic flow is a major goal of introducing eHMIs. To improve traffic flow, fast and correct intention recognition is crucial.

References [17, 18] developed different eHMI solutions in a user centered design process: one being a light-band wrapped around the vehicle, emitting different signals, and a further one being an external display, showing icons (Fig. 1). These eHMIs are able to transmit the intention of the AV through the messages "AV will give way" and "AV will pass" when encountering other traffic participants. Indeed, communicating the intention and awareness of the vehicle has been considered a more fruitful approach for eHMIs than showing commands of what the pedestrian should do [19–22].

HMIs have generally been found to differ depending on the cultural background. For instance, design patterns in websites were found to differ depending on the cultural background that the website was created in [23]. This suggests that mental models and expectations of HMIs differ between cultures. If external HMIs for AVs are introduced into different cultural contexts, there are two possible solutions: EHMIs are either not adapted by the manufacturer and must therefore work cross-culturally, or they are indeed tailored to the specific markets into which the AV is introduced. However, cultural influence might not be limited to explicit communication through eHMIs. Reference [4] states that driving behavior differs between the US, Europe, and China.

The author argues that in the US and Germany drivers show more consistent behavior than Chinese drivers, who are more prone to traffic violations. Indeed, the general rate of cars stopping to give way to pedestrians was found to be very low in

Fig. 1 eHMI Variants used in the study (from left: Baseline, Light-band, Icon)

China [24]. As expectancy is formed through experience, people from a Chinese cultural background might expect different stopping rates than Westerners. Furthermore, attitude towards automated systems has been found to differ between Western and Asian cultures. In a study run in the aviation context [25] Asian and Western pilots differed in their preferences and enthusiasm towards automated systems, with Asian pilots being more enthusiastic towards automation than pilots from Western societies.

To examine the feasibility of transferring eHMIs cross-culturally, we conducted a virtual reality (VR) study on the influences of two different eHMIs on pedestrians' intention recognition in the US, Germany, and China. We expect eHMI to improve intention recognition across all scenarios compared to baseline. As the eHMIs used in this study [17, 18] have been developed in Germany, we expect both eHMIs to generally work best in the Western and especially German background compared to a baseline without eHMI. One of the two eHMI solutions included icons, which are known to be culturally dependent [26, 27]. The other solution was purely light-based, featuring a light-band – a solution that resembles already cross-culturally used features such as turn indicators – which we thought should be more universally understood. We therefore expected the light-based eHMI to perform in a more stable way across the three studies.

2 Method

2.1 VR Pedestrian Simulator

The study took place at test studios in Mountain View (CA, USA), Mannheim (Germany) and Shanghai (China). In all three surveys, the same BMW research pedestrian simulator was used (Fig. 2). The pedestrian simulator consists of a standard HTC Vive Pro VR setup (head mounted display, and two infrared trackers, as well as the HTC VIVE's remote control) and a computer, running the simulation software which is based on Unity 3D. During the simulation, participants were immersed in an urban environment, standing on the sidewalk of a street and encountering an AV, a BMW i3.

Fig. 2. BMW research pedestrian simulator

2.2 Study Design and Measures

The intention of the AV was manipulated by simulating different driving behaviors. When the AV's intention was to give way to the pedestrian, it started to decelerate 20 m before the pedestrian at a constant deceleration of -3.5 m/s^2; When the AV's intention was to pass the pedestrian, it continued onwards at constant speed at 25 mph (about 40 km/h).

For the intention "Give Way", a 3×3 factorial study plan was used (Table 1), including three priorities (AV, HRU, undefined) and three eHMI solutions (none, icon, light-band). For the intention "Pass", a 2×3 factorial study plan with the factor's priority (AV, undefined) and eHMI solutions (none, icon, light-band) was used (Table 2). We did not include this intention for the HRU priority condition, because this condition would imply a violation of the traffic code, which might have influenced the overall study results.

Independent Variables

External Human Machine Interface
To account for the effects of the technologies and potential differences in the comprehensibility of eHMIs across cultures, two eHMI solutions [17, 18] were used in this study. Furthermore, a baseline without eHMI was included.

The first eHMI consisted of an exterior display, mounted behind the windscreen (see Fig. 3) which displayed two different icons. For the AV intention "Give Way" there was an icon displaying a car with a "stop" line in front of the car to symbolize a stopping car (Fig. 3, right). For the intention "Pass" an icon, showing an open hand (Fig. 3, left) was displayed to communicate that the pedestrian should stay back. The second eHMI was a light-band integrated into the chassis of the AV showing two different states for the two different intentions of the AV. The light-band pulsed slowly to communicate the intention "Give Way" and pulsed rapidly when communicating the intention "Pass".

In the baseline condition, no eHMI was displayed and participants had to derive the AV's intentions solely from the driving behavior. The icons as well as light-band states were displayed in white to prevent attentional biases due to color [27]. The signals were presented 6 s after the start of each trial, at a distance of 40 m from the pedestrian.

Priority
Three different traffic scenes with different priority regulations were included in the experimental setup: a street with a zebra crossing, where the HRU has priority (HRU), a 2-lane street with no cross markings, where the AV has priority (AV), and a parking space scenario with undefined priorities (Undefined). Speed and longitudinal and lateral distances to the pedestrian at the relative points in time were identical in all trials.

In the HRU priority, the pedestrian was standing at the curb waiting at a zebra crossing. The street was an urban two-lane street with no middle lane markings. In the US, a traffic guard was additionally placed at the zebra crossing, ready to stop the approaching car. In the AV priority condition, the pedestrian was placed at the sidewalk of the same two-lane street as in the zebra crossing condition, but distant from the zebra crossing. In the Undefined priority condition, participants were standing in a parking lot next to a parked car. The parked car was placed next to the pedestrian to create the same physical barriers as in the other conditions while not impeding visibility. The parking lot was large with multiple parked cars and no lane markings except the parking spots.

Dependent Variables
The intention recognition time (IRT) was measured from the moment participants recognized the AV's intention [9] and pressed the button of HTC Vive's remote control. By means of a short interview included after each trial, two variables were measured in all three countries: correct recognition of the AV's intention was measured by asking participants to judge the intent of the AV (give way or pass). We furthermore measured participants' certainty of choice (very uncertain to very certain on a scale from 1 to 5).

Table 1. Study plan for the intention "Give Way".

		Priority		
		AV	HRU	Undefined
eHMI	No eHMI			
	Icon			
	Light-band			

Table 2. Study plan for the intention "Pass".

		Priority	
		AV	Undefined
eHMI	No eHMI		
	Icon		
	Light-band		

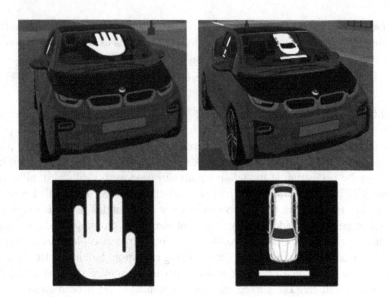

Fig. 3. Icons for intentions (left: "Pass" with open hands, right: "Give Way" with stopping car)

2.3 Procedure

After filling out their demographic data, participants were introduced to the pedestrian simulator and familiarized with the experimental setup. Participants were instructed that they would be participating in the study as a pedestrian encountering an AV in an urban environment. Following this, participants put on the head mounted display and ran three practice trials to become familiar with the setup, and the remote control they were holding in their right hands, as well as with the rating scales.

Participants were placed in identical starting positions for each trial. The study consisted of three study blocks which differed in terms of priority. Altogether 15 trials were executed. In the simulation, participants stood on the side of the street (on the sidewalk or next to a parked car, depending on the traffic scene). Each trial started with the AV driving at a constant speed of 25 mph (about 40 km/h). Participants were instructed that an AV would be approaching from the left, and were asked to press a button on the remote control once they decided that they had understood the intention of the AV. The moment they pressed the button the simulation froze. The IRT was measured and the experimenter completed a short interview. Afterwards the next trial started. In the end, short debriefing interviews regarding the interpretation of the eHMIs were conducted.

2.4 Participants

Across all countries N = 82 participants took part in this study. Table 3 shows the description of the US, German (GER), and Chinese (CN) samples. Participants' ages ranged between 20 and 65 years. They either had no visual impairment or corrected vision such as contact lenses or glasses. In the US, Germany, and China participants were recruited externally and received compensation.

Table 3. Sample description.

	USA (US)	Germany (GER)	China (CN)
N	n = 29	n = 30	n = 23
Age	M = 36 years (SD = 16 years)	M = 43 years (SD = 13 years)	M = 42 years (SD = 13 years)
Sex	16 males/13 females	14 males/16 females	11 males/12 females

2.5   Data Analysis

Data was analyzed using SPSS statistical software, version 23. As the goal was to compare the mechanisms of different eHMIs for different AV's intentions six repeated measures ANOVA were run: one for each country (US, GER, CN) and each intention of the AV (pass, give way). In case sphericity was violated, degrees of freedom were corrected according to Huynh-Feldt. The main effects of the two factors, eHMI (baseline, icon, light-band) and priority (AV, HRU, undefined), are reported separately. For the factor eHMI, two-tailed planned contrasts were conducted between baseline and the eHMI conditions in combination with effect size r. For the factor priority, post-hoc tests were conducted. Contrasts and post-hoc tests were Bonferroni adjusted.

3   Results

3.1   External Human Machine Interface (eHMI)

Correct Interpretation of Vehicle Intention

Intention "Give Way"
In the US ($F_{(2, 56)} = 7.95$, p = .001) and in Germany ($F_{(2, 56)} = 4.07$, p = .022), the type of eHMI had a significant main effect on correct interpretation rates of the vehicle intention (Fig. 4). In both countries, the light-band significantly enhanced the number of correct interpretations, improving the rate of understanding in the US from 59% (baseline) to 87% ($F_{(1, 28)} = 12.72$, p = .003, r = .56) and in Germany from 64% to 86% ($F_{(1, 29)} = 6.35$, p = .035, r = .42). For the Chinese sample, the type of eHMI did not influence the correct interpretation significantly ($F_{(2, 42)} = 0.58$, p = .563).

Intention "Pass"
Overall, eHMI influenced the rate of correct interpretation in an opposite way than for the intention "Give Way" and decreased the rate of understanding (Fig. 5). In the US ($F_{(2,56)} = 6.38$, p = .003), the icon eHMI significantly decreased the correct interpretation rate from 93% in the baseline to 69% ($F_{(1, 28)} = 10.98$, p = .005, r = .53). For the German sample, no significant main effect can be reported ($F_{(1.65, 47.98)} = 2.73$, p = .082). However, contrasts revealed that a significantly lower number of participants detected the intention of an AV with a light-band correctly (83%) compared to baseline (97%), ($F_{(1, 29)} = 7.86$, p = .018, r = .46). The significant main effect of the type of eHMI on correct interpretation rates in China ($F_{(1.44,}$

30.18) = .409, p = .041) could not reveal a significant Bonferroni-adjusted contrast. The intention recognition rate for light-band (82%) and icon (80%) did not significantly differ from the baseline (97%).

Intention Recognition Time (IRT)

Intention "Give Way"
In the US (F(2, 56) = 7.49, p = .001) and in China (F(2, 42) = 8.32, p = .001), the main effects of eHMI showed that responses were faster with an eHMI present (Fig. 6). Contrasts revealed that in both countries the icon (US: F(1, 28) = 13.04, p = .002, r = .56; CN: F(1, 21) = 11.26, p = .006, r = .59) as well as the light-band (US: F(1, 28) = 6.23, p = .037, r = .4; CN: F(1, 21) = 9.28, p = .012, r = .55) led to faster IRTs than the baseline condition. In Germany the IRT did not significantly differ between eHMIs (F(1.74, 48.75) = 1.80, p = .174).

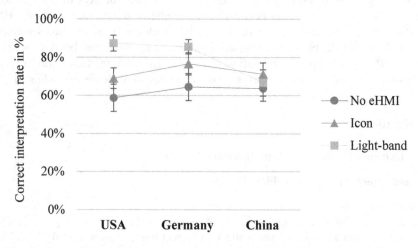

Fig. 4. Correct interpretation across countries and eHMI for the intention "Give Way"

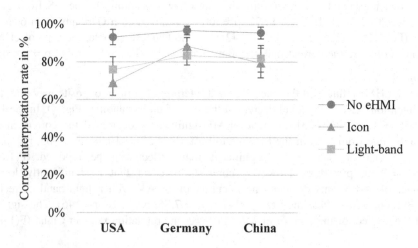

Fig. 5. Correct interpretation across countries and eHMI for the intention "Pass"

Intention "Pass"
In the US, results were similar to the intention "Give Way", with eHMI conditions leading to faster IRTs ($F(2, 56) = 7.74, p = .009$) (Fig. 7). The icon ($F(1, 28) = 11.93, p = .004, r = .55$) and the light-band ($F(1, 28) = 10.46, p = .006, r = .52$) differed significantly from the baseline condition. In China ($F(2, 42) = 3.13, p = .054$) as well as in Germany ($F(1.39, 40.33) = 3.26, p = .065$) eHMI did not influence IRT significantly when the AV's intention was to pass. In Germany, however, planned contrasts revealed that responses were significantly faster with an icon present as compared to the baseline condition ($F(1, 29) = 8.24, p = .015, r = .47$).

Certainty of Choice

Intention "Give Way"
When the AV's intention was to give way, participants in all three countries did not differ in their certainty of choice between eHMI variants (US: $F(1.62, 45.38) = 0.07, p = .899$; GER: $F(2, 56) = 0.42, p = .582$; CN: $F(2, 42) = 0.02, p = .984$).

Intention "Pass"
For the intention "Pass" there were significant effects of the eHMIs in the US ($F(2, 56) = 5.41, p = .007$) and in Germany ($F(2, 56) = 4.06, p = .022$) which can be attributed to the light-band leading to a lower certainty of choice (US: $F(1, 28) = 11.73, p = .004, r = .54$; GER: $F(1, 29) = 6.28, p = .036, r = .42$). In China, certainty did not differ significantly between eHMIs ($F(2, 40) = 1.20, p = .311$).

3.2   Priority

Intention "Give Way"
Across all countries and variables, there were three significant main effects of the factor priority. In the US, priority influenced the correct interpretation ($F(2, 56) = 3.33, p = .043$) and the certainty of choice ($F(2, 56) = 3.97, p = .024$). For the German sample, priority did have an influence on IRT ($F(1.52, 42.61) = 3.63, p = .046$). However, for all three effects, Bonferroni-adjusted post hoc tests revealed no significant differences between the type of priority. In China, no main effects of priority can be reported.

Intention "Pass"
For the intention "Pass" there was only one significant main effect of priority on correct interpretation rates in China ($F(1, 21) = 6.83, p = .016$). The post hoc test revealed that the parking scene with undefined priority led to a better understanding of the AV's intention than the two-lane street condition where the AV had priority ($M_{diff} = -0.14$ (95% CI[$-0.25--0.03$]), $p = .016$).

3.3   Interaction Effect (eHMI and Priority)

Intention "Give Way"
In total, three significant interaction effects between eHMI and priority occurred, being one effect per country.

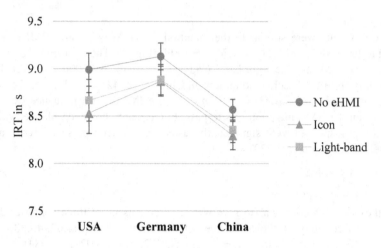

Fig. 6. IRT across countries and eHMI for the intention "Give Way"

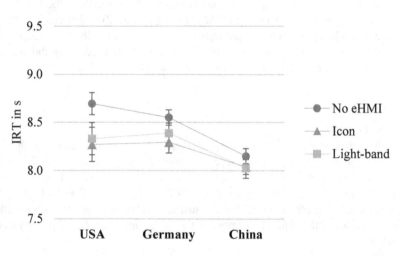

Fig. 7. IRT across countries and eHMI for the intention "Pass"

In the US ($F(2, 56) = 3.71$, $p = .007$) as well as in China ($F(4, 84) = 2.60$, $p = .041$), there were significant interactions associated with the correct interpretation rates. For the US sample, the light-band eHMI had a consistently high rate of correct interpretation for all traffic scenarios, while the icon eHMI only improved the correct ratings in the pedestrian priority situation and did not differ from baseline in the other scenarios. In China, the correct understanding for the light-band eHMI and the icon eHMI were rather stable for all traffic scenarios, while the baseline condition differed strongly between the priorities (best for HRU priority and worst for undefined priority).

In Germany, there was a significant interaction effect regarding the certainty of choice ($F(4, 112) = 2.79$, $p = .030$) which can be attributed to the HRU priority condition in which the icon and the light-band improved certainty compared to the baseline for the AV priority and undefined priority condition. EHMIs and baseline did not differ.

Intention "Pass"
When the AV was passing, no significant interaction effects between the eHMIs and priority occurred.

4 Discussion

This pedestrian VR simulator study investigated the influence of eHMI on pedestrians' intention recognition across three studies and three cultural backgrounds.

We predicted that an eHMI would generally improve intention recognition across all scenarios compared to the baseline. Additionally, since the eHMIs were developed in Germany, we expected them to generally work best in a Western and especially German environment. One of the two eHMI solutions included icons which are known to be culturally dependent [26, 27]. The other solution was a light-band. The light-band solution resembles common cross-culturally used features such as turn indicators, which we predicted to be more universally understandable. We therefore expected the light-based eHMI to perform in a more stable way across the three studies.

4.1   Main Findings: eHMI and Intentions of the AV

We expected eHMIs to improve intention recognition across all priorities and all cultures. We found that priority did not have systematic influences across all variables. Results show that eHMIs improved correct intention recognition rates in the US and Germany when the AV's intention was to give way. These effects were caused by the light-band eHMI, while the icon eHMI did not differ in terms of intention recognition from the baseline across all cultures. IRT was lowered in the US, while all participants in all three studies felt equally confident in their selection. EHMIs, however, led to deteriorated intention recognition when the AVs intention was not to yield. Intention recognition was most accurate in the baseline condition for all three countries (US: 93%, DE: 97%, CN: 95%), when the car did not yield to the pedestrian. Certainty of choice was lowered in the US and Germany when an eHMI was present and the AV passed and did not yield.

These findings suggest that using an eHMI to communicate that the AV will not yield to a pedestrian is not beneficial, rather potentially confusing or even detrimental. The event of a car approaching at 40 km/h itself seems to be dissuasive enough for the pedestrians to decide not to cross in front of the arriving AV. Showing eHMIs in this condition seems to influence pedestrians' safe legacy behavior, which could ultimately lead to an inappropriate or dangerous decision to cross the street. As scenarios were stopped at the point in time participants made their decision and pressed the button, we cannot conclude that participants would ultimately have walked into the street. The lower error rate without the eHMI, however, suggests that it would be beneficial to refrain from showing any eHMI signals unless the car is engaged in an already safe interaction with vulnerable road users – such as stopping.

When the AV was about to yield, eHMIs provided significant benefits for the pedestrians compared to baseline in the US and Germany. It could therefore be beneficial to use eHMIs to reinforce the AV's intention to let the pedestrian cross. This finding is in accordance with other findings of significant benefits for communicating the intention to yield to pedestrians [28, 29].

4.2   EHMIs and Culture

We expected the light-based eHMI to be less susceptible to cultural influence, thus showing the most stable effects across all three samples. While the German and US sample showed improved intention recognition with an eHMI when the AV was about to yield, Chinese participants did not profit from an eHMI in this scenario. IRTs were shortened in the US and China with an eHMI present, while remaining constant in Germany. The fact that the icon eHMI did not show any significant benefits in intention recognition rates might be caused by the shortcomings of the specific eHMI design, as the icon eHMI might be more difficult to process or was visible at a later point in time and both icons thus difficult to distinguish.

The Chinese participants did not seem to fully profit from any eHMI showing the AV's intention to yield. While they responded significantly more quick, intention recognition was not improved. Insights from the post study interviews suggest that this might be due to misinterpretations of the eHMI. Participants associated the slow pulsing with warning signals or even interpreted the eHMI as a design element without any further meaning. Thus, there are indications that the specific eHMI type that showed positive effects in Germany and the US was not suitable for Chinese pedestrians.

Cultural differences regarding the general traffic behavior in China might be a further explanation for this study's results. The traffic scenarios used were rated by all study participants as very suitable for the Chinese market at the study debriefing. However, the typical behavior of Chinese drivers encountering pedestrians in these scenarios might differ fundamentally from German or US drivers in equal scenarios. For instance, the general stopping rate of cars to give way to pedestrians was found to be very low in China [24]. This habitual behavior of traffic participants might lead to a very low expectancy of Chinese pedestrians that any car, manually driven or automated, will yield to them, thus influencing overall probabilities that an eHMI will be considered in their decisions. The interaction effect of improved intention recognition with eHMI in the parking space scene found in this study further supports this hypothesis.

4.3   Limitations

This study's results are inherently limited since very controlled traffic scenarios were used in order to isolate differences in the comprehensibility of the eHMI. Therefore, the results cannot be directly transferable to real traffic. All trials in this study included only one pedestrian and one AV, which only represents an excerpt of actual traffic. Furthermore, the findings have to be limited due to constraints caused by the VR setting, such as resolution, brightness, or angle of view, which might also have had an impact on effects such as the difference between a light-band and an icon eHMI. The visibility of the icon eHMI might have been reduced due to a lack of sufficient resolution or a limited field of view. In addition, the pedestrians' perception of the AV's braking behavior in VR might be different from the one they have for manually driven vehicles. If braking behavior is less predictable in real traffic, for instance, due to other influencing factors, eHMIs could have a greater impact than in VR. Thus, specific VR effects resulting in reduced visibility opposed and to the real world experience with eHMI should lead to reduced effects for the eHMIs compared to the baseline.

4.4 Future Research

To generalize this study's findings, more complex scenarios which also include additional traffic participants besides the study participant and the AV should be investigated. In addition, different speeds for the approaching AV should be included as this might profoundly impact the perception and interpretation of the eHMI. EHMIs should be tested on the real road to overcome the technological shortcomings of VR or other simulators. Furthermore, different methodological setups to evaluate eHMIs should be part of future research. Besides IRTs, critical gap acceptance [30] as an indicator of traffic efficiency or actual crossing behavior, such as crossing initiation time, might be of interest. Even in the scenarios in this study, which yielded significant benefits, negative side effects such as pedestrian distraction and the lack of safety glances at other vehicles present might be observed by using eye tracking in different setups. Pedestrians' attention might be captured by an eHMI solution and lead to neglect of the road traffic around them. This should therefore be investigated further.

No "one size fits all" eHMI solution was found in this study. It seems questionable to just deploy existing eHMI solutions to differing cultures. In further research and development, eHMIs should be adapted to their respective markets, such as the Chinese one, by means of new and improved design concepts by and for the respective market. The interaction effect between priority and eHMIs found in the Chinese sample suggests that eHMIs have further potential for Chinese pedestrians. Once developed, these localized eHMIs have to undergo thorough evaluation in different, culturally adapted traffic scenarios, also considering that we cannot know the cultural background of the recipient of the AVs message.

5 Conclusion

The authors conclude from the results presented that from a safety point of view it is not necessary or may even be counterproductive to display eHMIs in situations that might be harmful when the eHMI is misinterpreted. We therefore argue that situations in which an eHMI is displayed should be selected cautiously and benefits and potential problems should be studied carefully.

From a cultural point of view, the results of this study might have several implications. First, it might be concluded that eHMIs have to be localized and adapted to the respective market and the expectancy of the traffic participants in the culture they are introduced into. It will, however, be challenging to deploy localized eHMI solutions to different markets as, unlike in most other HMIs, an AV does not know the cultural background of the recipient of its messages. For instance, a European pedestrian encountering the same AV type in Europe and then China might be confused if the same vehicle interacts with him or her in a different way in each country.

Second, traffic scenarios in which an eHMI is used might differ between cultures. It might for instance be suitable to use eHMI in the Chinese market only when interacting in shared space scenarios, as interactions might be resolved equally well without eHMI in other traffic scenarios.

Third, traffic scenarios used for testing eHMIs have to be selected carefully taking into account cultural differences. These scenarios not only have to be comparable in terms of measurable context factors such as priorities, distances, or velocities [11], they also should take into account habitual behavioral patterns of current traffic and the expectations of road users as to what happens in such situations.

Acknowledgement. Some work is a part of the interACT project. interACT has received funding from the European Union's Horizon 2020 research & innovation program under grant agreement no 723395. Content reflects only the authors' view and European Commission is not responsible for any use that may be made of the information it contains.

References

1. SAE International: Automated driving: levels of driving automation are defined in new SAE International standard J3016 (2016). http://www.sae.org/misc/pdfs/automated_driving.pdf
2. Papadimitratos, P., De La Fortelle, A., Evenssen, K., Brignolo, R., Cosenza, S.: Vehicular communication systems: enabling technologies, applications, and future outlook on intelligent transportation. Comm. Mag. **47**(11), 84–95 (2009). https://doi.org/10.1109/MCOM.2009.5307471
3. Müller, L., Risto, M., Emmenegger, C.: The social behavior of autonomous vehicles. In: Proceedings of the 2016 ACM International Joint Conference on Pervasive and Ubiquitous Computing Adjunct - UbiComp 2016 (2016). https://doi.org/10.1145/2968219.2968561
4. Färber, B.: Communication and communication problems between autonomous vehicles and human drivers. In: Maurer, M., Gerdes, J.C., Lenz, B., Winner, H. (eds.) Autonomous Driving: Technical, Legal and Social Aspects, pp. 125–144. Springer, Heidelberg (2016). https://doi.org/10.1007/978-3-662-48847-8_7
5. Clamann, M., Aubert, M., Cummings, M. L.: Evaluation of vehicle-to-pedestrian communication displays for autonomous vehicles. In: TRB 96th Annual Meeting Compendium of Papers (2017)
6. Rosenbloom, T.: Crossing at a red light: behaviour of individuals and groups. Transp. Res. Part F Traffic Psychol. Behav. **12**(5), 389–394 (2009). https://doi.org/10.1016/j.trf.2009.05.002
7. Lundgren, V.M., et al.: Will there be new communication needs when introducing automated vehicles to the urban context? In: Stanton, N.A., Landry, S., Bucchianico, G., Vallicelli, A. (eds.) dvances in Human Aspects of Transportation: Proceedings of the AHFE 2016 International Conference on Human Factors in Transportation, July 27–31, 2016, Walt Disney World®, Florida, USA, pp. 485–497. Springer, Cham (2017). https://doi.org/10.1007/978-3-319-41682-3_41

8. Röse, K.: Methodik zur Gestaltung interkultureller Mensch-Maschine-Systeme in der Produktionstechnik. Kaiserslautern, Univ. (2002). https://portal.dnb.de/opac.htm?method= simpleSearch&cqlMode=true&query=idn%3D964834855

9. Fuest, T., Michalowski, L., Träris, L., Bellem, H., Bengler, K.: Using the driving behavior of an automated vehicle to communicate intentions – a wizard of Oz study. In: 21st IEEE International Conference on Intelligent Transportation Systems (2018). https://doi.org/10. 1109/itsc.2018.8569486

10. Nathanael, D., Portouli, E., Papakostopoulos, V., Gkikas, K., Amditis, A.: Naturalistic observation of interactions between car drivers and pedestrians in high density urban settings. In: Bagnara, S., Tartaglia, R., Albolino, S., Alexander, T., Fujita, Y. (eds.) Proceedings of the 20th Congress of the International Ergonomics Association, vol. 823, pp. 389–397. Springer, Cham (2019). https://doi.org/10.1007/978-3-319-96074-6_42

11. Fuest, T., Sorokin, L., Bellem, H., Bengler, K.: Taxonomy of traffic situations for the interaction between automated vehicles and human road users BT. In: Stanton, N.A. (ed.) Advances in Human Aspects of Transportation, pp. 708–719. Springer, Cham (2018)

12. Rothenbucher, D., Li, J., Sirkin, D., Mok, B., Ju, W.: Ghost driver: a field study investigating the interaction between pedestrians and driverless vehicles. In: 25th IEEE International Symposium on Robot and Human Interactive Communication, RO-MAN 2016 (2016). https://doi.org/10.1109/roman.2016.7745210

13. Petzoldt, T., Schleinitz, K., Banse, R.: Laboruntersuchung zur potenziellen Sicherheitswirkung einer vorderen Bremsleuchte in PKW. Zeitschrift für Verkehrssicherheit 63(1), 19–24 (2017)

14. Song, Y.E., Lehsing, C., Fuest, T., Bengler, K.: External HMIs and their effect on the interaction between pedestrians and automated vehicles. In: Karwowski, W., Ahram, T. (eds.) IHSI 2018. AISC, vol. 722, pp. 13–18. Springer, Cham (2018). https://doi.org/10. 1007/978-3-319-73888-8_3

15. Benderius, O., Berger, C., Lundgren, V.M.: The best rated human-machine interface design for autonomous vehicles in the 2016 grand cooperative driving challenge. IEEE Trans. Intell. Transp. Syst. 19(4), 1302–1307 (2018). https://doi.org/10.1109/TITS.2017.2749970

16. Camara, F., Belotto, N., Cosar, S.: Pedestrian Models for autonomous driving: a literature review. IEEE Trans. Intell. Transp. Syst. (submitted)

17. Sorokin, L., Hofer, M.: A new traffic participant and its language. In: Proceedings of the 12 th International Symposium on Automotive Lightning, pp. 565–574. Herbert Utz Verlag GmbH, Darmstadt (2017)

18. Sorokin, L., Chadowitz, R., Kauffmann, N.: A change of perspective: designing the automated vehicle as a new social actor in a public space. In: CHI Conference on Human Factors in Computing Systems Extended Abstracts (CHI 2019 Extended Abstracts) (2019, accepted)

19. Mahadevan, K., Somanath, S., Sharlin, E.: Can interfaces facilitate communication in autonomous vehicle-pedestrian interaction? In: Companion of the 2018 ACM/IEEE International Conference on Human-Robot Interaction - HRI 2018 (2018). https://doi.org/ 10.1145/3173386.3176909

20. Mahadevan, K., Somanath, S., Sharlin, E.: Communicating awareness and intent in autonomous vehicle-pedestrian interaction. In: Proceedings of the 2018 CHI Conference on Human Factors in Computing Systems, pp. 429:1–429:12. ACM, New York (2018). https:// doi.org/10.1145/3173574.3174003

21. Habibovic, A., et al.: Communicating intent of automated vehicles to pedestrians. Front. Psychol. 9, 1336 (2018). https://doi.org/10.3389/fpsyg.2018.01336

22. Andersson, J., Habibovic, A., Klingegård, M., Englund, C., Malmsten-Lundgren, V.: Hello Human, can you read my mind? ERCIM News (109), 36–37 (2017). http://urn.kb.se/resolve?urn=urn:nbn:se:ri:diva-29618

23. Alexander, R., Thompson, N., Murray, D.: Towards cultural translation of websites: a large-scale study of Australian, Chinese, and Saudi Arabian design preferences. Behav. Inf. Technol. **36**(4), 351–363 (2017). https://doi.org/10.1080/0144929X.2016.1234646

24. Zhuang, X., Wu, C.: Pedestrian gestures increase driver yielding at uncontrolled mid-block road crossings. Accid. Anal. Prev. **70**, 235–244 (2014). https://doi.org/10.1016/j.aap.2013.12.015

25. Helmreich, R.L., Merritt, A.C.: National culture and flight deck automation: results of a multination survey AU - Sherman, Paul. J. Int. J. Aviat. Psychol. **7**(4), 311–329 (1997). https://doi.org/10.1207/s15327108ijap0704_4

26. Rau, P.-L.P., Gao, Q., Liang, S.-F.M.: Good computing systems for everyone – how on earth? Cultural aspects. Behav. Inf. Technol. **27**(4), 287–292 (2008). https://doi.org/10.1080/01449290701761250

27. Romberg, M., Röse, K., Zühlke, D.: Global demands of non-european markets for the design of user-interfaces. IFAC Proc. Volumes **31**(26), 137–141 (1998). https://doi.org/10.1016/S1474-6670(17)40082-6

28. Langström, T., Lundgren, V.M.: AVIP – Autonomous Vehicles' Interaction with Pedestrians. Chalmers University of Tehnology, Gothenborg, Sweden (2015)

29. Dietrich, A., Willrodt, J.-H., Wagner, K., Bengler, K.: Projection-based external human machine interfaces-enabling interaction between automated vehicles and pedestrians. In: Proceedings of the DSC 2018 Europe VR (2018)

30. Rodríguez Palmeiro, A., van der Kint, S., Vissers, L., Farah, H., de Winter, J.C.F., Hagenzieker, M.: Interaction between pedestrians and automated vehicles: a Wizard of Oz experiment. Transp. Res. Part F Traffic Psychol. Behav. **58**, 1005–1020 (2018). https://doi.org/10.1016/j.trf.2018.07.020

Author Index

Printed in the United States
By Bookmasters

Printed in the United States
By Bookmasters